URSYN

DANCE CODE

DANCE STEPS AS A CODE

-URSYN.COM

Cover design and images Moonlit Manifestation (front cover), Stellar Questions (back cover), and images © 2025 by Anna Ursyn. The codes, with resulting outputs and musical notations, are used with permissions and are duly credited.

First published 2026
by CRC Press
2385 NW Executive Center Drive, Suite 320, Boca Raton FL 33431

and by CRC Press
4 Park Square, Milton Park, Abingdon, Oxon, OX14 4RN

CRC Press is an imprint of Taylor & Francis Group, LLC

Library of Congress Cataloging-in-Publication Data

ISBN: 9781032972565 (hbk)
ISBN: 9781032971629 (pbk)
ISBN: 9781003592952 (ebk)

DOI: 10.1201/9781003592952

Typeset in Times New Roman and Herculanum

Publisher's note: This book has been prepared from printer-ready copy provided by the author.

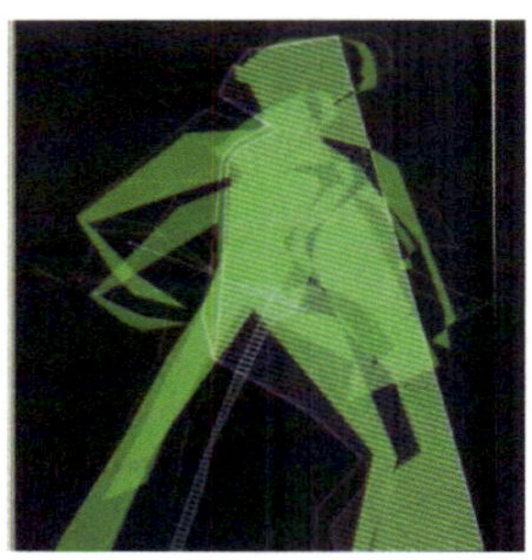

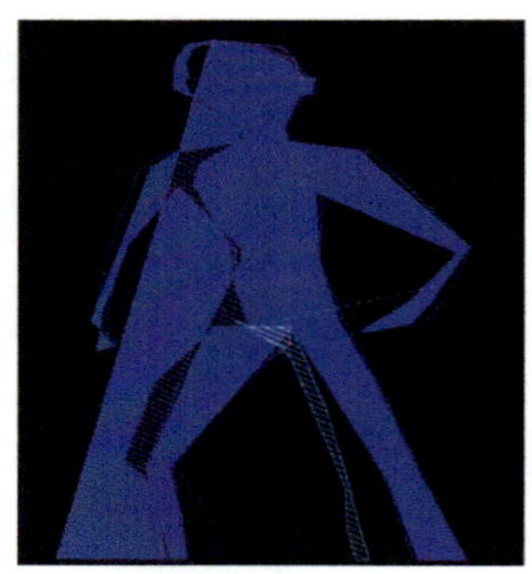

The scripts with resulting outcomes, and videos are available to those willing to learn at available the website:

Ursyn.com/art-and-programming

URSYN.COM

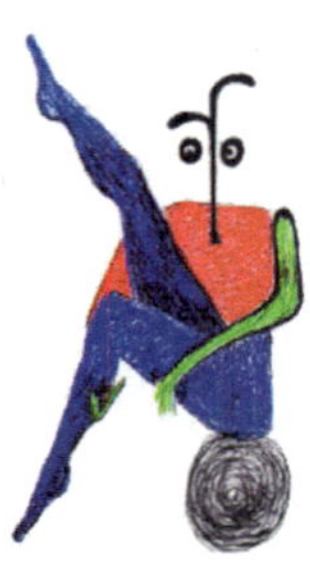

For M and M, with Much Love.

A Series of Books: Knowledge Through the Arts
Playful Books About Serious Things

This series consists of the following volumes:

Dance Code: Dance Steps as a Code, Taylor & Francis
New Storytelling: Learning Through Metaphors, Taylor & Francis
Code Appreciation: Reshaping Knowledge, Taylor & Francis
Nature Appreciation: Knowledge as Art, Taylor & Francis

Several common themes are woven throughout these books. They provide crucial insights into coding and the ongoing connections between programming and various fields of knowledge, areas of specialization, courses of study, and a diverse range of subjects. The books playfully explore knowledge domains and technologies within the framework of everyday situations faced by characters who endeavor to express their challenges through coding. These volumes also examine a collective understanding of nature. In each book, text and illustrations are accompanied by artworks from the author and other artists, visual outputs from computer programs, and sketches of imaginary characters.

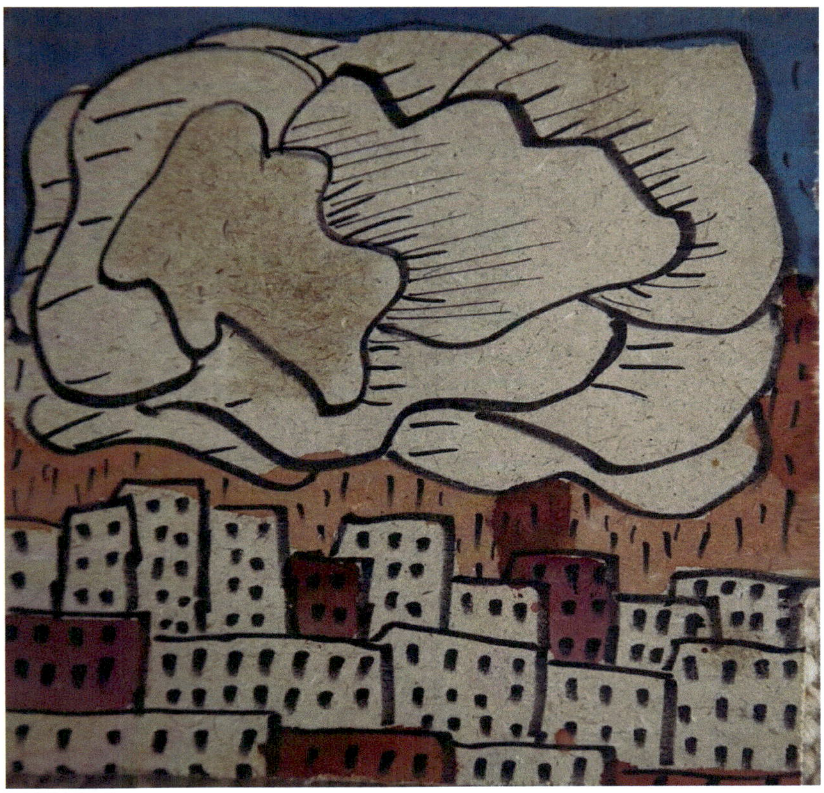

One City. © 2025 by Anna Ursyn

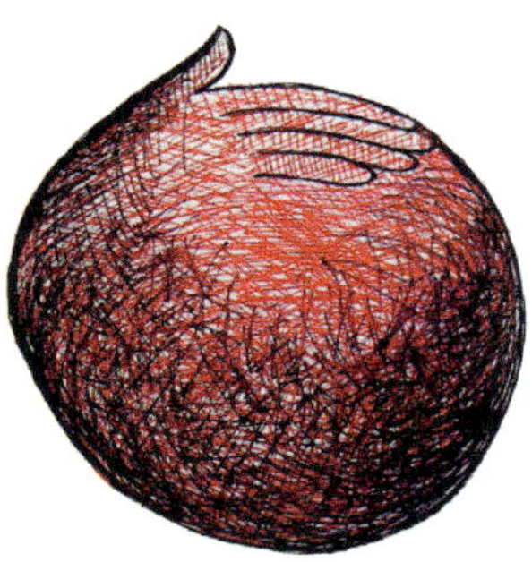

Labudo: enjoys every meal.

Labudo: enjoys every meal.

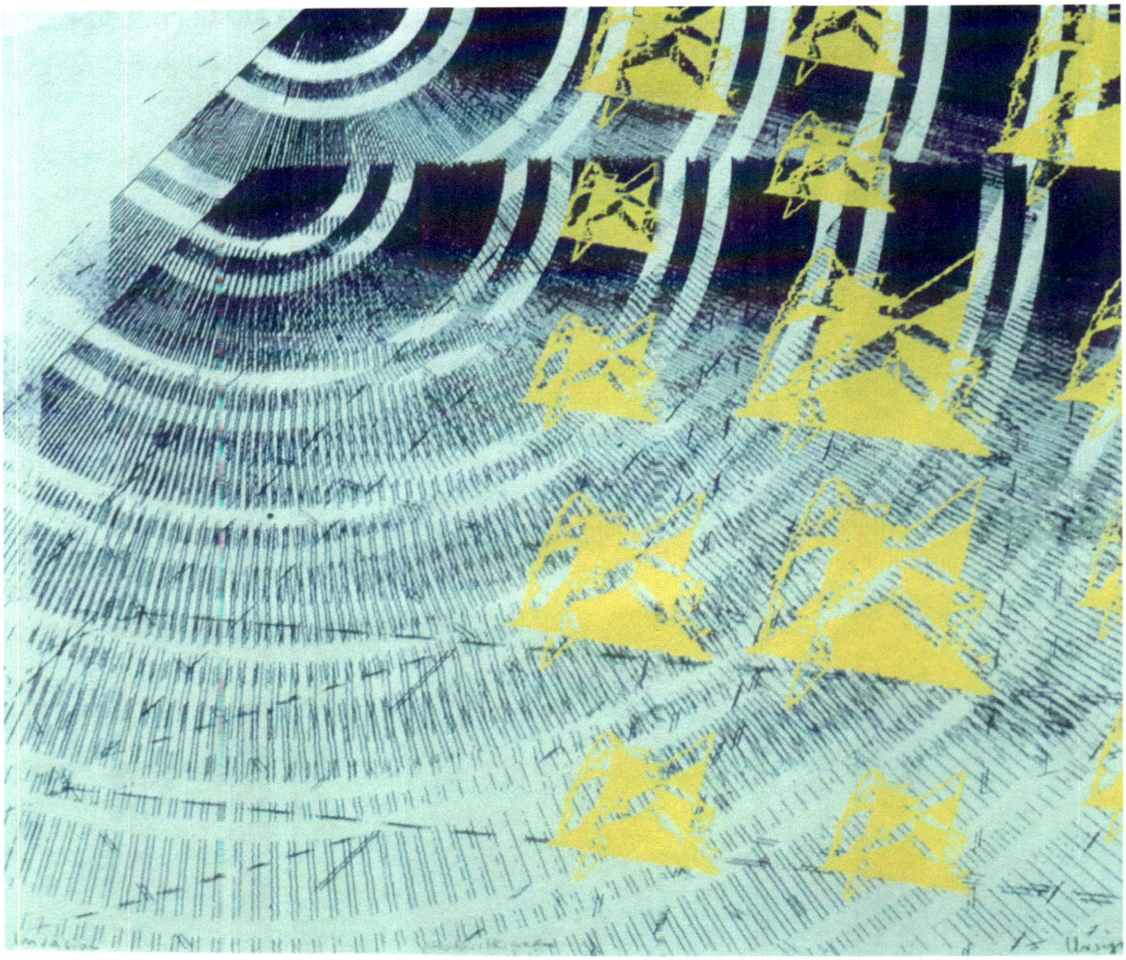

Dancing, Photolithograph after computer code. © 2025 by Anna Ursyn

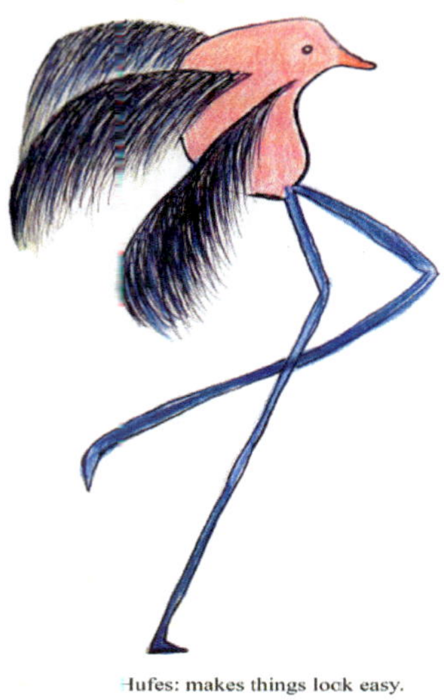

Hufes: makes everything easy.

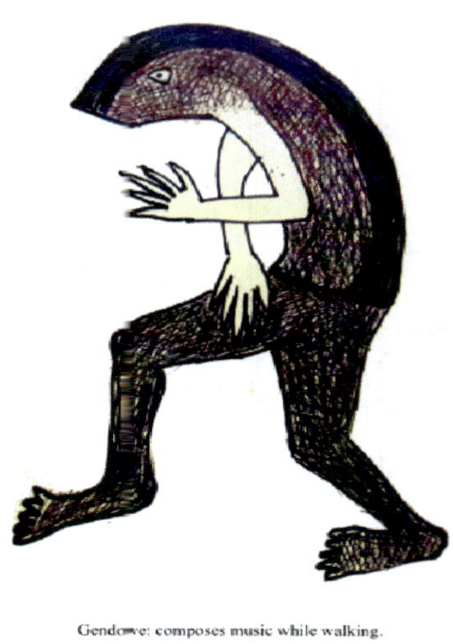

Gendowe: composes music when walking.

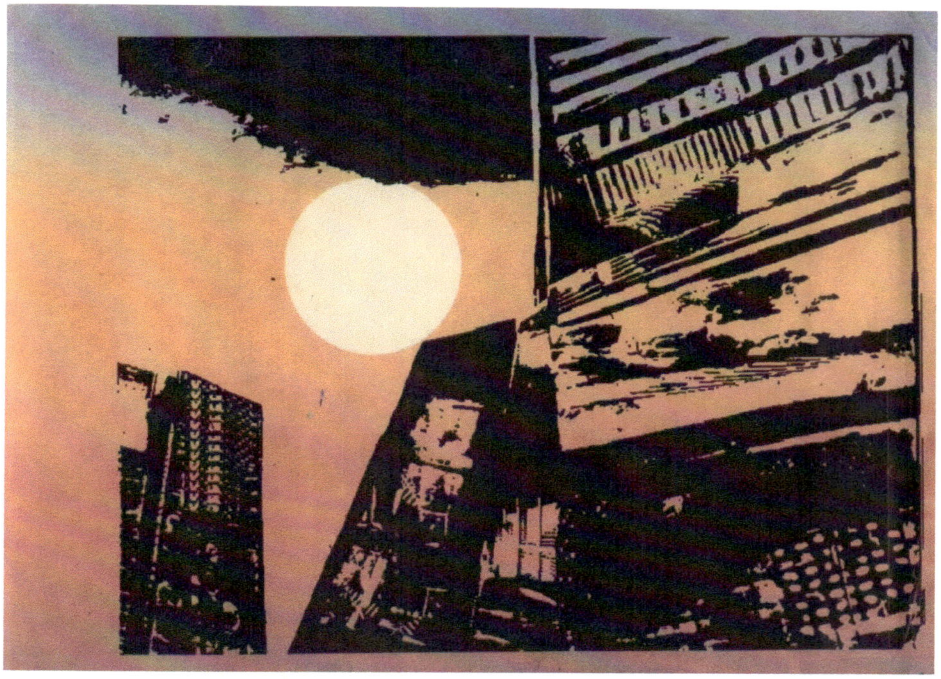

Moon. Photolithograph on an airbrushed paper. © 2025 by Anna Ursyn

TABLE OF CONTENTS

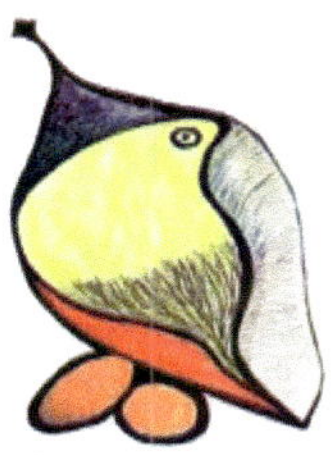

Jerlon: can sense you coming.

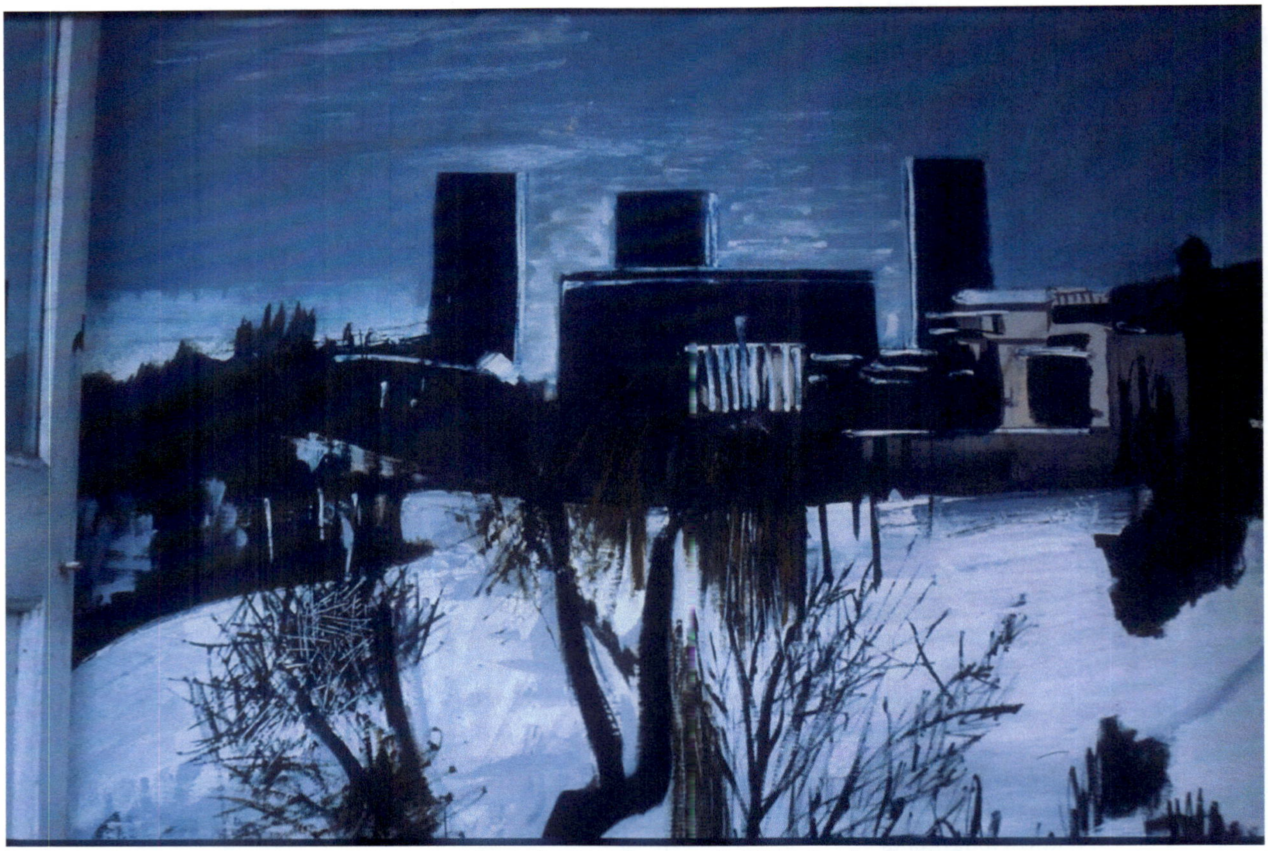

Into the Landscape. © *2025 by Anna Ursyn*

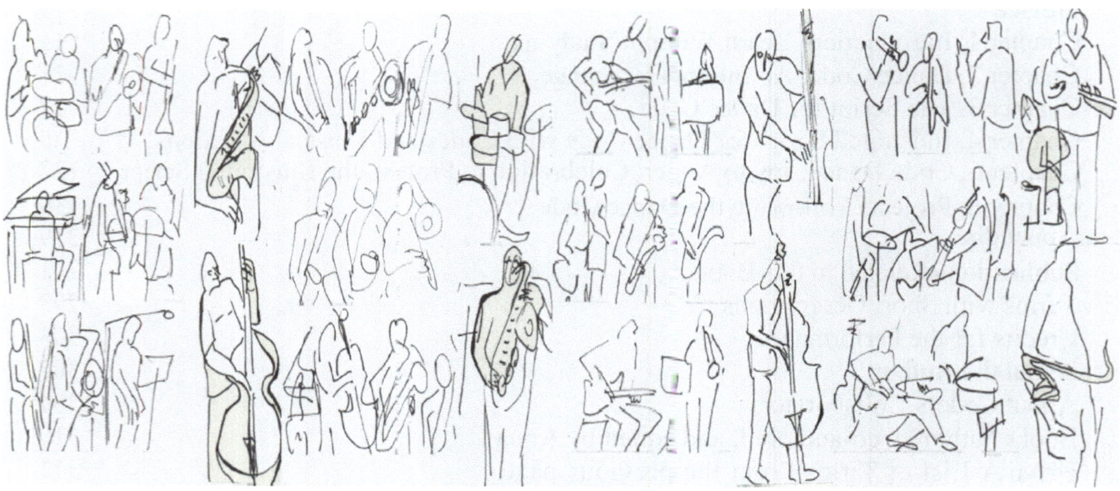

Jazz Players. © *2025 by Anna Ursyn*

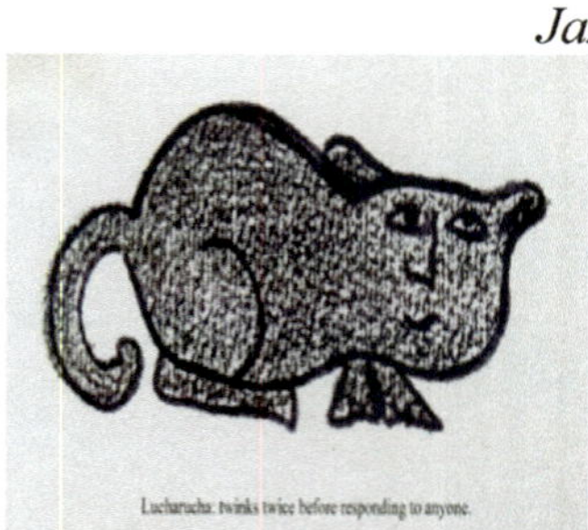

Lucharucha: thinks twice before responding to anyone.

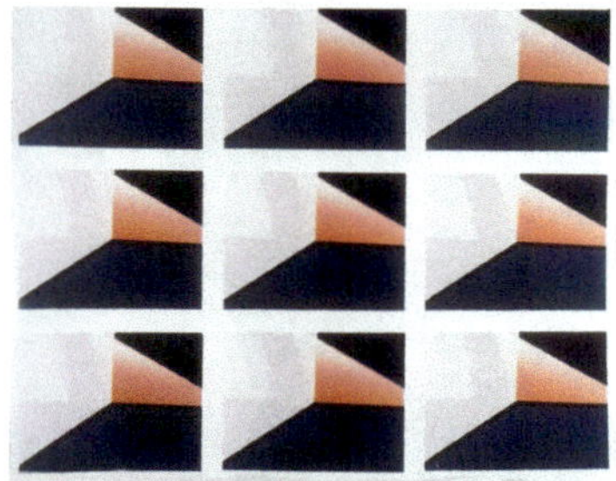

Windows. The output from the computer code. © 2025 by Anna Ursyn

ACKNOWLEDGMENTS

The UNC President, Board of Trustees members, Provost and members of the Provost Office, Dean of the Performance and Visual Arts, Chair of the School of Art and Design, and colleagues, have been supporting me in my research and performances based on my books. My great thanks go to all of them and everybody else involved.

My gratitude goes to Dr. Irwing Stuart Smith, Department of Music, and Department of Computer and Information Sciences, UMass, Lowell, Dr. Mohammad Majid al-Rifaie, Associate Professor in Artificial Intelligence, Goldsmiths College, University of London, Md. Fahimul Islam, JPMorgan Chase & Co., Chicago, IL, and Md. Asifur Rahman, Infozillion Teletech BD, Dhaka, Bangladesh, for creating codes with their outputs and for Stu for composing music.

I want to thank the Members of the Taylor & Francis Group team, especially Randi Slack for her cheerful and personal guidance for this book series, and Heba Kusseibati for her assistance in shaping the publications.

Many thanks to the Reviewers for helping to make this book happen.

As for the resulting performances, my thanks go to theater production members, Professor Brian Luedloff, Director of the Opera Theatre, Associate Professor of Dance, Kim Clay, all the students from the School of Theater Arts & Dance, the School of Art and Design at the University of Northern Colorado (UNC), as noted in the Credits section, for their collaboration, and input in the performance, Cristina Goletti, the Dean of the PVA College, and Brandon Ingold, Technical Director for making the performance possible.

Thank you to the committees' members for awarding grants enabling the performance:

- The Provost Research Dissemination and Faculty Development (RDFD) Competition, Office of Sponsored Programs, (OSP), University of Northern Colorado
- The grant from the Office of Undergraduate Research (OUR), University of Northern Colorado, to my Digital Media student assistant at the School of Art & Design.
- The College of Performing and Visual Arts grants *Through Lines,* securing the performance.

My appreciation goes to Matylda Czarnecka and Joohyun Pyune for the photographs and the artwork adorning the books. For books and performances, software and apps were used, including music and visual music among them, Grammarly, ChatGPT, WordQuill, Wordtune, Pexels, and Udio. Images are by the author and collaborators, as credited.

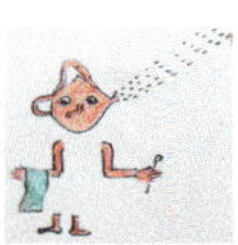

Surevi: always ready to listen.

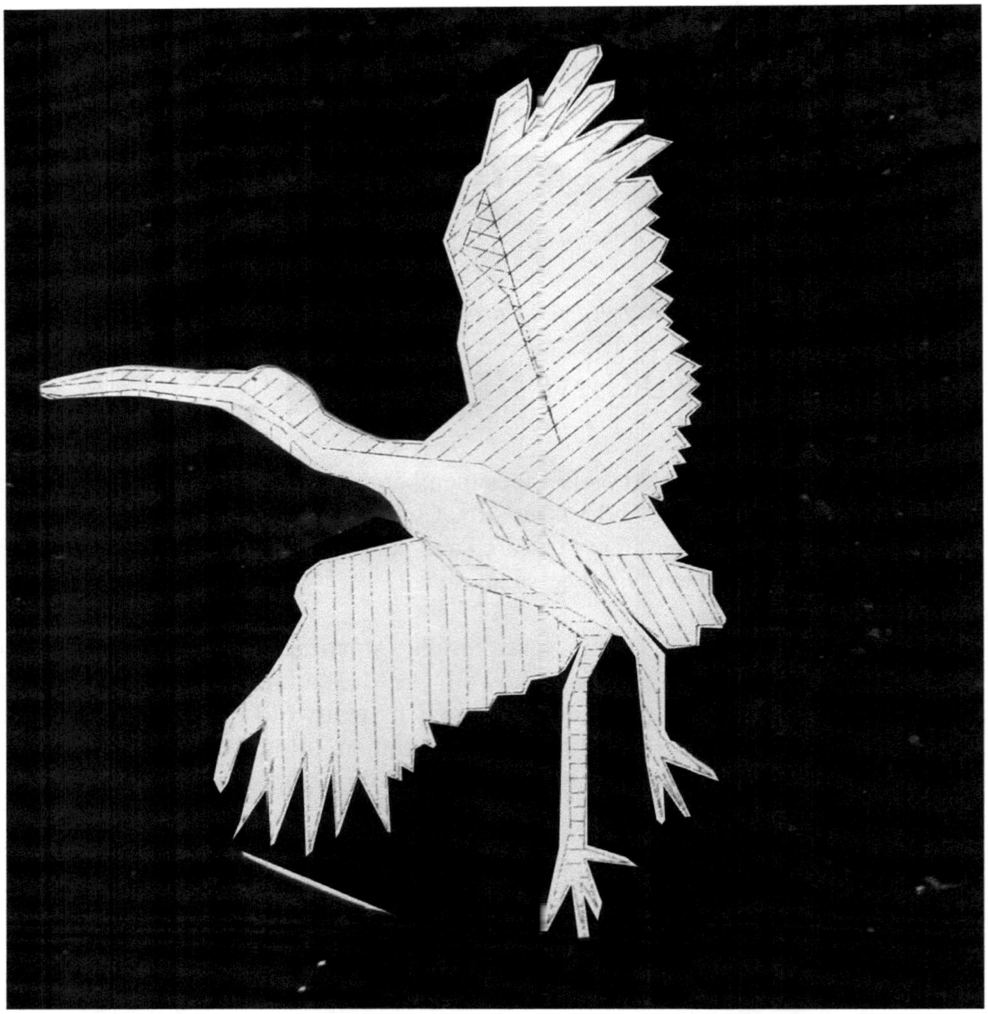

A Takeoff. The output from the computer code. © *2025 by Anna Ursyn*

KEYWORDS

New Storytelling, Metaphors, Analogies, Art, Coding, Programming, Music, Dance, Video, Mathematical Art, Stage Design, Art and Science Connection, Professions, Computer Graphics, Digital Media, Knowledge Visualization, Knowledge Transfer through Art, Character Creation, Concept Art, Time and Space-based Digital Illustration, Cognitive and Abstract, and Non-Linear Thinking, Types of Intelligences, Talents, Understanding Programming for Better Prompts, Information through New Choreography, Interdisciplinary Collaboration, Preparing for a Job, Performance as Knowledge Transfer, New Storytelling-Learn without Studying, Buffered image, OCR, Digital image processing, Pattern Recognition, Artificial Intelligence, Machine Learning, Image Acquisition, Greyscale Conversion.

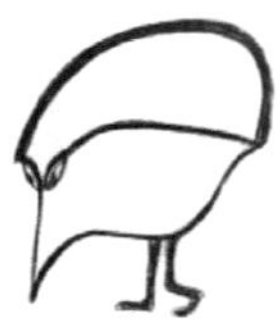

Fusto: finds all lost items.

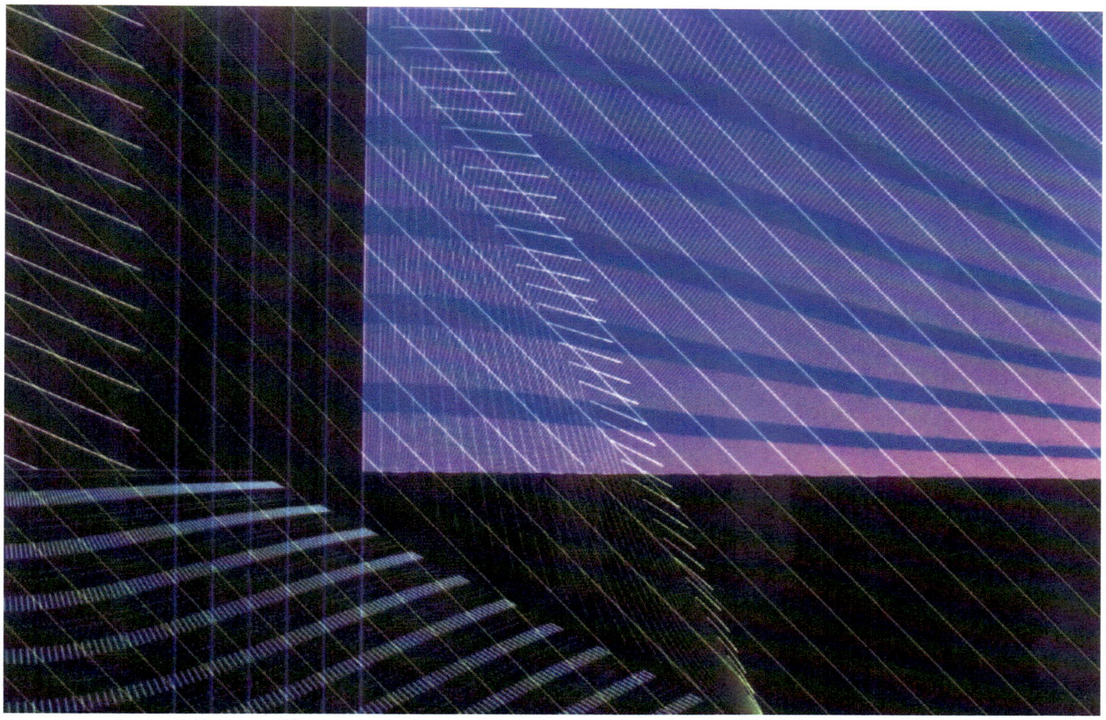

A Scene. The output from the computer code. © 2025 by Anna Ursyn

Dialog is in this font.

Characters and some directions are in italics.

`Exposition in this font (Information for the audience's benefit).`

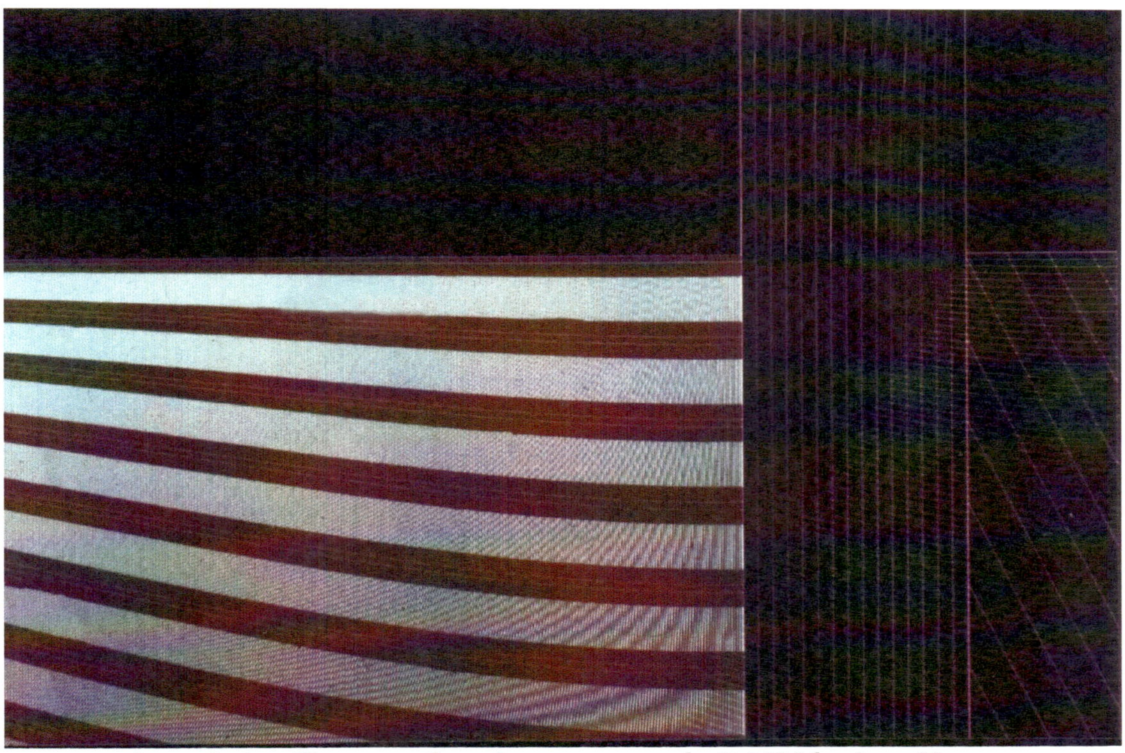

Intermission. The output from the computer code. © 2025 by Anna Ursyn

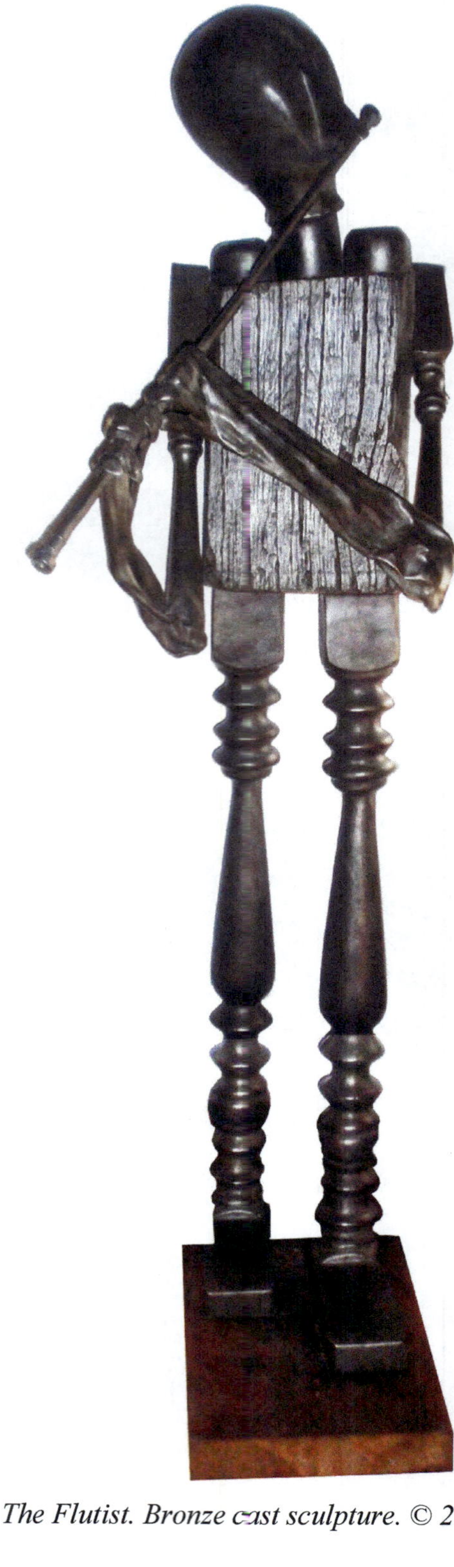

The Flutist. Bronze cast sculpture. © 2025 by Anna Ursyn

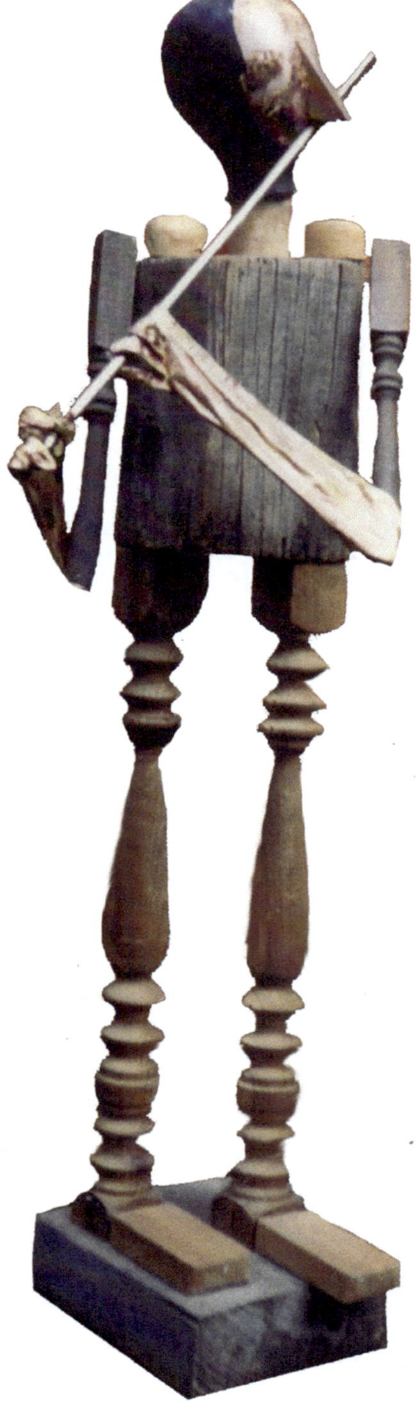

The Flutist. Wooden sculpture. © 2025 by Anna Ursyn

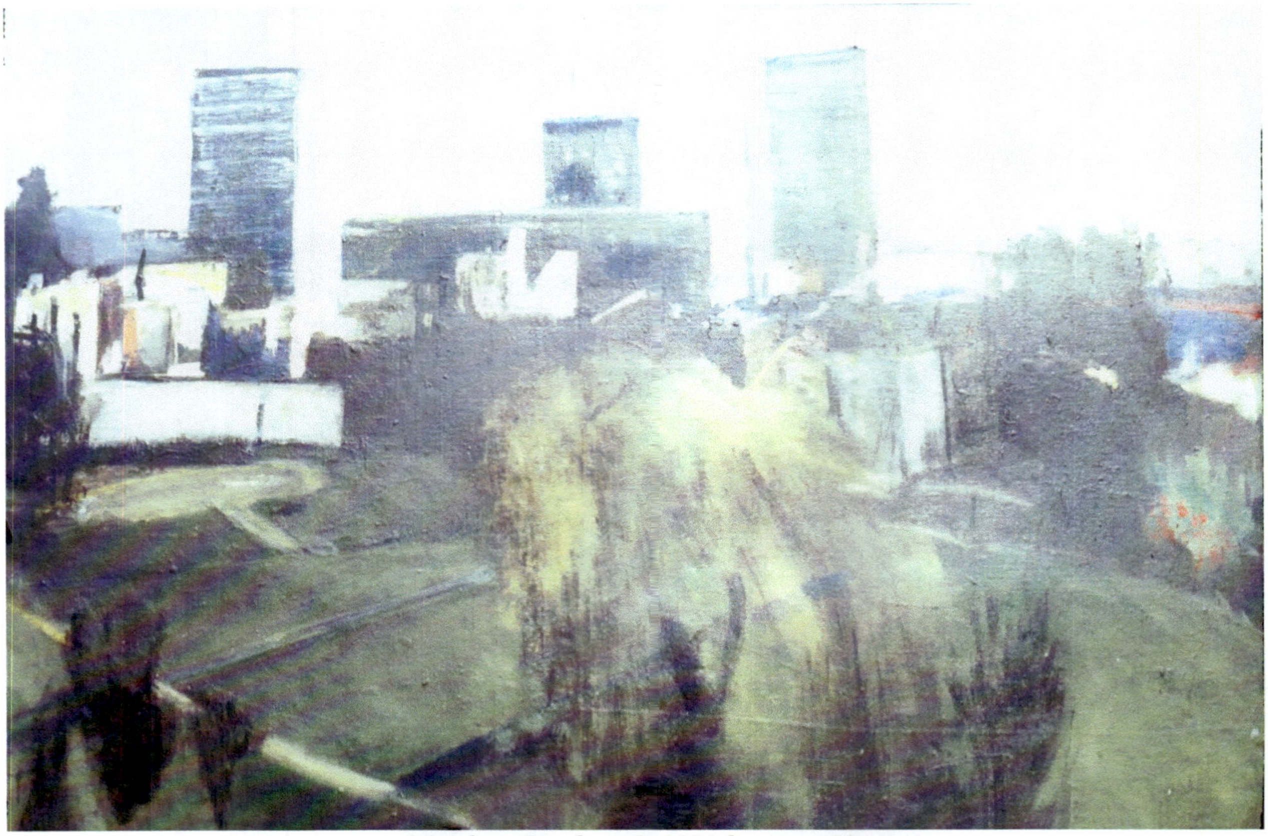

Out of my Window. © 2025 by Anna Ursyn

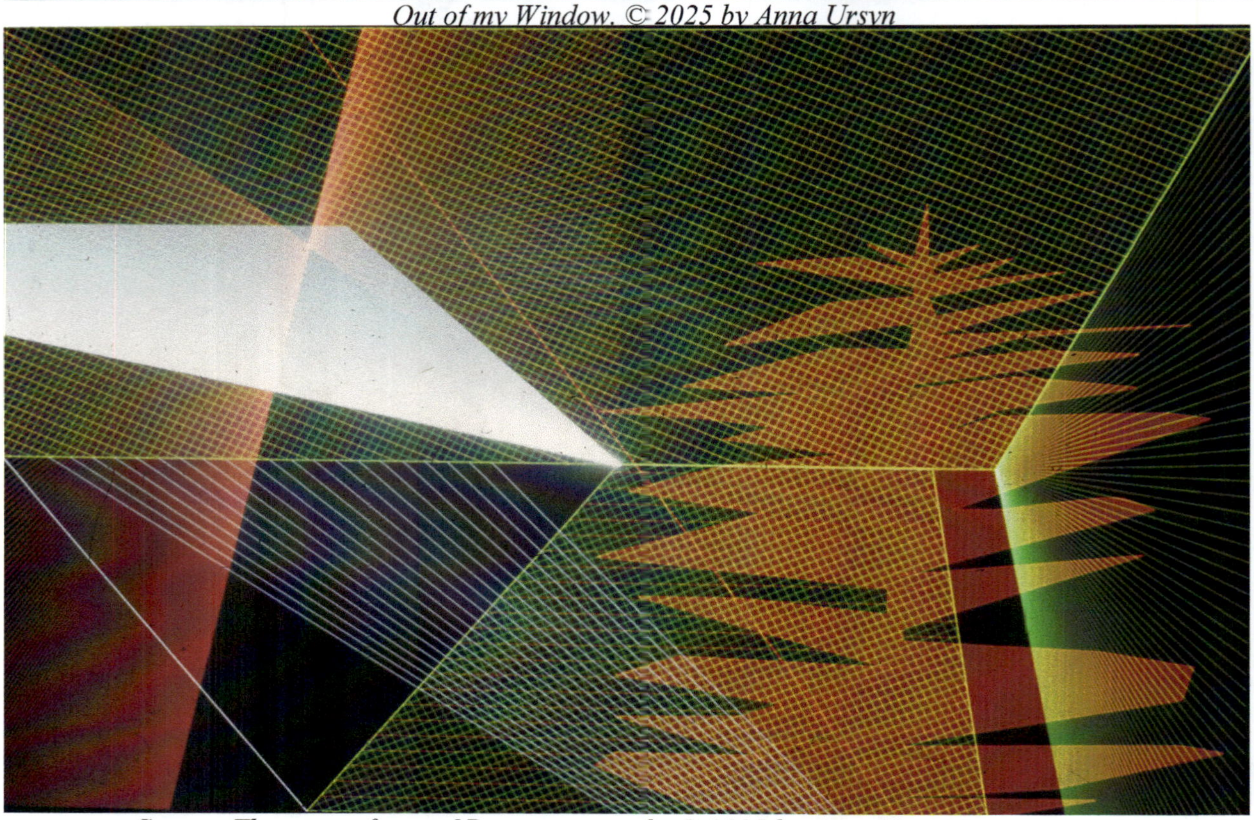

Spruce. The output from a 3D computer code. © 2025 by Anna Ursyn

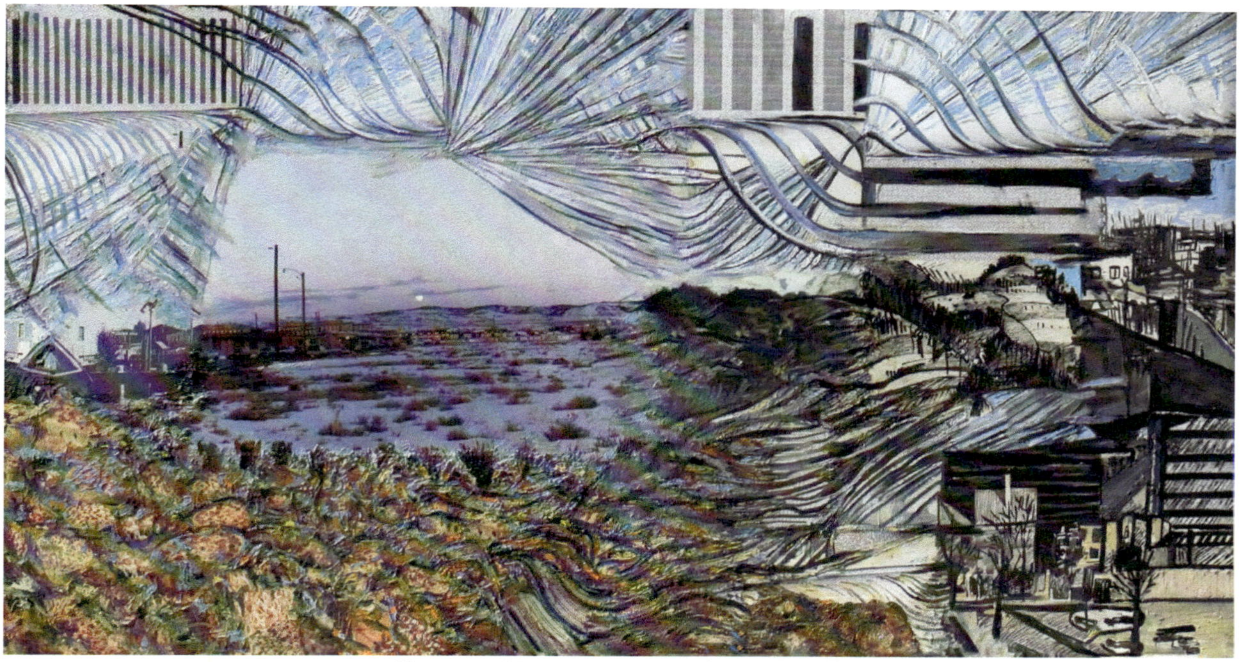

Change of Matter. © *2025 by Anna Ursyn.*

PREFACE: OUR WORLD IS RUN BY PATTERNS

Programming helps to convert data into information, and visualization pictorially transforms information. Understood and interpreted information set in some context becomes knowledge. We use patterns and gain insight for making informed decisions.

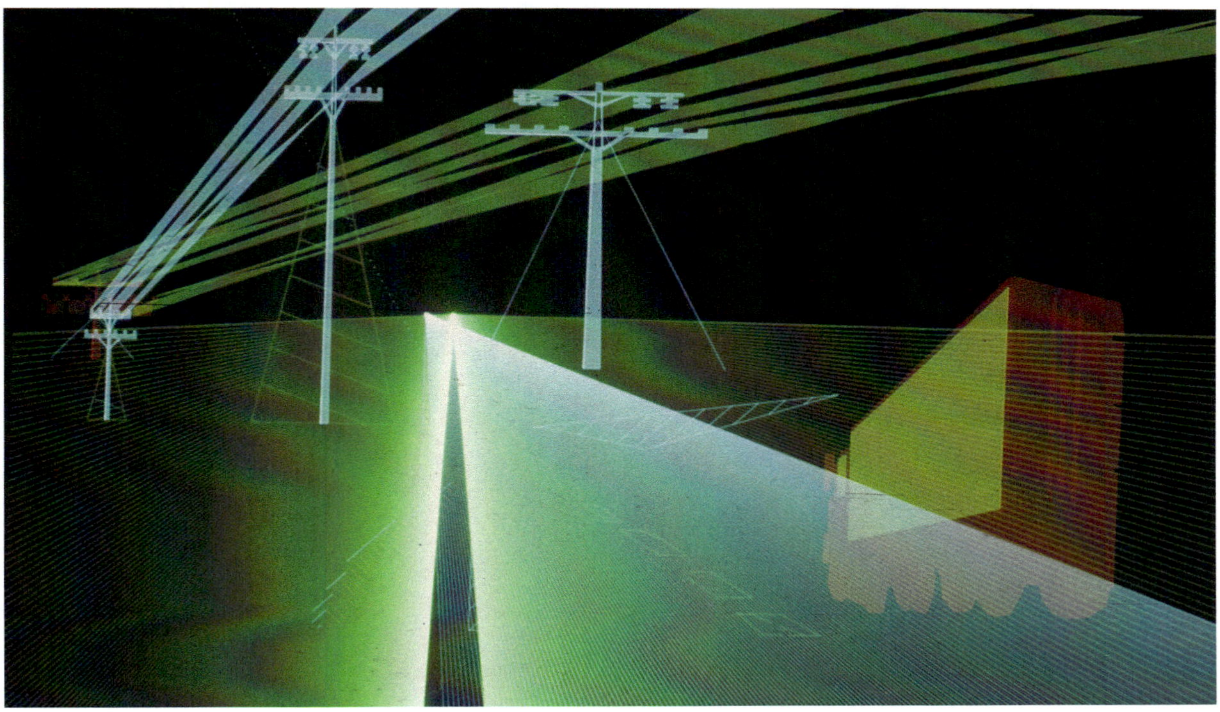

Energies. The output from a 3D computer code. © *2025 by Anna Ursyn*

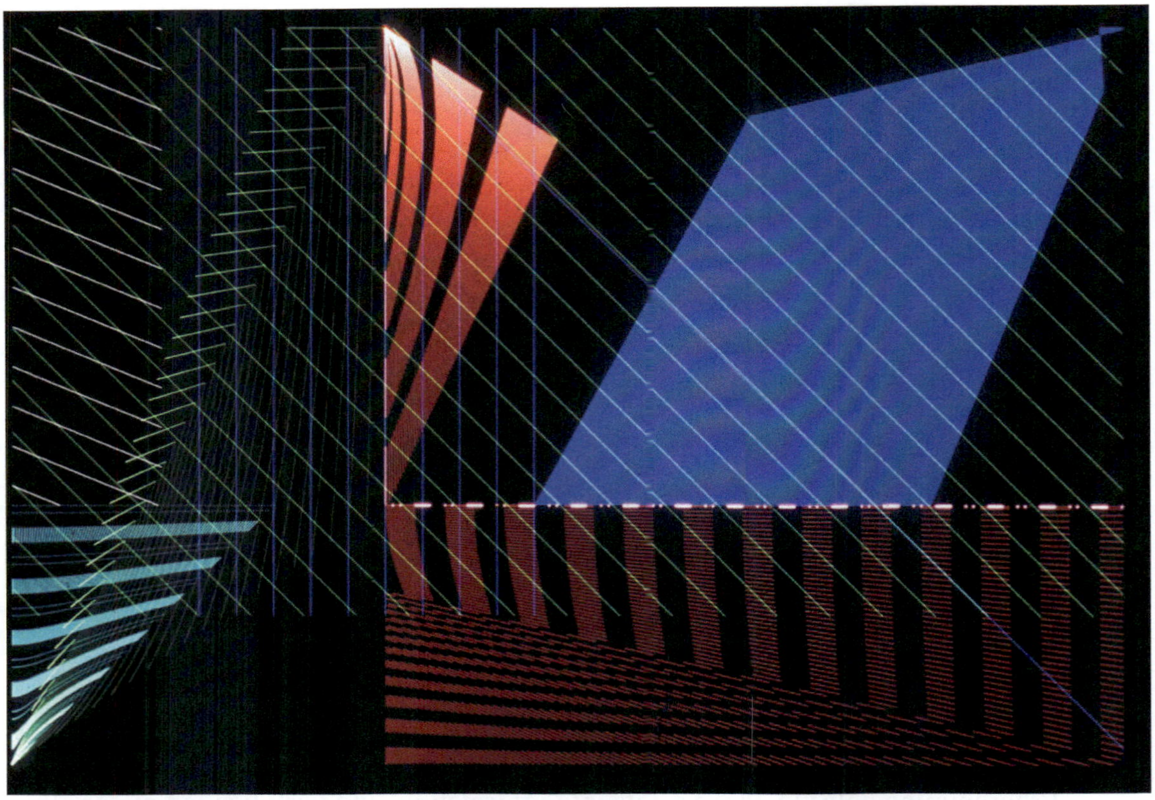

Report from Colorado. The output from the computer code. *© 2025 by Anna Ursyn*

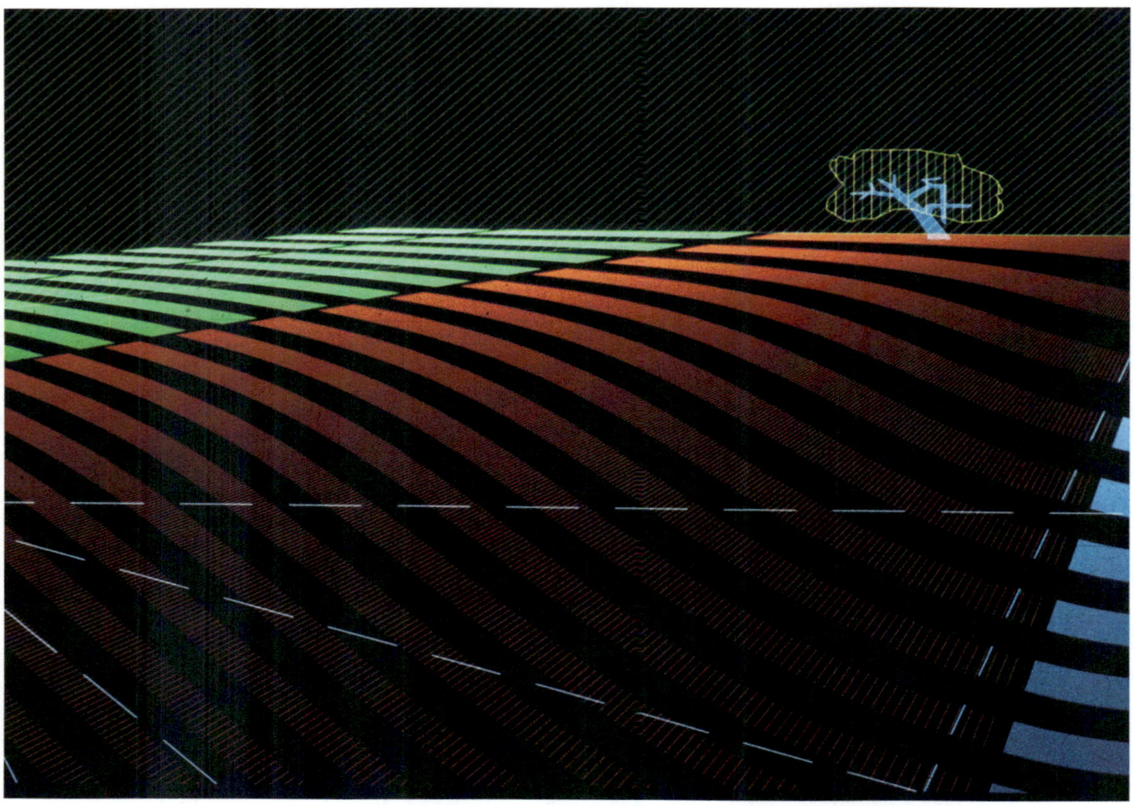

Useful Patterns The output from the computer code. *© 2025 by Anna Ursyn*

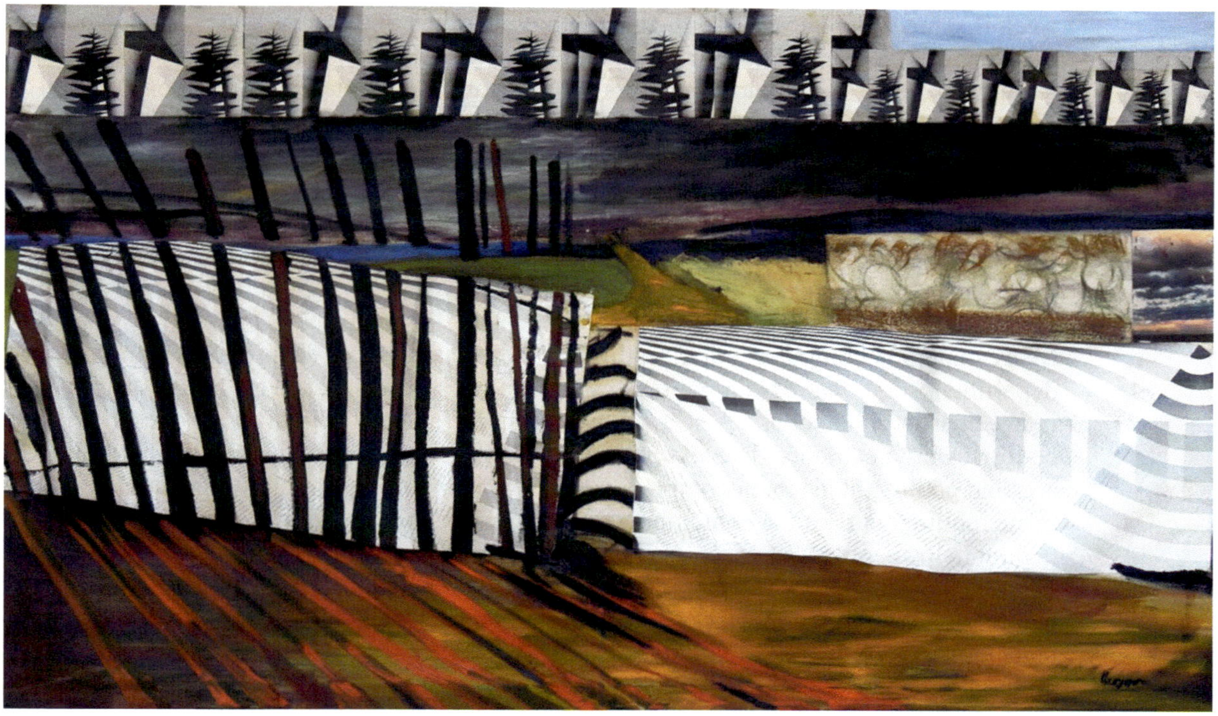

No Man No Shadow. © 2025 by Anna Ursyn

CHAPTER 1
INTRODUCTION: LEARN WITHOUT STUDYING

Dancing and coding have some similar features. Dances have unique structures, and so do computer codes. In both disciplines, some steps are applied in a sequence to achieve a certain goal. Dancing is structured to follow patterns. Both dances and codes are guided by rules and restrictions. The rules obey condition

The *Dance Code* script aims to make coding less feared by readers when talking with coders on the job and typing better prompts when using artificial intelligence. As coding rules our environment, we need a comfortable way of learning about technology. In this book, a dramatized, choreographed story unfolds technical information about coding and dance.

The *Dance Code* script tells the story of an online interaction between a coder and a prima ballerina, resulting in a shared understanding of their respective fields. An exchange between a coder and a dancer may inspire new ways to look at visually presenting knowledge through dancing, performing, or choreographed movement. Hence, the audience learns without knowing that they do: learn without studying.

Many people relax when coding is introduced as a language, rather than math The concept of using programming for artistic creation is not a new one. Artists started creating art with computers by writing codes before any software for art was available. The script was used as a basis for collaborative, international, and interdisciplinary performances involving computer science colleagues, faculty, staff, and students from music, dance, and the arts worldwide. It was performed by faculty and student dancers at the University of Northern Colorado (UNC). Credits are at the end of this book.

Even if someone creates alone, the technology involved in the process has already been developed by somebody else or needs more expertise. Even one person's business requires professional input from others. Collaborations are often performed online with participating parties living across some seas, oceans, mountains, or forests.

This book mixes experiences in art, coding, music, dance, choreography, video, mathematical art, stage design, web development, and Nature. Computers allow for space and time exploration, and our work is time-based. Technology allows us to put our multisensory experiences into visual ways.

Many majors in computer graphics, digital media, and graphic design studying at the art department some fear of coding. However, employers see that skill as a necessary bridge between artists and coders. A scare-free way is offered here to familiarize oneself with coding using metaphors. The text makes the readers familiar with principles that link the arts and coding, helps them comprehend basic principles in programming, and meets some requirements of the contemporary job market. Understanding basic concepts about coding helps people write prompts to obtain the most interesting solution.

Continuing with experimentation in multimedia informs and enriches our learning, teaching, service, and community outreach. This book uses a story about dance, music, and performance as an analogy of coding to understand how technologies affect us in our work. It offers an innovative way of transferring knowledge about programming to instructional methods. It introduces the audience to the ideas behind basic concepts in programming through metaphors, with movement in mind. The script for a *Dance Code* shows how coding is necessary, related to the work, and may even have similar traits regarding structure, rules, or concepts behind creations in another discipline. Thus, this book provides materials for people working in dance, music, choreography, and theater.

Dancing and coding

Many students are active, love sports, and enjoy dancing. There is a long list of existing dances, such as waltz, tango, foxtrot, charleston, rumba, mamba, ballet, tap, hip-hop, jazz/lyrical, pointe and creative movement, of course, jazz, and much more. Dances are very structured, with rules to obey, and restrictions. They produce beauty, all just like coding. When writing a code, certain steps to a sequence are applied to achieve a goal. Dancing a particular dance is also based on following sequences according to a pattern. This project facilitates learning how to write programs by translating selected dance performances into computing codes. Each dance has its own composition described through the structure of a code.

Videos and graphics serve as visual tools for introducing programming: to carry information with explanatory power. First, a look at conditions, restrictions, and ideas related to coding, then dance, shows how they overlap, and we can see similarities. For the initial video, the focus was on sequences, conditions (if statements), and repetitions, (loops, for, or while statements). Sequences of dancers' steps were converted into computer programs – lists of instructions written to tell a computer what task to perform. In this project, more than one computer language is involved.

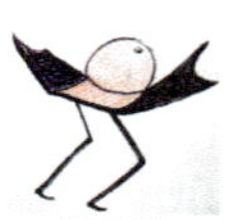

Fahuo: knows how to fly but prefers walking.

Metaphors as a tool for knowledge transfer

The visual approach to coding uses metaphors to link a code to another discipline. Using metaphors is a powerful and relatively easy way for a novice to comprehend these links. Both disciplines deliver beauty and elegance combined with emotions. We can see some similarities between the work of the coders and dancers. The use of metaphors is considered a powerful and relatively easy way to comprehend for a novice. Learning coding through any dance of one's choice creates comparative ways to approach the task.

The analogy between coding and dancing

We can see rules in coding and in dancing through visual objects, which can move according to the rules. This story explores techniques behind various dances and in a symbolic way introduces basic concepts beyond coding. Dances are accompanied by codes with comments. See *What is Coding: Coding 101* tutorial (2024): https://www.bitdegree.org/tutorials/what-is-coding/

Descriptions of programming languages may make one feel comfortable with coding. Several computer languages were introduced, and music was composed for various dances for these projects. An open-ended approach to help visual people learn about programming for visual outputs can be seen from any direction and fit anyone's interests, passions, fun time, and more.

Methods and procedures

Sets of coding describe the selected dances. Sequences of dancers' steps were recorded and converted into computer programs – lists of instructions written to tell a computer what task it should perform. Students performed traditional dances, while the codes and music notations were projected behind them. The results of this phase were translated back into dance performances. After that, students improvised.

Online learning materials

Several codes and videos are on an accompanying website for those willing to try the actual codes used for the visuals in the production during the performance after reading the tutorials provided. The link is listed at the beginning and the end of this book.

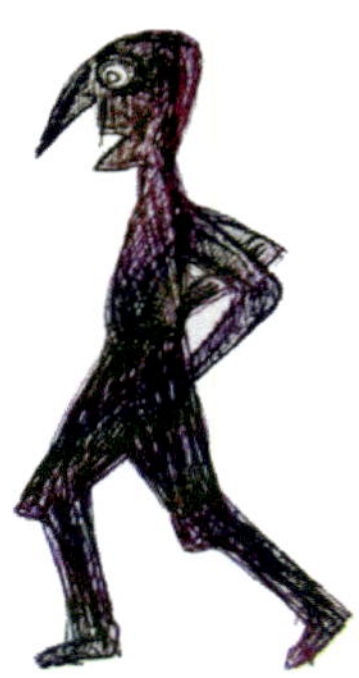

Gutco: holds some snakes in your pockets.

Gutco: holds some snakes in his pockets.

Enhancement of one's career: educational approach

Continuing with experimentation in multimedia informs and enriches our learning, teaching, service, and community outreach. Walter Isaacson wrote about a scientist Emmanuelle Charpentier: "There were lessons from the arts, she would discover, that applied to science. The methodology is important in both, she says. You also must know the basics and master the methods. That requires persistence – repeating experiments and repeating them perfecting how to prepare the DNA when you clone a gene, and then doing it over and over again. It's part of the training, just like the hard work of a ballet dancer, repeating all day long the same moves and methods." Also like the arts, once a scientist has mastered the basic routine, she must combine it with creativity. "You have to be rigorous and disciplined, Charpentier explains, "but also know when to let yourself loose and blend in a creative approach. I found in biological research the right combination of persistence and creativity." (Isaacson, 2021)

Learning about codes and dance in a visual way adds knowledge transfer methods to the existing instructional strategies with the STEAM

(with S for science, T for technology, E for education, A for Art, and M for mathematics (STEAM, 2024; STEM to STEAM: https://www.edutopia.org/article/STEM-resources-downloads, 2023). It follows the current educational trend: STREAM (Science, Technology, Reading, Engineering, Art, and Math; Holoworld, The Basics of STREAM Education, 2024). The scope of this book contains, explanations, with an interest in collaboration with the spectrum of these movements.

The goal of this project is to use dance as a metaphor for coding so as a result a dancer feels ready to code, and the coder feels ready for dancing. The use of metaphors offers a different, more comfortable way of learning (Ursyn, & Mostowfi, 2015). The focus is on teaching through metaphors first (Ursyn, al-Rifaie, & Md Islam, 2015) and then focus on choreography. For example, a visual object can be moved, based on the dance rule we provide. One may start with a dance following the code, then the code will somehow evolve out of the dance.

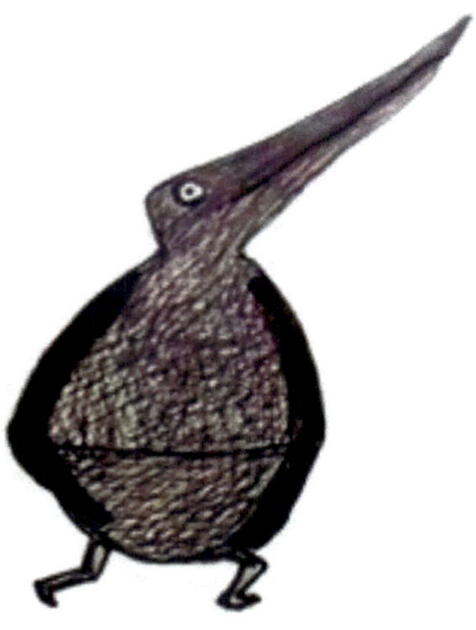

Rusuna: can whistle, but never does

Rusuna: can whistle but never does.

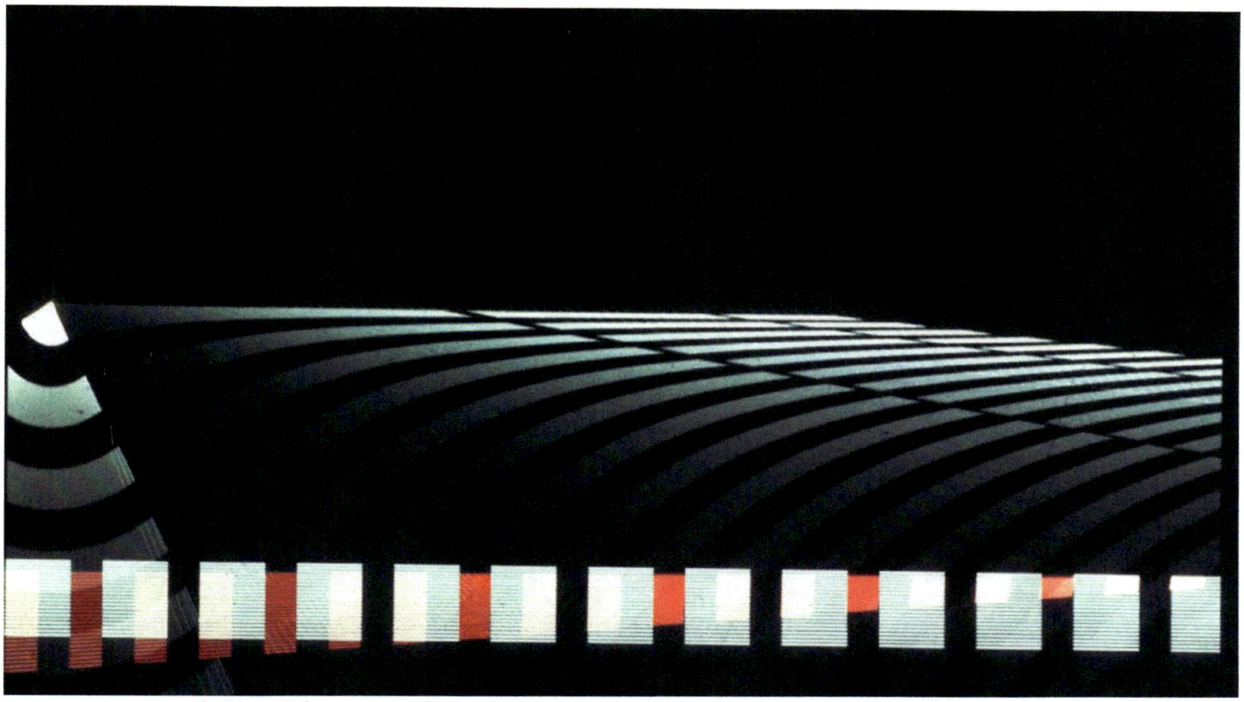

Rhythm. The output from the computer code. © 2025 by Anna Ursyn

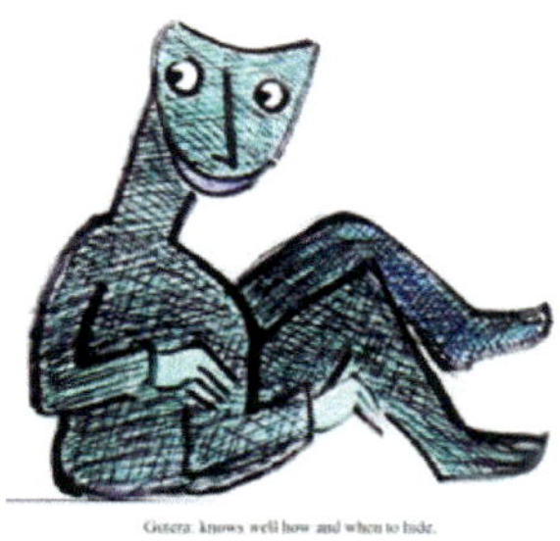

Gutera: knows well how and when to hide.

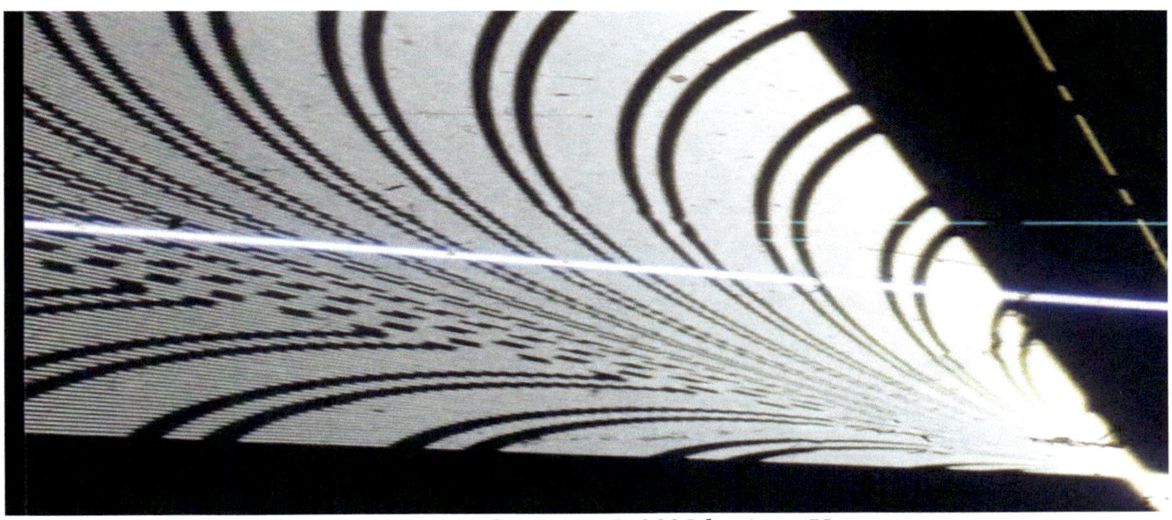

The Turns. © 2025 by Anna Ursyn

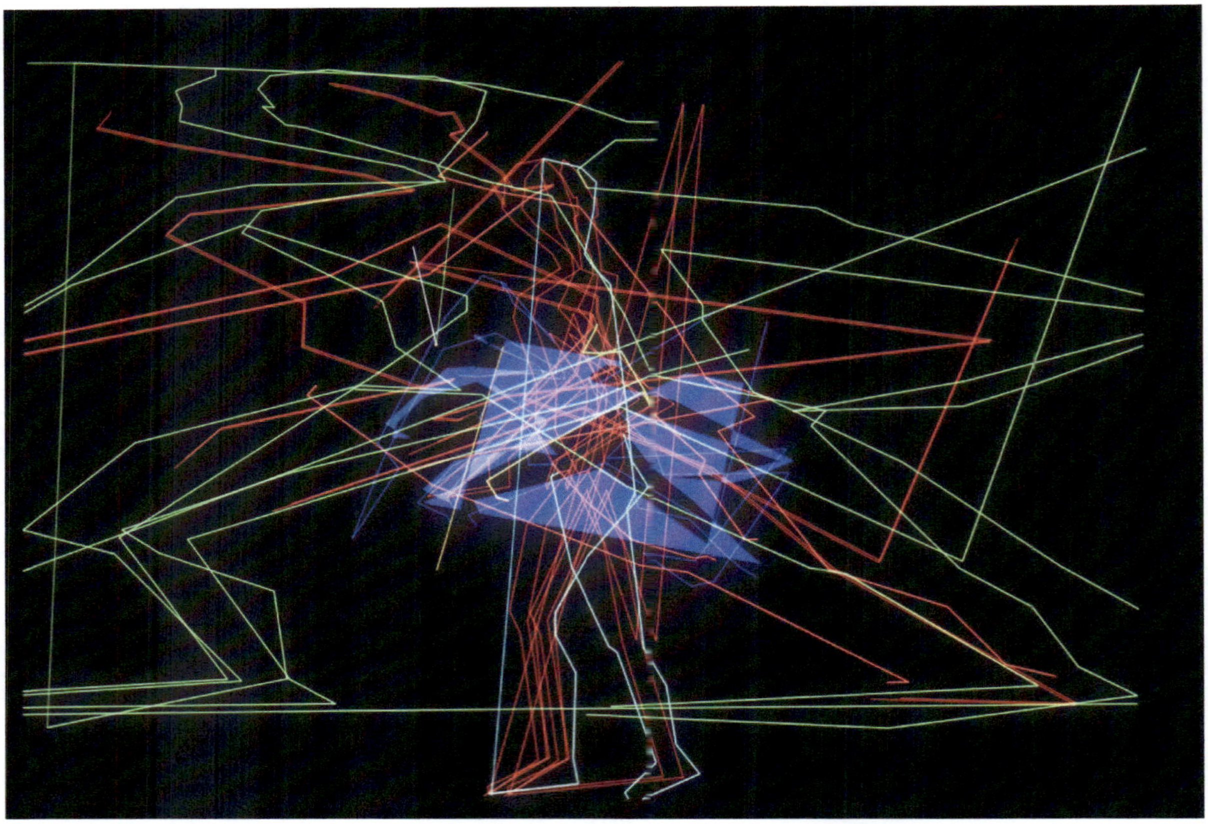

People. The output from the computer code. © 2025 by Anna Ursyn.

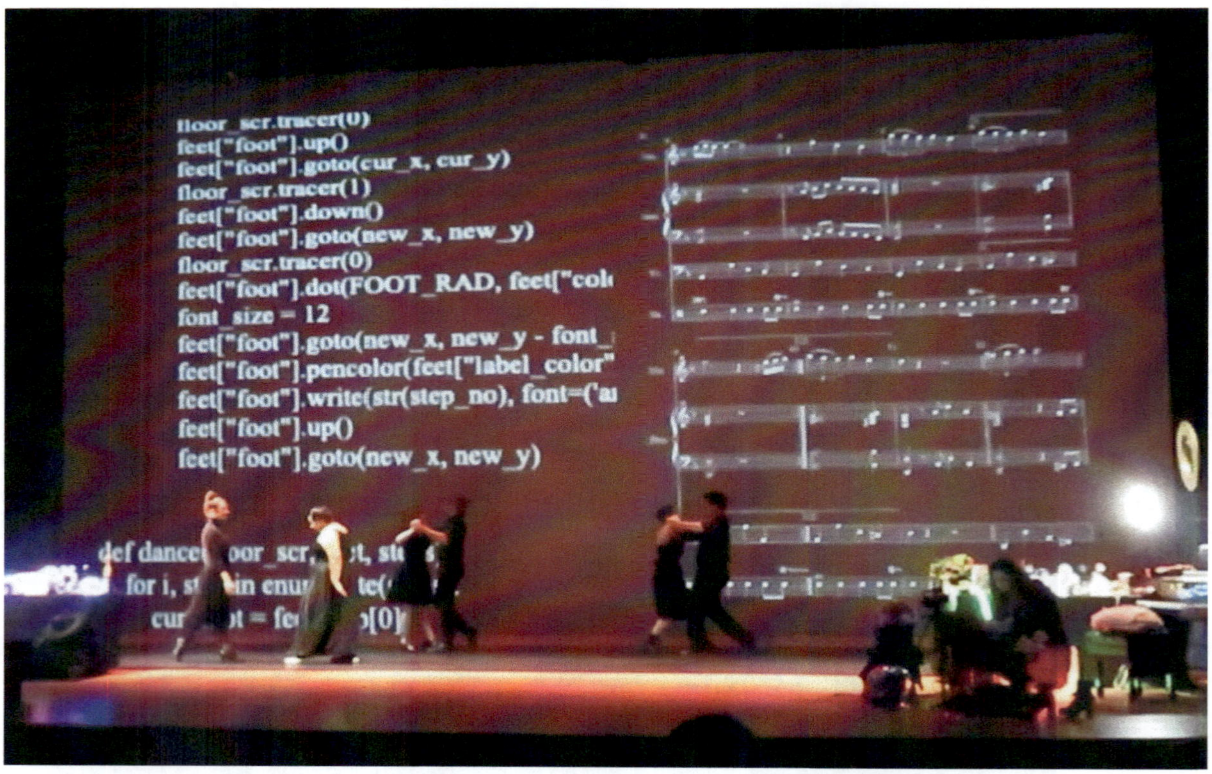

Dancers, projected codes, and music notation – a scene from a theater performance.

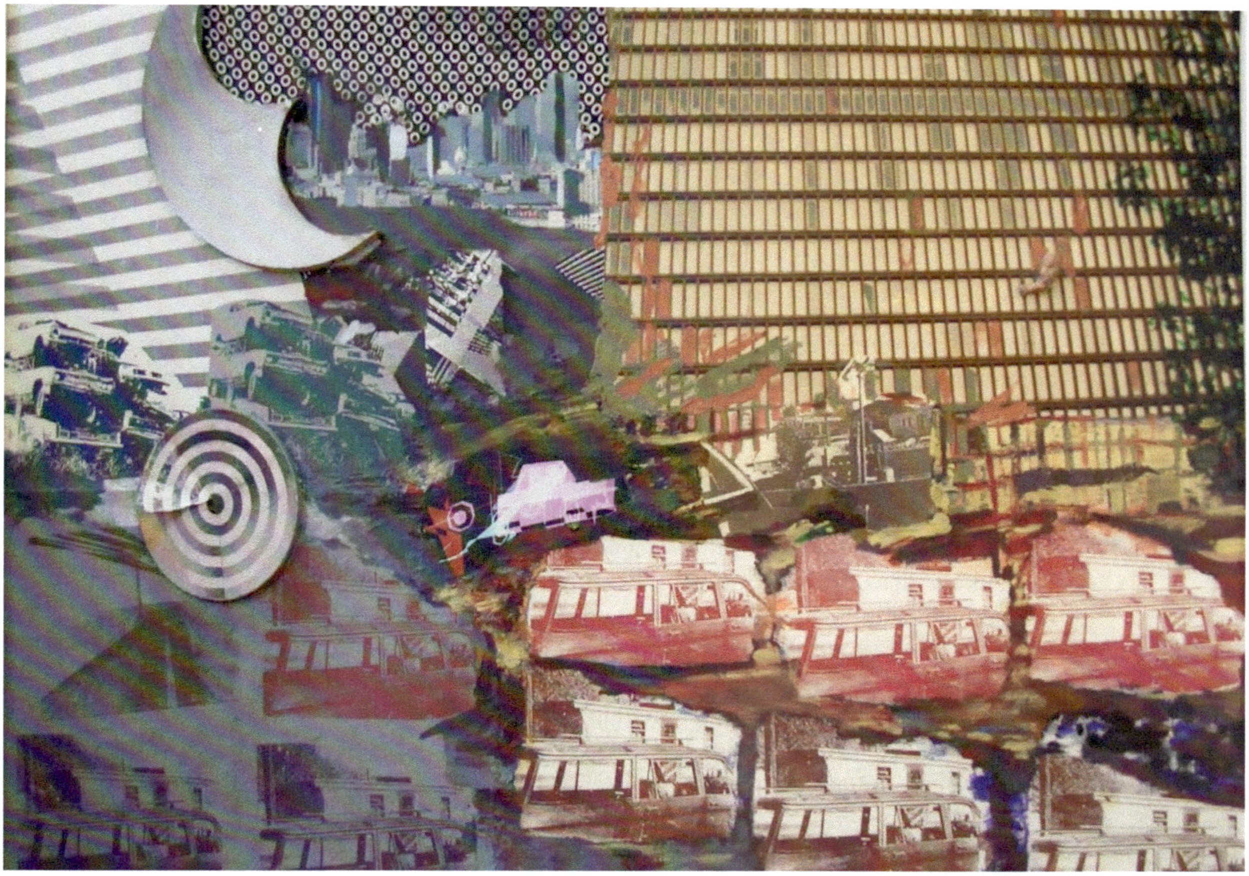

Two Moons. © 2025 by Anna Ursyn

CHAPTER 2

DANCE CODE – INTERNET EXCHANGE

This chapter tells the story of an online interaction between a coder and a prima ballerina that resulted in a shared understanding of their respective fields. The script was converted into theatrical performances through an international collaboration with respective codes, available on the accompanying website for people to explore further.

This story encourages readers to become more open to coding. In a short internet exchange between the two professionals, dancing is a metaphor for programming, so the dancer is ready to code, and the coder starts dancing. Thus, the project attempts to facilitate learning how to comprehend programs by loosely translating dance performances into codes. Hopefully, readers and viewers will start thinking about computing as something doable. Thus, this story-based learning project links dance performances and music with computing codes. Scripts in several programming languages used for animated videos and performances present general ideas behind programming and are credited at the end of the book.

Several performances and dances are introduced in the following text, with various forms, steps, rules, and conditions juxtaposed with steps in coding. It simplifies introducing basic conditions, restrictions, and ideas related to coding and dance to someone unfamiliar with these areas.

There is a universal need to help people become fearless of coding, as it improves communication between them and the coders to bridge a gap between them on the job. The resulting materials serve students in computer graphics, digital media, or anyone interested in understanding the principles and logic behind coding, which hiring companies often request. They also help understand how machines are set when writing a prompt.

SYNOPSIS (A PLOT)

Dramata, a well-recognized, self-assured dancer, reads a letter from Kalizuko, a timid computer programmer who has recently watched her dance, which made him aware of parts missing in his life to be a whole one. Kalizuko, a modest coder, is afraid that the famous prima-ballerina could ignore him; thus, he acts strongly to schedule an online meeting. They discuss their fields, as they initially thought, as very different disciplines, finding out that both are based on steps. They soon discover that they have a lot in common. Dramata, a bit spoiled by fame and her success, begins to understand Kalizuko's ideas, so they compare the guiding rules, restrictions, and norms in their respective disciplines while explaining them one to each other. Self-aware Kalizuko becomes more secure after including arts, while somewhat bossy, rule-driven Dramata finds her interest in learning and discovering new ideas. They keep comparing and contrasting their disciplines, which helps the audience explore the hidden nature behind the rules that shape each field. The dance-based codes are provided on the website for the audience's further explorations.

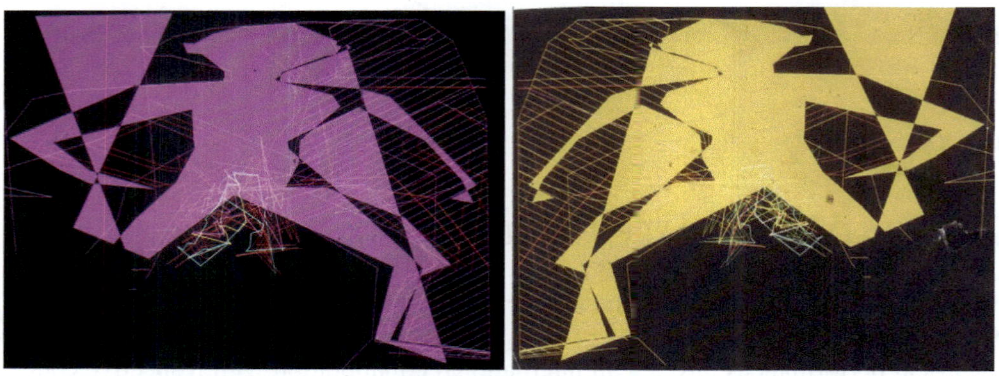

Men. Outputs from computer codes. © 2025 by Anna Ursyn

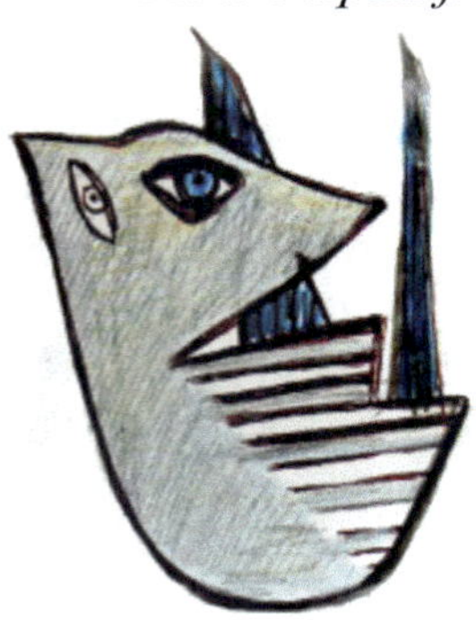

Laxura: sees it all.

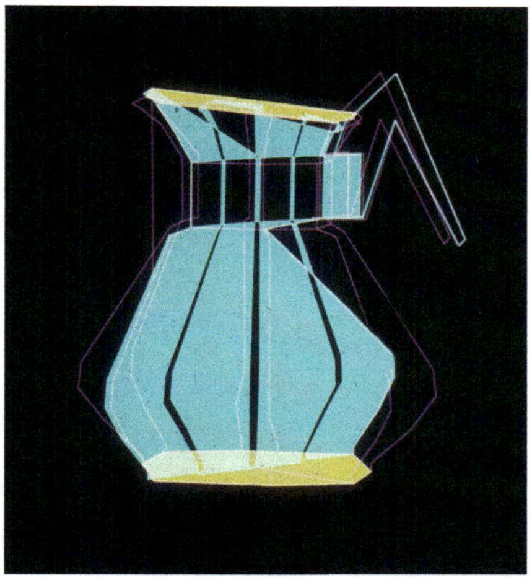

A Teapot. The output from the computer code. © 2025 by Anna Ursyn

CAST OF CHARACTERS

Kalizuko, a coder
Dramata, a dancer
The dancers

SUMMARY

Two characters sit in front of their laptops interacting, while a well-lit, colorful space on the stage allows the dancers to explain major coding and dance-related ideas in front of projected codes, visuals, and props. This performance involves art, coding, music, dance, choreography, video, mathematical art, stage design, and web development. In this interdisciplinary and collaborative project, we are working with colleagues from Computer Science, Dance, Music, and the Arts, as multidisciplinary collaborators in many countries. Many talents contributed with their research, codes, voices, characters, and art projects to this enterprise. There is some singing to the code on display, and also to a poem about a coder who taught a dancer about coding. The dancer does what's not permitted in programming, and the dancer who taught a coder dance sees what's not generally accepted when presented by their student.

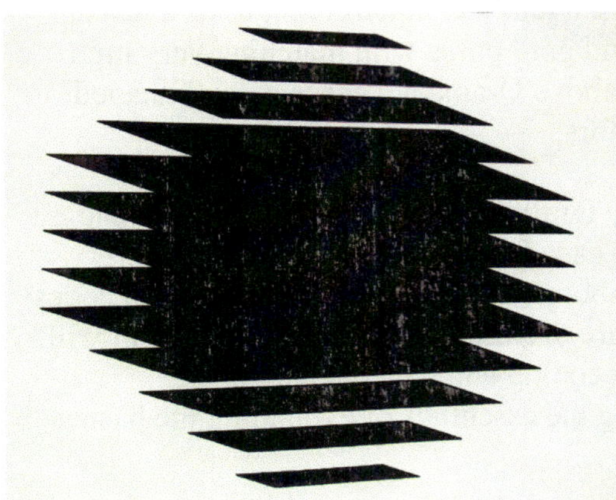

A Paradox. The output from the computer code. © 2025 by Anna Ursyn

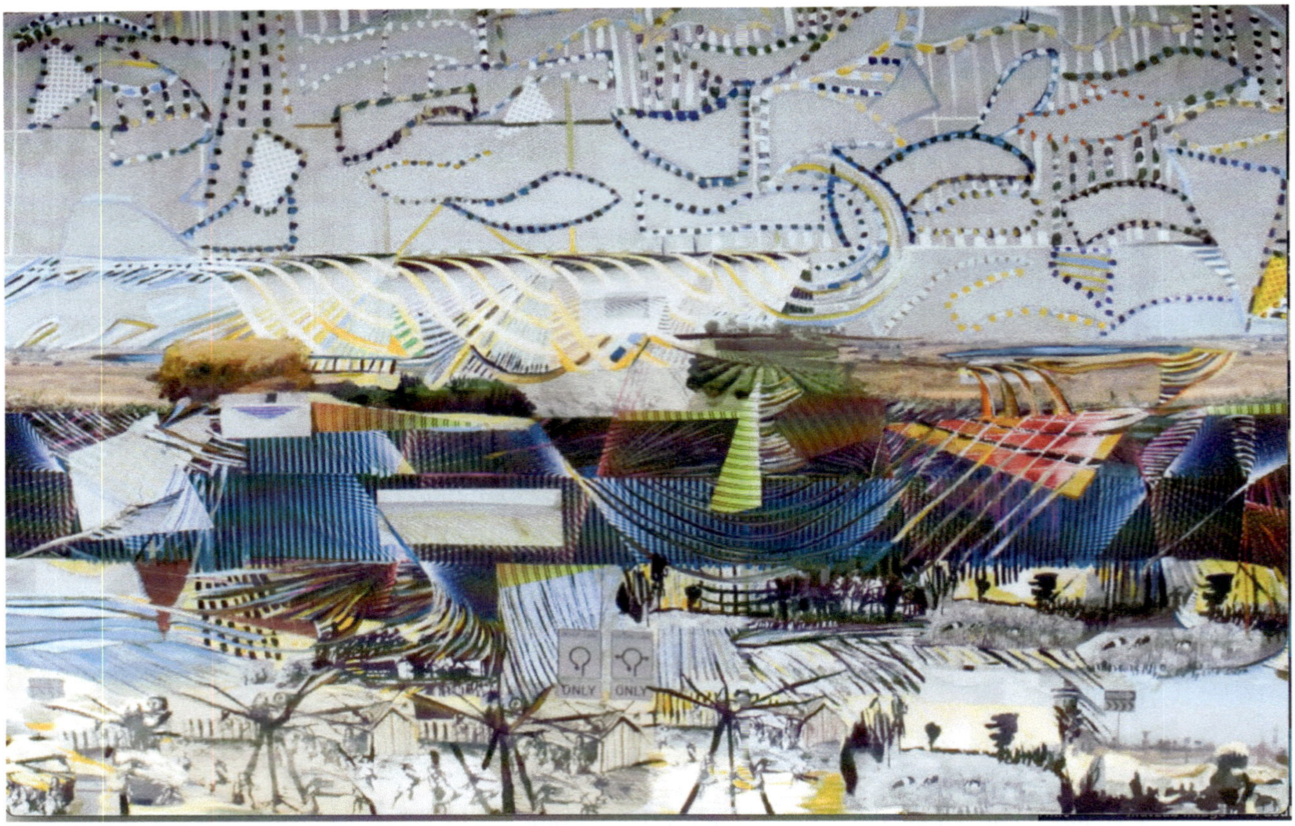

Rondo. © *2025 by Anna Ursyn*

STAGE DIRECTIONS

Kalizuko sits on a blue-lit cube in front of his laptop, while Dramata sits on a red-lit cube on the right side of the stage with the grand piano. Both corners are decorated with lights reacting with reflective surfaces. Their rooms are filled with typical props. Kalizuko has maps, clocks, computers with multiple screens, neon signs, an aquarium, an oversized black stuffed teddy bear, LED lights, sport-related props, messy shoes, shirts, etc. There is a large workshop-style mirror hanging. Balls and balloons are shaped like our planet, and the moon lies around while a large crescent silver moon hangs above Kalizuko. Dramata's space is organized. There are flowers, a jellyfish aquarium, masks, a teacup, instruments, a dressed-up dog, and her silver and gold purse with matching very high heels shoes. A butterfly-shaped balloon hangs above Dramata's space. Cloud-shaped balloons connect the space between the two actors.

The following text, a script for the *Dance Code* familiarizes the reader with the dialogs, an essential feature of the story that serves as a base for the performance. The text following this script will provide a detailed, elaborated, and illustrated script with codes written for dances that Dramata and Kalizuko are discussing. This enhanced script will offer the readers a more substantial view of the coding and dancing analogies, clarifications of codes, and guidance concerning the essentials of explaining the basics of writing codes through the metaphor of dance.

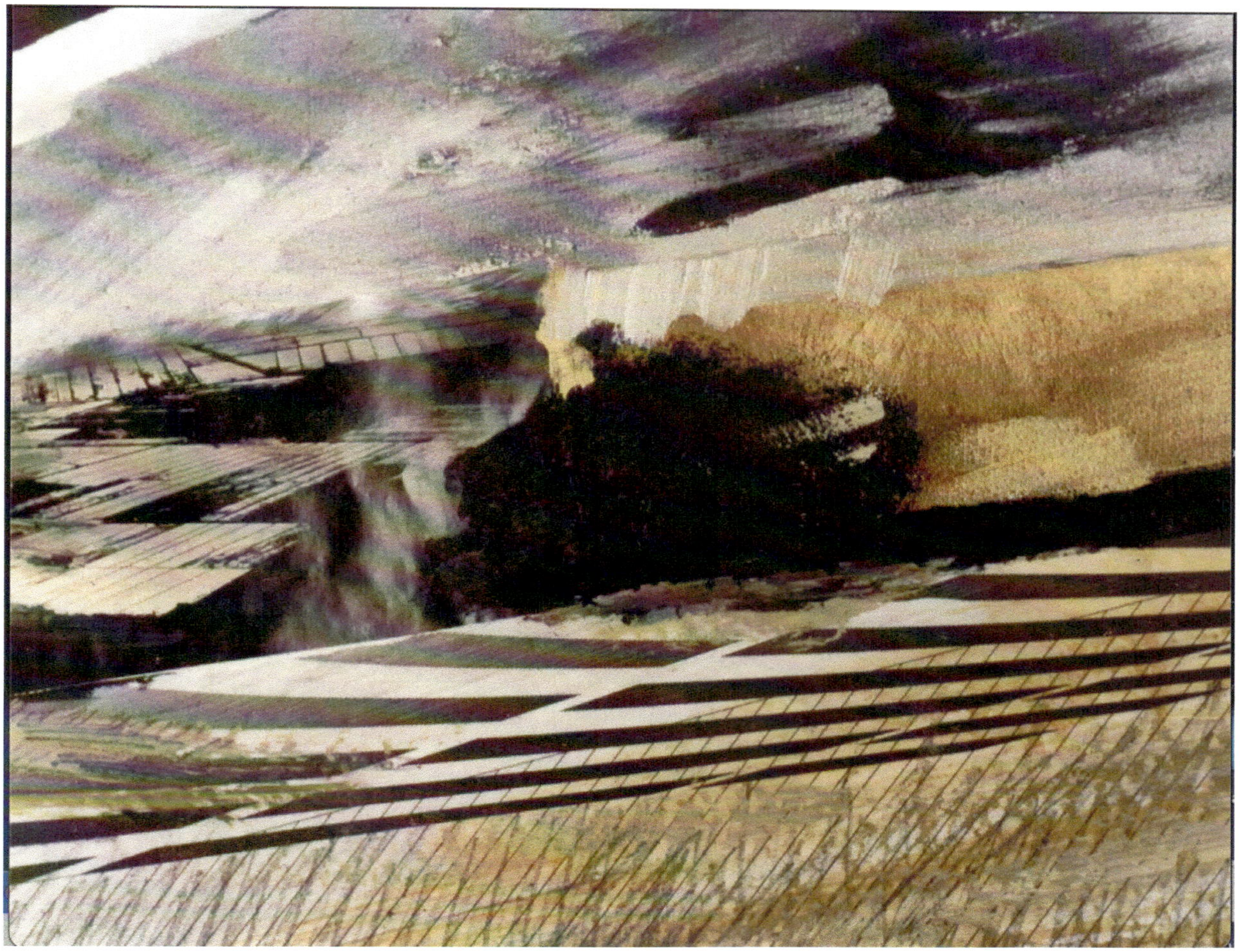

Goldview. © *2025 by Anna Ursyn*

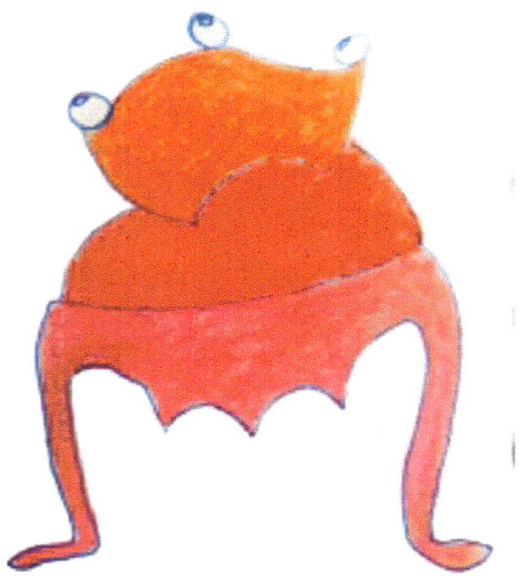

Husago: is happy, sad, or just is.

Hutago: the happy one.

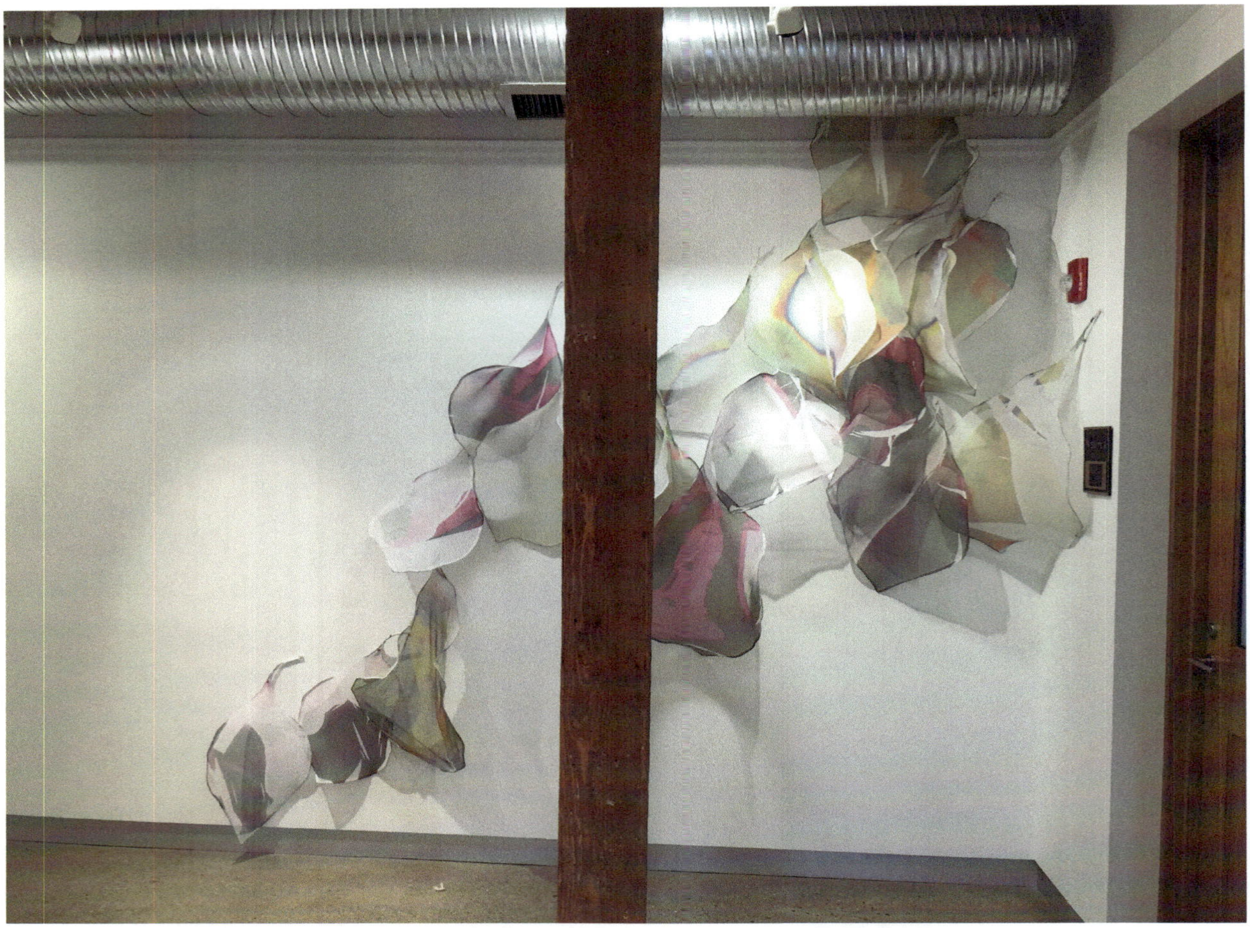

Artwork by Joohyun Pyune. (Used with permission).

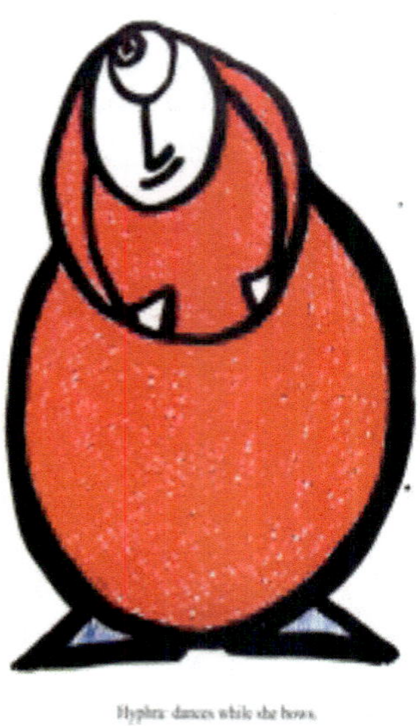

Hyphra: dances while bowing.

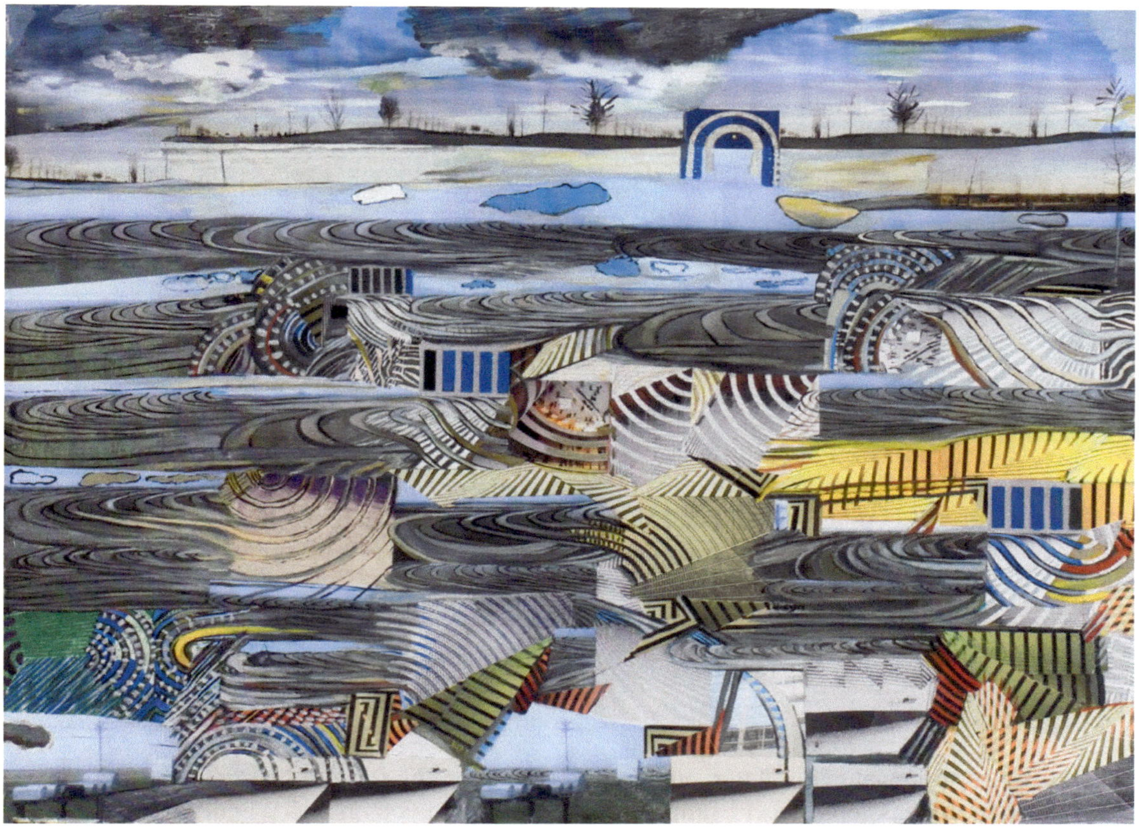

Fragile Balance. © 2025 by Anna Ursyn

CHAPTER 3

THE SCRIPT FOR DANCE CODE

The curtain goes up (or opens)

Kalizuko: "Dear Dramata,

Please forgive me for writing to you. I know it comes as unexpected, but I am simply urged to share my strong senses with you, so please read it unless you want me to stay alert, nervous, unhappy, and destructed. I am in your hands: I saw you on TV dancing one night and cannot shake the feeling off. The rhythm and structure caused some consequences and conditions embedded into your movement; it was all like some instructions would come up to you from above, causing you to create new solutions, an environment, and very special air to breathe. This night was such a multisensory experience, and I saw the arts in a totally new light, causing my brain to function on a new level. I cannot lose that experience; I cannot forget the unique organization my brain went through, which created a new order for me to cope with information, knowledge, beauty, and all connecting movements. I used to think about the world being 2-D, flat, 3D, sculptural, and 4D, time-based. I used to enjoy looking at images by changing my distance to see less to focus or more to appreciate and connect. I used to walk around sculptures to see their different dimensions, structures, or variations, or I enjoyed architecture by seeing it from all around, from the inside, or from above. I used to appreciate work based on time, from reading a book to listening to music. I enjoyed musicals combining images, voices, and motions.

I cannot name the number of "Ds" you added to my world last night just by dancing. Now, I can see it all in 4D, 5D, and even more. What I felt during your performance is hard to describe; elation might be the closest word, and I only experienced that once before, when I saw a butterfly for the first time.

I am begging you to meet with me via video conferencing so we can talk, and all my questions, excitement, and anxiety can be resolved one way or another. I do realize that you are a talented, well-known artist, and I am but an unknown coder, yet it causes me to think about your art and craft, pushing me to write this letter and beg you for a virtual meeting. We live so far apart, but I am hoping that when I explore some options with you, you yourself will ask me for yet another meeting. Please do not think of me as callous or self-important. I am the opposite. It is just that something happened to me, and I cannot resist contacting you and risking going on with my life without losing my sleep. We all know that might lead to some special circumstances we would all appreciate less, so please take it as a social duty to release my tensions."
Sincerely,
Kalizuko"

Dramata: "Dear Kalizuko,
I surely appreciate your letter, and despite coding or any form of computing going beyond my passions, I feel I simply have to meet with you online. Wednesday at 3 a.m. works for me. You need to set it up, as for me computers are for typing and answering excessively long and demanding messages. The same applies to the other arts: I cannot draw a stick figure.
Sincerely,
Dramata."

Kalizuko: "Dear Dramata,
Wednesday at 3 a.m. it is; ops, this is Wednesday, 3 a.m. already, so I am connecting us now. Bear with me. It will take just a minute.
P.S. I cannot dance. I never could, but I enjoy watching people dance very much.
Sincerely,
Kalizuko."

`Kalizuko and Dramata began interacting online via video conferencing, which Kalizuko had just set up for them. Their faces are projected on the big screens.`

Kalizuko, "Hello"
Dramata, "Hello"
Kalizuko, "How are you"
Dramata, "I'm fine. And you?"
Kalizuko, "Fine. Thank you for asking. Have you read my letter?"
Dramata, "Oh, yes. Therefore, I must be here."
Kalizuko, "Oh, that is such a relief. I am a simple coder."
Dramata, "A coder?"

Kalizuko, "A programmer. It is not so simple. I mean, it is quite complicated, but people are so afraid of my profession that it makes me feel smaller and less important. I do many good things that you benefit from in your daily life…"
Dramata, "Enough, I know, but it does not interest me. I am a dancer and only function in a step-by-step-based world. I think in steps. My life consists of steps, routines, tempos, values, and more. I need to know music and, of course, all the elements and principles of design. I need to know literature, as they often want me to sing texts while dancing, but I am a dancer. I always was, and I will always be!"
Kalizuko, "*I* know, but when you dance, why do I feel so connected to you and so involved? My job and passion are the routines, steps, conditions, and all the resulting consequences. I also think in steps. Maybe I am losing my sleep because I can see it all in your craft."
Dramata, "My craft is art! Do not offend me."
Kalizuko, "There is nothing wrong with craft. You sleep under a cover and have a keychain on your car keys."
Dramata, "Nope. I have a manufactured weighted blanket and open every door with my phone."
Kalizuko, "So nice to hear. You are immersed in technology, then."
Dramata, "I did not make it. I bought it, and my dancing provided for it."
Kalizuko, "So, I see, we have much more in common."
Dramata, "What?"
Kalizuko, "We both follow steps, routines, and conditions. We create new content or make the old content more appealing. We are both structured and organized but playful and imaginative."
Dramata, "But I only can think in a linear fashion."
Kalizuko, "Even when you improvise?
Dramata, "Well…."
Kalizuko, "See?"
Dramata, "Pardon me?"
Kalizuko, "Well, we are the same thinkers. We are both creators and developers, often bending the rules but guided by some."
Dramata, "Oh, I see."
Kalizuko, "Would you describe your biggest challenges for me, please?"
Dramata, "Yes, there is a tango or a waltz. Tango originated in Buenos Aires, Argentina, in the 19th century. Working-class women danced it and then became partner dancers. Argentine tango is highly improvisational; there is also an International Ballroom style. The basic step consists of a slow 1,2, slow 3,4, quick 5, quick 6, and slow 7,8 in an 8-count pattern where the lead begins forward on their left and the follow steps back on their right. They often have me dance it, but I simply do not follow it well. There are steps and all, but it does not appeal to me together."
Kalizuko, "Which way?"
Dramata, "*The* turns."
Kalizuko, "Here is a sequence of how they could be defined through visually recorded steps:"

`A photograph of a sequence with still frames of the dancers dancing tango appears, showing dancers performing informal tango dance frame by frame.`

Dramata, "I know what the dance looks like!"

Kalizuko, "I can simplify it for you, let's look at a graph."

A graph of an output from a computer code appears on the screen.

Dramata, "*But* it is an abstract!"
Kalizuko, "*There* is nothing wrong with abstracts, but let me show you how the codes work."

Kalizuko, "Typically, you'd need a partner to dance it with, so perhaps we can solve it by looking at a code. Give me a second. Make yourself a nice cup of tea, and I will show you the code: "

A dancer enters the stage carrying a giant teapot and tries to pour some tea for Dramata into an oversized teacup, but the dancer with the teacup runs away from the teapot-carrying dancer.

Dramata drinks her tea while Kalizuko works on a program. Then, Kalizuko shows her the code written in Python. The output of basic tango dance steps and the respective musical notes appear on the screen.

Dramata, "I cannot understand a single line."
Kalizuko, "It's all in English

A short video showing coded steps of the dance appears on the screen.

Dramata, "This I can refer to."

Kalizuko, "Let's see what coding is about."

Another animation of dance steps (from a code) appears on the screen).

Kalizuko, "Here's another example of a tango in a different language: C # (C Sharp) in Visual Studio."
Dramata, "Oh, I see."
Kalizuko, "Can we dance it now?"

Six dancers, dressed in black outfits, dance. (tango).
The actors dance it, each differently while looking at their screens.

Dramata, "Now I get it. OOOO, I made it. I made proper turns."
Kalizuko, "Can you make some unproper ones now?"
Dramata, "Beg your pardon?"
Kalizuko, "Oh, well, try to release your tension in old patterns and relax the rules a bit. Keep in tune with the general rules but try to listen to and focus on the music more and see what happens."

Dramata makes a big, indescribable sound.

Kalizuko, "See?"
Dramata, "See! Now, the waltz, please. Waltz is an 13th-century Bohemia, Bavaria, and Austrian word derived from the word Walzer (to roll or revolve). Dancers perform it in the closed position in ¾ time. Steps include a basic box step, natural turn, and reverse

turn danced counterclockwise around the room, with the lead stepping forward on the left foot and the follower stepping back on their right."
Kalizuko, "All right, have some freshly brewed coffee now while I work on the code."

```
A dancer carrying a giant stovetop espresso coffee maker is
chasing another who carries an oversized coffee mug across a
scene.
Six dancers dressed in black dance (waltz).
```

Kalizuko, "Here it is."

```
     A set of photographs showing sequential still frames of
dancers waltzing. Waltz code is projected with the respective
musical notes.
```

Dramata, "I know what the waltz looks like, but the code?"

```
A code for the waltz appears, along with the musical notes,
followed
by its animated output while the dancers perform it (waltz).
```

Kalizuko, "Now, we'll work on the cha-cha. Make yourself a nice cup of..."
Dramata, "I cannot possibly drink that much!"
Kalizuko, "Eat an apple!"

```
An actor carrying an apple crosses the stage in a dignified way.
Six dancers dressed in black dance (cha-cha).
`
```

Dramata, "Can I see the rumba now?"
Kalizuko, "Sure. Make yourself a cup of cacao, and I will be right back with the code. "

```
Two actors carrying two cacao pitchers approach from the two
sides of the stage, trapping the cacao cup as it tries to run
away.
```

Kalizuko, "Here it is, the graphical output, then rumba when danced simultaneously. Here is the code for you.

```
Rumba output from the code and the musical notes, followed by an
appearance of the six dancers dressed in black dance (rumba).
```

Dramata, "That makes sense. How about the foxtrot?"
Kalizuko, "Yes, of course. Please pour yourself a glass of water while I come up with a code."

```
A water pitcher is chasing a water glass which lovingly accepts
the liquid. The graph, the code, and the musical notes are
projected.
Six dancers dressed in black dance (foxtrot).
```

Dramata, "I feel ready to become a coder!"
Kalizuko, "Sure, you are."
Dramata, "So, how do I know which dance to code and how? Which language to use?"

Kalizuko, "For that, you will need a simulator: waltz, foxtrot, cha-cha, salsa, tango, rumba, samba, jive)."
Dramata, "Yes, swing (Lindy Hop). This dance originated in Harlem in the 1920s in the Savoy Ballroom. An aviator Charles Lindbergh 'hopped 'the Atlantic Ocean in 1927, thus the name of the dance. This dance has the 6 and 8-count patterns to 4/4 music, with the lead starting on their left foot and the follow starting on their right. An 8-count basic would consist of a rock step on 1, 2, triple steps 3&4, then rock, 5, 6, and triple steps 7&8. The 6-count basic would consist of rock steps 1, 2, followed by two triple steps, 3&4 and 5&6.
Kalizuko, "I will explore how to translate it into any other language. Please prepare some juice while I make it available for you."

Two dancers carrying juice pitchers run towards one carrying a juice glass to compete over who will fill it. Codes and music notations appear—six dancers dressed in black dance (swing).

Dramata, "I can code!
Would you describe your own challenges for me, please?"
Kalizuko, "That would be the conditions. I never know what to expect."
Dramata, "That should be easy; just dance a tango with two partners."

The three dancers dance the tango by alternating their partners.

Kalizuko, "I feel ready for a dancing career!"
Dramata, "I can code!"
Kalizuko, "I can dance!"
Kalizuko + *Dramata (together),* "Let's do both!"

Dramata sings to a computer code projected on the screen, while Kalizuko dances with a large stuffed bear.

Dramata, "You made me curious!"
Kalizuko, "You made me dance!"

The End.
The curtain goes down (or closes).

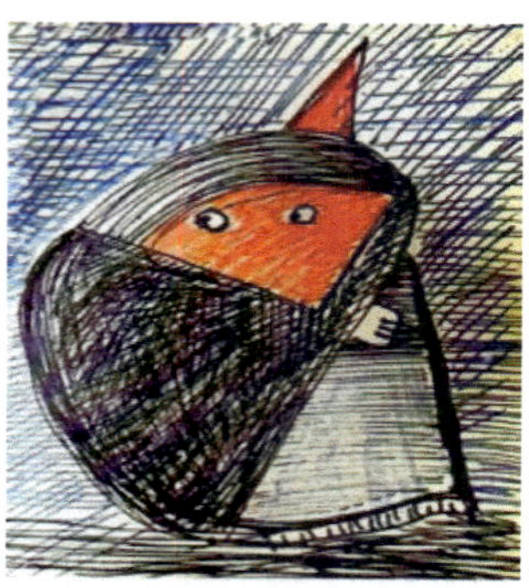

Okuda: grants one wish per day.

Okuda: grants as much as one wish per day.

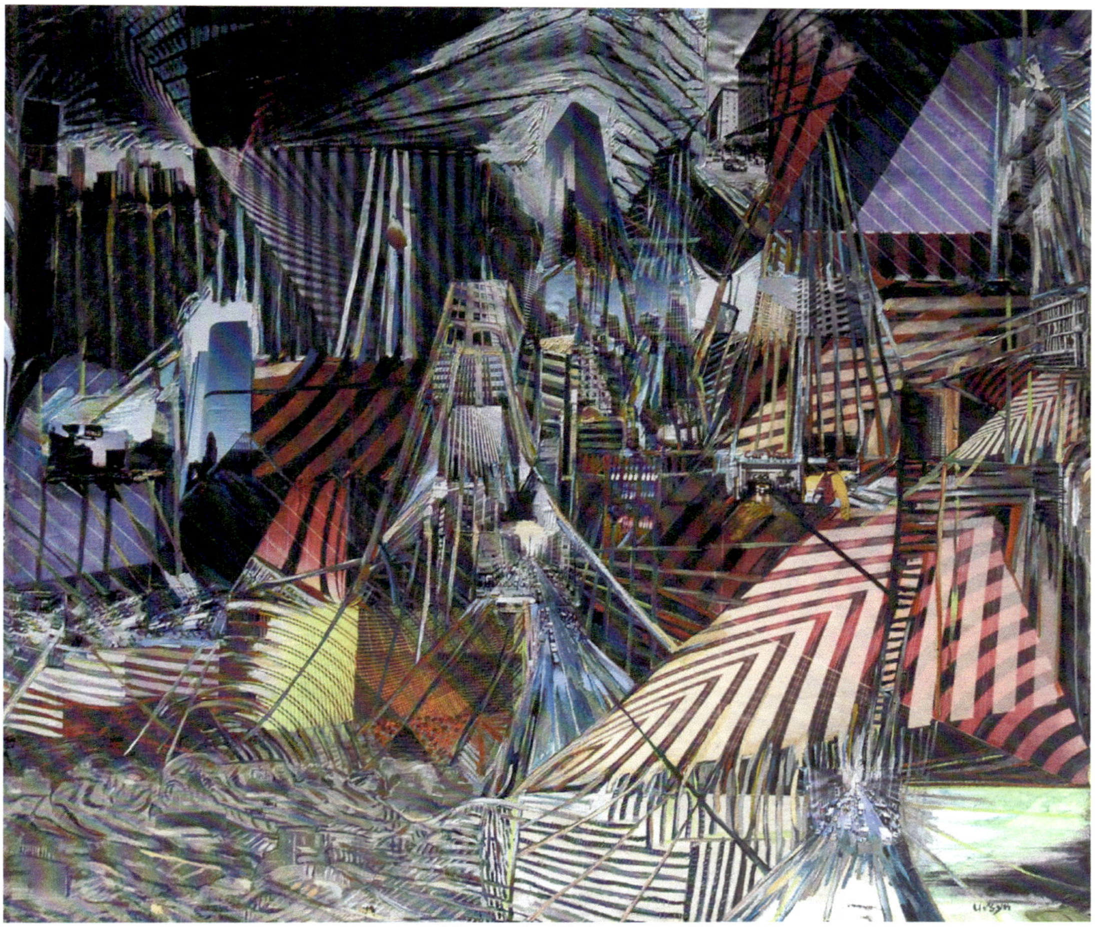

Timetable. © 2025 by Anna Ursyn

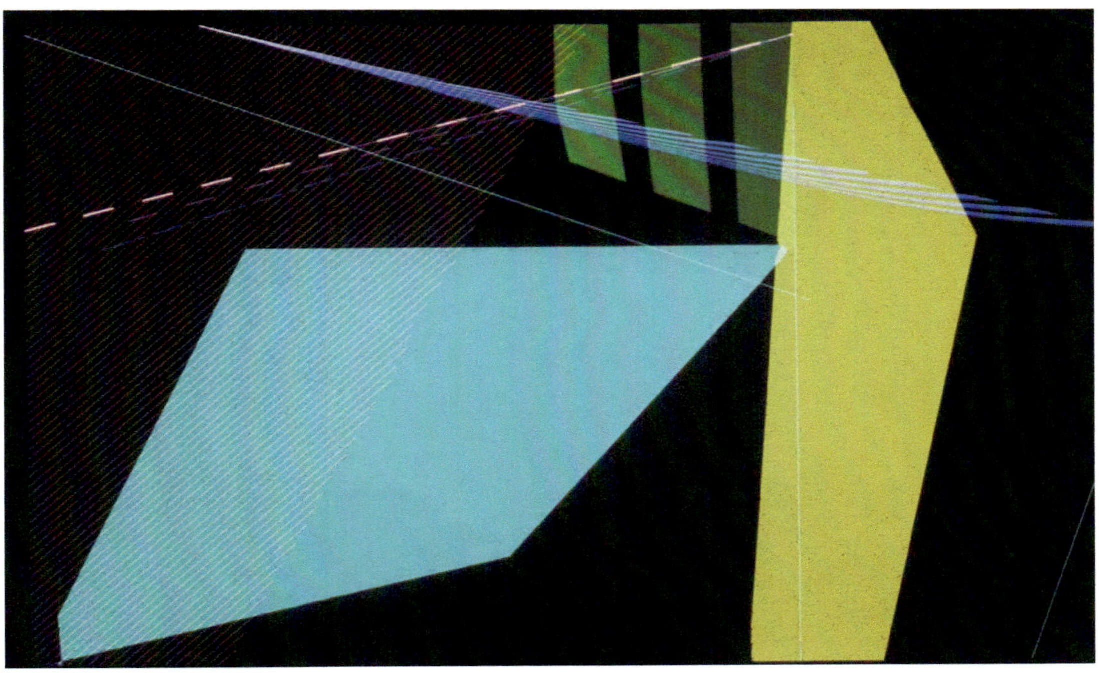

Movements. The output from the computer code. © 2025 by Anna Ursyn

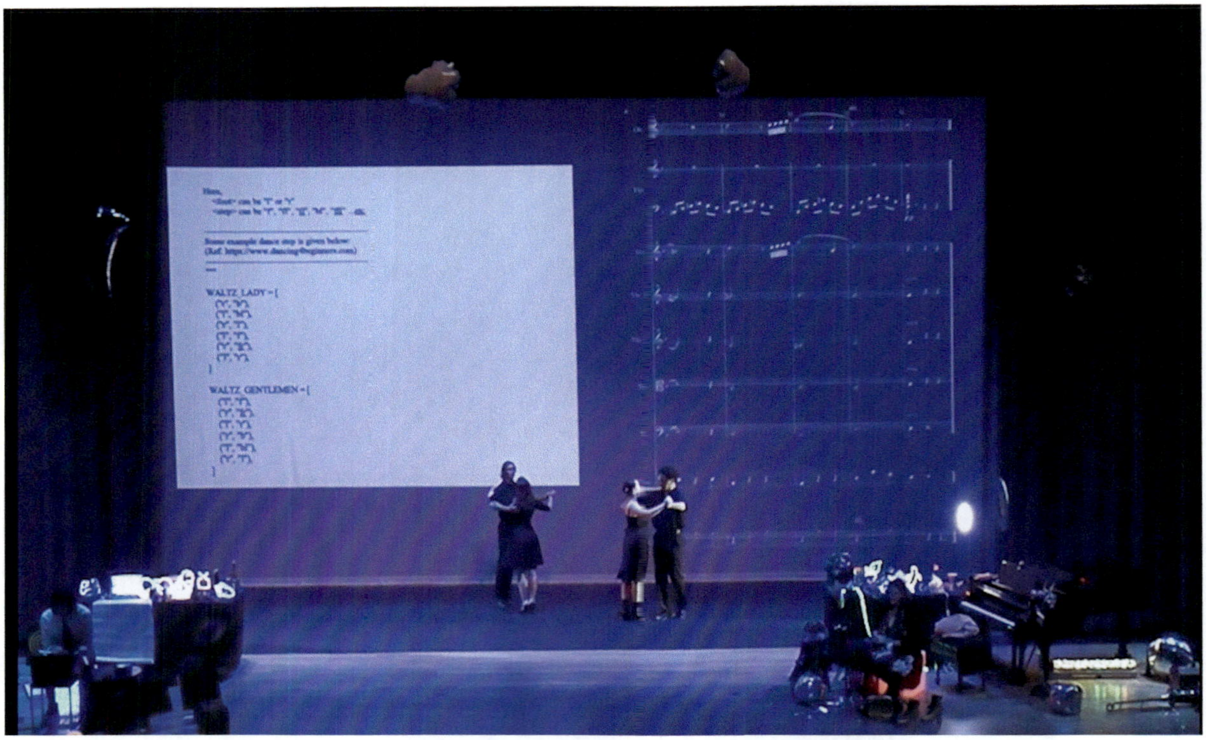

A scene from a theatrical performance.

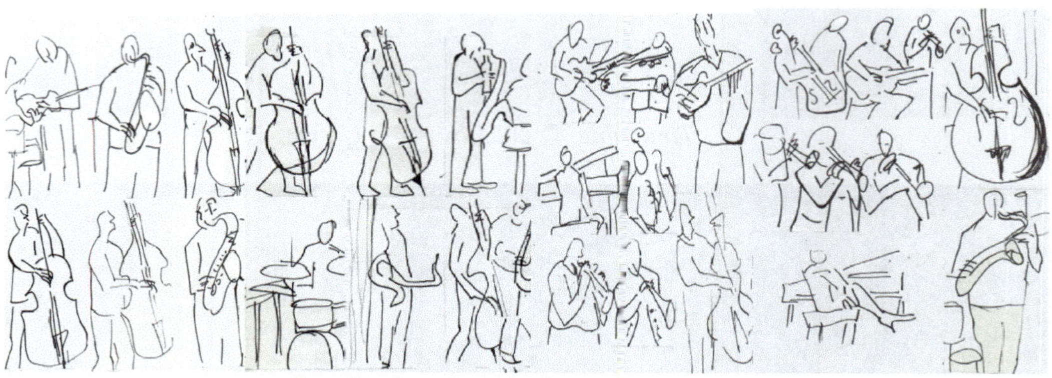

Jazz Studies. © 2025 by Anna Ursyn.

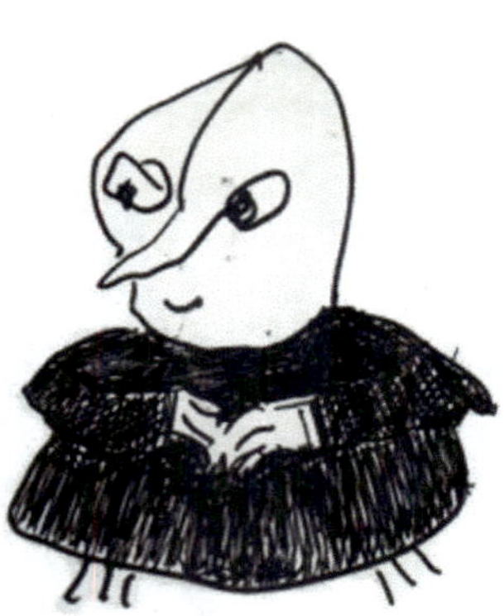

Hadalemel: knits while she walks.

Hadalemel: knits while she walks.

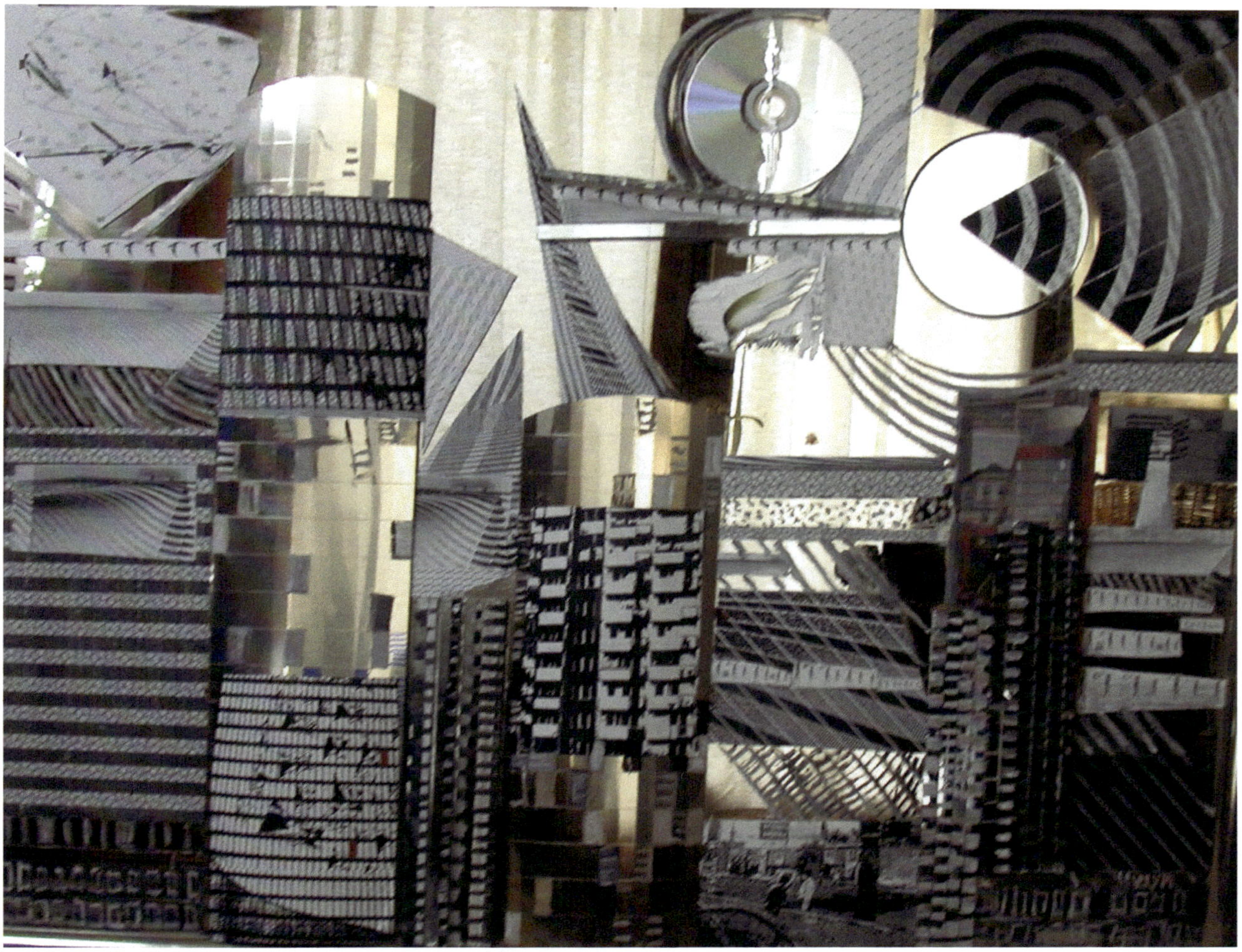

Data Mining. © 2025 by Anna Ursyn

CHAPTER 4

THE ILLUSTRATED SCRIPT FOR DANCE CODE WITH CODES, OUTPUTS, AND MUSICAL NOTATIONS

This version of the script for the performance is illustrated with artwork resulting from the codes along with their outputs, musical notations, and scenes from the actual performance. This way the reader can envision the action while examining the codes, or even copying them from the accompanying website and trying to work on them. Some instructions written by the code's authors are provided in the Appendix section of this book.

After introducing the script for the *Dance Code* to the readers and presenting its general features that serve as a base for the performance, the further text provides a detailed, elaborated, and illustrated script with codes and music notations for the dances credited to their creators. The readers can read the codes and dancing analogies, clarifications of codes, and explanations of the basics of writing codes through the dance metaphor, copy them from the website to try, and watch the resulting animations.

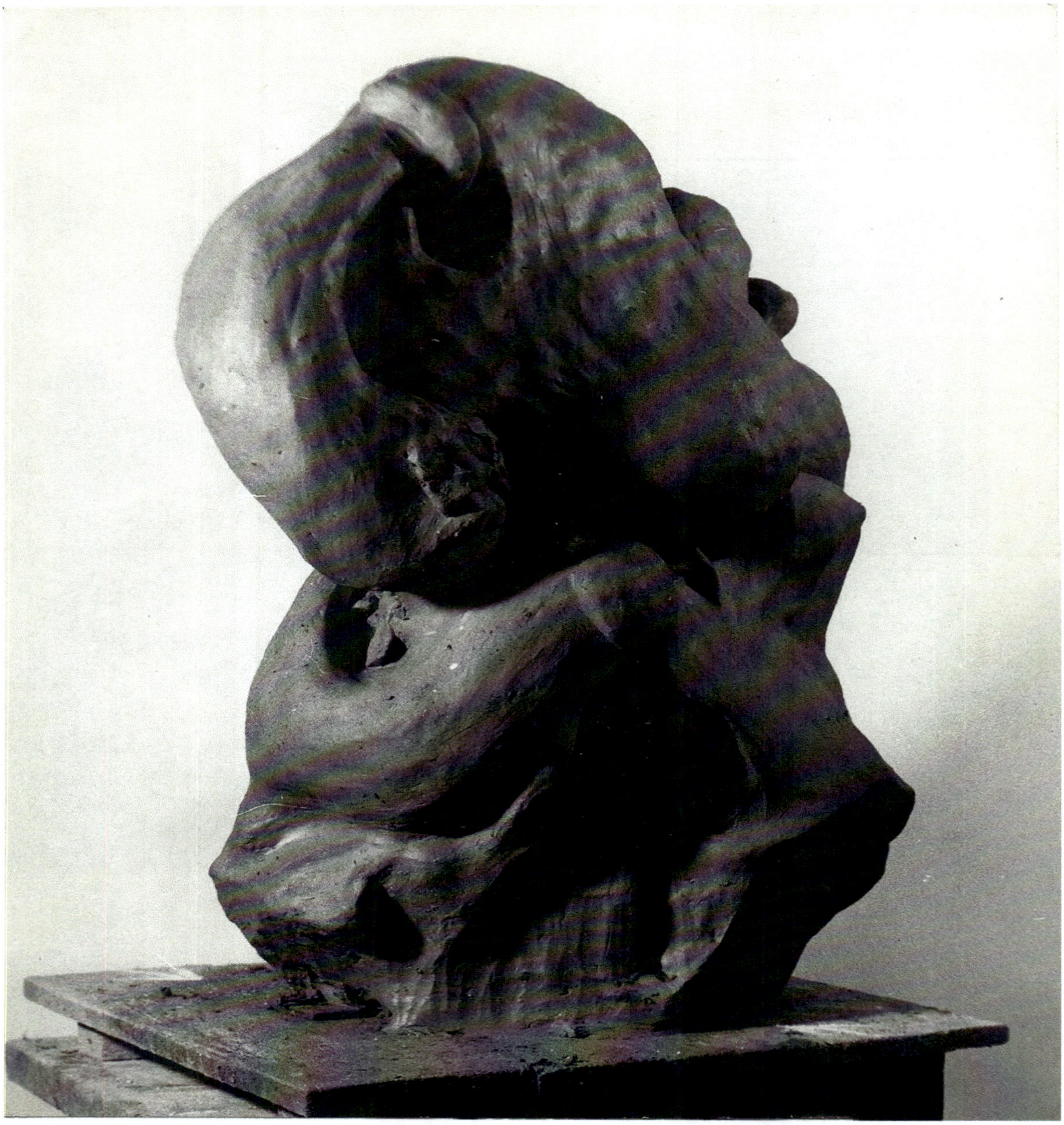

A Closed Form. © *2025 by Anna Ursyn*

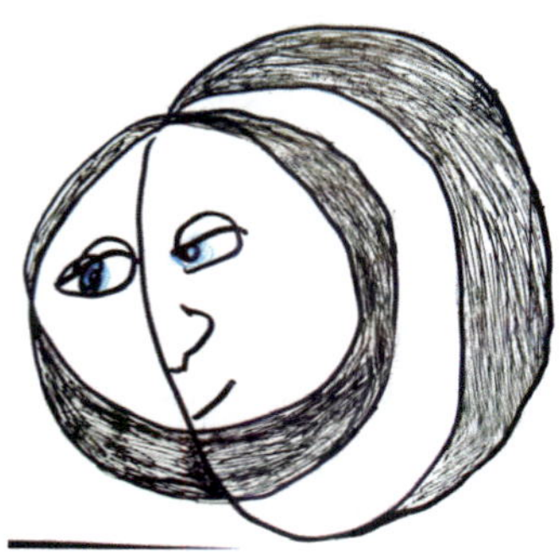

Oneva: translates it all.

Oneva: can do all sorts of translations.

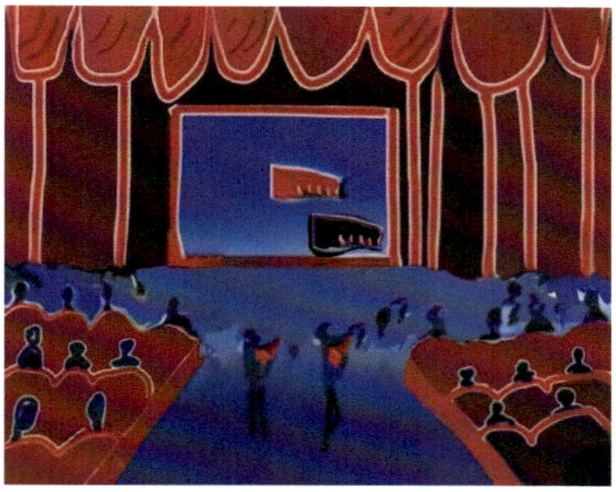

The curtain goes up (or opens)

Kalizuko: “Dear Dramata,

Please forgive me for writing to you. I know it comes as unexpected, but I am simply urged to share my strong senses with you, so please read it unless you want me to stay alert, nervous, unhappy, and destructed. I am in your hands: I saw you on TV dancing one night and cannot shake the feeling off. The rhythm and structure caused some consequences and conditions embedded into your movement; it was all like some instructions would come up to you from above, causing you to create new solutions, an environment, and very special air to breathe. This night was such a multisensory experience, and I saw the arts in a totally new light, causing my brain to function on a new level. I cannot lose that experience; I cannot forget the unique organization my brain went through, which created a new order for me to cope with information, knowledge, beauty, and all connecting movements. I used to think about the world being 2-D, flat, 3D, sculptural, and 4D, time-based. I used to enjoy looking at images by changing my distance to see less to focus or more to appreciate and connect. I used to walk around sculptures to see their different dimensions, structures, or variations, or I enjoyed architecture by seeing it from all around, from the inside, or from above. I used to appreciate work based on time, from reading a book to listening to music. I enjoyed musicals combining images, voices, and motions.

I cannot name the number of “Ds” you added to my world last night just by dancing. Now, I can see it all in 4D, 5D, and even more. What I felt during your performance is hard to describe; elation might be the closest word, and I only experienced that once before, when I saw a butterfly for the first time.

I am begging you to meet with me via video conferencing so we can talk, and all my questions, excitement, and anxiety can be resolved one way or another. I do realize that you are a talented, well-known artist, and I am but an unknown coder, yet it causes me to think about your art and craft, pushing me to write this letter and beg you for a virtual meeting. We live so far apart, but I am hoping that when I explore some options with you, you yourself will ask me for yet another meeting. Please do not think of me as callous or self-important. I am the opposite. It is just that something happened to me, and I cannot resist contacting you and risking going on with my life without losing my sleep. We all know that might lead to some special circumstances we would all appreciate less, so please take it as a social duty to release my tensions.”
Sincerely,
Kalizuko”

Dramata: "Dear Kalizuko,

I surely appreciate your letter, and despite coding or any form of computing going beyond my passions, I feel I simply have to meet with you online. Wednesday at 3 a.m. works for me. You need to set it up, as for me computers are for typing and answering excessively long and demanding messages. The same applies to the other arts: I cannot draw a stick figure.
Sincerely,
Dramata"

Kalizuko: "Dear Dramata,
Wednesday at 3 a.m. it is; ops, this is Wednesday, 3 a.m. already, so I am connecting us now. Bear with me. It will take just a minute.
P.S. I cannot dance. I never could, but I enjoy watching people dance very much.
Sincerely,
Kalizuko"

`Kalizuko and Dramata began interacting online via video conferencing, which he just set up for them. Their faces are projected on the big screen above the actors plaing their roles..`

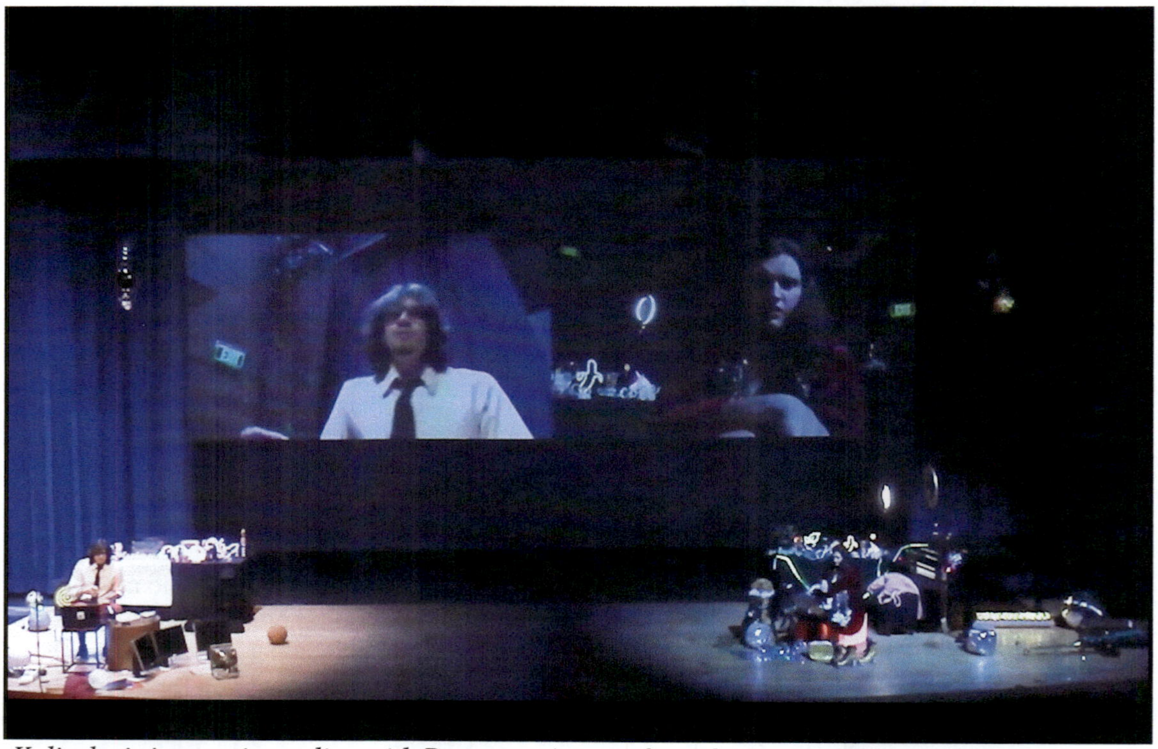

Kalizuko is interacting online with Dramata. A scene from the Dance Code performance.

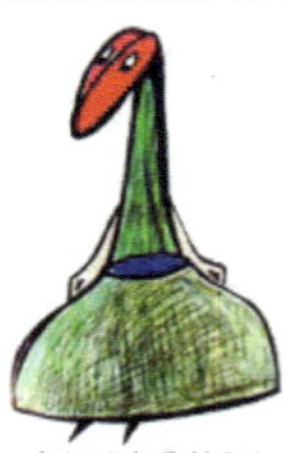

Jasatran: stays still when dancing.

Kalizuko, "Hello"
Dramata, "Hello"
Kalizuko, "How are you"
Dramata, "I'm fine. And you?"
Kalizuko, "Fine. Thank you for asking. Have you read my letter?"
Dramata, "Oh, yes. Therefore, I must be here."
Kalizuko, "Oh, that is such a relief. I am a simple coder."
Dramata, "A coder?"
Kalizuko, "A programmer. It is not so simple. I mean, it is quite complicated, but people are so afraid of my profession that it makes me feel smaller and less important. I do many good things that you benefit from in your daily life…"
Dramata, "Enough, I know, but it does not interest me. I am a dancer and only function in a step-by-step-based world. I think in steps. My life consists of steps, routines, tempos, values, and more. I need to know music and, of course, all the elements and principles of design. I need to know literature, as they often want me to sing texts while dancing, but I am a dancer. I always was, and I will always be!"
Kalizuko, "I know, but when you dance, why do I feel so connected to you and so involved? My job and passion are the routines, steps, conditions, and all the resulting consequences. I also think in steps. Maybe I am losing my sleep because I can see it all in your craft."
Dramata, "My craft is art! Do not offend me."
Kalizuko, "There is nothing wrong with craft. You sleep under a cover and have a keychain on your car keys."
Dramata, "Nope. I have a manufactured weighted blanket and open every door with my phone."
Kalizuko, "So nice to hear. You are immersed in technology, then."
Dramata, "I did not make it. I bought it, and my dancing provided for it."
Kalizuko, "So, I see, we have much more in common."
Dramata, "What?"
Kalizuko, "We both follow steps, routines, and conditions. We create new content or make the old content more appealing. We are both structured and organized but playful and imaginative."
Dramata, "But I only can think in a linear fashion."
Kalizuko, "Even when you improvise?
Dramata, "Well…."
Kalizuko, "See?"
Dramata, "Pardon me?"
Kalizuko, "Well, we are the same thinkers. We are both creators and developers, often bending the rules but guided by some."
Dramata, "Oh, I see."
Kalizuko, "Would you describe your biggest challenges for me, please?"
Dramata, "Yes, there is a tango or a waltz. Tango originated in Buenos Aires, Argentina, in the 19th century. Working-class women danced it and then became partner dancers. Argentine tango is highly improvisational; there is also an International Ballroom style. The basic step consists of a slow 1,2, slow 3,4, quick 5, quick 6, and slow 7,8 in an 8-count pattern where the lead begins forward on their left and the follow steps back on their right. They often have me dance it, but I simply do not follow it well. There are steps and all, but it does not appeal to me together."
Kalizuko, "Which way?"
Dramata, "The turns."

Kalizuko, "Here is a sequence of how they could be defined through visually recorded steps:"

`A sequence with still frames of the dancers dancing appears, showing dancers performing informal dance frame by frame (tango).`

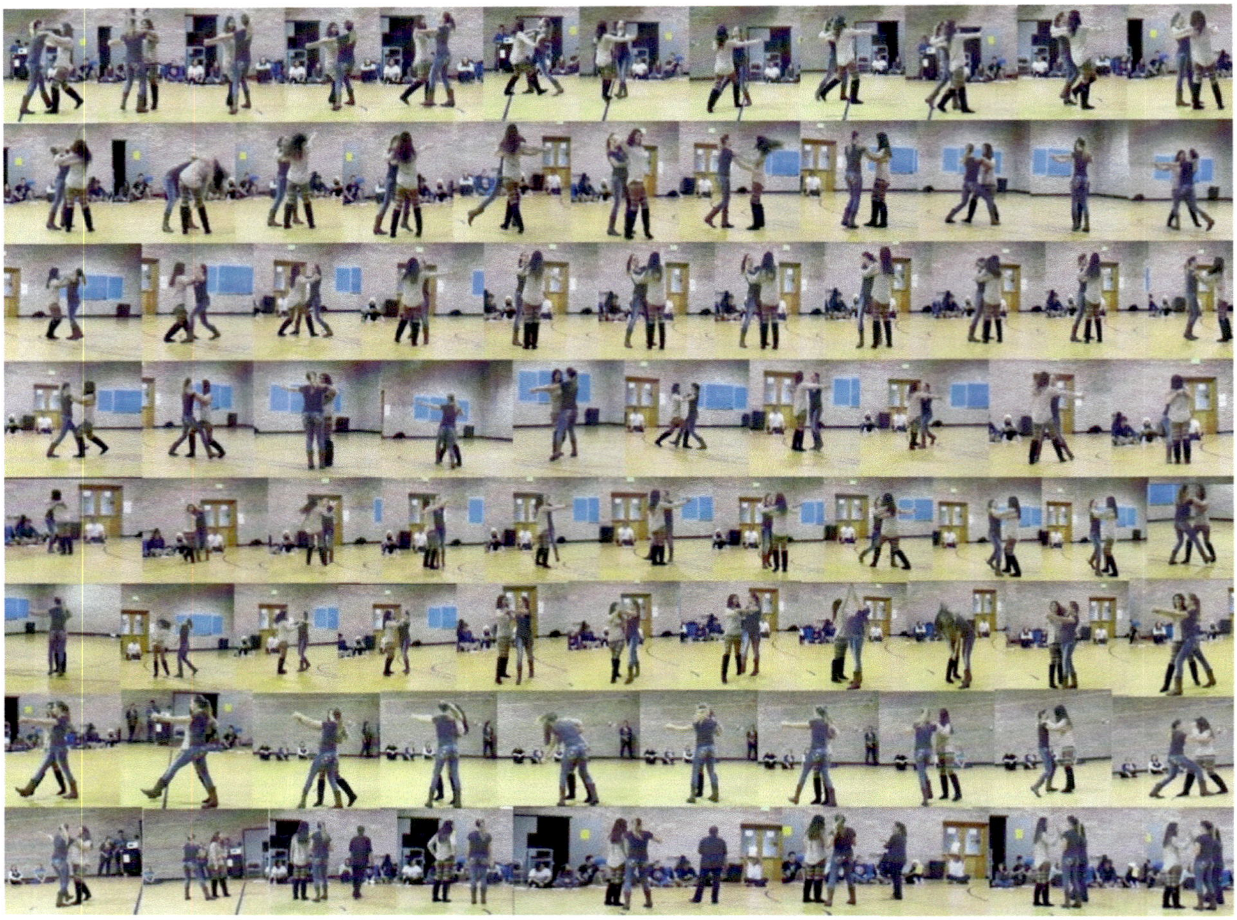

Social Dance: a tango. © 2025 by Anna Ursyn

Dramata, "I know what the dance looks like!"
Kalizuko: "Here is a simple graph showing the footwork."

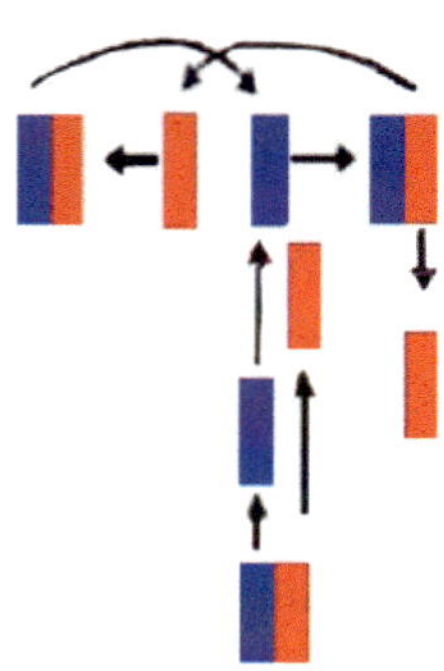

A graph showing a dancing foot pattern.

Dramata: That's a dancing foot pattern.

Kalizuko, "I can simplify it for you, let's look at a graph:"

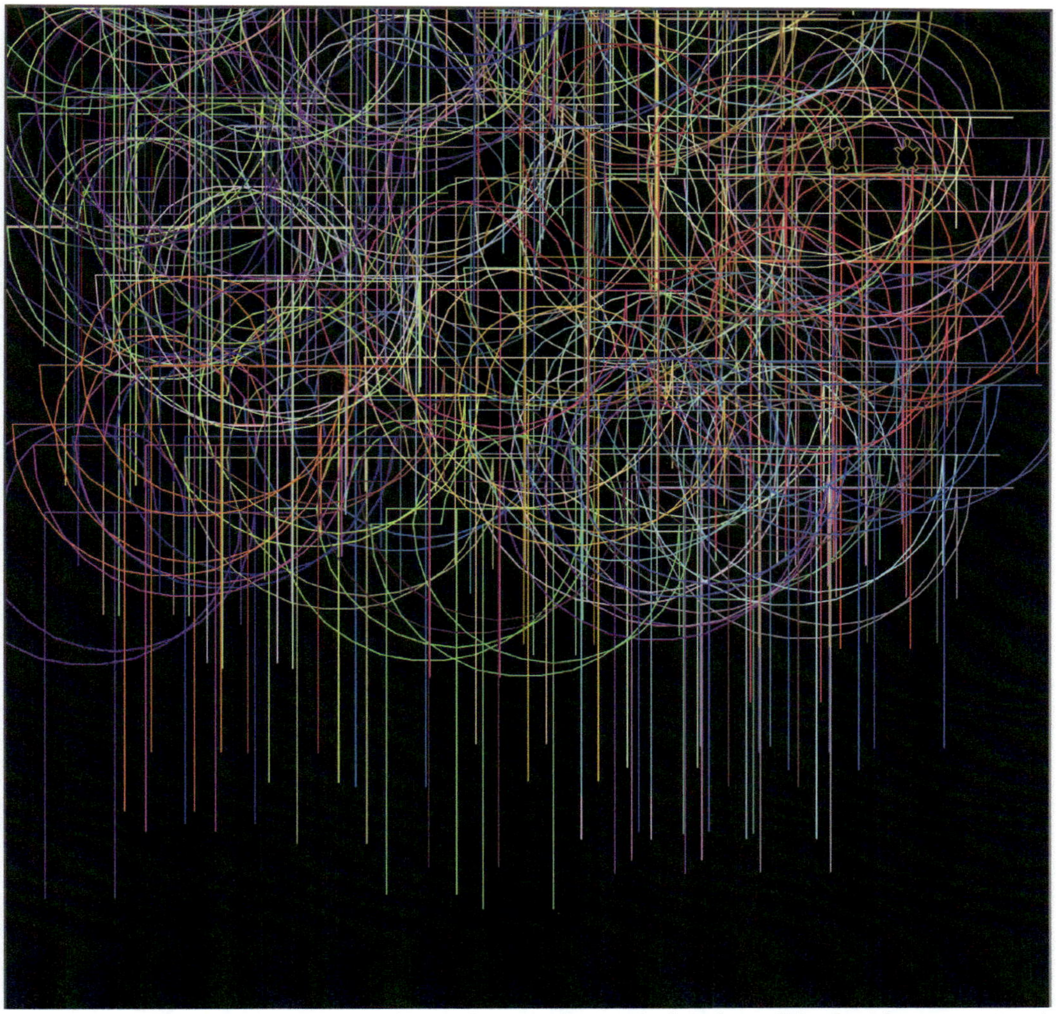

100 cycles of tango steps © 2015 by Mohammad Majid al-Rifaie. (Used with permission.)

Dramata, "*But* it is an abstract!"

Kalizuko, "*There* is nothing wrong with abstracts, but let me do it for you: I'll show you how the codes work."

Kalizuko, "Typically, you'd need a partner to dance it with, so perhaps we can solve it by looking at a code. Give me a second. Make yourself a nice cup of tea, and I will show you the code: "

```
A dancer enters the stage carrying a giant teapot trying to pour
tea for Dramata into an oversized teacup. The dancer holding the
teacup runs away from a teapot-carrying dancer.
```

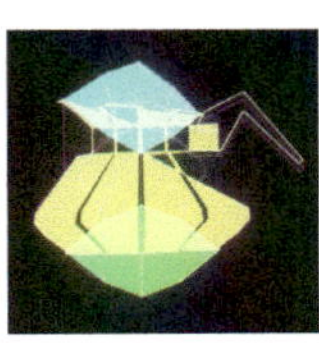

A Teapot. The output from the computer code. © 2025 by Anna Ursyn

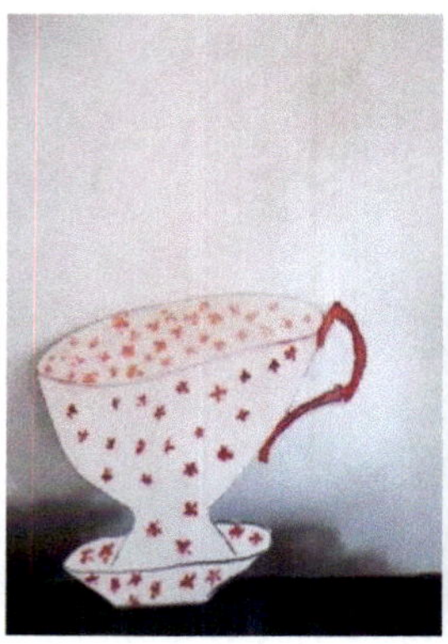

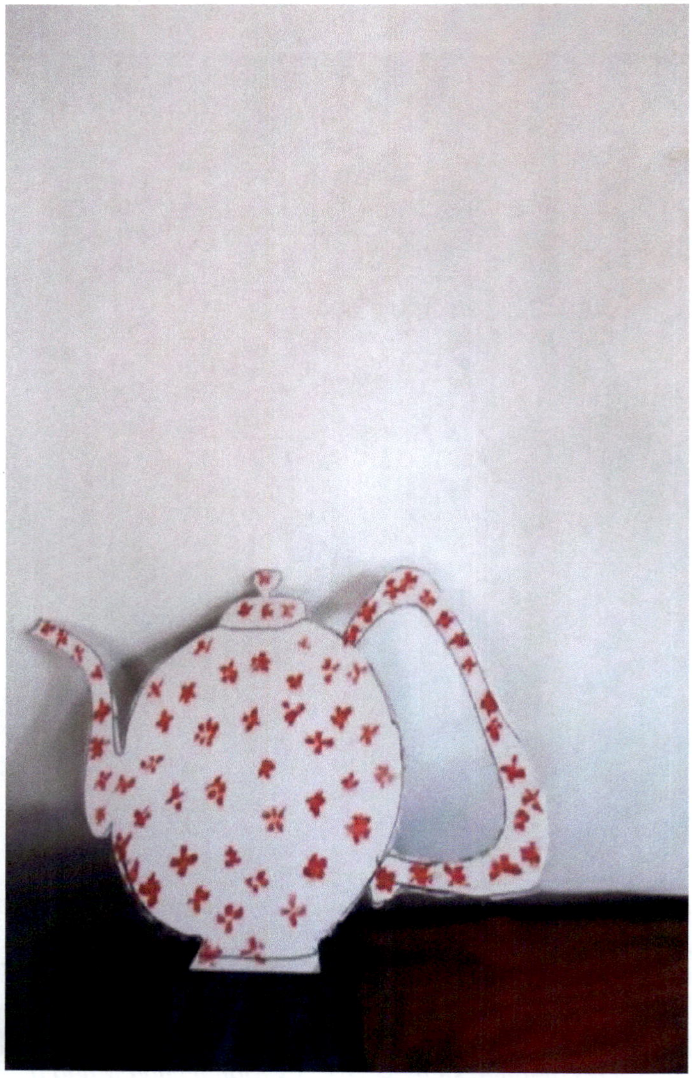

Dramata drinks her tea while Kalizuko works on a program. Then, Kalizuko shows her the code written in Python. (Please see the code on the website listed at the beginning of the book).

Kalizuko: " I simplified it for you: here is a Python code for basic tango dance steps performed by a solo dancer. This dance is repeated 100 times by 100 different solo dancers. Each time, the dance starts at a different position on the dance floor. Here is its visual output of the 100 cycles: "*
Tango code © by Mohammad Majid al-Rifaie. (Used with permissions.) *

```
import turtle   # this line makes sure the turtle object can be used
import random   # this lines allows the program to generate random numbers when needed
            # e.g. in order to decide the step size

screen = turtle.Screen()
screen.screensize(300,300)
screen.colormode(255)
screen.bgcolor(0,0,0)

# the screen is made to be 300x300 with black background
```

```
l = turtle.Turtle() # l for left foot
r = turtle.Turtle() # r for right foot

# hiding the left and right foot until they are in their positions
l.hideturtle()
r.hideturtle()

l.penup()
r.penup()

# adjusting the speed and rhythom of the dance
speed = 50
l.speed(speed)
r.speed(speed)

# setting the shape of the feet to turtles
l.shape("turtle")
r.shape("turtle")

# initially the turtle is headed east,
# so they are rotated 90 degrees anti-clockwise to head upward (north)
# and go back 100 unit to be placed in the initial position
l.left(90); l.backward(100)
r.left(90); r.backward(100)

# here the right foot is given a distance from the left one
# (so they don't overlap)
r.right(90); r.forward(30); r.left(90)

# Now that each foot is in its place, they are made visible
l.showturtle()
r.showturtle()

# and the pen is down, so the user see the traces of their movements
l.pendown()
r.pendown()

# a stamp allows the trace of the feet to be printed whenever required
#l.stamp()
#r.stamp()

# Whenever this functin is called, one full basic tango movement is executed
def tango():

    # a random colour is generated for the traces of the feet
    red   = random.randint(0,255)
    green = random.randint(0,255)
    blue  = random.randint(0,255)

    l.pencolor(red,green,blue)
    r.pencolor(red,green,blue)

    # No 1 & 2
    l.forward(80)   # left moves forward 80 units
    r.forward(200)  # right moves forward 200 units

    #l.stamp()
    #r.stamp()

    # No 3
    l.forward(200)
    #l.stamp()

    # No 4
    r.forward(80)
    r.right(90)     # right foot is rotated to the right (clockwise) for 90 degrees
    r.forward(100)  # then it moves forward 100 units
    r.left(90)      # then rotated 90 degrees anti-clockwise
    #r.stamp()
```

```
# No 5
l.right(90)
l.forward(100)
l.left(90)
#l.stamp()

# No 6
#r.left(90)
#r.forward(150)
#r.right(90)
r.circle(65, 180)  # this lines allows a circle to be drawn showing the arc made during the dance
r.right(180)
#r.stamp()

# No 7
#l.left(90)
#l.forward(200)
#l.right(90)
l.right(180)
l.circle(-100, 180)
#l.stamp()

# No 8
r.left(90)
r.forward(50)
r.right(90)
#r.stamp()

# No 9
#l.right(90)
#l.forward(100)
#l.left(90)
l.circle(-65,180)
l.left(180)
#l.stamp()

# No 10
#r.right(90)
#r.forward(200)
#r.left(90)
r.left(180)
r.circle(110,180)
#r.stamp()

# No 11
l.right(90)
l.forward(100)
l.left(90)
#l.stamp()

# No 12
#r.right(180)
#r.forward(80)
#r.right(180)
r.backward(80)
#r.stamp()


# the FOR loop

 below makes sure that the lines are called n (n=100) times  for i in
range(100):
   #l.stamp();    r.stamp()
   tango()    # this line calls all the steps of tango dance

   # once the first round is complete:
   # the pen is lifted
   l.penup(); r.penup()

   # a new dance starts at a random position
   # and feet left and right are places within the range of -300 to 300 which
```

```
    # is the within the dance floor
    randX = random.randint(-300,300)
    randY = random.randint(-300,300)
    l.goto( randX, randY)
    r.goto( randX-50, randY)

    # the pen is put down again to show the the movement of the feet
    l.pendown(); r.pendown()


# saving the final image as a pdf file (without the background colour)
screen.getcanvas().postscript(file="Tango.pdf")
```

Dramata: "Stop it! I cannot understand a single line."
Kalizuko, It's all in English…This is what it looks like when outputted as a graph:

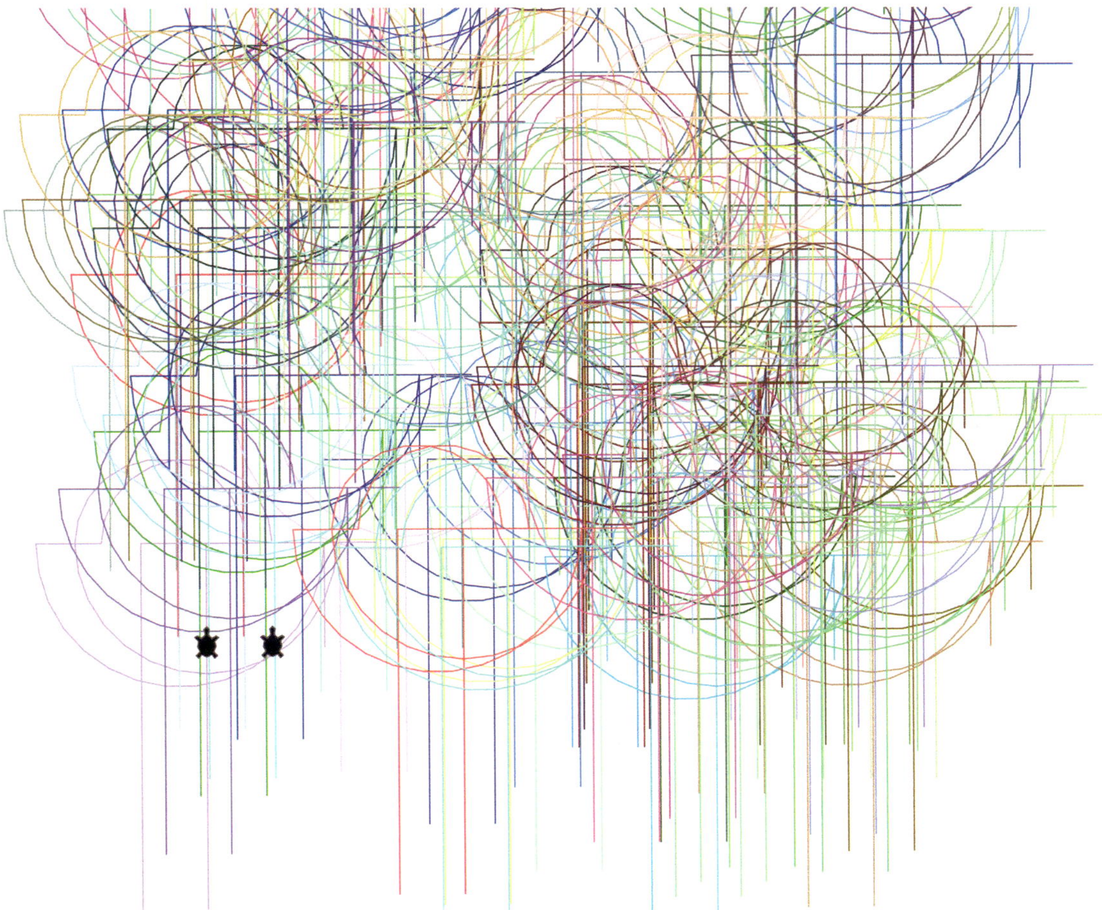

Tango, the output of a code © by Mohammad Majid al-Rifaie. (Used with permissions.)

`Animated tango steps with musical notation appear on the screen.`

Dramata, "This I can refer to."

Kalizuko, " Here's another example of a tango, but in a different language: C # (C Sharp). First, I quickly sketched the basic steps to create images.

```
using System;
using System.Collections.Generic;
```

```
using System.ComponentModel;
using System.Data;
using System.Drawing;
using System.Linq;
using System.Text;
using System.Windows.Forms;


namespace DanceGraphics
{
  public partial class Form1 : Form
  {

    int L = 0;
    int R = 1;

    int x = 0;
    int y = 1;

    int Step = 0;

    int[, ,] xy = new int[,,]{{{250,400},{310,400}},
                    {{250,274},{310,400}},
                    {{250,274},{310,190}},
                    {{250,76},{310,190}},
                    {{250,76},{450,84}},
                    {{400,76},{450,84}},
                    {{400,76},{200,76}},
                    {{40,66},{200,76}},
                    {{40,66},{100,76}},
                    {{250,76},{100,76}},
                    {{250,76},{450,84}},
                    {{400,76},{450,84}},
                    {{400,76},{450,220}}};

    public Form1()
    {
      InitializeComponent();
      this.Text = "Tango Dance";
    }

    private void Form1_Load(object sender, EventArgs e)
    {
      this.DoubleBuffered = true;
      this.Paint += new PaintEventHandler(Form1_Paint);
    }

    private void Form1_Paint(object sender, System.Windows.Forms.PaintEventArgs e)
    {
      e.Graphics.DrawImage(Properties.Resources.LeftFoot, xy[Step, L, x],xy[Step, L, y],40,84);
      e.Graphics.DrawImage(Properties.Resources.RightFoot, xy[Step, R, x], xy[Step, R, y],40,84);

      Step++;
      if(Step==13)Step=0;
    }

    private void timer1_Tick(object sender, EventArgs e)
    {
      this.Refresh();
    }
  }

}
```

Kalizuko, "And this is a shortened record of the animation it generated."

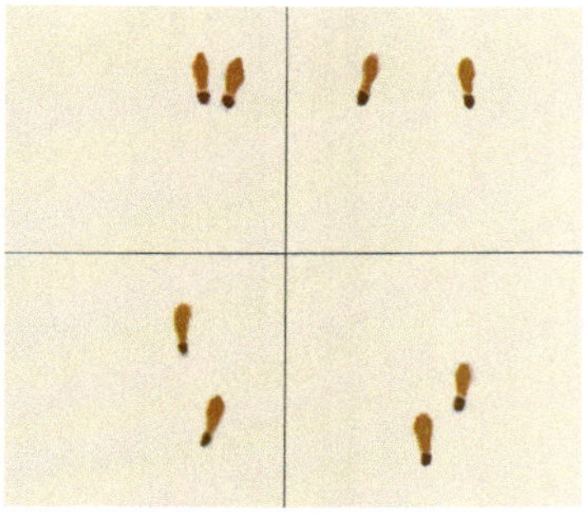

An animated footwork appears on the screen as a code's output.

Dramata, "Interesting. I can see the pattern."
Kalizuko, "But let's get more sophisticated by having the code draw abstract art out of the code in the same C# language as an output, forming a rhythmic graphical pattern without any drawing by hand: "

*Tango in C # © by Md Fahimul Islam (Used with permissions). **

```
using System;
using System.Collections.Generic;
using System.ComponentModel;
using System.Data;                                  Default class library
using System.Drawing;
using System.Linq;
using System.Text;
using System.Windows.Forms;
using System.Reflection;          // for brushes property retrieval


namespace DanceGraphics
{
  public partial class Form1 : Form
  {

    int L = 0;
    int R = 1;

    int x = 0;
    int y = 1;

    int Step = 0;


        int[, ,] xy = new int[,,]{{{250,400},{310,400}},
                 {{250,274},{310,400}},
                 {{250,274},{310,190}},
                 {{250,76},{310,190}},
                 {{250,76},{450,84}},
                 {{400,76},{450,84}},
                 {{400,76},{200,76}},
                 {{40,66},{200,76}},
                 {{40,66},{100,76}},
                 {{250,76},{100,76}},
                 {{250,76},{450,84}},
                 {{400,76},{450,84}},
                 {{400,76},{450,220}}};

    Random rnd = new Random();
    int xrnd = 0;
```

```
        Brush LBrush;
        Brush RBrush;


        public Form1()
        {
            InitializeComponent();
            this.Text = "Tango Dance";
            LBrush = RandomBrush();
            RBrush = RandomBrush();

        }

        private void Form1_Load(object sender, EventArgs e)
        {
            this.DoubleBuffered = true;
            this.Paint += new PaintEventHandler(Form1_Paint);
        }

        private void Form1_Paint(object sender, System.Windows.Forms.PaintEventArgs e)
        {


            e.Graphics.FillRectangle(LBrush, xy[Step, L, x] + xrnd, xy[Step, L, y], 11, 84);
            e.Graphics.FillRectangle(LBrush, xy[Step, L, x] + xrnd+5, 0, 1, 600);


            e.Graphics.FillRectangle(RBrush, xy[Step, R, x] + xrnd, xy[Step, R, y], 11, 84);
            e.Graphics.FillRectangle(RBrush, xy[Step, R, x] + xrnd+5, 0, 1, 600);


            Step++;
            if (Step == 13)
            {
                Step = 0;
                xrnd = rnd.Next(-300,300);
                do
                {
                    LBrush = RandomBrush();
                    RBrush = RandomBrush();
                } while (LBrush == RBrush);
            }
        }

        Brush RandomBrush()
        {
            PropertyInfo[] brushInfo = typeof(Brushes).GetProperties();
            Brush[] brushList = new Brush[brushInfo.Length];
            for (int i = 0; i < brushInfo.Length; i++)
            {
                brushList[i] = (Brush)brushInfo[i].GetValue(null, null);
            }
            Random randomNumber = new Random(DateTime.Now.Millisecond);
            return brushList[randomNumber.Next(1, brushList.Length)];
        }


        private void timer1_Tick(object sender, EventArgs e)
        {
            this.Refresh();
        }
    }

}
```

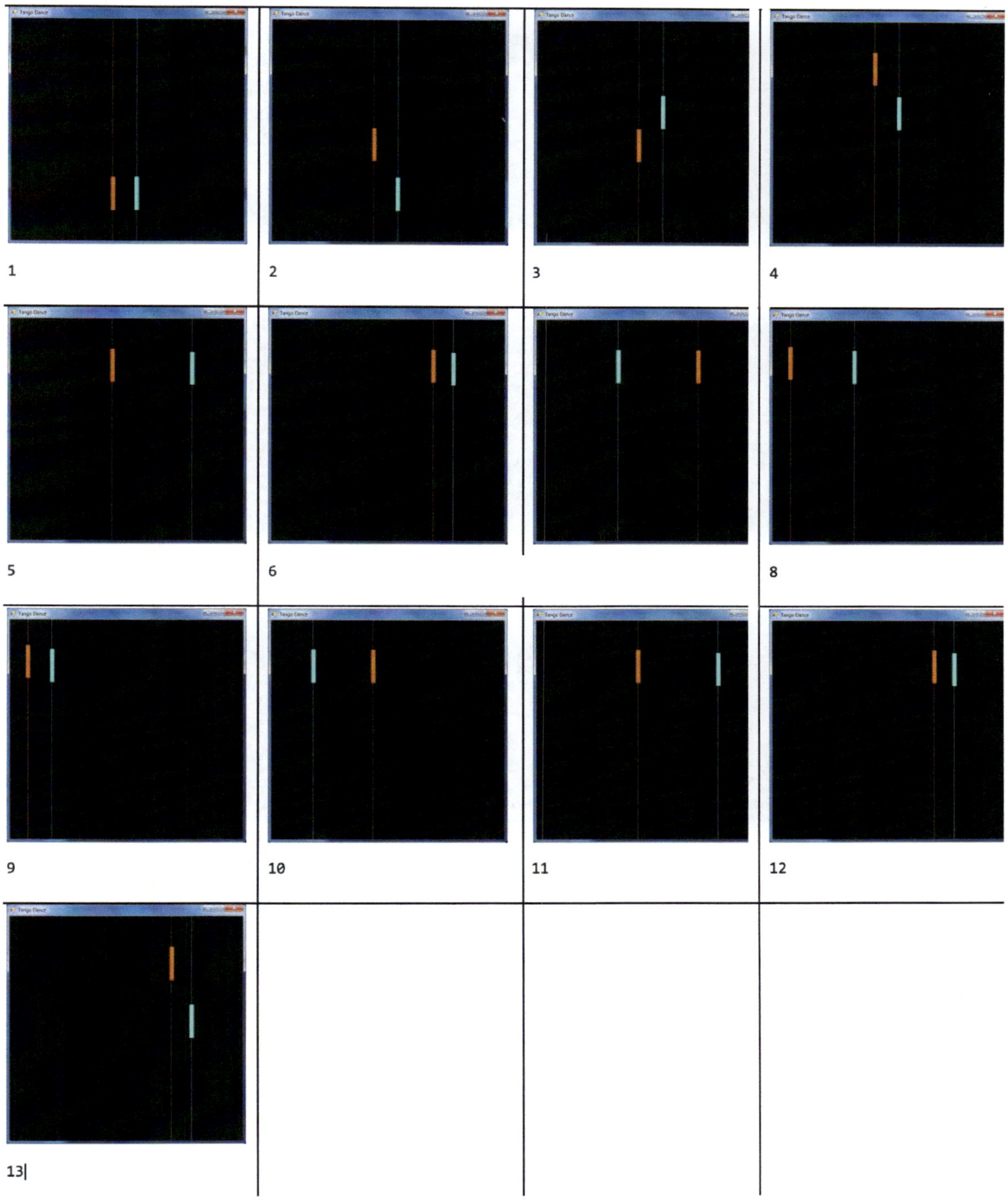

Tango from the computer code. © by Md Fahimul Islam. (Used with permissions.) *

Kalizuko, "And here is the output as a sequence of steps."

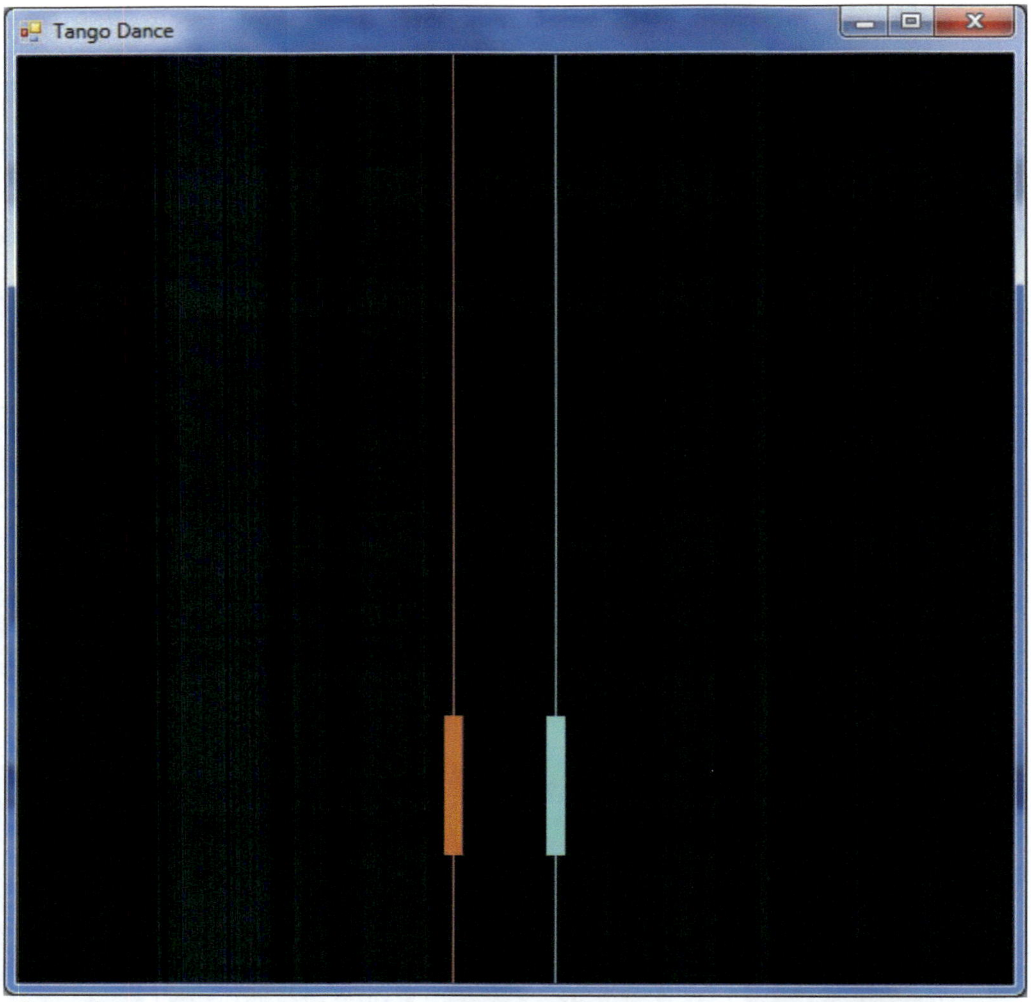

The code's output: tango. © 2021 by Md Fahimul Islam. (Used with permission).

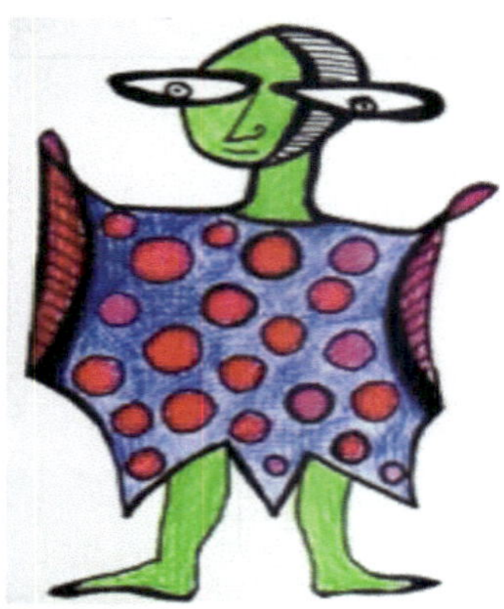

Nasati: he knows more tricks than he can perform.

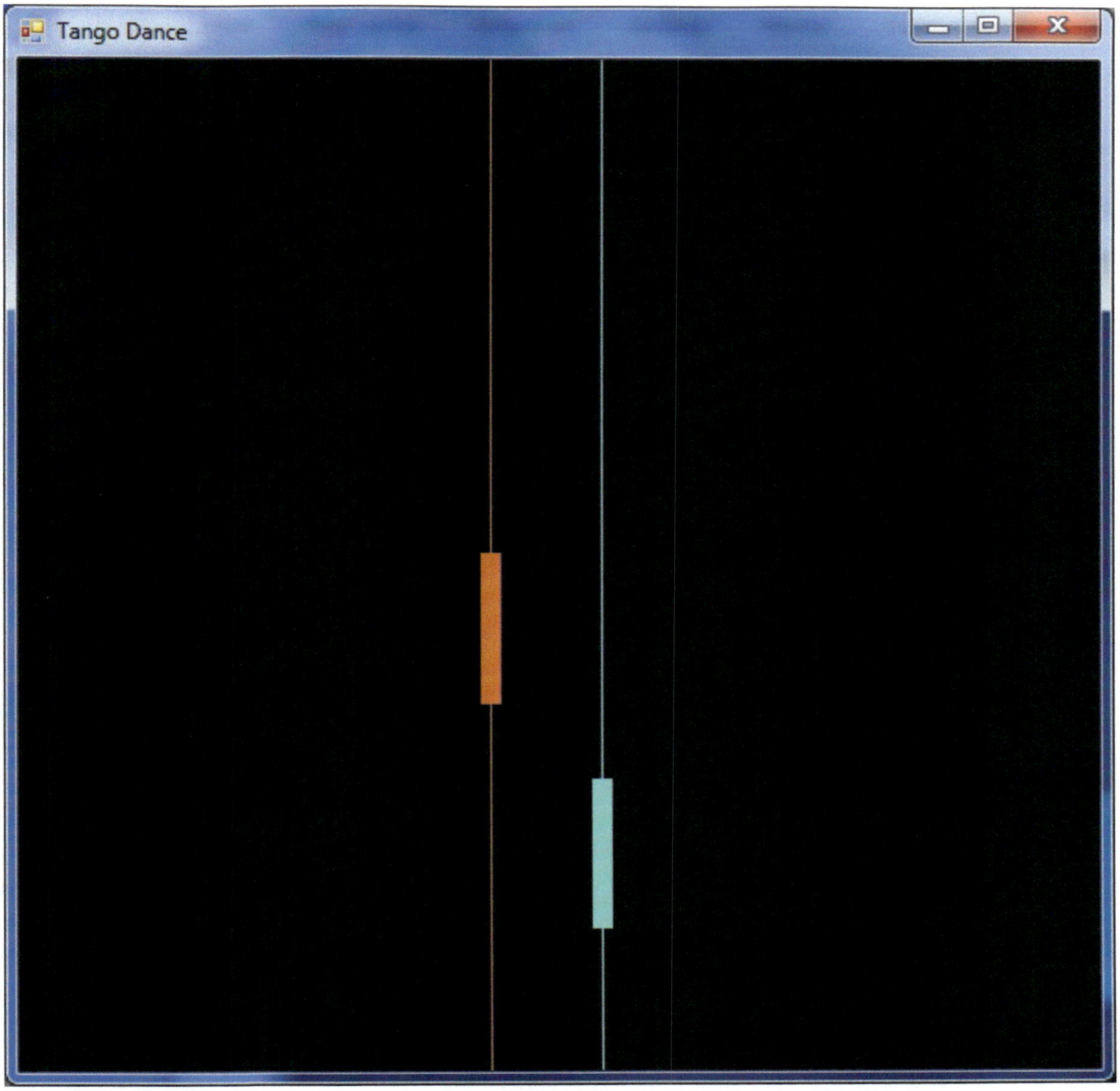

The output from the code: tango. © 2021 by Md Fahimul Islam. (Used with permission).

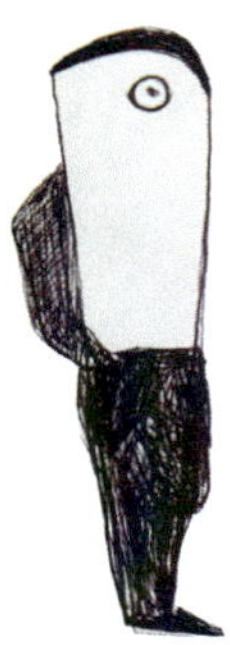

Hendo: collects small objects.

Hendo: collects all that's small.

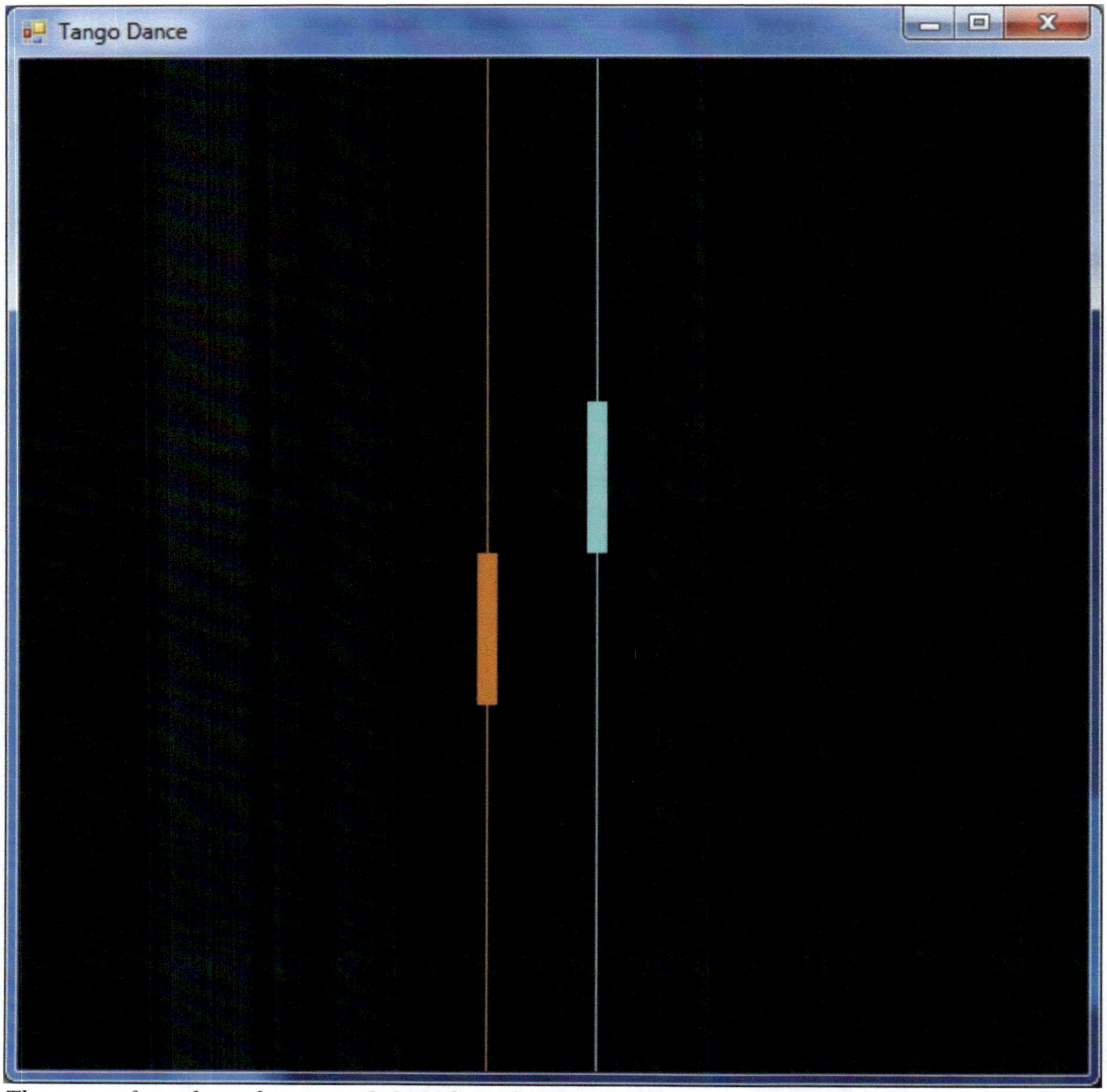

The output from the code: tango. © 2021 by Md Fahimul Islam. (Used with permission).

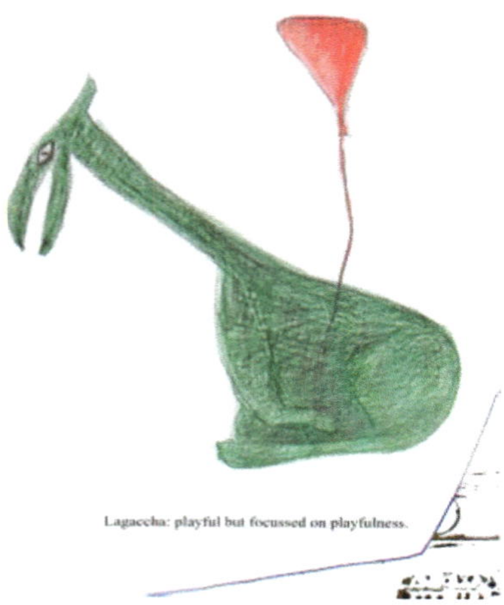

Lagaccha: playful but focused on playfulness.

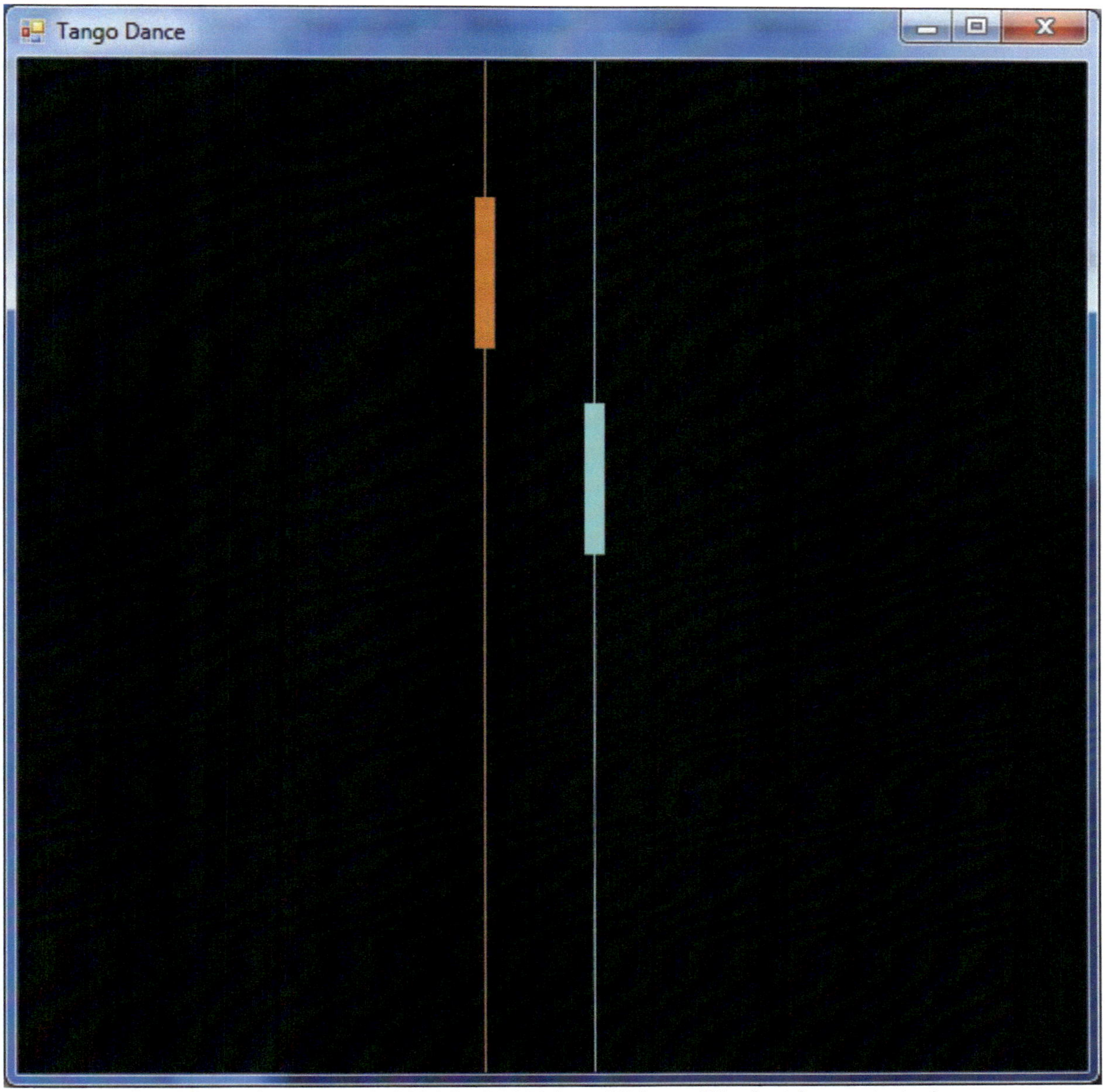

The output from the code: tango. © 2021 by Md Fahimul Islam. (Used with permission).

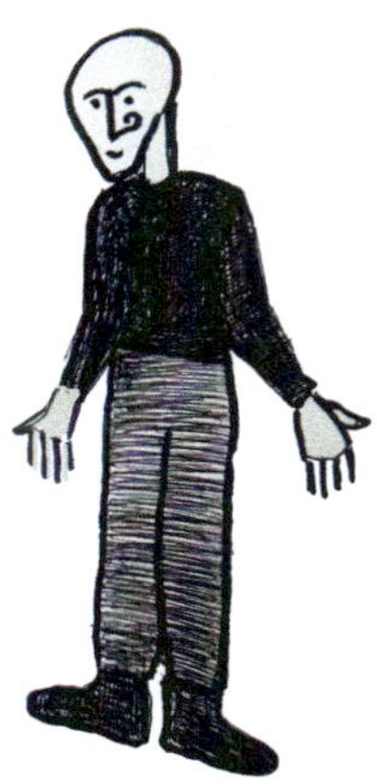

Bundo: never answers why questions.

Bundo: never answers any question.

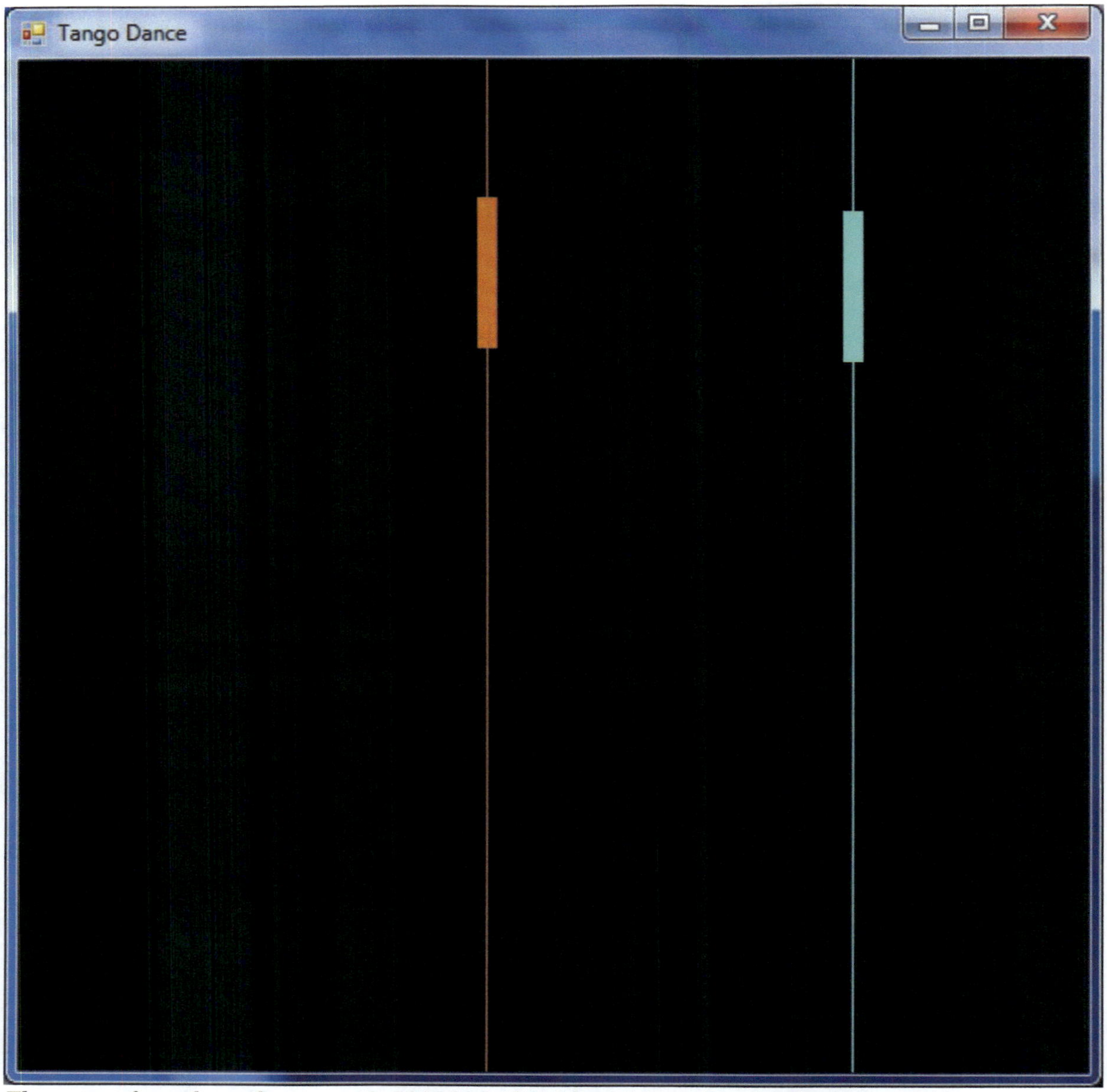

The output from the code: tango. © 2021 by Md Fahimul Islam. (Used with permission).

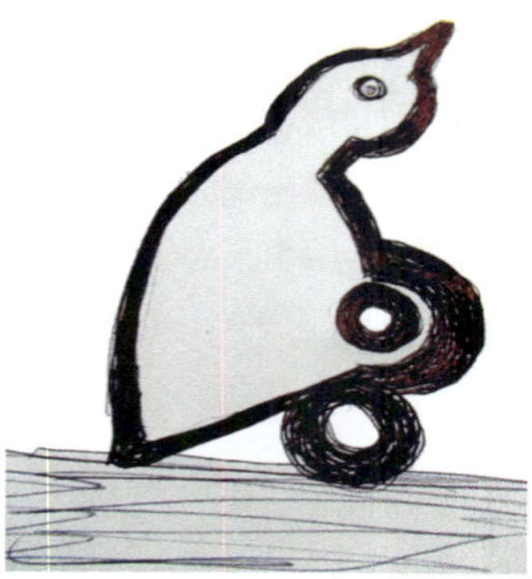

Bugerta: can drive on one wheel when nobody is watching.

Bugerta: can drive on one wheel when no one is watching.

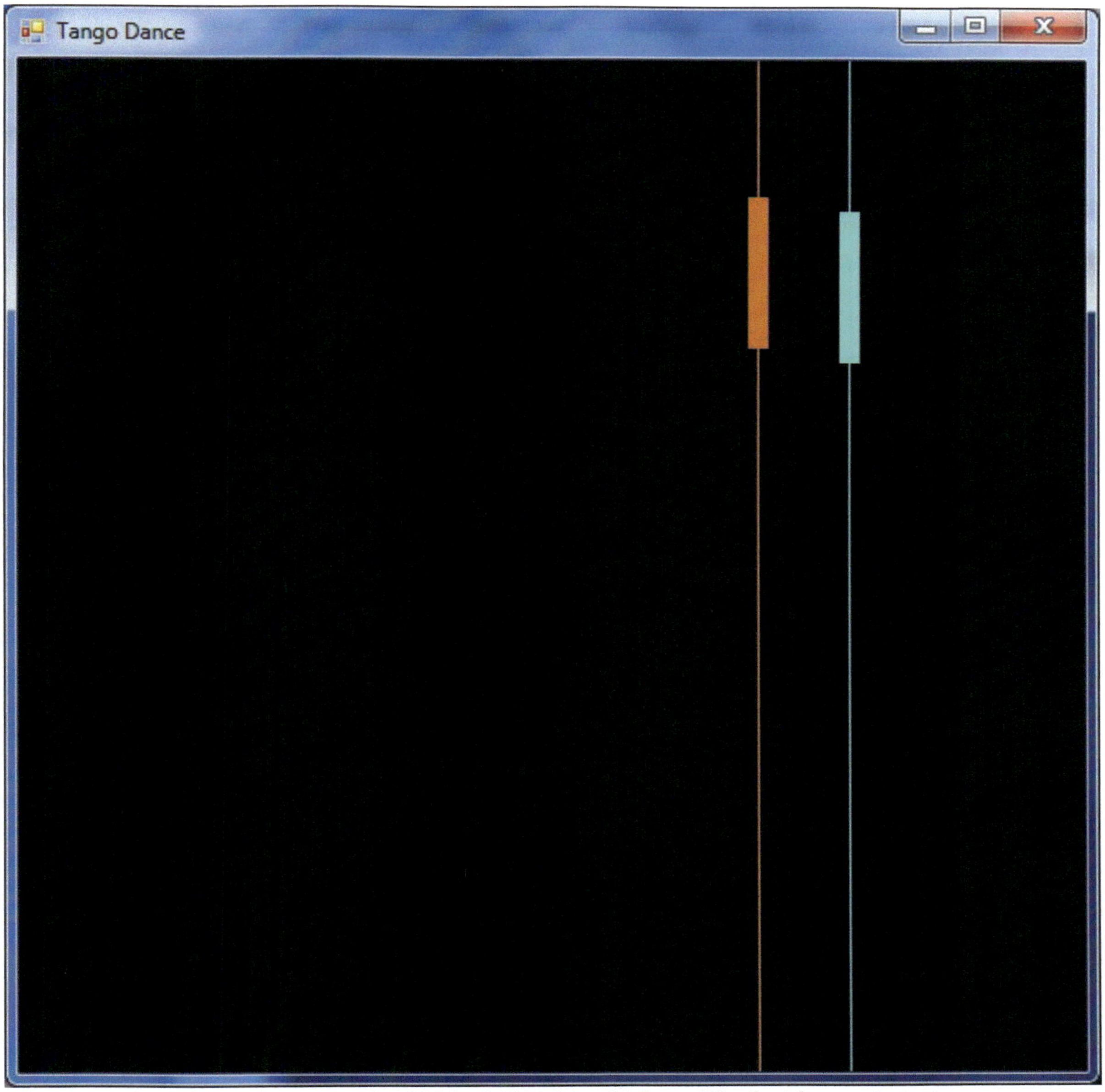

The output from the code: tango. © 2021 by Md Fahimul Islam. (Used with permission).

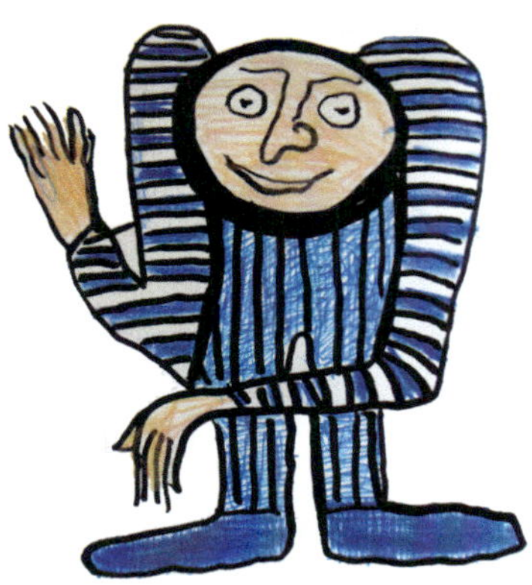

Baduran: knows everyone, if not, sends an invite.

Baduran: knows everyone; if not, sends an invite.

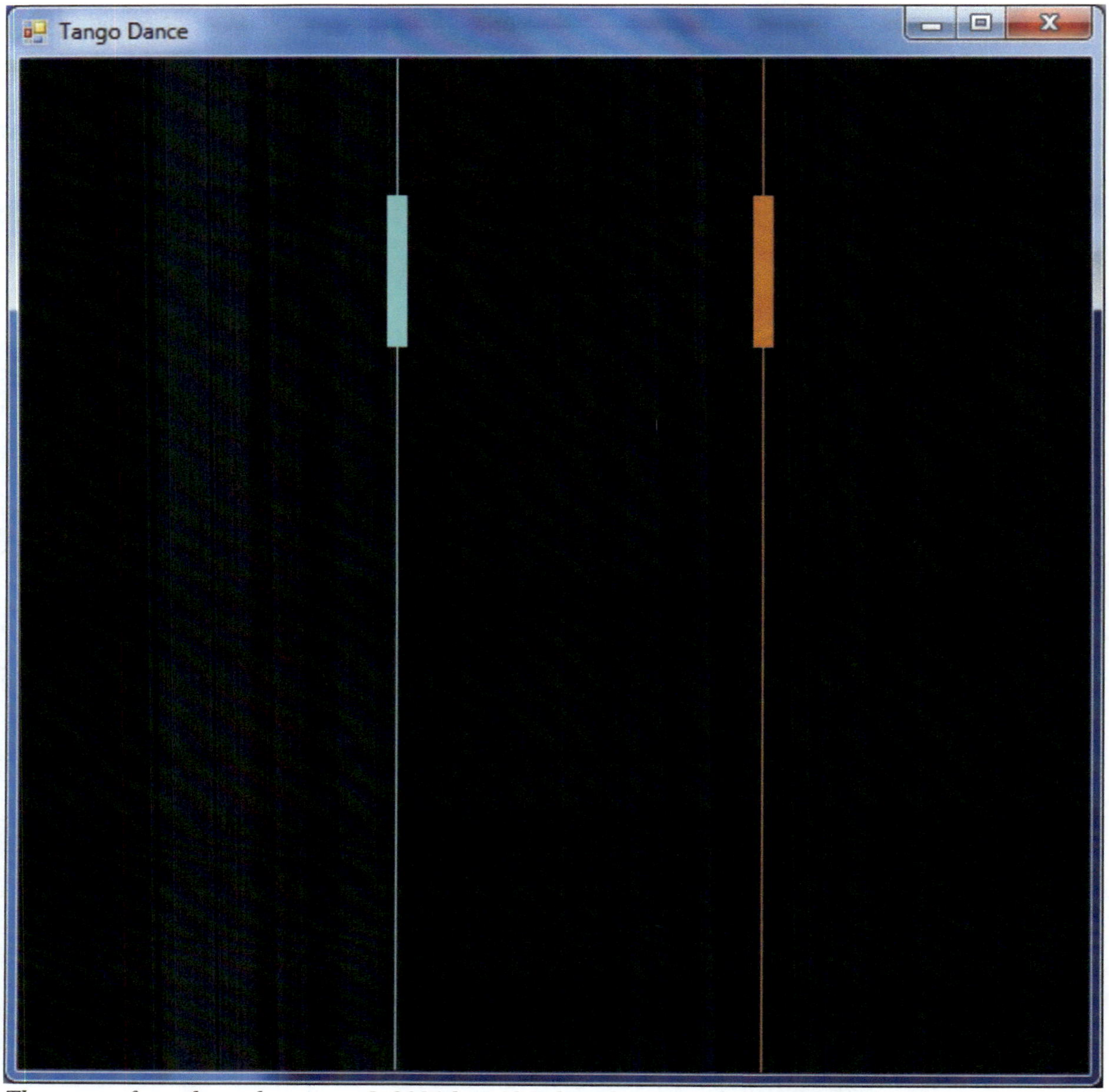

The output from the code: tango. © 2021 by Md Fahimul Islam. (Used with permission).

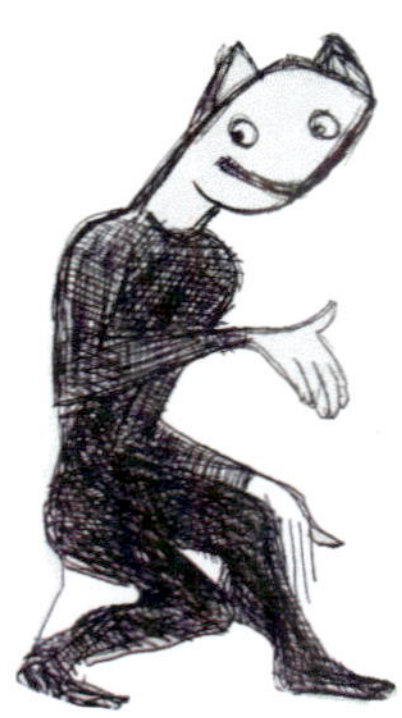

Bungo: knows his way around.

Bungo: knows his way around.

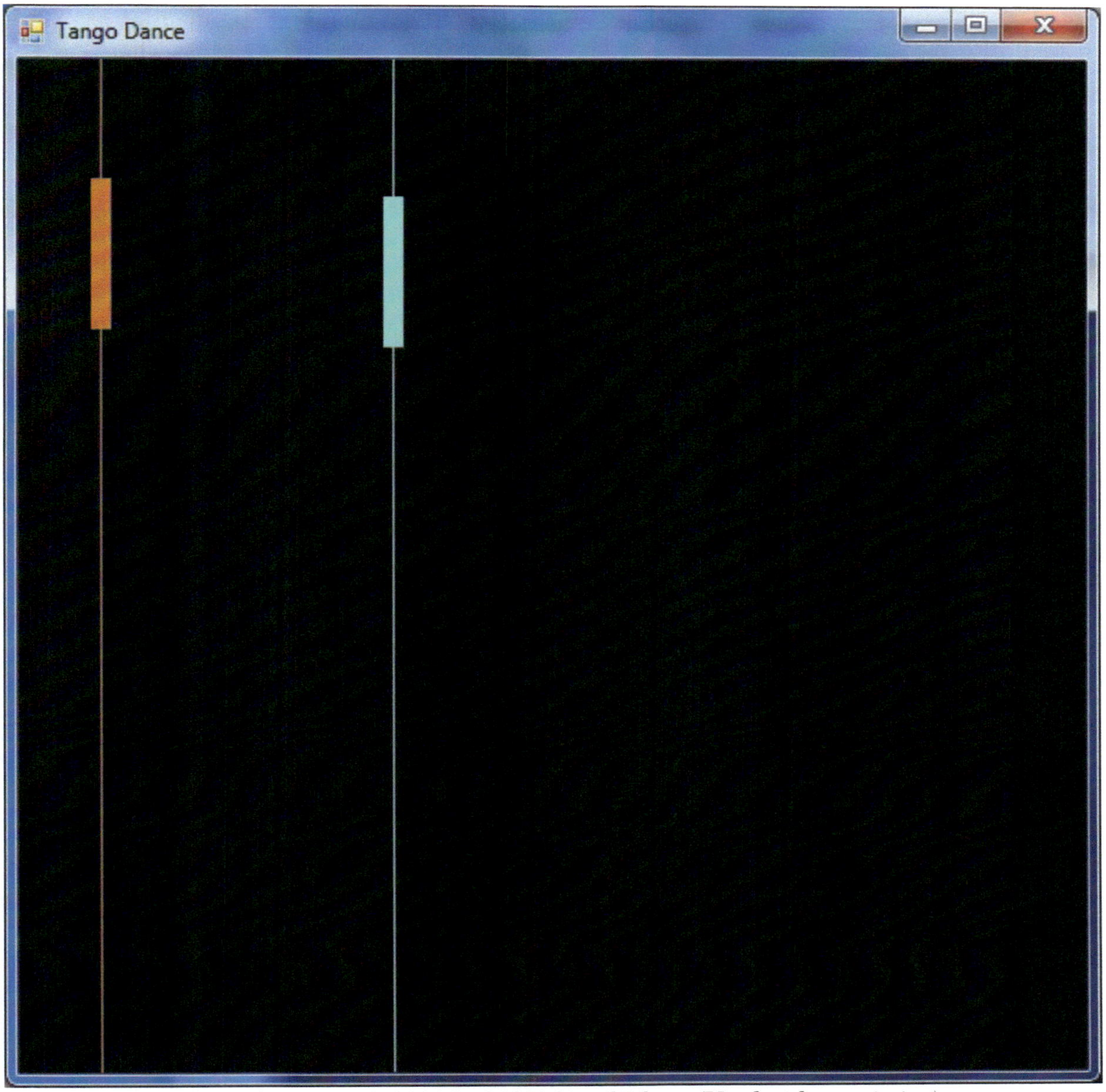

The output from the code: tango. © 2021 by Md Fahimul Islam. (Used with permission).

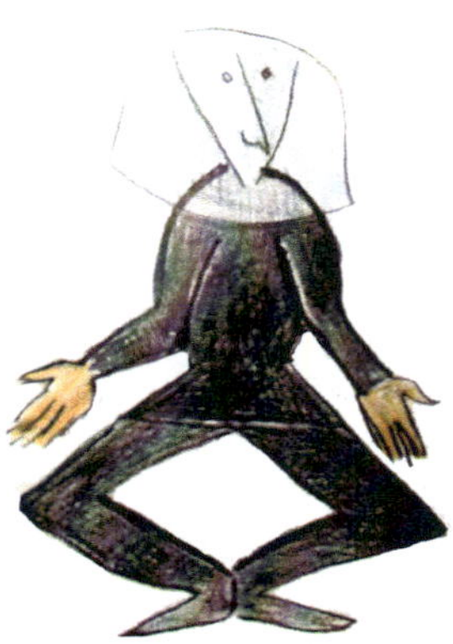

Bujdhor: knows nothing about it.

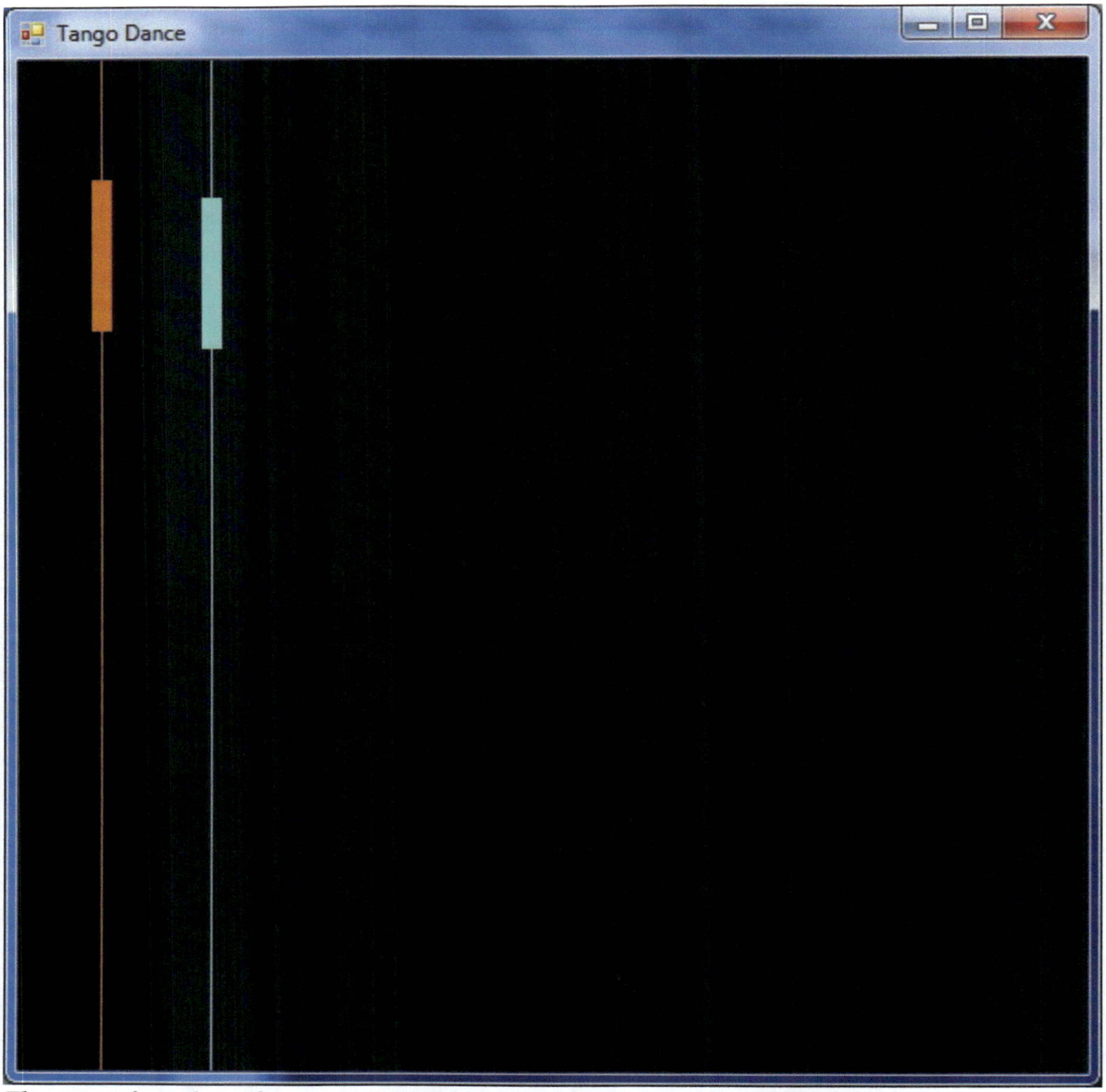

The output from the code: tango. © 2021 by Md Fahimul Islam. (Used with permission).

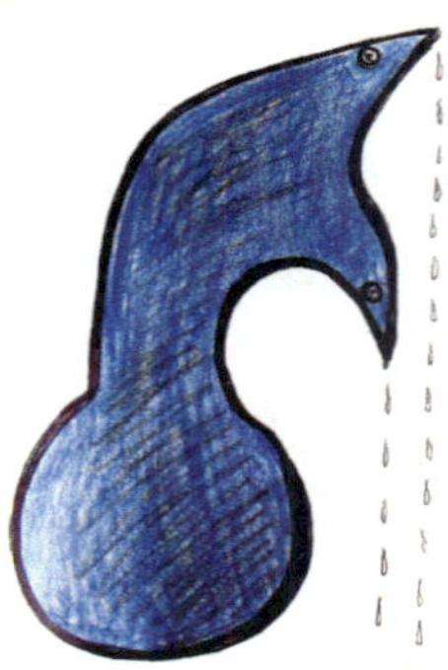

Felango: can understand distributions.

Felango shared how distributions can be visualized.

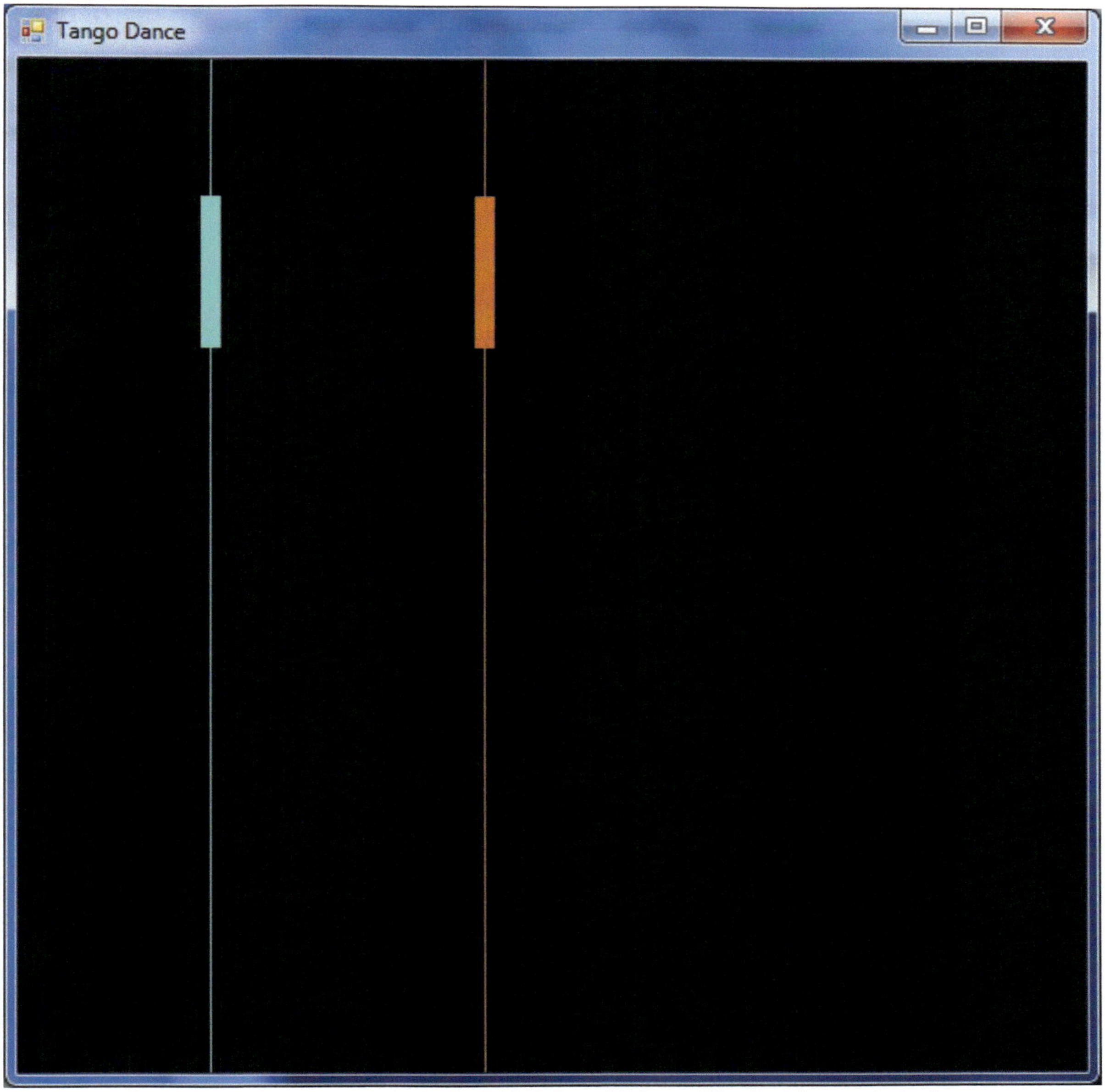

The output from the code: tango. © 2021 by Md Fahimul Islam. (Used with permission).

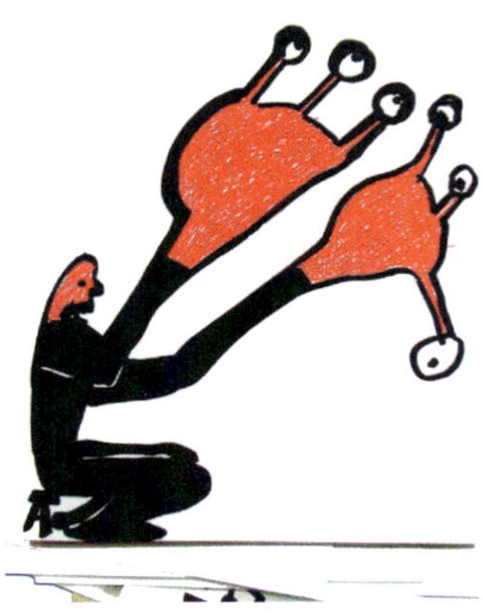

Pazeto: can see in any scale, any direction.

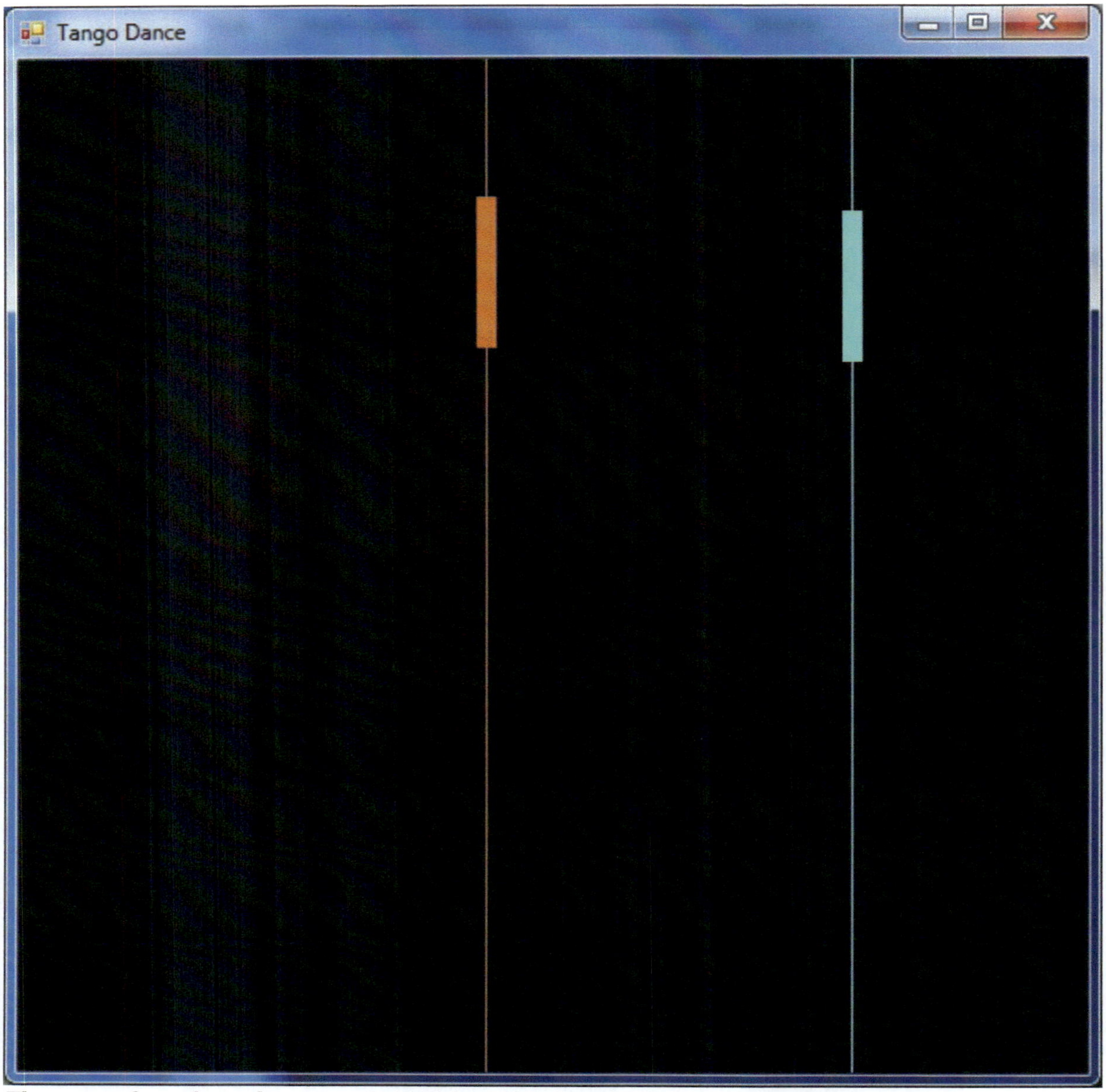

The output from the code: tango. © 2021 by Md Fahimul Islam. (Used with permission).

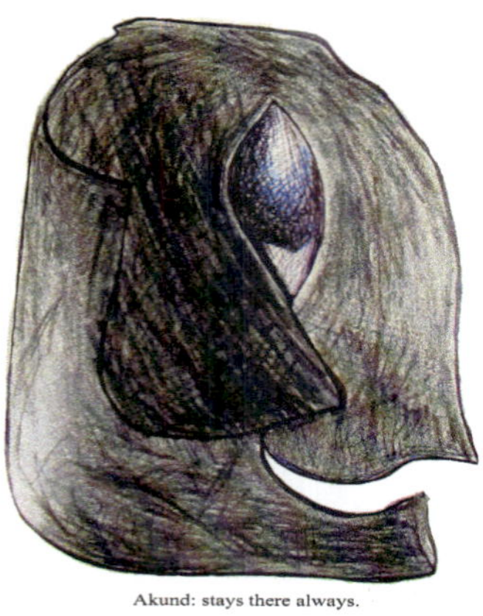

Akund: stays there always.

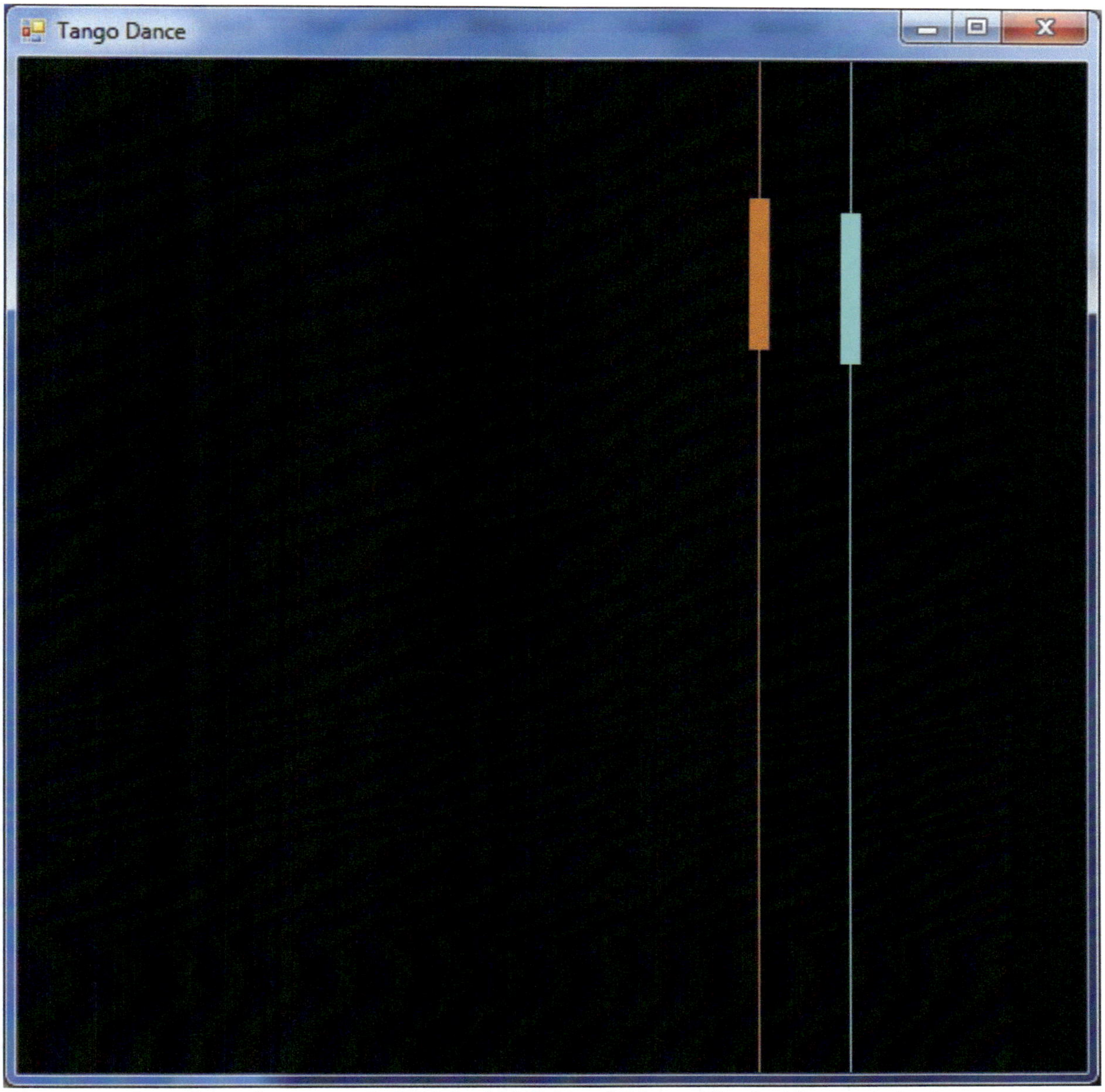

The output from the code: tango. © 2021 by Md Fahimul Islam. (Used with permission).

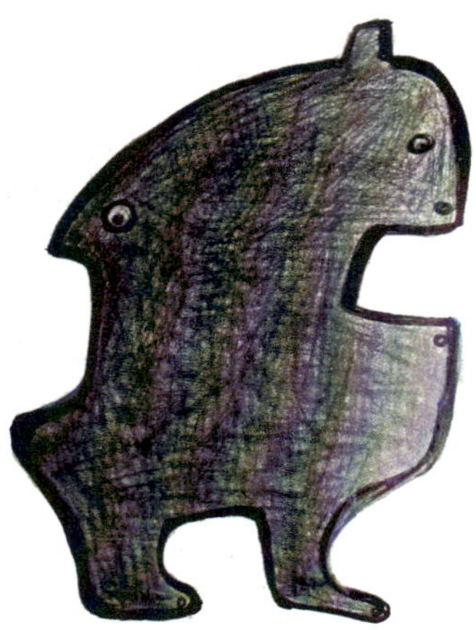

Badoj: always hungry.

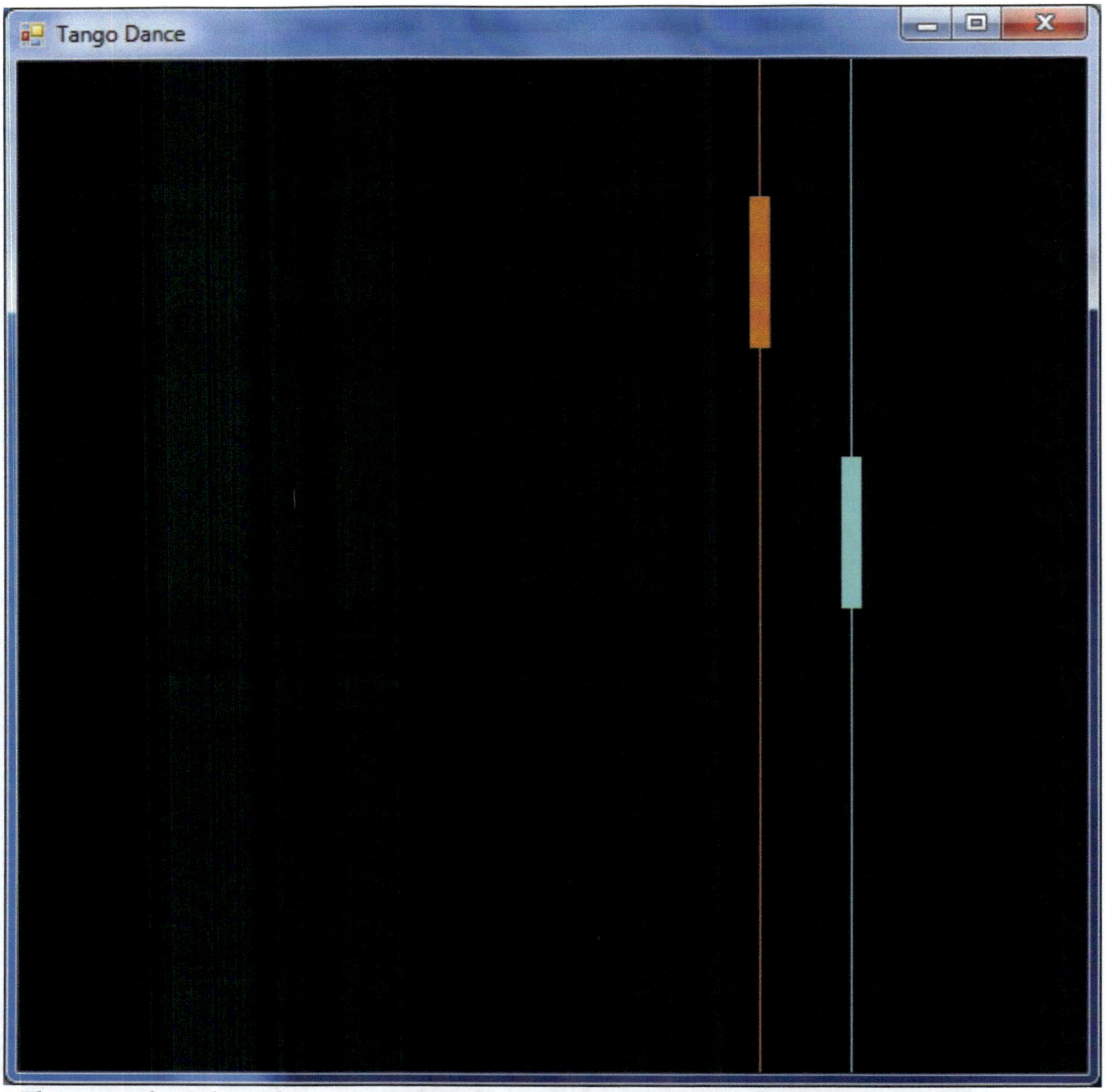

The output from the code: tango. © 2021 by Md Fahimul Islam. (Used with permission).

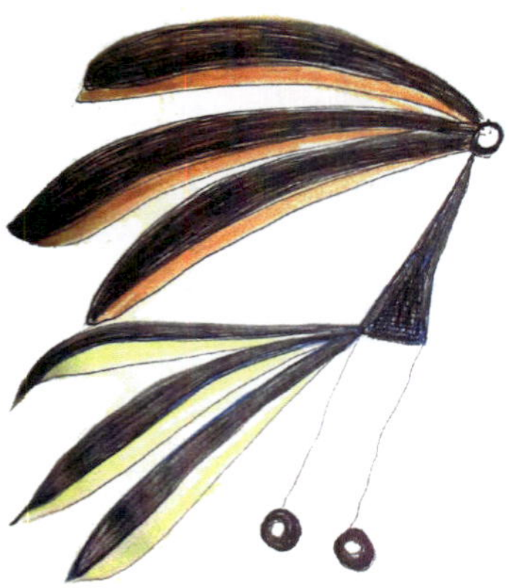

Fuftu: makes only right turns.

Fuftu: wanted to do good things only.

Musical notations appear on the screen (tango), along with the codes. Six dancers, dressed in black outfits, dance (tango).

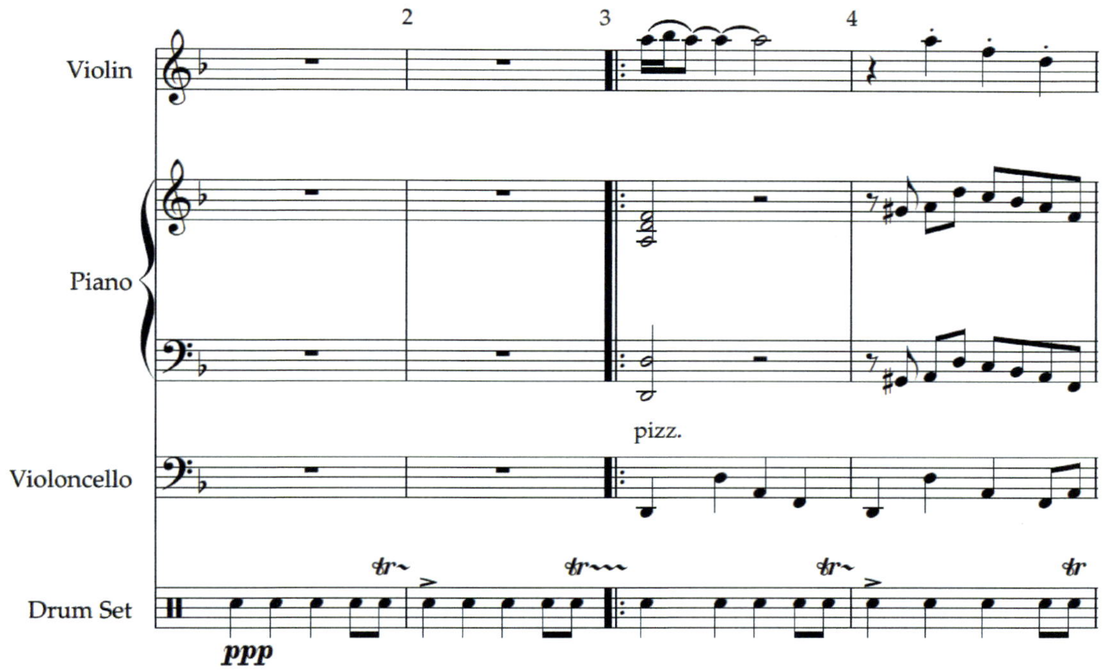

Musical Notation – A tango. © 2021 by Irwin Stuart Smith. (Used with permission).

Dramata, "Oh, I see."
Kalizuko, "In terms of Python, the easiest way is to use IDLE and run code. "

More information about Python and how to use it can be found in the Appendix section of this book.

Dramata, "That makes some sense to me now."
Kalizuko, "Can we dance it now?"

They both dance it, each differently while looking at the screen.

Dramata, "Now I get it. OOOO, I made it. I made proper turns."
Kalizuko, "Can you make some unproper ones now?"
Dramata, "Beg your pardon?"
Kalizuko, "Oh, well, try to release your tension, and old patterns, and relax the rules a bit. Keep in tune with the general rules, but try to listen to the music more, focus on the music, and see what happens."

Dramata makes a big, indescribable sound.

Kalizuko, "See?

Dramata, "See! Now the waltz, please. Waltz came from 18th-century Bohemia, Bavaria, and Austria, from the German word Walzer, meaning to roll or revolve. Steps include a basic box step, natural turn, and reverse turn danced in a counterclockwise direction around the room, with the lead stepping forward on the left foot and the follow stepping back on their right, ¾ time in the closed position.
Kalizuko, "The waltz looks like that:"

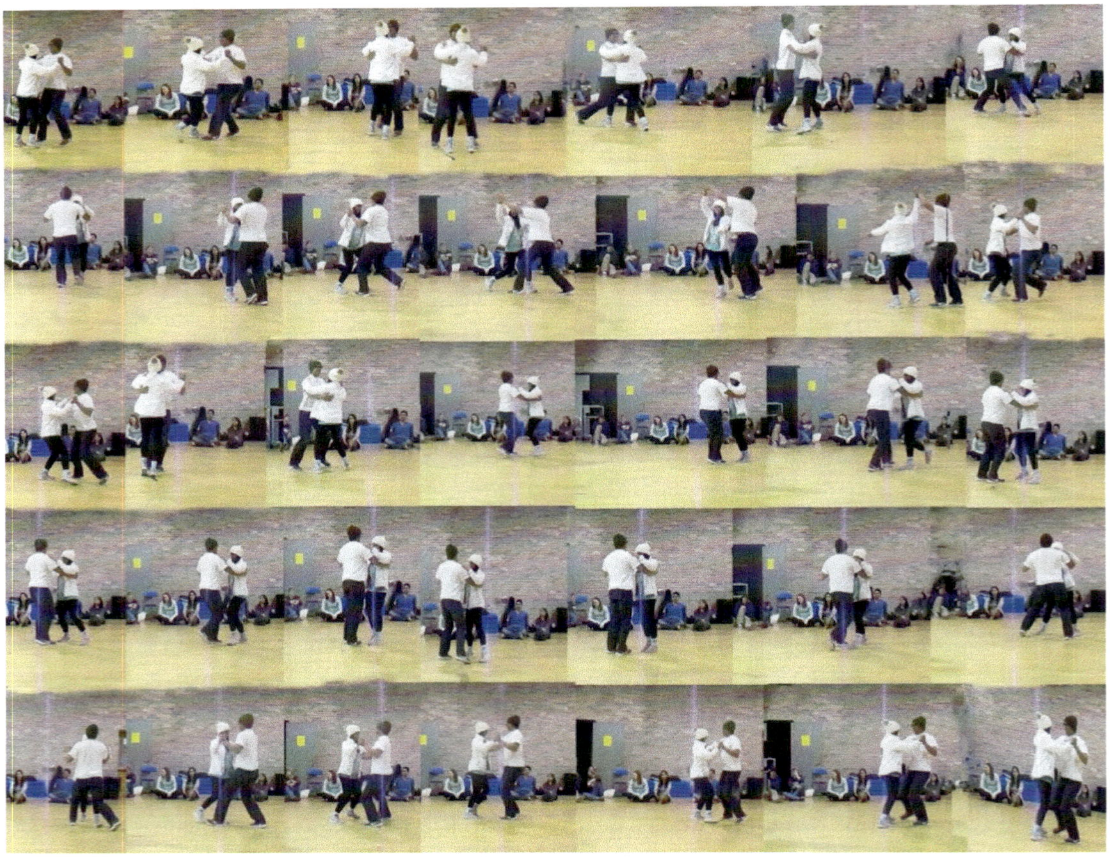

Social Dance: waltz. © 2015, by Anna Ursyn

Dramata, "I know what the waltz looks like, but the code?"
Kalizuko, "All right, have some freshly brewed coffee now, while I work on the code.

```
An actor holding a stovetop espresso maker is chasing another
one, running away with a coffee mug. That is the wrong cup to
drink this coffee from!
```

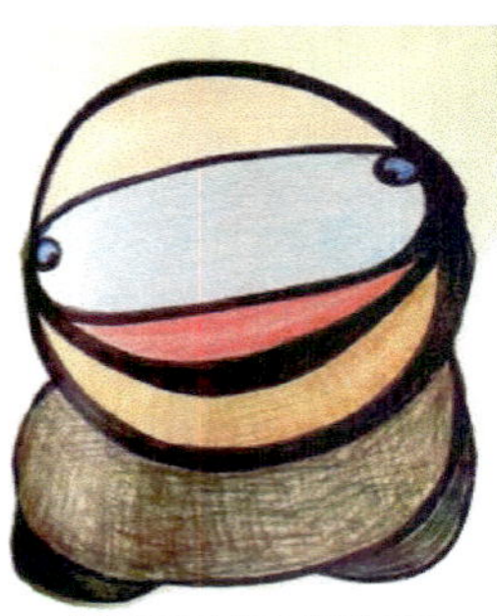

Pukute: can make anything happen.

Pukute: he can make anything happen.

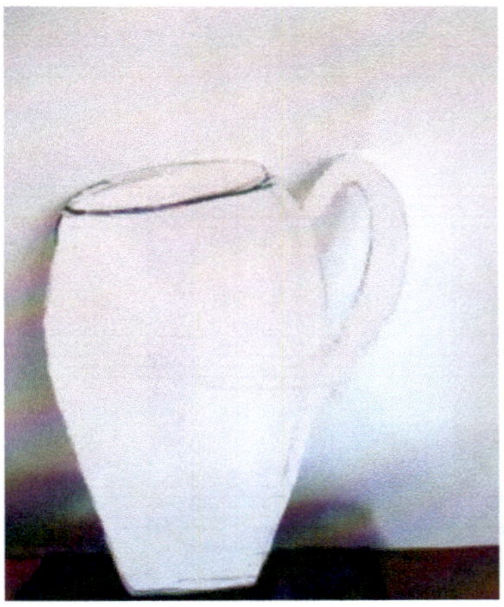
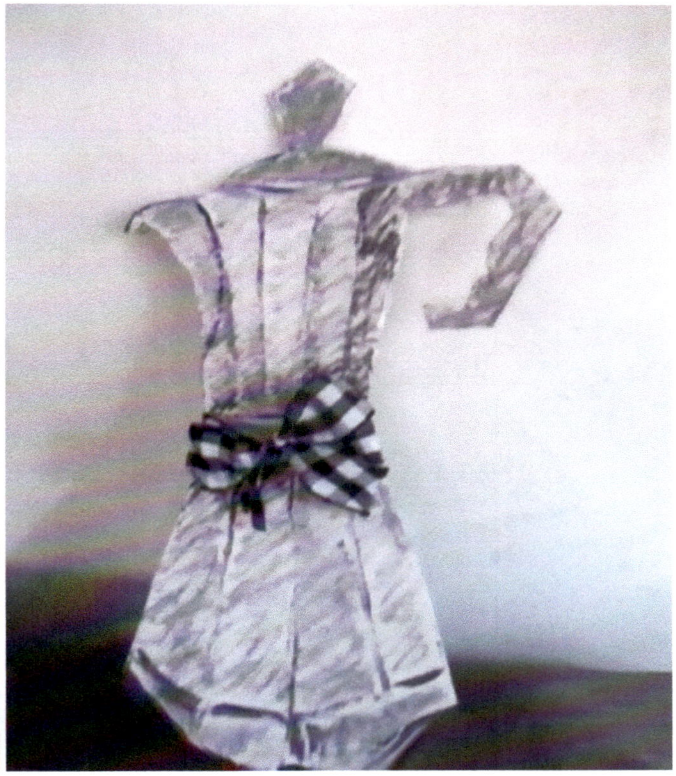

The coded steps for the waltz along with musical notations appear on the screen, while the dancers perform it (waltz). This is not the proper mug to drink this type of coffee from!

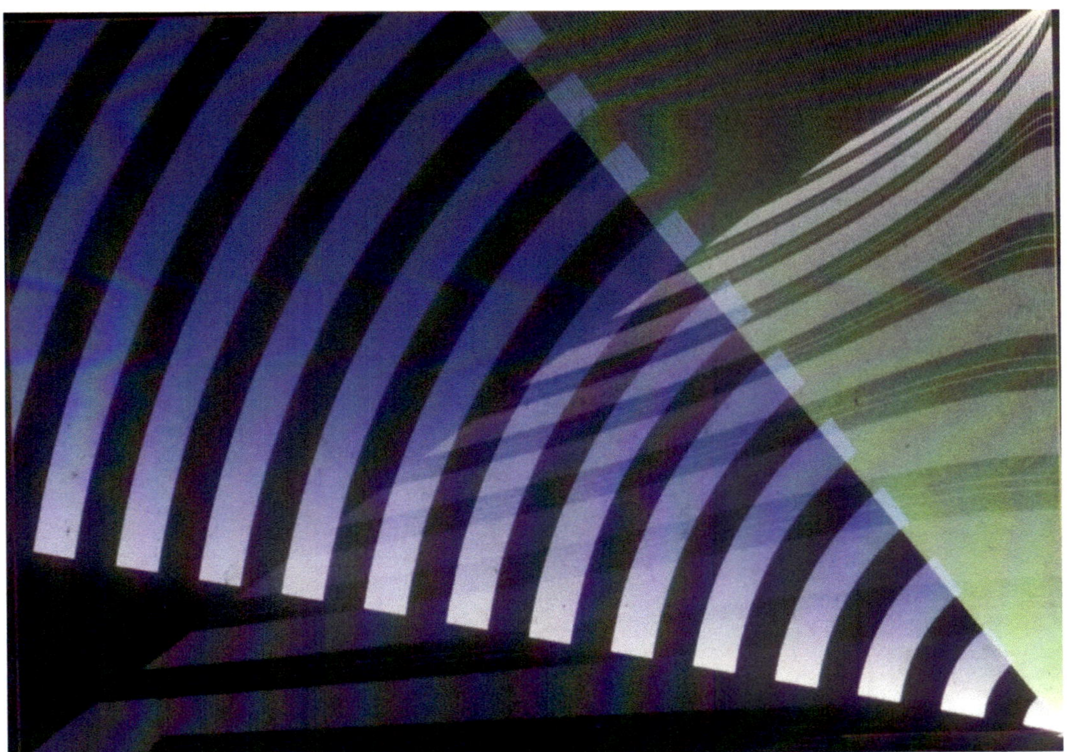

The Moves. The output from the computer code. © 2025 by Anna Ursyn.

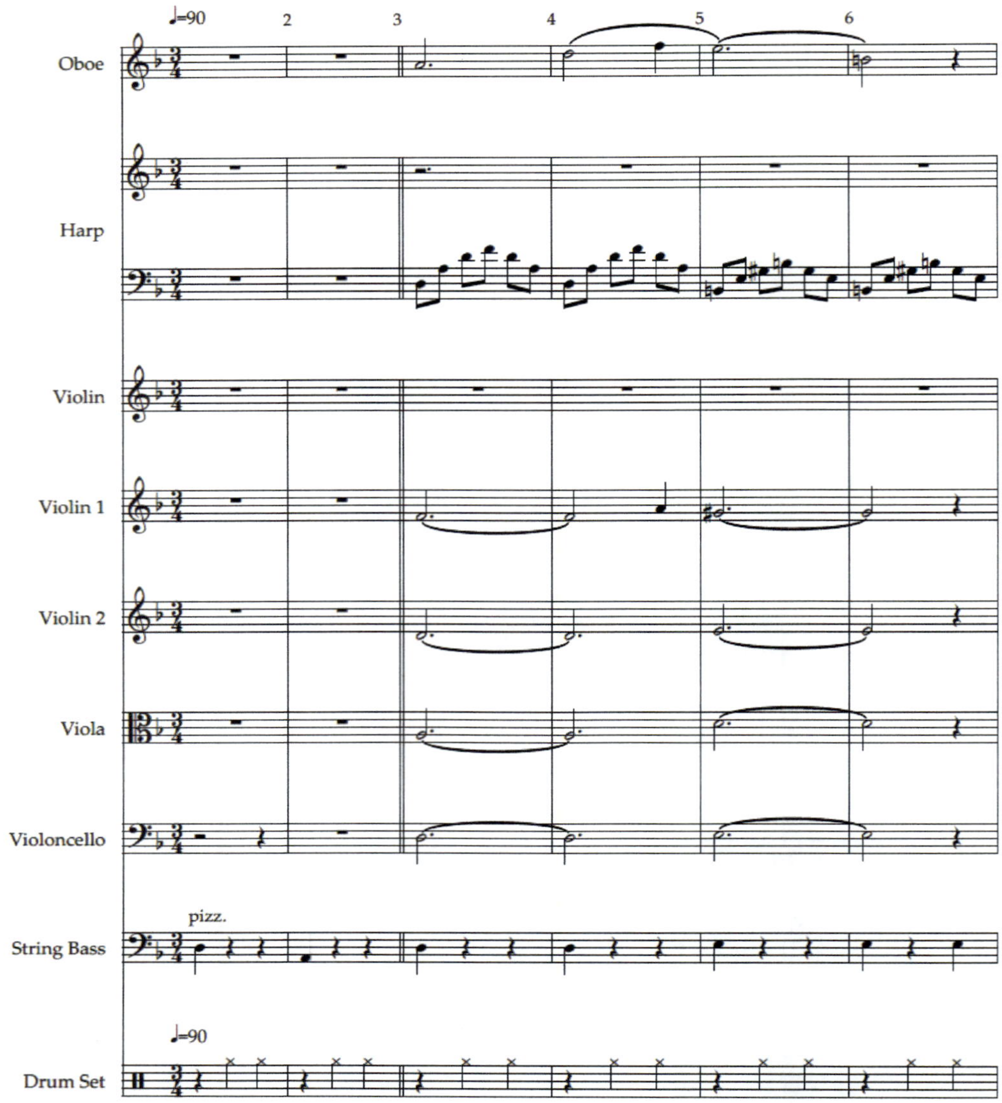

Musical Notations – a waltz. © 2021 by Irwin Stuart Smith. (Used with permission).

Kalizuko, “Here it is:”

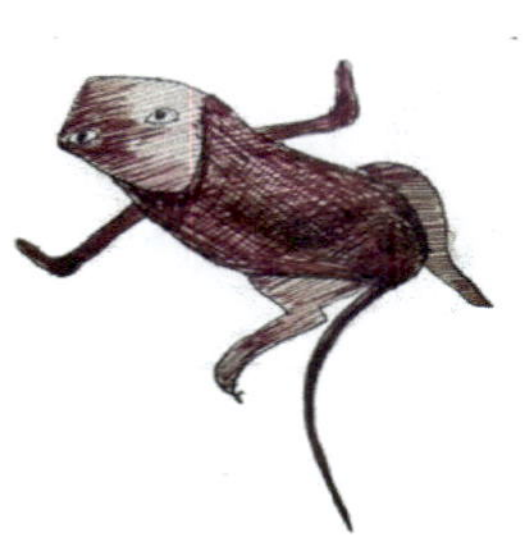

Pufat: always lands on his feet.

Pufat: always lands on its feet.

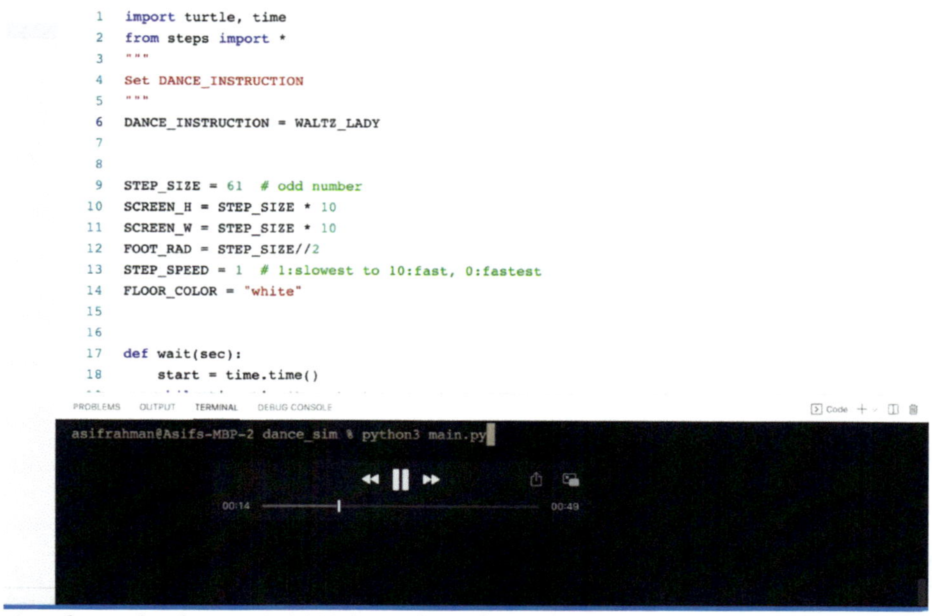

A code: waltz dance. © 2022 by Md. Asifur Rahman. (Used with permission).

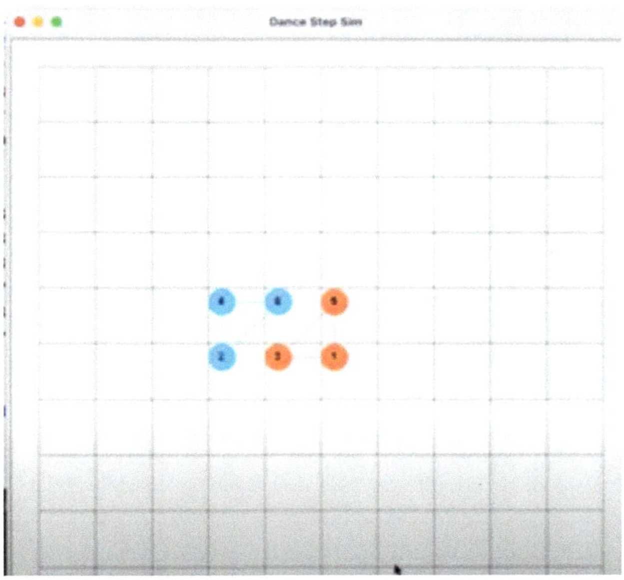

The output from the code. © 2022, Md. Asifur Rahman (Used with permission).

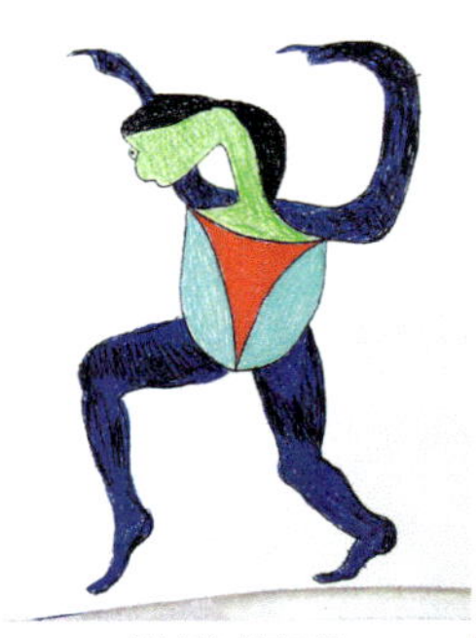

Gubala: feels the wind and records it.

Kalizuko, "Now, we'll work on the cha-cha."
Make yourself a nice cup of...
Dramata, "I cannot drink that much!"
Kalizuko, "Eat an apple!"

An actor carrying an oversized apple crosses the stage in a dignified way.

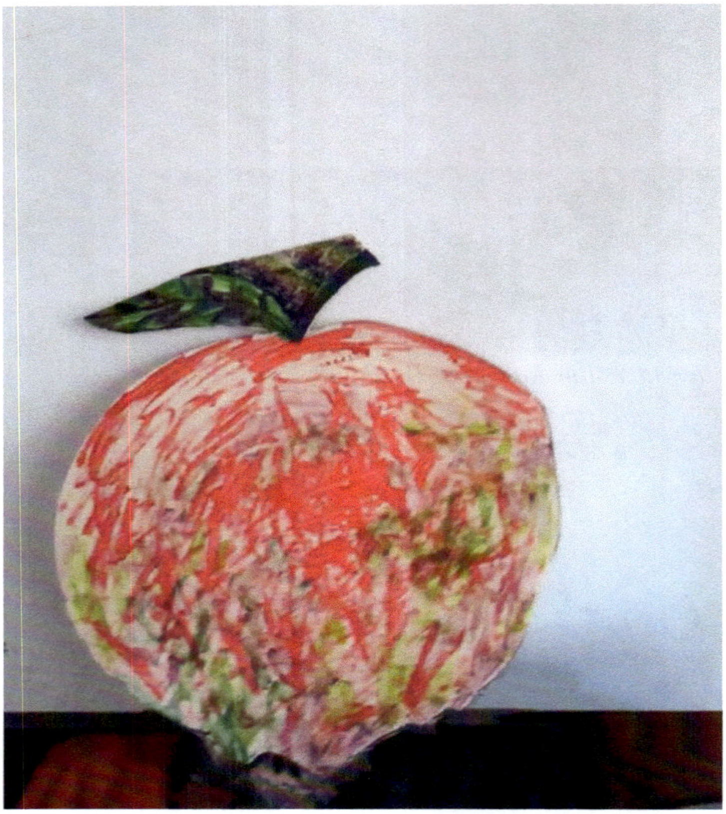

Kalizuko, " And this is cha-cha."

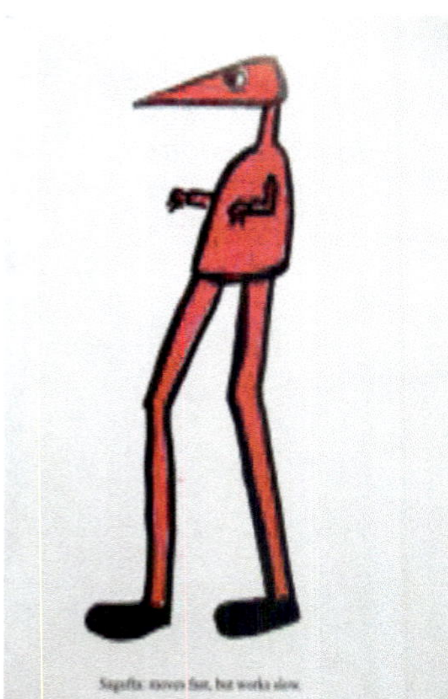

Sagufta: moves fast but walks slow.

```
import time

def cha_cha_dance():
    for _ in range(2):
        print("  .-\"\"\"-.      .-\"\"\"-.      .-\"\"\"-.")
        print(" /      /    /      /    /      /")
        print("|  ~ ~ |   |  ~ ~ |   |  ~ ~ |")
        print(" \    _\\    \    _\\    \    _\\")
        print("  `\"\"\"-'      `\"\"\"-'      `\"\"\"-'")
        time.sleep(0.5)
        print("  .-\"\"\"-.      .-\"\"\"-.      .-\"\"\"-.")
        print(" /      /    /      /    /      /")
        print("|  ~ ~ |   |  ~ ~ |   |  ~ ~ |")
        print(" \    _\\    \    _\\    \    _\\")
        print("  `\"\"\"-'      `\"\"\"-'      `\"\"\"-'")
        time.sleep(0.5)
        print("  .-\"\"\"-.      .-\"\"\"-.      .-\"\"\"-.")
        print(" /      /    /      /    /      /")
        print("|  ~ ~ |   |  ~ ~ |   |  ~ ~ |")
        print(" \    _\\    \    _\\    \    _\\")
        print("  `\"\"\"-'      `\"\"\"-'      `\"\"\"-'")
        time.sleep(0.5)
        print("  .-\"\"\"-.      .-\"\"\"-.      .-\"\"\"-.")
        print(" /      /    /      /    /      /")
        print("|  ~ ~ |   |  ~ ~ |   |  ~ ~ |")
        print(" \    _\\    \    _\\    \    _\\")
        print("  `\"\"\"-'      `\"\"\"-'      `\"\"\"-'")
        time.sleep(0.5)


if __name__ == "__main__":
    cha_cha_dance()
```

Cha-cha. © 2022 by Md. Asifur Rahman. (Used with permission).

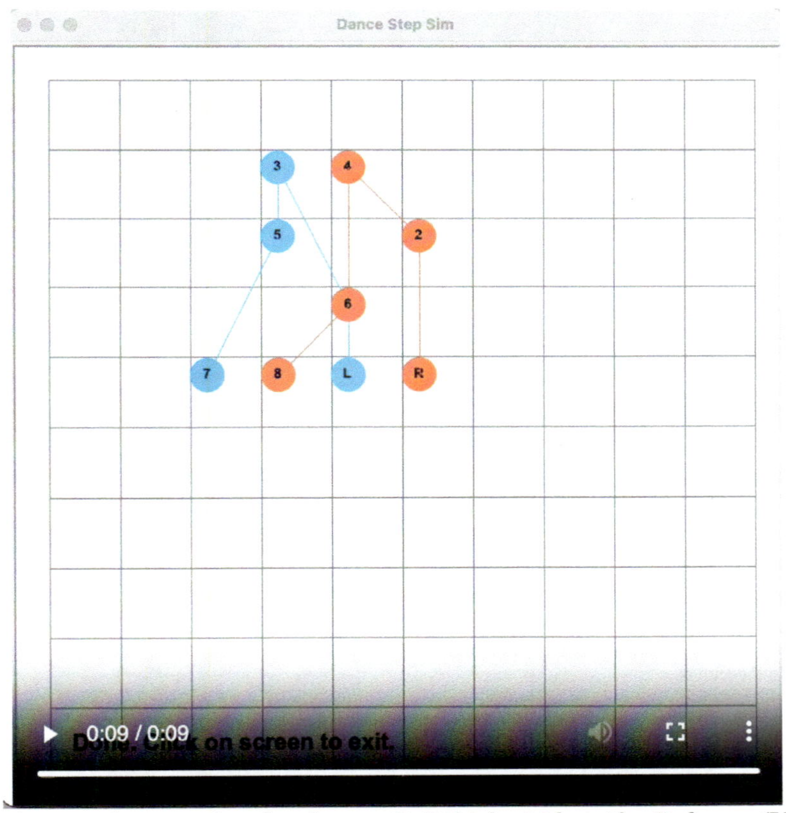

Animated steps: cha-cha dance. © 2022 by Md. Asifur Rahman (Used with permission)

`Six dancers dressed in black dance (cha-cha).`

Dramata, “Can I see the rumba, now? Rumba was a synonym for party. In northern Cuba, in the late 19th century it was danced for the complex of secular music, and in the 20th century, it became to be a ballroom dance. The basic rhythm in rumba is known as slow-quick-quick, with the first step taking two beats and the last two steps each taking one beat, a box step.”

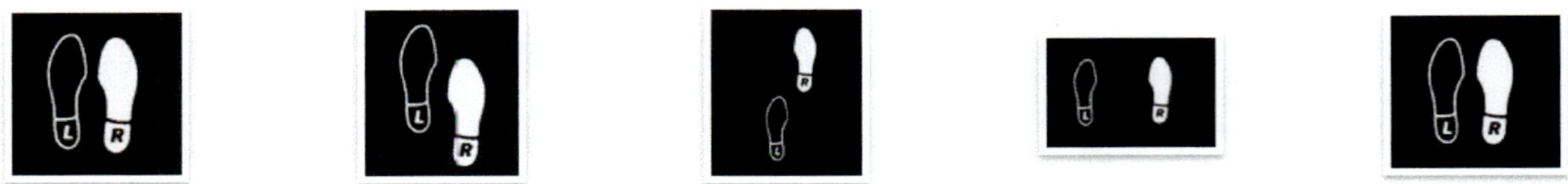

Box step. © 2021 by Irwin Stuart Smith. (Used with permission).

Kalizuko, “That’s just like the waltz!”
Dramata, “Bravo!”
Kalizuko, “Make yourself a cup of cacao, and I will be right back with the code.

`Two actors, carrying two cacao pitchers, approach from two directions, trapping the cacao cup as it tries to run away.`

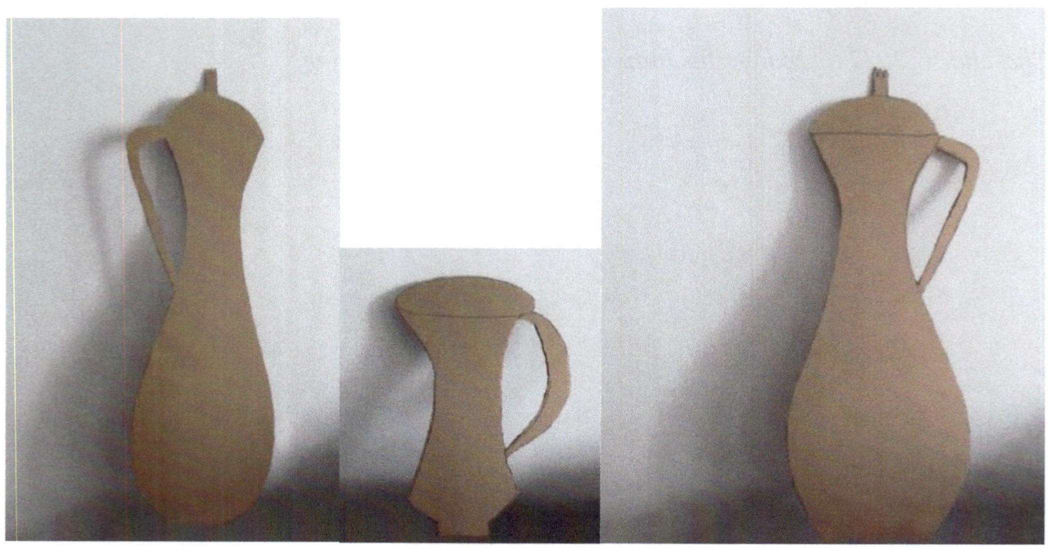

Kalizuko, “Here it is, first, the code for you, followed by the graphical output.”

Rumba code by Mohammad Majid al-Rifaie. (Used with permissions): *

```
import turtle   # this line makes sure the turtle object can be used
import random   # this lines allows the program to generate random numbers when needed
             # e.g. in order to decide the step size

screen = turtle.Screen()
screen.screensize(300,300)
screen.colormode(255)
screen.bgcolor(0,0,0)

# a function that takes the name of the turtle, the angle at which
# it should be rotated and the distance that it should move
def move(turtle, angle, distance):
   turtle.left(angle)
   turtle.forward(distance)
```

```
# this function takes the turtle name as well as
# the x and y distances that the turtle should move
def moveXY(turtle, dx, dy):
   turtle.setpos(turtle.xcor()+dx,
            turtle.ycor()+dy)

# four turtles are created
# lM for left foot of the man
# rM for right foot of the man
lM = turtle.Turtle()
rM = turtle.Turtle()

# lW for left foot of the woman
# rW for right foot of the woman
lW = turtle.Turtle()
rW = turtle.Turtle()

lM.hideturtle()
rM.hideturtle()

lW.hideturtle()
rW.hideturtle()

# the pens are lifted until the feet are in the starting position
lM.penup()
rM.penup()
lW.penup()
rW.penup()

# the man's feet are blue and woman's feet are red
lM.fillcolor("blue")
rM.fillcolor("blue")

lW.fillcolor("red")
rW.fillcolor("red")

lM.pencolor("blue")
rM.pencolor("blue")

lW.pencolor("red")
rW.pencolor("red")

lM.shape("turtle")
rM.shape("turtle")

lW.shape("turtle")
rW.shape("turtle")

lM.showturtle()
rM.showturtle()

lW.showturtle()
rW.showturtle()

screen.delay(10)

# adjust the direction of the foot to make sure that
# the man's feet are facing the woman's feet
lM.setpos(200,200)
rM.setpos(150,200)

move(lM, -90, 0)
move(rM, -90, 0)

rW.setpos(200,150)
lW.setpos(150,150)

move(lW, 90, 0)
move(rW, 90, 0)
```

```
stepSize = 200

# once the feet at their initial position, the pen of each foot is down
# and ready to show the traces of their feet's movements
lM.pendown()
rM.pendown()
lW.pendown()
rW.pendown()

screen.delay(1)

# this function takes two numbers (m and n) and generate a random integer number between m and n
def rand(a,b):
   return random.randint(a,b)

# Whenever this functin is called, one full basic rumba movement is executed
def rumba():

   # a random colour is generated for the man
   r = rand(0,255); b = rand(0,255); g = rand(0,255)
   lM.pencolor(r,g,b)#;   rM.fillcolor(r,g,b)
   rM.pencolor(r,g,b)#;      rM.fillcolor(r,g,b)

   # a random colour is generated for the the woman
   r = rand(0,255); b = rand(0,255); g = rand(0,255)
   lW.pencolor(r,g,b)#;   rM.fillcolor(r,g,b)
   rW.pencolor(r,g,b)#;      rM.fillcolor(r,g,b)

   stepSize = rand(50,300)      # a random step size is generated for each new cycle
   r = 2;                       # this number offsets the location of the feet byt a small distance

   # no. 0
   move(rW, 720, 0)          # turn woman's right foot 720 degrees anti-clockwise
                        # this is used as an indicated showing that the dance is about to start

   # no. 1
   # here woman's right foot and then man's left foot are moved along the y axis
   moveXY(rW, 0, -stepSize+rand(-r,r))
   moveXY(lM, 0, -stepSize+rand(-r,r))

   # no. 2
   moveXY(rM, -stepSize+rand(-r,r), -stepSize+rand(-r,r) )
   moveXY(lW, -stepSize+rand(-r,r), -stepSize+rand(-r,r) )
   #move(rM, -45, 100)
   #move(lW, 180-45, 100)

   # no. 3
   moveXY(lM, -stepSize+rand(-r,r), 0)
   moveXY(rW, -stepSize+rand(-r,r), 0)

   # no. 4
   moveXY(rM, 0, stepSize+rand(-r,r) )
   moveXY(lW, 0, stepSize+rand(-r,r) )

   # no. 5
   moveXY(lM, stepSize+rand(-r,r), stepSize+rand(-r,r))
   moveXY(rW,  stepSize+rand(-r,r),  stepSize+rand(-r,r))

   # no. 6
   moveXY(rM, stepSize+rand(-r,r), 0 )
   moveXY(lW, stepSize+rand(-r,r), 0 )
   #rM.stamp(); lM.stamp(); rW.stamp(); lW.stamp()

# MAIN
# Here, 100 cycle of the rumba basic step dance is executed
for i in range(100):
   rumba()


# saving the final image as a pdf file (without the background colour)
```

```
screen.getcanvas().postscript(file="Rumba_couple.pdf")
```

Rumba output from the code and the musical notes show up on the screen. The six dancers dressed in black are dancing (rumba).

The output: rumba. by Mohammad Majid al-Rifaie. (Used with permission).

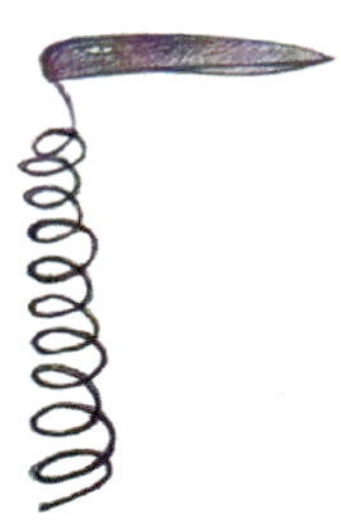

Duvi: can cut anything you cannot reach out to.

Kalizuko, "The same rumba but the dancers move simultaneously, with no offsets between the location of their steps. There is an overwrite of the trace of the second dancer's steps. The artwork of the 100 cycles is here. Here comes the code for you:"

The code for simultaneous rumba by Mohammad Majid al-Rifaie. (Used with permissions): *

```
import turtle   # this line makes sure the turtle object can be used
import random   # this lines allows the program to generate random numbers when needed
            # e.g. in order to decide the step size

screen = turtle.Screen()
screen.screensize(300,300)
screen.colormode(255)
screen.bgcolor(0,0,0)

# a function that takes the name of the turtle, the angle at which
# it should be rotated and the distance that it should move
def move(turtle, angle, distance):
   turtle.left(angle)
   turtle.forward(distance)

# this function takes the turtle name as well as
# the x and y distances that the turtle should move
def moveXY(turtle, dx, dy):
   turtle.setpos(turtle.xcor()+dx,
              turtle.ycor()+dy)


# four turtles are created
# lM for left foot of the man
# rM for right foot of the man
lM = turtle.Turtle()
rM = turtle.Turtle()

# lW for left foot of the woman
# rW for right foot of the woman
lW = turtle.Turtle()
rW = turtle.Turtle()

lM.hideturtle()
rM.hideturtle()

lW.hideturtle()
rW.hideturtle()

# the pens are lifted until the feet are in the starting position
lM.penup()
rM.penup()
lW.penup()
rW.penup()

# the man's feet are blue and woman's feet are red
lM.fillcolor("blue")
rM.fillcolor("blue")

lW.fillcolor("red")
rW.fillcolor("red")

lM.pencolor("blue")
rM.pencolor("blue")

lW.pencolor("red")
rW.pencolor("red")

lM.shape("turtle")
```

```
rM.shape("turtle")

lW.shape("turtle")
rW.shape("turtle")

lM.showturtle()
rM.showturtle()

lW.showturtle()
rW.showturtle()

screen.delay(10)

# adjust the direction of the foot to make sure that
# the man's feet are facing the woman's feet
lM.setpos(200,200)
rM.setpos(150,200)

move(lM, -90, 0)
move(rM, -90, 0)

rW.setpos(200,150)
lW.setpos(150,150)

move(lW, 90, 0)
move(rW, 90, 0)

stepSize = 200

# once the feet at their initial position, the pen of each foot is down
# and ready to show the traces of their feet's movements
lM.pendown()
rM.pendown()
lW.pendown()
rW.pendown()

screen.delay(0.00001)

# this function takes two numbers (m and n) and generate a random integer number between m and n
def rand(a,b):
   return random.randint(a,b)

# Whenever this functin is called, one full basic rumba movement is executed
def rumba():

   # a random colour is generated for the man
   r = rand(0,255); b = rand(0,255); g = rand(0,255)
   lM.pencolor(r,g,b)#;   rM.fillcolor(r,g,b)
   rM.pencolor(r,g,b)#;    rM.fillcolor(r,g,b)

   # a random colour is generated for the the woman
   r = rand(0,255); b = rand(0,255); g = rand(0,255)
   lW.pencolor(r,g,b)#;   rM.fillcolor(r,g,b)
   rW.pencolor(r,g,b)#;    rM.fillcolor(r,g,b)

   stepSize = rand(50,300)      # a random step size is generated for each new cycle

   # no. 0
   move(rW, 720, 0)             # turn woman's right foot 720 degrees anti-clockwise
                          # this is used as an indicated showing that the dance is about to start

   # no. 1
   for i in range(stepSize):    # the FOR loop allows the relevant feet to move together one unit at a time
      moveXY(rW, 0, -1)
      moveXY(lM, 0, -1)

   # no. 2
   for i in range(stepSize):
      moveXY(rM, -1, -1 )
      moveXY(lW, -1, -1 )
```

```
    # no. 3
    for i in range(stepSize):
        moveXY(lM, -1, 0)
        moveXY(rW, -1, 0)

    # no. 4
    for i in range(stepSize):
        moveXY(rM, 0, 1 )
        moveXY(lW, 0, 1 )

    # no. 5
    for i in range(stepSize):
        moveXY(lM, 1, 1)
        moveXY(rW, 1, 1)

    # no. 6
    for i in range(stepSize):
        moveXY(rM, 1, 0 )
        moveXY(lW, 1, 0 )
    #rM.stamp(); lM.stamp(); rW.stamp(); lW.stamp()


# MAIN
# Here, 100 cycle of the rumba basic step dance is executed
for i in range(100):
    rumba()


# saving the final image as a pdf file (without the background colour)
screen.getcanvas().postscript(file="Rumba_Couple_simultaneous.pdf")
```

Please see Mohammad Majid al-Rifaie's code on the website listed at the beginning and at the end of the book in case you would like to copy the code to try it. If so, please read Stuart Smith Python introduction at the Appendix section. Rumba output and the musical notes are projected while the six dancers dressed in black are dancing (rumba).

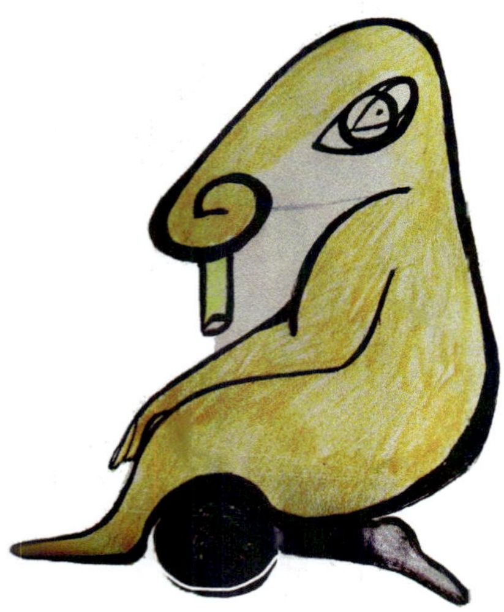

Iufafu: has no questions.

Iufafu: has no questions.

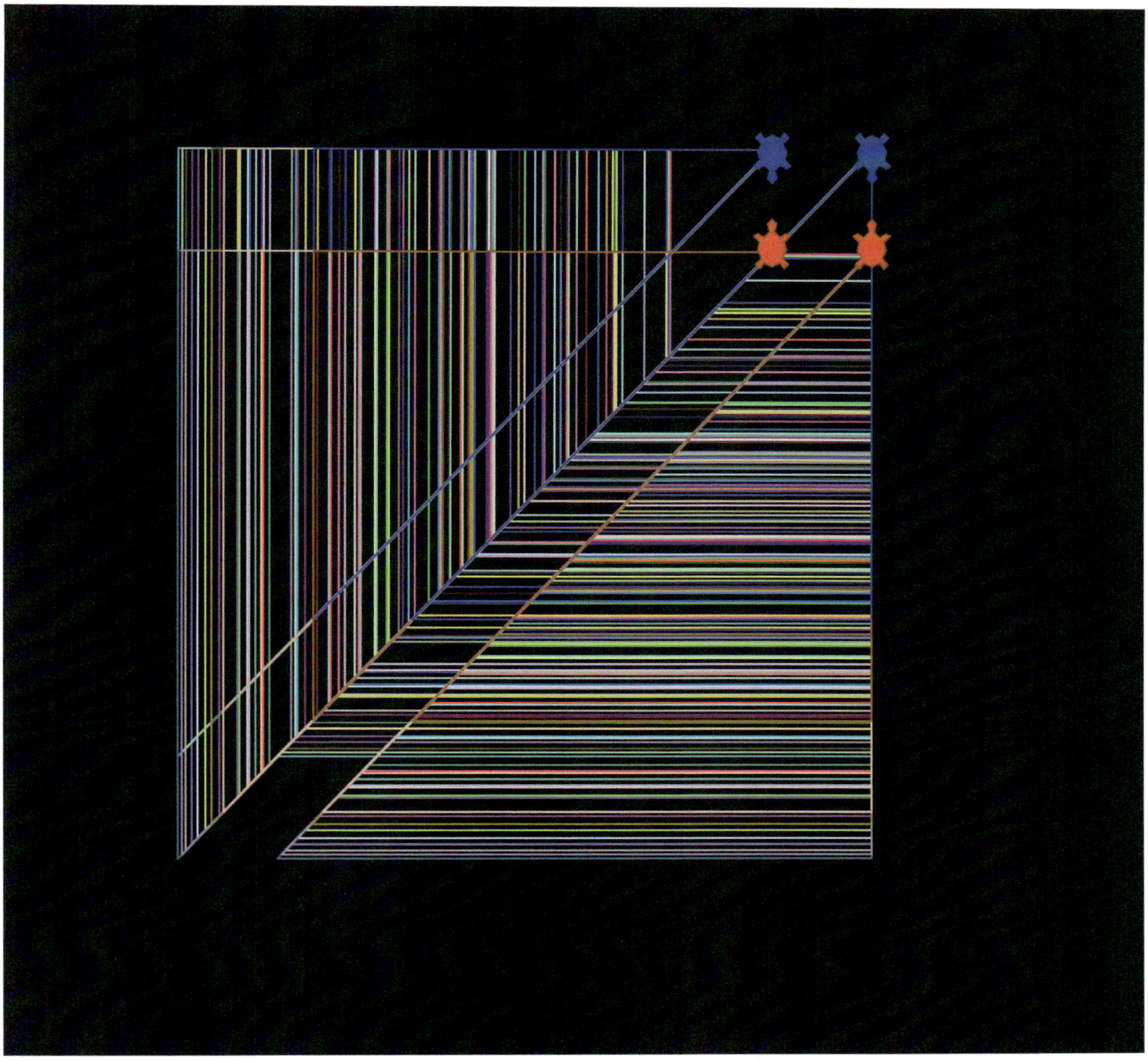

Turtles dance simultaneous rumba by Mohammad Majid al-Rifaie. (Used with permission.)

Dramata, "That's a beautiful artwork."

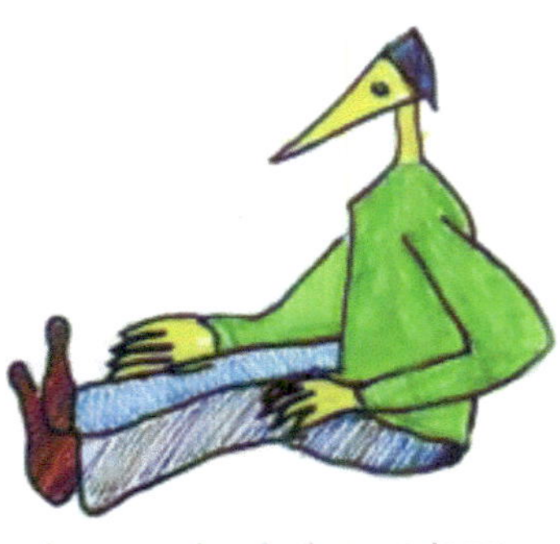

Lezeter: remembers when there was no Internet.

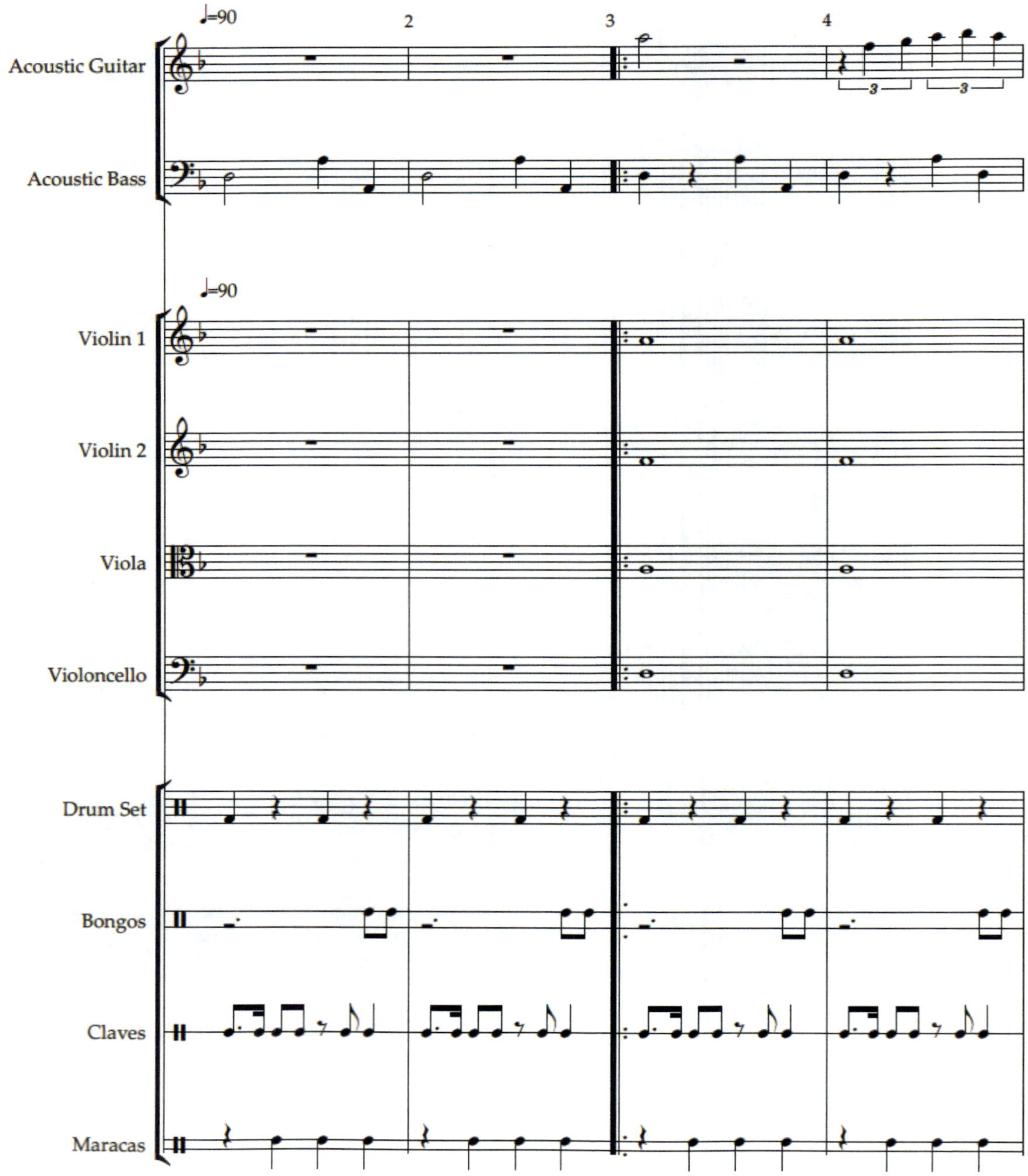

Musical Notation: rumba. © 2021 by Irwin Stuart Smith. (Used with permission).

Dramata: "That makes sense. How about the foxtrot? It originated in the 1910s in America, was first danced by African Americans, believed to have been named after Vaudeville actor Harry Fox. A progressive dance to a 4/4-time signature, with a basic foot pattern of slow, slow, quick, quick, with the lead smoothly stepping forward on their left and the follow stepping back on their right."

Kalizuko, "Yes, of course, please pour yourself a glass of water, while I come up with a code. "

```
A water pitcher is chasing a water glass which lovingly accepts
the liquid. The six dancers are dancing (foxtrot).
```

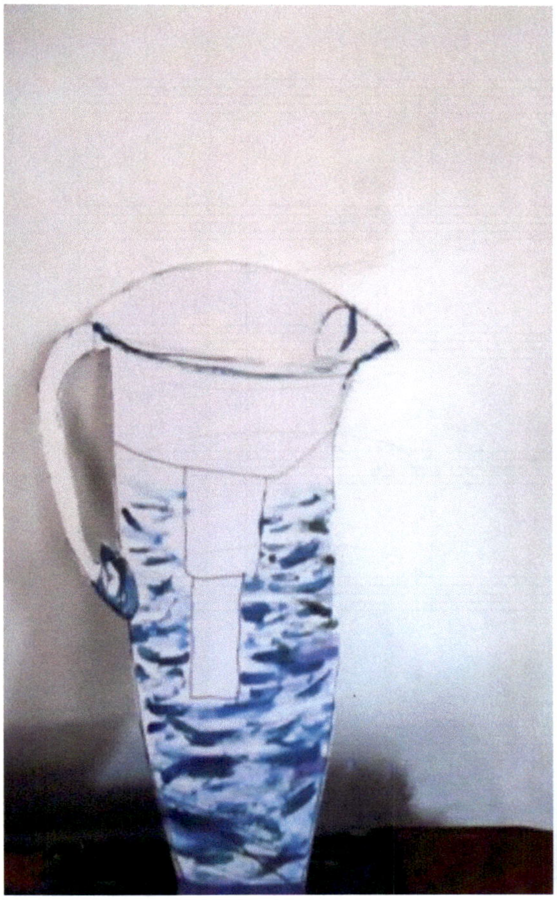

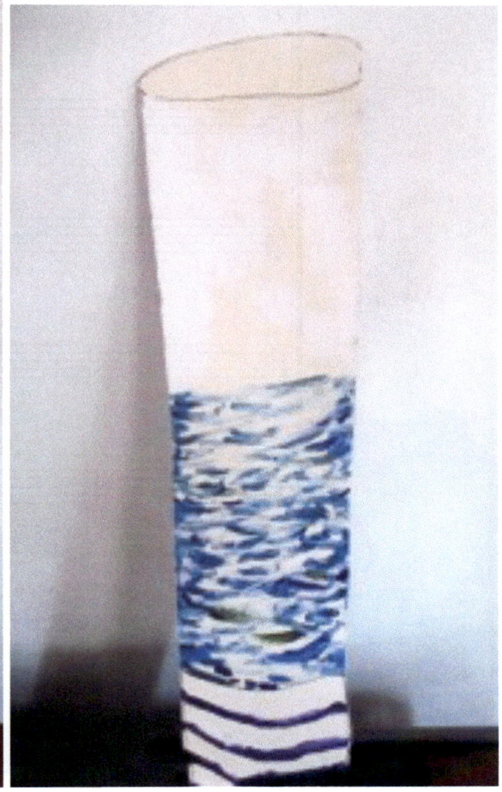

```
The musical notes followed by the code with the outputted artwork
are projected. Six dancers dressed in black dance (foxtrot).
```

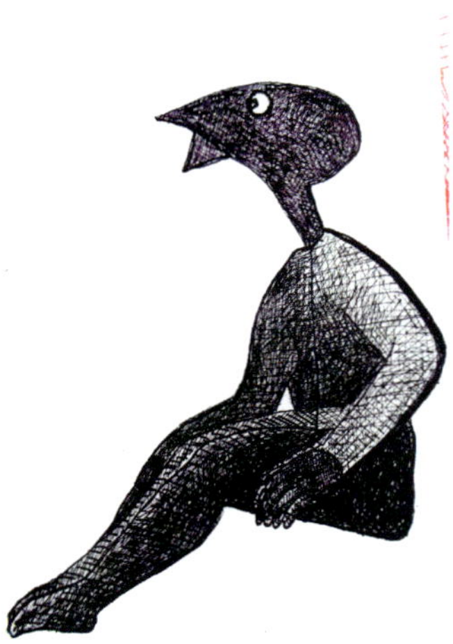

Gatae: depends on his imagination.

Gatae: depends on his imagination.

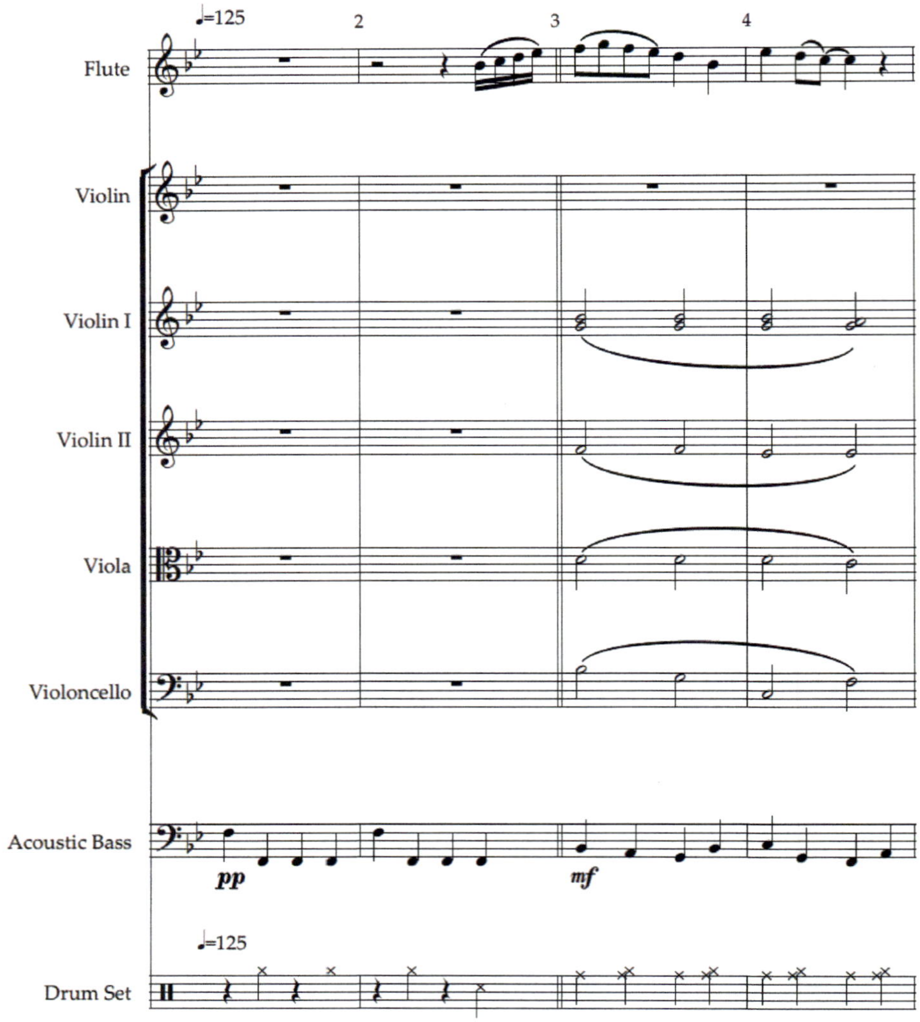

Musical Notations: a foxtrot. © 2021 by Irwin Stuart Smith. (Used with permission).

Kalizuko, "Here it is: first the code in Python, followed by the graphical output."

The code for a foxtrot. © by Mohammad Majid al-Rifaie. (Used with permissions): *

```
import turtle   # this line makes sure the turtle object can be used
import random   # this lines allows the program to generate random numbers when needed
            # e.g. in order to decide the step size
import time

screen = turtle.Screen()
screen.screensize(300,300)
screen.colormode(255)
screen.bgcolor(0,0,0)
```

```
sleepTime = 0.3

# a function that takes the name of the turtle, the angle at which
# it should be rotated and the distance that it should move
def move(turtle, angle, distance):
   turtle.left(angle)
   turtle.forward(distance)

# this function takes the turtle name as well as
# the x and y distances that the turtle should move
def moveXY(turtle, dx, dy):
   turtle.setpos(turtle.xcor()+dx,
              turtle.ycor()+dy)

# four turtles are created
# lM for left foot of the man
# rM for right foot of the man
lM = turtle.Turtle()
rM = turtle.Turtle()

# lW for left foot of the woman
# rW for right foot of the woman
lW = turtle.Turtle()
rW = turtle.Turtle()

lM.hideturtle()
rM.hideturtle()

lW.hideturtle()
rW.hideturtle()

lM.shape("turtle")
rM.shape("turtle")

lW.shape("turtle")
rW.shape("turtle")

lM.showturtle()
rM.showturtle()

lW.showturtle()
rW.showturtle()

# adjust the direction of the foot to make sure that
# the man's feet are facing the woman's feet
move(lW, 90, 0)
move(rW, 90, 0)

move(lM, -90, 0)
move(rM, -90, 0)

screen.delay(10)

# this function takes two numbers (m and n) and generate a random integer number between m and n
def rand(m,n):
   random.randint(m,n)

# this function places the feet of the dances in the starting position
def startingPosition():

   # the initial position of the feet are marked and the pen is lifted
   lM.stamp();   lM.penup()
   rM.stamp();   rM.penup()
   lW.stamp();   lW.penup()
   rW.stamp();   rW.penup()

   # a random colour is generated for the man
   r = rand(0,255); b = rand(0,255); g = rand(0,255)
   lM.pencolor(r,g,b)#;   rM.fillcolor(r,g,b)
   rM.pencolor(r,g,b)#;   rM.fillcolor(r,g,b)
```

```
    # a random colour is generated for the the woman
    r = rand(0,255); b = rand(0,255); g = rand(0,255)
    lW.pencolor(r,g,b)#;     rM.fillcolor(r,g,b)
    rW.pencolor(r,g,b)#;     rM.fillcolor(r,g,b)

    # initial position of the feet is randomly selected on the dance floor
    initX = random.randint(-200,300)
    initY = random.randint(-200,300)

    # lowe and upper difference allows the feet to be slightly unsynched
    # rather than precisely opposite each other
    diffLower = 5
    diffUpper = 10

    lM.goto(initX,initY)
    rM.goto(initX-50,initY)

    rW.goto(initX+rand(diffLower,diffUpper),
            initY-50+rand(diffLower,diffUpper))

    lW.goto(initX-50+rand(diffLower,diffUpper),
            initY-50+rand(diffLower,diffUpper))

    # once the feet at their initial position, the pen of each foot is down
    # and ready to show the traces of their feet's movements
    lM.pendown()
    rM.pendown()
    lW.pendown()
    rW.pendown()


# setting the the speed at which the dance should take place
screen.delay(0.1)

# this function checks if the feet inside the dance floor
# if not, a new position is generated and the dance starts over
def checkLoc():
    if ( rM.ycor() > 300 or rM.ycor() < -300 or
        lM.ycor() > 300 or lM.ycor() < -300 or
        rW.ycor() > 300 or rW.ycor() < -300 or
        lW.ycor() > 300 or lW.ycor() < -300):

        startingPosition()


def rand(a,b):
    return random.randint(a,b)

# this function performs one full cycle of the basic foxtrot basic dance steps
def foxtrot():

    stepSize = rand(100,200)  # a random step size is generated for each new cycle

    # no. 1
    for i in range(stepSize):  # the FOR loop allows the relevant feet to move together one unit at a time
        moveXY(rW, 0, -1)
        moveXY(lM, 0, -1)

    checkLoc()     # once done, the position of the feet are checked to see that
                   # they are inside the dance floor, otherwise a new dance starts

    time.sleep(sleepTime)  # this allows the dancers to stop for a little while to show that once step is complete
    #move(lM, 0, 100)      # these line allows the fast movement of each foot separately
    #move(rW, 0, -100)

    # no. 2
    for i in range(int(stepSize/2)):
        moveXY(rM, 0, -1 )
        moveXY(lW, 0, -1 )
```

```
    checkLoc()
    time.sleep(sleepTime)
    #move(rM, -45, 100)
    #move(lW, 180-45, 100)

    # no. 3
    for i in range(int(stepSize/4)):
        moveXY(lM, 0, -1)
        moveXY(rW, 0, -1)

    for i in range(stepSize):
        moveXY(lM, -1, 0)
        moveXY(rW, -1, 0)

    checkLoc()
    time.sleep(sleepTime)

    # no. 4
    for i in range(int(stepSize*3/4)):
        moveXY(rM, 0, -1 )
        moveXY(lW, 0, -1 )

    for i in range(int(stepSize)):
        moveXY(rM, -1, 0 )
        moveXY(lW, -1, 0 )

    checkLoc()
    time.sleep(sleepTime)


# ==================================
# initially the feet are located in an initial position
startingPosition()

# then the dance starts, in this case 100 cycle of the foxtrot basic step dance
for i in range(100):
    foxtrot()


# saving the final image as a pdf file (without the background colour)
screen.getcanvas().postscript(file="Foxtrot.pdf")
```

Please find the code by Mohammad Majid al-Rifaie on the website provided at the beginning and the end of the book if you want to try it. Use the instructions on how to use IDLE by Stuart Smith to run this code and see its output on your screen. You may alter some variables to see the changes in a new output after you select "Run" from the menu.

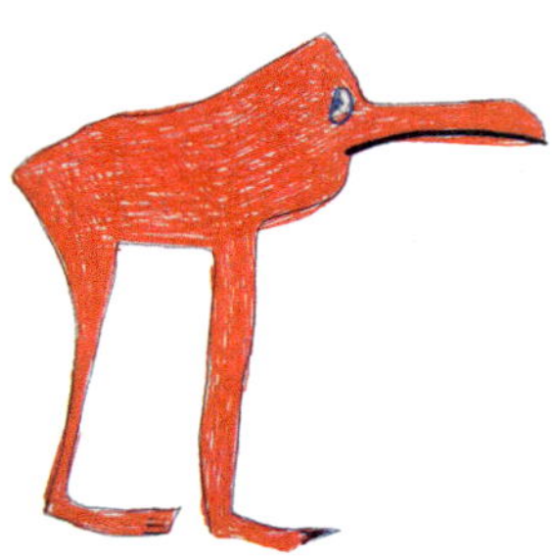

Vuturandu: knows when to act.

Vuturandu: knows when to act.

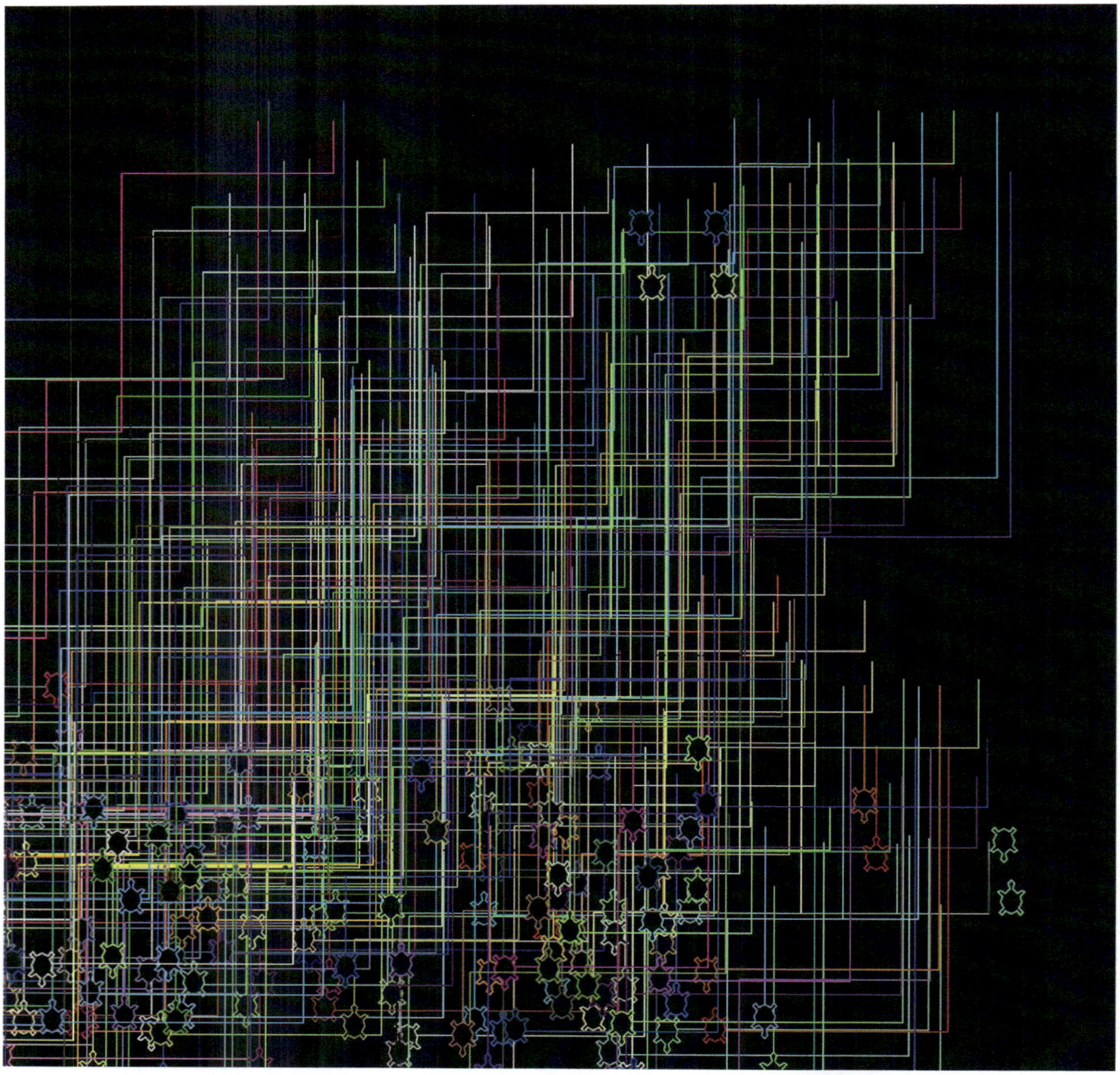

The output from a foxtrot code. © by Mohammad Majid al-Rifaie. (Used with permission).

Dramata, "I feel ready to become a coder!"

Kalizuko, "Sure, you are."

Dramata, " So, how do I know which dance to code and how? Which language to use?"

Kalizuko, "For that, you will need a simulator: waltz, foxtrot, cha-cha, salsa, tango, rumba, samba, jive, and swing."

```
/*
Dance Simulator in JavaScript.
Author: Md. Asifur Rahman.
NB: Set variable 'dance_instruction' at Line 156 for a different dance.
*/
```

```
/* Some Example Dance steps/instructions */
const WALTZ_MEN = [
    ["l", "f"],
    ["r", "fr"],
    ["l", "r"],
    ["r", "b"],
    ["l", "bl"],
    ["r", "l"],
];
const WALTZ_LADY = [
    ["r", "b"],
    ["l", "bl"],
    ["r", "l"],
    ["l", "f"],
    ["r", "fr"],
    ["l", "r"],
];
const FOXTROT_MEN = [
    ["l", "f"],
    ["r", "ff"],
    ["l", "ffl"],
    ["r", "fl"],
    ["l", "b"],
    ["r", "bb"],
    ["l", "bbl"],
    ["r", "bl"],
];
const FOXTROT_LADY = [
    ["r", "b"],
    ["l", "bb"],
    ["r", "bbr"],
    ["l", "br"],
    ["r", "f"],
    ["l", "ff"],
    ["r", "ffr"],
    ["l", "fr"],
];
const CHA_CHA_MEN = [
    ["l", "l"],
    ["r", "bl"],
    ["r", "fr"],
    ["l", "r"],
    ["r", "r"],
    ["l", "fr"],
    ["l", "bl"],
    ["r", "l"],
];

const CHA_CHA_LADY = [
    ["r", "r"],
    ["l", "fr"],
    ["l", "bl"],
    ["r", "l"],
    ["l", "l"],
    ["r", "bl"],
    ["r", "fr"],
    ["l", "r"],
];
const SALSA_MEN = [
    ["l", "f"],
    ["l", "b"],
    ["r", "b"],
    ["r", "f"],
];
const SALSA_LADY = [
    ["r", "b"],
    ["r", "f"],
    ["l", "f"],
    ["l", "b"],
];
```

```
const TANGO_MEN = [
  ["l", "f"],
  ["r", "ff"],
  ["l", "ff"],
  ["r", "fr"],
  ["l", "fr"],
];
const TANGO_LADY = [
  ["r", "b"],
  ["l", "bb"],
  ["r", "bb"],
  ["l", "bl"],
  ["r", "bl"],
];
const RUMBA_MEN = [
  ["l", "f"],
  ["r", "fr"],
  ["l", "r"],
  ["r", "b"],
  ["l", "bl"],
  ["r", "l"],
];
const RUMBA_LADY = [
  ["r", "b"],
  ["l", "bl"],
  ["r", "l"],
  ["l", "f"],
  ["r", "fr"],
  ["l", "r"],
];
const SAMBA_MEN = [
  ["l", "f"],
  ["r", "f"],
  ["r", "b"],
  ["l", "b"],
];
const SAMBA_LADY = [
  ["r", "b"],
  ["l", "b"],
  ["l", "f"],
  ["r", "f"],
];
const JIVE_MEN = [
  ["l", "b"],
  ["l", "fl"],
  ["r", "l"],
  ["l", "l"],
  ["l", "r"],
  ["r", "r"],
];
const JIVE_LADY = [
  ["r", "b"],
  ["r", "fr"],
  ["l", "r"],
  ["r", "r"],
  ["r", "l"],
  ["l", "l"],
];
const SWING_MEN = [
  ["l", "l"],
  ["r", "l"],
  ["l", "l"],
  ["l", "r"],
  ["r", "r"],
  ["l", "br"],
];
const SWING_LADY = [
  ["r", "r"],
  ["l", "r"],
  ["r", "r"],
```

```
    ["r", "l"],
    ["l", "l"],
    ["r", "bl"],
];


main(async function () {
    let ctx = await init();
    let dance_instruction = TANGO_MEN; /* <----- Set instruction variable */
    setTimeout(async function () {
        await dance(ctx, dance_instruction);
    }, 1200);
    window.document.getElementById("reloadButton").addEventListener("click",
        function () {
            window.location.reload();
        }
    );
});

// the entire canvas will be devided into equal sized square blocks
const BLOCK_COUNT = 10; // number of blocks in the canvas
const BLOCK_SIZE = 60; // length of a side of the square block in pixel
const HALF_BLOCK_SIZE = parseInt(BLOCK_SIZE / 2);
const CANVAS_HEIGHT = BLOCK_COUNT * BLOCK_SIZE;
const CANVAS_WIDTH = BLOCK_COUNT * BLOCK_SIZE;
const CANVAS_COLOR = "white";
const GRID_COLOR = "silver";
const TRAIL_COLOR = "white";
const FOOT_ICON_RADIUS = parseInt(BLOCK_SIZE / 4);
const STEP_SPEED = 15; // in milli sec. higher the number, slower the speed
const FOOT_ICON_ATTRIBUTES = {
    "l": {
        iconColor: "yellow",
        labelColor: "navy",
        label: "L",
    },
    "r": {
        iconColor:  "green",
        labelColor: "white",
        label: "R",
    },
};
let currentFootOnBlock = {
    "l": {
        h: Math.ceil(BLOCK_COUNT / 2),
        v: Math.ceil(BLOCK_COUNT / 2),
        step: 'L',
    },
    "r": {
        h: Math.ceil(BLOCK_COUNT / 2) + 1,
        v: Math.ceil(BLOCK_COUNT / 2),
        step: 'R',
    },
};

function blockToPixel(blockNo) {
    return blockNo * BLOCK_SIZE;
}

function main(callback) {
    if (document.readyState != "loading") {
        callback();
    } else if (document.addEventListener) {
        document.addEventListener("DOMContentLoaded", callback);
    } else {
        document.attachEvent("onreadystatechange", function () {
            if (document.readyState == "complete") callback();
        });
    }
```

```
}

async function initiateHtmlElements() {
  return new Promise(function (resolve) {
    let reloadButton = window.document.createElement("button");
    reloadButton.innerHTML = "Reload";
    reloadButton.id = "reloadButton";
    window.document.body.prepend(reloadButton);
    window.document.body.prepend(window.document.createElement("br"));
    let danceSimCanvas = window.document.createElement("canvas");
    danceSimCanvas.id = "danceSimCanvas";
    window.document.body.prepend(danceSimCanvas);
    danceSimCanvas.width = CANVAS_WIDTH;
    danceSimCanvas.height = CANVAS_HEIGHT;
    danceSimCanvas.style = "border:2px solid " + GRID_COLOR + ";margin:10px;";
    let ctx = danceSimCanvas.getContext("2d"); // canvas context
    ctx.fillStyle = CANVAS_COLOR; // set background color
    ctx.fillRect(0, 0, CANVAS_WIDTH, CANVAS_HEIGHT);
    danceSimCanvas.insertAdjacentHTML("beforebegin", "<h1>Dance Simulator</h1>");
    resolve(ctx);
  });
}

function drawGrid(ctx) {
  return new Promise(function (resolve) {
    ctx.strokeStyle = GRID_COLOR;
    ctx.lineWidth = 1;
    for (let block = 1; block < BLOCK_COUNT; block++) {
      // vertical lines
      ctx.beginPath();
      ctx.moveTo(blockToPixel(block), 0);
      ctx.lineTo(blockToPixel(block), CANVAS_HEIGHT);
      ctx.stroke();
      // horizontal lines
      ctx.beginPath();
      ctx.moveTo(0, blockToPixel(block));
      ctx.lineTo(CANVAS_WIDTH, blockToPixel(block));
      ctx.stroke();
    }
    resolve(true);
  });
}

async function drawTrail(ctx, currentFootOnBlock, toHorizontalBlockNo, toVerticalBlockNo, trailColor) {
  return new Promise(function (resolve) {
    let startX = blockToPixel(currentFootOnBlock.h) - HALF_BLOCK_SIZE;
    let startY = blockToPixel(currentFootOnBlock.v) - HALF_BLOCK_SIZE;
    let endX = blockToPixel(toHorizontalBlockNo) - HALF_BLOCK_SIZE;
    let endY = blockToPixel(toVerticalBlockNo) - HALF_BLOCK_SIZE;
    currentFootOnBlock.h = toHorizontalBlockNo;
    currentFootOnBlock.v = toVerticalBlockNo;
    let wayPointCount = 50;
    let wayPoints = [];
    let dx = endX - startX;
    let dy = endY - startY;
    for (let wC = 0; wC < wayPointCount; wC++) {
      let x = startX + dx * wC / wayPointCount;
      let y = startY + dy * wC / wayPointCount;
      wayPoints.push({
        x: x,
        y: y
      });
    }
    let wayPointNo = 1;
    let intervalHandler = setInterval(function () {
      ctx.strokeStyle = trailColor;

      ctx.beginPath();
      ctx.lineWidth = 2;
      ctx.setLineDash([30, 15]);
```

```
            ctx.moveTo(wayPoints[wayPointNo - 1].x, wayPoints[wayPointNo - 1].y);
            ctx.lineTo(wayPoints[wayPointNo].x, wayPoints[wayPointNo].y);
            ctx.stroke();
            wayPointNo = wayPointNo + 1;
            if (wayPointNo >= wayPointCount) {
                clearInterval(intervalHandler);
                resolve(true);
            }
        }, STEP_SPEED);
    });
}

async function drawFootSymbol(ctx, horizontalBlockNo, verticalBlockNo, symbolColor, symbolLabel, labelColor) {
    return new Promise(function (resolve) {
        ctx.fillStyle = symbolColor;
        ctx.strokeStyle = "white";
        ctx.lineWidth = 1;
        ctx.beginPath();
        ctx.arc(
            blockToPixel(horizontalBlockNo) - BLOCK_SIZE / 2,
            blockToPixel(verticalBlockNo) - BLOCK_SIZE / 2,
            FOOT_ICON_RADIUS,
            0,
            360 * Math.PI / 180
        );
        ctx.fill();
        ctx.stroke();
        ctx.font = "16px Arial";
        ctx.textAlign = "center";
        ctx.fillStyle = labelColor;
        ctx.fillText(
            symbolLabel,
            blockToPixel(horizontalBlockNo) - BLOCK_SIZE / 2,
            blockToPixel(verticalBlockNo) - BLOCK_SIZE / 2 + 5
        );
        resolve(true);
    });
}

function nextTrailColor() {
    let colors = ["teal", "blue", "olive", "maroon", "green"];
    if (typeof nextTrailColor.colorNo == 'undefined') nextTrailColor.colorNo = 0;
    nextTrailColor.colorNo = nextTrailColor.colorNo == colors.length - 1 ? 0 : nextTrailColor.colorNo + 1;
    return colors[nextTrailColor.colorNo];
}

async function move(ctx, stepNo, foot, currentFootOnBlock, destHBlock, destVBlock) {
    let footOrigin = {};
    Object.assign(footOrigin, currentFootOnBlock[foot]);
    await drawTrail(
        ctx,
        currentFootOnBlock[foot],
        destHBlock, destVBlock,
        nextTrailColor()
    );
    await drawFootSymbol(
        ctx,
        footOrigin.h,
        footOrigin.v,
        FOOT_ICON_ATTRIBUTES[foot].iconColor,
        footOrigin.step,
        FOOT_ICON_ATTRIBUTES[foot].labelColor
    );
    await drawFootSymbol(
        ctx,
        destHBlock,
        destVBlock,
        FOOT_ICON_ATTRIBUTES[foot].iconColor,
        stepNo,
        FOOT_ICON_ATTRIBUTES[foot].labelColor
```

```
  );
  currentFootOnBlock[foot].step = stepNo;
}

async function dance(ctx, instruction) {
  for (let stepNo = 0; stepNo < instruction.length; stepNo++) {
    let foot = instruction[stepNo][0];
    let direction = instruction[stepNo][1];
    let destHBlock = currentFootOnBlock[foot].h;
    let destVBlock = currentFootOnBlock[foot].v;
    for (let dN = 0; dN < direction.length; dN++) {
      if (direction[dN] == "l") destHBlock = destHBlock - 1;
      if (direction[dN] == "r") destHBlock = destHBlock + 1;
      if (direction[dN] == "f") destVBlock = destVBlock - 1;
      if (direction[dN] == "b") destVBlock = destVBlock + 1;
    } // next direction
    await move(
      ctx,
      stepNo + 1,
      foot,
      currentFootOnBlock,
      destHBlock,
      destVBlock
    );
  } // next instruction
}

async function init() {
  let ctx = await initiateHtmlElements();
  await drawGrid(ctx);
  // draw foot symbol at initial position
  await drawFootSymbol(
    ctx,
    currentFootOnBlock.l.h,
    currentFootOnBlock.l.v,
    FOOT_ICON_ATTRIBUTES.l.iconColor,
    FOOT_ICON_ATTRIBUTES.l.label,
    FOOT_ICON_ATTRIBUTES.l.labelColor
  );
  await drawFootSymbol(
    ctx,
    currentFootOnBlock.r.h,
    currentFootOnBlock.r.v,
    FOOT_ICON_ATTRIBUTES.r.iconColor,
    FOOT_ICON_ATTRIBUTES.r.label,
    FOOT_ICON_ATTRIBUTES.r.labelColor
  );
  return ctx;
}
```

Dramata, “What kind of a language is this?”
Kalizuko, “Those scripts are written in JavaScript and Python languages to support all the dances mentioned. You just need to change the input-instruction variable. I already have written instructions for all the dances. Just set the desired variable name that holds the instruction set. “
Dramata, “That’s a translator!”
Kalizuko, “And now, let’s see the more visual outcome of a waltz and then, on a foxtrot; those are, like nonverbal instructions:”

Please watch the Md. Asifur Rahman’s video:

https://youtu.be/zhuMGF8_Whw

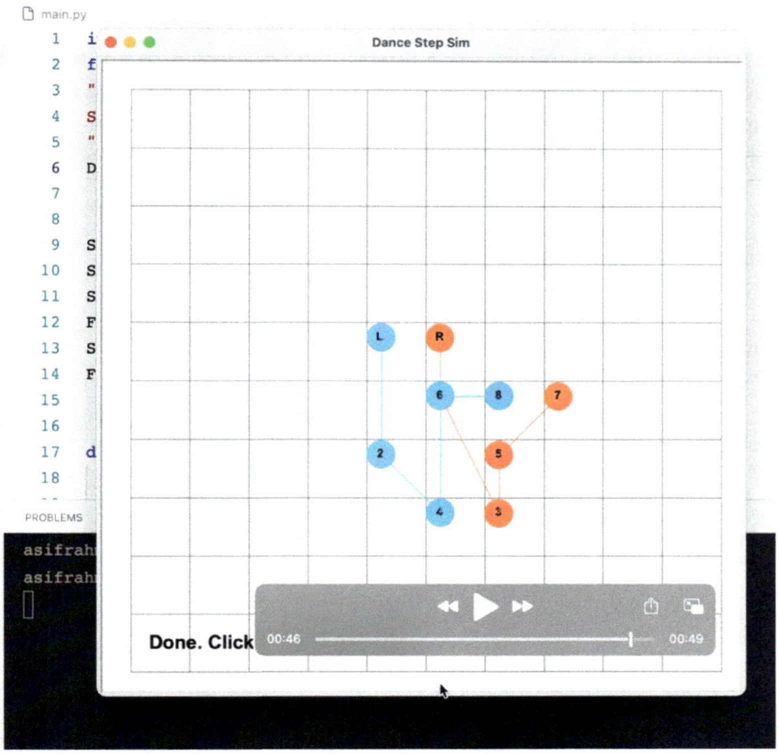

Dance Simulator © 2022 Md. Asifur Rahman (Used with permission).

Kalizuko, "And another visual example:"

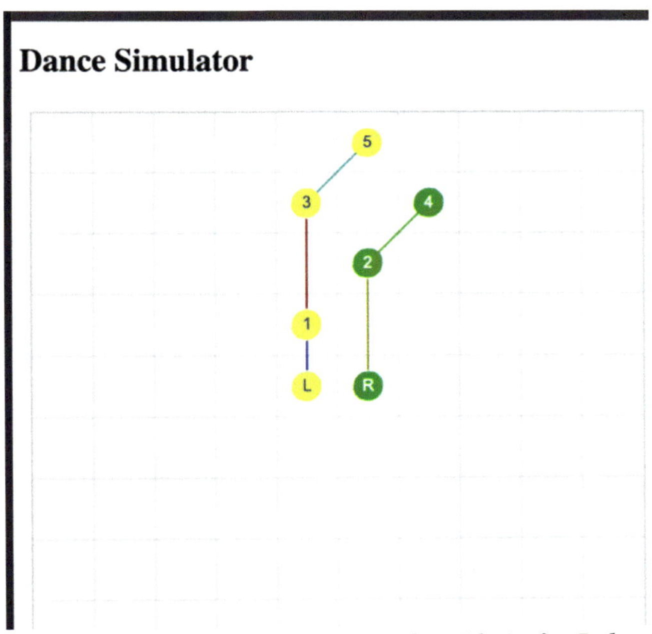

Another Dance Simulator © 2022 by Md. Asifur Rahman. (Used with permission).

Kalizuko, "Now, I'll do both, salsa and samba, in Python. I'll use the " #s" for comments, so you and other coders can see them, but the computer won't as it will ignore them".

Dramata, "First, the salsa. It originated in 1900s Eastern Cuba, combining elements and rhythms from Cuban and Afro-Cuban rumba. In salsa music, the beat is 4/4, with three weight changes in each measure. It can be danced in linear or circular motion. The basic step is counted as quick, quick, slow, with the lead stepping forward on their left and follow stepping back on their right."
Kalizuko, "There you go:"

Code for a salsa by Mohammad Majid al-Rifaie. (Used with permission).

```
import turtle  # This line makes sure the turtle object can be used
import time    # This allows a pause before starting the dance

screen = turtle.Screen()
screen.setup(500, 500) # Set the size of the dance floor
screen.colormode(255) # The colour of the dance floor

""" ########################################## """
""" Create the dancers AND set the dance floor """
""" ########################################## """

ml = turtle.Turtle() # Create male's left foot
mr = turtle.Turtle() # Create male's right foot

fl = turtle.Turtle() # Create female's left foot
fr = turtle.Turtle() # Create female's right foot

# Leave no trace behind
ml.penup(); mr.penup()
fl.penup(); fr.penup()

# Setting the shape of the feet to turtles
ml.shape("arrow"); mr.shape("arrow")
fl.shape("classic"); fr.shape("classic")

# Initially the turtle is headed east,
# so the male feet are rotated 90 degrees anti-clockwise to head upward (north)
# and go back 50 unit to be placed in the initial position
ml.left(90); ml.backward(50)
mr.left(90); mr.backward(50)

# Here, for male dancer, the right foot is moved away from the left one
# (so they don't overlap)
mr.right(90); mr.forward(50); mr.left(90)

# The same is repeated for the female dancer now.
fl.forward(50) # Move away the left foot from the right
fl.right(90); fr.right(90) # Turn the feet to face south

# set the speed of the dance
speed = 1
ml.speed(speed); mr.speed(speed)
fl.speed(speed); fr.speed(speed)

# set the shoe colour of the male (black) and female dancer (red)
ml.pencolor(0,0,0) # RED, GREEN, BLUE
mr.pencolor(0,0,0)

fl.pencolor(255,0,0) # RED, GREEN, BLUE
fr.pencolor(255,0,0)

# Give the dancers a few seconds and let the dance begin!
time.sleep(3)


""" ################## """
```

```
""" Create Salsa Steps """
""" ################## """

# Whenever this function is called, one full Salsa dance routine is executed
def salsa():

  ml.forward(50)  # Move male left foot forward 50 units
  fr.backward(50) # Move female right foot backward 50 units

  shift = 45      # Magnitude of "in place rotation" representing weight shift
  mr.left(shift)  # Rotate male right foot 45 degrees
  fl.right(shift) # Rotate female left foot 45 degrees
  mr.right(shift)
  fl.left(shift)

  ml.backward(50)
  fr.forward(50)
  time.sleep(.5)  # pause on beat

  mr.backward(50)
  fl.forward(50)

  ml.right(shift)
  fr.left(shift)
  ml.left(shift)
  fr.right(shift)

  mr.forward(50)
  fl.backward(50)
  time.sleep(.5)  # pause on beat

""" ################### """
""" LET THE DANCE BEGIN """
""" ################### """

for i in range(10): # Repeat 10 times
  salsa() # Calls the steps of salsa
```

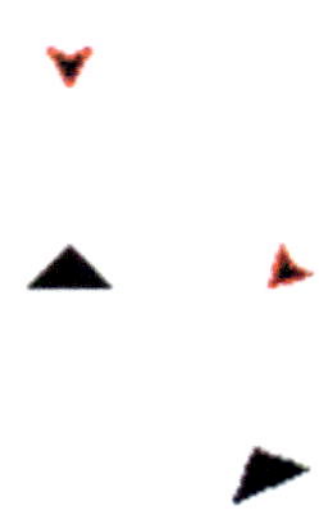

Salsa Dance © by Mohammad Majid al-Rifaie. (Used with permission).

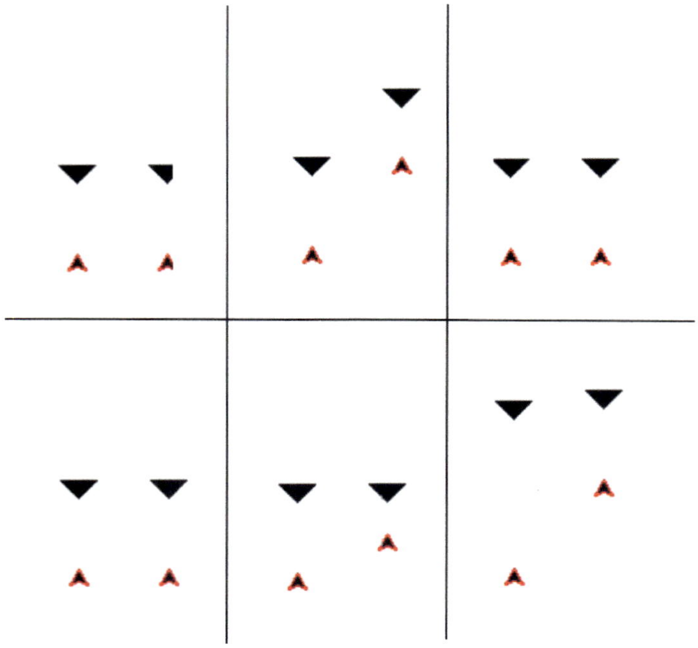

Salsa dance animated.© by Mohammad Majid al-Rifaie. (Used with permission).

Dramata: “And now, the samba code, please. It originated in Bahia, Brazil, then expanded to Rio de Janeiro, in the 19th century. Brazilian Samba is a set of dances, danced during a festival, or carnival, they differ from the ballroom style samba, with the basic step being a slow, quick quick, or, 1 and 2. The beat value is ¾, ¼, one for every three steps, and the lead stepping on their left and the follow stepping on their right.” *Kalizuko*, “I just emailed you the Python code, but you will find two files attached, you will need to add the ".py" extension yourself when the file is downloaded as ".py" files are not accepted as attachments). I also attached the video of the code in action.”

The code for a samba. © by Mohammad Majid al-Rifaie. (Used with permission).

```
import turtle  # This line makes sure the turtle object can be used
import time    # This allows a pause before starting the dance

screen = turtle.Screen()
screen.setup(500, 500) # Set the size of the dance floor
screen.colormode(255)  # The colour of the dance floor

""" ########################################## """
""" Create the dancers AND set the dance floor """
""" ########################################## """
time.sleep(3)
ml = turtle.Turtle() # Create male's left foot
mr = turtle.Turtle() # Create male's right foot

fl = turtle.Turtle() # Create female's left foot
fr = turtle.Turtle() # Create female's right foot

# Leave no trace behind
ml.penup(); mr.penup()
fl.penup(); fr.penup()

# Setting the shape of the feet to turtles
ml.shape("arrow");  mr.shape("arrow")
fl.shape("classic"); fr.shape("classic")
```

```
# Initially the turtle is headed east,
# so the male feet are rotated 90 degrees anti-clockwise to head upward (north)
# and go back 50 unit to be placed in the initial position
ml.left(90); ml.backward(50)
mr.left(90); mr.backward(50)

# Here, for male dancer, the right foot is moved away from the left one
# (so they don't overlap)
mr.right(90); mr.forward(50); mr.left(90)

# The same is repeated for the female dancer now.
fl.forward(50) # Move away the left foot from the right
fl.right(90); fr.right(90) # Turn the feet to face south

# set the speed of the dance
speed = 1
ml.speed(speed); mr.speed(speed)
fl.speed(speed); fr.speed(speed)

# set the shoe colour of the male (black) and female dancer (red)
ml.pencolor(0,0,0) # RED, GREEN, BLUE
mr.pencolor(0,0,0)

fl.pencolor(255,0,0) # RED, GREEN, BLUE
fr.pencolor(255,0,0)

# Give the dancers a few seconds and let the dance begin!
time.sleep(3)


""" ################## """
""" Create Samba Steps """
""" ################## """

# Whenever this function is called, one full samba dance routine is executed
def samba():

    ml.forward(50)  # Move male left foot forward 50 units
    fr.backward(50)  # Move female right foot 50 units
    mr.forward(50)
    fl.backward(50)

    shift = 45     # Magnitude of "in place rotation" representing weight shift
    ml.right(shift) # Rotate male left foot 45 degrees
    fr.left(shift)  # Rotate female right foot 45 degrees
    ml.left(shift)
    fr.right(shift)

    mr.backward(50)
    fl.forward(50)
    ml.backward(50)
    fr.forward(50)

    mr.left(shift)
    fl.right(shift)
    mr.right(shift)
    fl.left(shift)

""" ################## """
""" LET THE DANCE BEGIN """
""" ################## """

for i in range(10): # Repeat 10 times
    samba() # Calls the steps of samba
```

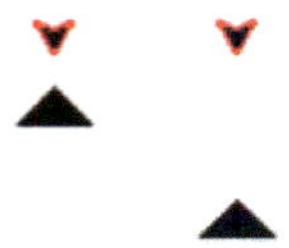

The output: samba © 2022 by Mohammad. Majid al-Rifaie (Used with permission).

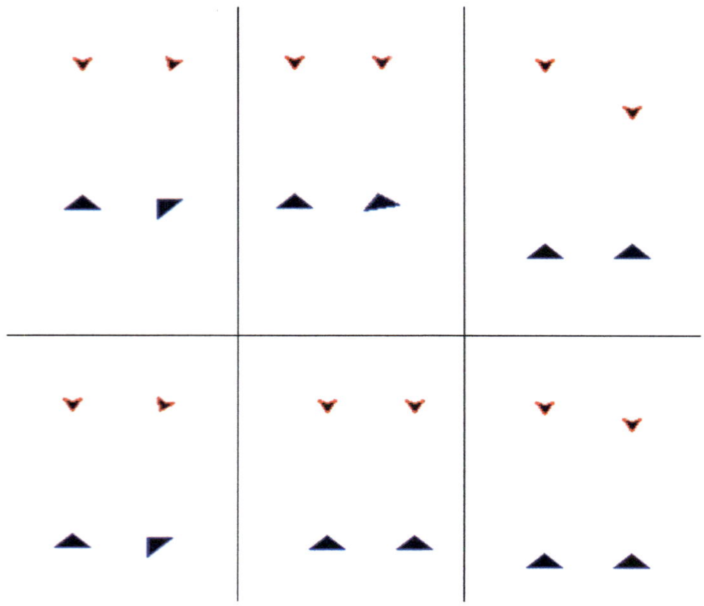

Samba dance animated. © 2022 by Mohammad Majid al-Rifaie. (Used with permission).

Dramata: "Yes it is all working, and the code starts to make some sense to me, but…"
Kalizuko: "Let me do it for you: I'll show you how the codes work in general.
1. Conditional: Python code for the "conditional execution" section of the video
2. Looping: Python code for the "looping" section of the video
3. Numeric: Python code for the "numeric computation" section of the video
4. Sequence: Python code for the "sequence" section of the video"

`The following codes with various components and videos and are available on the accompanying website as listed in the book.`

Kalizuko: "Here they come:"

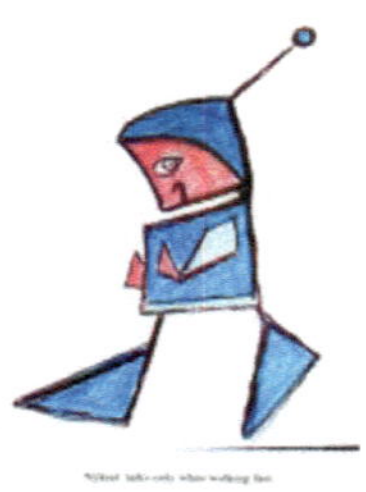

Nykret: talks only when moving fast.

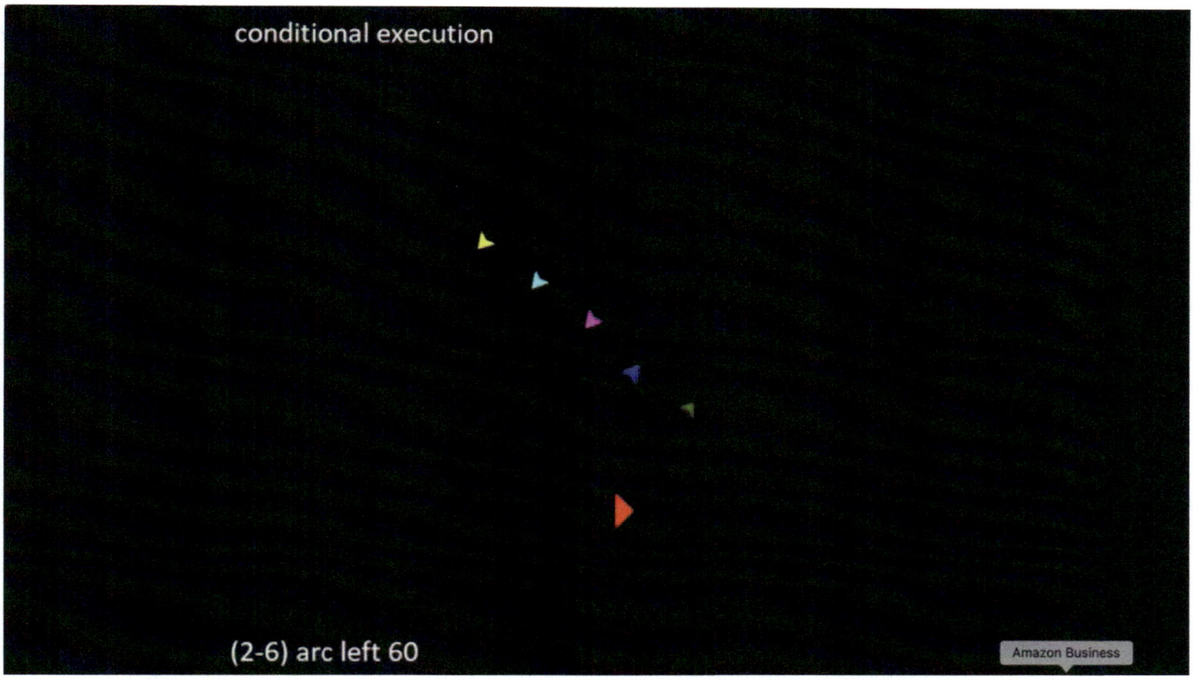

Conditional execution. © 2021 by Irwin Stuart Smith (Used with permission).

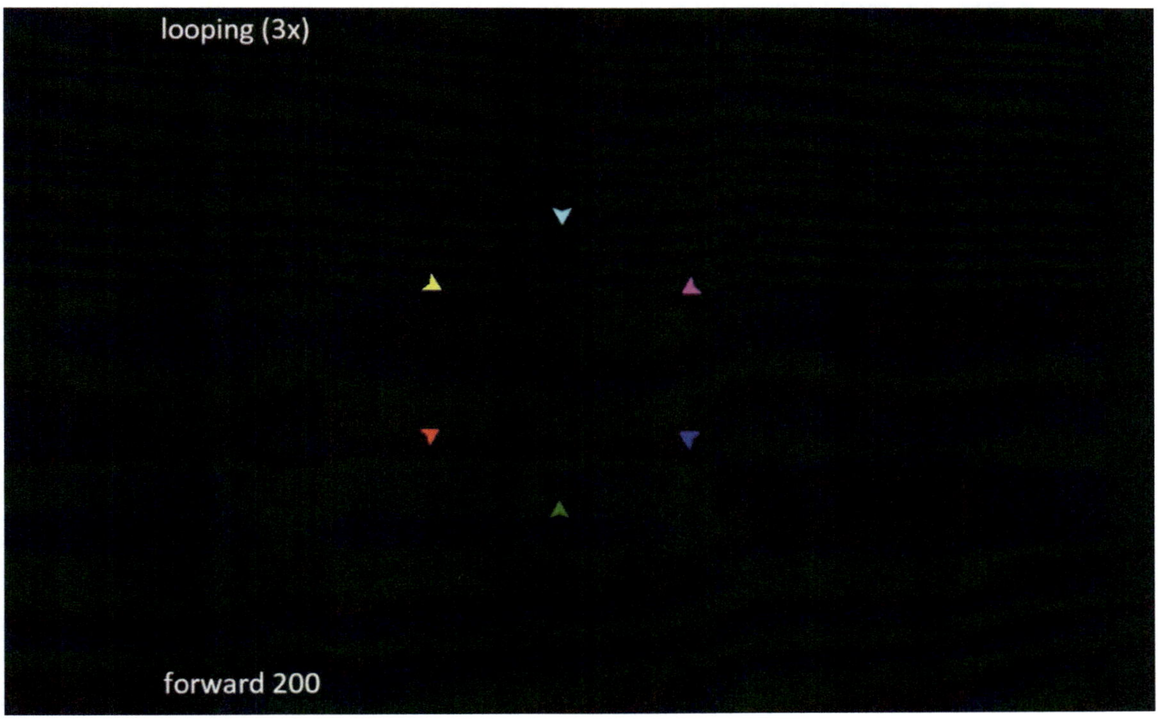

Looping. © 2021 by Irwin Stuart Smith (Used with permission).

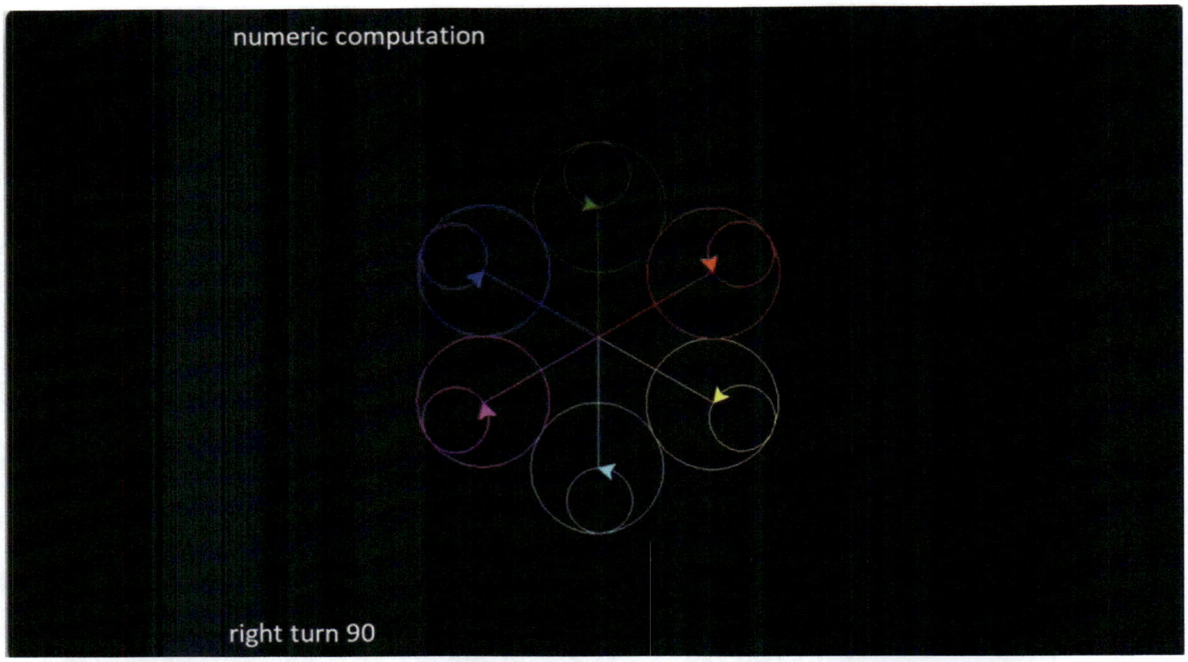

Numeric Computation. © *2021 by Irwin Stuart Smith (Used with permission).*

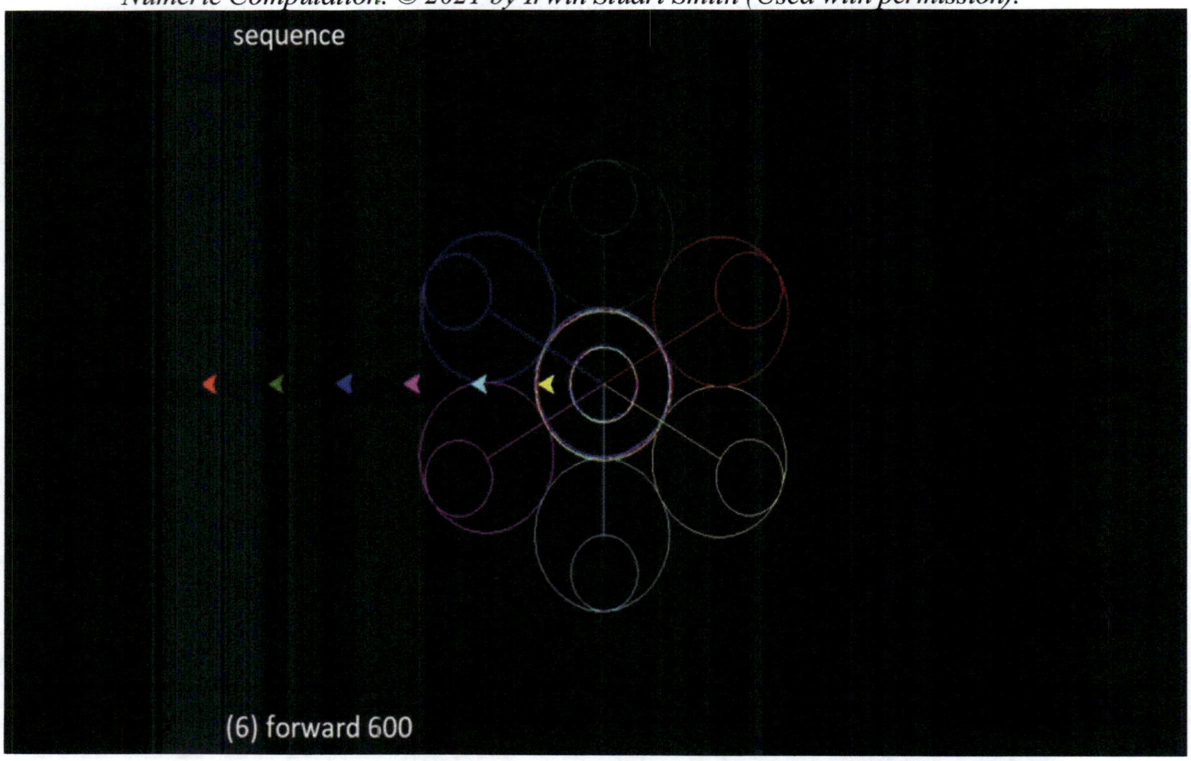

Sequence. © *2021 by Irwin Stuart Smith (Used with permission).*

Kalizuko, "Those are general rules, of course, but if you'd like to see the codes behind them, I'll show you:"

The following codes were written by Irwin Stuart Smith, (Used with permission).

```
# CONDITIONAL EXECUTION

from turtle import *
from math import *
import time

class Dancer(Turtle):

    def __init__(self,x,y):
        Turtle.__init__(self)
        self.shapesize(2,2,1)
        self.pu()
        self.ht()
        self.speed(0)

    def arcright(self,r,a):
        theta = 2*pi*(360-2*a)/360
        c=2*r*sin(theta/2)
        self.rt(a)
        self.fd(c)

    def arcleft(self,r,a):
        theta = 2*pi*(360-2*a)/360
        c=2*r*sin(theta/2)
        self.lt(a)
        self.fd(c)

def main():
    t1 = Dancer(0,0)
    t2 = t1.clone()
    t3 = t1.clone()
    t4 = t1.clone()e
    t5 = t1.clone()
    t6 = t1.clone()
    ts = t1.getscreen()
    ts.bgcolor("black")
    d = ts.delay(1)
#
    t1.color("red")
    t2.color("green")
    t3.color("blue")
    t4.color("magenta")
    t5.color("cyan")
    t6.color("yellow")
#
    t1.goto(500,0)
    t2.goto(500,0)
    t3.goto(500,0)
    t4.goto(500,0)
    t5.goto(500,0)
    t6.goto(500,0)
#
    t1.seth(180)
    t2.seth(180)
    t3.seth(180)
    t4.seth(180)
    t5.seth(180)
    t6.seth(180)
#
    t1.st()
    t2.st()
    t3.st()
    t4.st()
    t5.st()
    t6.st()
#
    t1.goto(440,0)
```

```
    t2.goto(520,0)
    t3.goto(600,0)
    t4.goto(680,0)
    t5.goto(760,0)
    t6.goto(840,0)
#
    t1.seth(180)
    t2.seth(180)
    t3.seth(180)
    t4.seth(180)
    t5.seth(180)
    t6.seth(180)
#
    t1.st()
    t2.st()
    t3.st()
    t4.st()
    t5.st()
    t6.st()
#
#  1. All enter single-file stage left, 640 steps to starting positions.
#
    for i in range(160):
        t1.fd(4)
        t2.fd(4)
        t3.fd(4)
        t4.fd(4)
        t5.fd(4)
        t6.fd(4)
    time.sleep(0.5)
#
# 2. t1-t3 turn left 90, t4-t6 turn right 90
#
    for i in range(30):
        t1.lt(3)
        t2.lt(3)
        t3.lt(3)
        t4.rt(3)
        t5.rt(3)
        t6.rt(3)
    time.sleep(0.5)
#
# 3. Circle clockwise.
#
    for i in range(30):
        t1.arcleft(100,3)
        t2.arcleft(60,3)
        t3.arcleft(20,3)
        t4.arcleft(20,3)
        t5.arcleft(60,3)
        t6.arcleft(100,3)
    time.sleep(0.5)
#
# 4. All face forward
#
    for i in range(30):
        t1.rt(3)
        t2.rt(3)
        t3.rt(3)
        t4.lt(3)
        t5.lt(3)
        t6.lt(3)
#
# 5. t1 becomes "traffic cop."
#
    for i in range(60):
        t1.rt(3)
    t1.shape("arrow")
    time.sleep(0.5)
    for i in range(30):
```

```
      t1.lt(3)
   time.sleep(0.5)
#
# 6. Turn right 90.
#
   for i in range(30):
      t2.rt(3)
      t3.rt(3)
      t4.lt(3)
      t5.lt(3)
      t6.lt(3)
   time.sleep(0.5)
#
# 7. Circle clockwise 60.
#
   for i in range(20):
      t2.arcright(60,3)
      t3.arcright(20,3)
      t4.arcright(20,3)
      t5.arcright(60,3)
      t6.arcright(100,3)
   time.sleep(0.5)
   t1.color("black")
#
# 8. Wait for traffic signal.
#
   t1.seth(0)
   time.sleep(1)
   t1.color("red")
#
# 9. Turn right 180.
#
   for i in range(60):
      t2.rt(3)
      t3.rt(3)
      t4.lt(3)
      t5.lt(3)
      t6.lt(3)
   time.sleep(0.5)
#
# 10. Circle left 60.
#
   for i in range(20):
      t2.arcleft(60,3)
      t3.arcleft(20,3)
      t4.arcleft(20,3)
      t5.arcleft(60,3)
      t6.arcleft(100,3)
   time.sleep(0.5)
   t1.color("black")
#
# 11. Wait for traffic signal.
#
   t1.seth(0)
   time.sleep(1)
   t1.color("red")
#
# 12. Continue counterclockwise.
#
   for i in range(20):
      t2.arcleft(60,3)
      t3.arcleft(20,3)
      t4.arcleft(20,3)
      t5.arcleft(60,3)
      t6.arcleft(100,3)
   time.sleep(0.5)
   t1.color("black")
#
# 13. Wait for traffic signal.
#
```

```
    t1.seth(180)
    time.sleep(1)
    t1.color("red")
#
# 14. Turn right 180.
#
    for i in range(60):
        t2.rt(3)
        t3.rt(3)
        t4.lt(3)
        t5.lt(3)
        t6.lt(3)
    time.sleep(0.5)
#
# 15. Circle right 60.
#
    for i in range(20):
        t2.arcright(60,3)
        t3.arcright(20,3)
        t4.arcright(20,3)
        t5.arcright(60,3)
        t6.arcright(100,3)
    time.sleep(0.5)
    t1.color("black")
#
# 16. t1 rejoins the ensemble.
#
    t1.seth(90)
    t1.shape("classic")
    t1.color("red")
    for i in range(30):
        t1.lt(3)
#
# 17. Circle right back into starting line.
#
    for i in range(30):
        t1.arcright(100,3)
        t2.arcright(60,3)
        t3.arcright(20,3)
        t4.arcright(20,3)
        t5.arcright(60,3)
        t6.arcright(100,3)
    time.sleep(0.5)
#
# 18. Turn to exit heading.
#
    t1.seth(180)
    t2.seth(180)
    t3.seth(180)
    t4.seth(180)
    t5.seth(180)
    t6.seth(180)
#
# 19. Exit single-file in order stage right.
#
    ts.tracer(4,100)
    for i in range(80):
        t1.fd(1)
    for i in range(80):
        t1.fd(1)
        t2.fd(1)
    for i in range(80):
        t1.fd(1)
        t2.fd(1)
        t3.fd(1)
    for i in range(80):
        t1.fd(1)
        t2.fd(1)
        t3.fd(1)
        t4.fd(1)
```

```
    for i in range(80):
        t1.fd(1)
        t2.fd(1)
        t3.fd(1)
        t4.fd(1)
        t5.fd(1)
    for i in range(1200):
        t1.fd(1)
        t2.fd(1)
        t3.fd(1)
        t4.fd(1)
        t5.fd(1)
        t6.fd(1)
#
if __name__=="__main__":
    msg=main()
    print("DONE!")
    mainloop()
```

Kalizuko, *"Now*, let's look at Looping:"

```
# LOOPING

from turtle import *
from math import *
import time

class Dancer(Turtle):

    def _init_(self,x,y):
        Turtle.__init__(self)
        self.shapesize(2,2,1)
        self.pu()
        self.ht()
        self.speed(10)

    def arcright(self,r,a):
        theta = 2*pi*(360-2*a)/360
        c=2*r*sin(theta/2)
        self.rt(a)
        self.fd(c)

    def arcleft(self,r,a):
        theta = 2*pi*(360-2*a)/360
        c=2*r*sin(theta/2)
        self.lt(a)
        self.fd(c)

def main():
    t1 = Dancer(0,0)
    t2 = t1.clone()
    t3 = t1.clone()
    t4 = t1.clone()
    t5 = t1.clone()
    t6 = t1.clone()
    ts = t1.getscreen()
    ts.bgcolor("black")
    d = ts.delay(1)
#
    t1.color("red")
    t2.color("green")
    t3.color("blue")
    t4.color("magenta")
    t5.color("cyan")
    t6.color("yellow")
#
    t1.goto(500,0)
    t2.goto(500,0)
    t3.goto(500,0)
    t4.goto(500,0)
```

```
    t5.goto(500,0)
    t6.goto(500,0)
#
    t1.seth(180)
    t2.seth(180)
    t3.seth(180)
    t4.seth(180)
    t5.seth(180)
    t6.seth(180)
#
    t1.st()
    t2.st()
    t3.st()
    t4.st()
    t5.st()
    t6.st()
#
    t1.goto(440,0)
    t2.goto(520,0)
    t3.goto(600,0)
    t4.goto(680,0)
    t5.goto(760,0)
    t6.goto(840,0)
#
    t1.seth(180)
    t2.seth(180)
    t3.seth(180)
    t4.seth(180)
    t5.seth(180)
    t6.seth(180)
#
    t1.st()
    t2.st()
    t3.st()
    t4.st()
    t5.st()
    t6.st()
#
#  1. All enter single-file stage left, 640 steps to starting positions.
#
    for i in range(160):
        t1.fd(4)
        t2.fd(4)
        t3.fd(4)
        t4.fd(4)
        t5.fd(4)
        t6.fd(4)
    time.sleep(0.5)
#
# 2. t1-t3 about face right.
#
    for i in range(90):
        t1.rt(2)
        t2.rt(2)
        t3.rt(2)
    time.sleep(0.5)
#
# 3. Gather at center.
#
    for i in range(40):
        t3.fd(1)
        t4.fd(1)
    for i in range(120):
        t2.fd(1)
        t5.fd(1)
    for i in range(200):
        t1.fd(1)
        t6.fd(1)
    time.sleep(0.5)
#
```

```
# 4. Turn to initial dance headings
#
   t1.seth(30)
   t2.seth(90)
   t3.seth(150)
   t4.seth(210)
   t5.seth(270)
   t6.seth(330)
#
   t1.pd()
   t2.pd()
   t3.pd()
   t4.pd()
   t5.pd()
   t6.pd()
#
# 5. Begin loop (3 repetitions)
#
   ts.tracer(2,50)
   for i in range( 3 ):
#      5a. Move out from center.
      for j in range(200):
         t1.fd(1)
         t2.fd(1)
         t3.fd(1)
         t4.fd(1)
         t5.fd(1)
         t6.fd(1)
      # 5b. Turn right 90.
      for j in range(30):
         t1.rt(3)
         t2.rt(3)
         t3.rt(3)
         t4.rt(3)
         t5.rt(3)
         t6.rt(3)
      # 5c. Advance 1/6 circle.
      for j in range(60):
         t1.arcright(100,1)
         t2.arcright(100,1)
         t3.arcright(100,1)
         t4.arcright(100,1)
         t5.arcright(100,1)
         t6.arcright(100,1)
      # 5d. Turn right back toward center.
      for j in range(30):
         t1.rt(3)
         t2.rt(3)
         t3.rt(3)
         t4.rt(3)
         t5.rt(3)
         t6.rt(3)
      # 5e. Move back into center
      for j in range(200):
         t1.fd(1)
         t2.fd(1)
         t3.fd(1)
         t4.fd(1)
         t5.fd(1)
         t6.fd(1)
      # 5f. Face back out from center.
      for j in range(60):
         t1.rt(3)
         t2.rt(3)
         t3.rt(3)
         t4.rt(3)
         t5.rt(3)
         t6.rt(3)
#
# End loop
```

```
#
# 6. Turn to exit heading.
#
    t1.seth(180)
    t2.seth(180)
    t3.seth(180)
    t4.seth(180)
    t5.seth(180)
    t6.seth(180)
#
    t1.speed(1)
    t2.speed(1)
    t3.speed(1)
    t4.speed(1)
    t5.speed(1)
    t6.speed(1)
#
# 7. Exit single-file in order stage right.
#
    for i in range(80):
        t1.pu()
        t1.fd(1)
    for i in range(80):
        t2.pu()
        t1.fd(1)
        t2.fd(1)
    for i in range(80):
        t3.pu()
        t1.fd(1)
        t2.fd(1)
        t3.fd(1)
    for i in range(80):
        t4.pu()
        t1.fd(1)
        t2.fd(1)
        t3.fd(1)
        t4.fd(1)
    for i in range(80):
        t5.pu()
        t1.fd(1)
        t2.fd(1)
        t3.fd(1)
        t4.fd(1)
        t5.fd(1)
    for i in range(1200):
        t6.pu()
        t1.fd(1)
        t2.fd(1)
        t3.fd(1)
        t4.fd(1)
        t5.fd(1)
        t6.fd(1)
#
if __name__=="__main__":
    msg=main()
    print("DONE!")
    mainloop()
```

Kalizuko, " And now, Numeric:"

```
# NUMERIC COMBINED

from turtle import *
from math import *
import time

class Dancer(Turtle):

    def __init__(self,x,y):
```

```
        Turtle.__init__(self)
        self.shapesize(2,2,1)
        self.pu()
        self.ht()
        self.speed(0)

    def arcright(self,r,a):
        theta = 2*pi*(360-2*a)/360
        c=2*r*sin(theta/2)
        self.rt(a)
        self.fd(c)

    def arcleft(self,r,a):
        theta = 2*pi*(360-2*a)/360
        c=2*r*sin(theta/2)
        self.lt(a)
        self.fd(c)

def main():
    t1 = Dancer(0,0)
    t2 = t1.clone()
    t3 = t1.clone()
    t4 = t1.clone()
    t5 = t1.clone()
    t6 = t1.clone()
    ts = t1.getscreen()
    ts.bgcolor("black")
    d = ts.delay(1)
#
    t1.color("red")
    t2.color("green")
    t3.color("blue")
    t4.color("magenta")
    t5.color("cyan")
    t6.color("yellow")
#
    t1.goto(500,0)
    t2.goto(500,0)
    t3.goto(500,0)
    t4.goto(500,0)
    t5.goto(500,0)
    t6.goto(500,0)
#
    t1.seth(180)
    t2.seth(180)
    t3.seth(180)
    t4.seth(180)
    t5.seth(180)
    t6.seth(180)
#
    t1.st()
    t2.st()
    t3.st()
    t4.st()
    t5.st()
    t6.st()
#
    t1.goto(440,0)
    t2.goto(520,0)
    t3.goto(600,0)
    t4.goto(680,0)
    t5.goto(760,0)
    t6.goto(840,0)
#
    t1.seth(180)
    t2.seth(180)
    t3.seth(180)
    t4.seth(180)
    t5.seth(180)
    t6.seth(180)
```

```
#
   t1.st()
   t2.st()
   t3.st()
   t4.st()
   t5.st()
   t6.st()
#
#  1. All enter single-file stage left.
#
   for i in range(160):
      t1.fd(4)
      t2.fd(4)
      t3.fd(4)
      t4.fd(4)
      t5.fd(4)
      t6.fd(4)
   time.sleep(0.5)
#
# 2. t1-t3 about face right.
#
   for i in range(90):
      t1.rt(2)
      t2.rt(2)
      t3.rt(2)
   time.sleep(0.5)
#
# 3. Gather at center.
#
   for i in range(40):
      t3.fd(1)
      t4.fd(1)
   for i in range(120):
      t2.fd(1)
      t5.fd(1)
   for i in range(200):
      t1.fd(1)
      t6.fd(1)
   time.sleep(0.5)
#
# 4. Turn to initial dance headings
#
   t1.seth(30)
   t2.seth(90)
   t3.seth(150)
   t4.seth(210)
   t5.seth(270)
   t6.seth(330)
   time.sleep(0.5)
#
   t1.pd()
   t2.pd()
   t3.pd()
   t4.pd()
   t5.pd()
   t6.pd()
#
# 5.Move out from center.
#
   for j in range(80):
      t1.fd(1)
      t2.fd(1)
      t3.fd(1)
      t4.fd(1)
      t5.fd(1)
      t6.fd(1)
#
# 6. Turn left 90.
#
   for j in range(30):
```

```
        t1.lt(3)
        t2.lt(3)
        t3.lt(3)
        t4.lt(3)
        t5.lt(3)
        t6.lt(3)
    time.sleep(0.5)
#
# 7. Circle right 360.
#
    for i in range(120):
        t1.arcright(40,3)
        t2.arcright(40,3)
        t3.arcright(40,3)
        t4.arcright(40,3)
        t5.arcright(40,3)
        t6.arcright(40,3)
    time.sleep(0.5)
#
# 8. Turn right(90)
#
    for i in range(30):
        t1.rt(3)
        t2.rt(3)
        t3.rt(3)
        t4.rt(3)
        t5.rt(3)
        t6.rt(3)
    time.sleep(0.5)
#
# 9. Move out 80.
#
    for i in range(80):
        t1.fd(1)
        t2.fd(1)
        t3.fd(1)
        t4.fd(1)
        t5.fd(1)
        t6.fd(1)
    time.sleep(0.5)
#
# 10. Turn right 90
#
    for i in range(30):
        t1.rt(3)
        t2.rt(3)
        t3.rt(3)
        t4.rt(3)
        t5.rt(3)
        t6.rt(3)
    time.sleep(0.5)
#
# 11. Circle left 360.
#
    for i in range(120):
        t1.arcleft(40,3)
        t2.arcleft(40,3)
        t3.arcleft(40,3)
        t4.arcleft(40,3)
        t5.arcleft(40,3)
        t6.arcleft(40,3)
    time.sleep(0.5)
#
# 12. Turn right 90.
#
    for i in range(30):
        t1.rt(3)
        t2.rt(3)
        t3.rt(3)
        t4.rt(3)
```

```
        t5.rt(3)
        t6.rt(3)
    time.sleep(0.5)
#
# 13. Forward towards center.
#
    for i in range(80):
        t1.fd(1)
        t2.fd(1)
        t3.fd(1)
        t4.fd(1)
        t5.fd(1)
        t6.fd(1)
    time.sleep(0.5)
#
# 14. Turn right 90.
#
    for i in range(30):
        t1.rt(3)
        t2.rt(3)
        t3.rt(3)
        t4.rt(3)
        t5.rt(3)
        t6.rt(3)
    time.sleep(0.5)
#
# 15. Circle left 360.
#
    for i in range(120):
        t1.arcleft(40,3)
        t2.arcleft(40,3)
        t3.arcleft(40,3)
        t4.arcleft(40,3)
        t5.arcleft(40,3)
        t6.arcleft(40,3)
    time.sleep(0.5)
#
# 16. Turn left 90.
#
    for i in range(30):
        t1.lt(3)
        t2.lt(3)
        t3.lt(3)
        t4.lt(3)
        t5.lt(3)
        t6.lt(3)
    time.sleep(0.5)

#
# 17. Forward into center.
#
    for i in range(80):
        t1.fd(1)
        t2.fd(1)
        t3.fd(1)
        t4.fd(1)
        t5.fd(1)
        t6.fd(1)
    time.sleep(0.5)
#
# 18. Turn to exit heading.
#
    t1.seth(180)
    t2.seth(180)
    t3.seth(180)
    t4.seth(180)
    t5.seth(180)
    t6.seth(180)
    time.sleep(0.5)
#
```

```
# 19. Exit single-file in order stage right.
#
    for i in range(80):
        t1.pu()
        t1.fd(1)
    for i in range(80):
        t2.pu()
        t1.fd(1)
        t2.fd(1)
    for i in range(80):
        t3.pu()
        t1.fd(1)
        t2.fd(1)
        t3.fd(1)
    for i in range(80):
        t4.pu()
        t1.fd(1)
        t2.fd(1)
        t3.fd(1)
        t4.fd(1)
    for i in range(80):
        t5.pu()
        t1.fd(1)
        t2.fd(1)
        t3.fd(1)
        t4.fd(1)
        t5.fd(1)
    for i in range(1200):
        t6.pu()
        t1.fd(1)
        t2.fd(1)
        t3.fd(1)
        t4.fd(1)
        t5.fd(1)
        t6.fd(1)
#
if __name__=="__main__":
    msg=main()
    print("DONE!")
    mainloop()
```

Kalizuko, “And now, time for the Sequence:”

```
# SEQUENCE

from turtle import *
from math import *
import time

class Dancer(Turtle):

    def _init_(self,x,y):
        Turtle.__init__(self)
        self.shapesize(2,2,1)
        self.pu()
        self.speed(0)

    def arcright(self,r,a):
        theta = 2*pi*(360-2*a)/360
        c=2*r*sin(theta/2)
        self.rt(a)
        self.fd(c)

    def arcleft(self,r,a):
        theta = 2*pi*(360-2*a)/360
        c=2*r*sin(theta/2)
        self.lt(a)
        self.fd(c)
```

```
def main():
    t1 = Dancer(0,0)
    t2 = t1.clone()
    t3 = t1.clone()
    t4 = t1.clone()
    t5 = t1.clone()
    t6 = t1.clone()
    ts = t1.getscreen()
    ts.bgcolor("black")
    d = ts.delay(1)
#
    t1.ht()
    t2.ht()
    t3.ht()
    t4.ht()
    t5.ht()
    t6.ht()
#
    t1.color("red")
    t2.color("green")
    t3.color("blue")
    t4.color("magenta")
    t5.color("cyan")
    t6.color("yellow")
#
    t1.goto(440,0)
    t2.goto(520,0)
    t3.goto(600,0)
    t4.goto(680,0)
    t5.goto(760,0)
    t6.goto(840,0)
#
    t1.seth(180)
    t2.seth(180)
    t3.seth(180)
    t4.seth(180)
    t5.seth(180)
    t6.seth(180)
#
    t1.st()
    t2.st()
    t3.st()
    t4.st()
    t5.st()
    t6.st()
#
#   1. All enter single-file stage left, 640 steps to starting positions.
#
    for i in range(160):
        t1.fd(4)
        t2.fd(4)
        t3.fd(4)
        t4.fd(4)
        t5.fd(4)
        t6.fd(4)
    time.sleep(0.5)
#
# 2. All turn left 90.
#
    ts.tracer(1,10)
    for i in range(30):
        t1.lt(3)
        t2.lt(3)
        t3.lt(3)
        t4.lt(3)
        t5.lt(3)
        t6.lt(3)
    time.sleep(0.5)
#
# 3. Circle outward 90, t1-t3 right, t4-t6 left.
```

```
#
    for i in range(90):
        t1.arcright(40,1)
        t2.arcright(80,1)
        t3.arcright(120,1)
        t4.arcleft(120,1)
        t5.arcleft(80,1)
        t6.arcleft(40,1)
    time.sleep(0.5)
#
# 4. All about face, 180.
#
    ts.tracer(1,10)
    for i in range(60):
        t1.lt(3)
        t2.lt(3)
        t3.lt(3)
        t4.rt(3)
        t5.rt(3)
        t6.rt(3)
    time.sleep(0.5)
#
# 5. All move towards each other.
#
    for i in range(10):
        t1.fd(1)
        t2.fd(1)
        t3.fd(1)
        t4.fd(1)
        t5.fd(1)
        t6.fd(1)
    time.sleep(0.5)
#
# 6. Turn back inward 90, t1-t3 left, t4-t6 right.
#
#   ts.tracer(1,10)
    for i in range(90):
        t1.arcleft(40,1)
        t2.arcleft(80,1)
        t3.arcleft(120,1)
        t4.arcright(120,1)
        t5.arcright(80,1)
        t6.arcright(40,1)
    time.sleep(0.5)
#
# 7. All turn left 90, facing stage right.
#
    ts.tracer(1,10)
    for i in range(30):
        t1.lt(3)
        t2.lt(3)
        t3.lt(3)
        t4.lt(3)
        t5.lt(3)
        t6.lt(3)
#
# 8. All exit single-file stage right.
#
    for i in range(375):
        t1.fd(4)
        t2.fd(4)
        t3.fd(4)
        t4.fd(4)
        t5.fd(4)
        t6.fd(4)
#
if __name__=="__main__":
    msg=main()
    print("DONE!")
    mainloop()
```

Kalizuko, "So now, let's look at the whole code and how it generates the explanatory video posted on the accompanying website. It has music, and there is also another version of that project, with a dancer performing while following the trajectories. You may copy the codes and run them on your computer, but you need to have Python. There are instructions on how to do it if you scroll down to the end. You may always google terms used if they are new to you or not explained enough:"

```
# PROJECT DANCE

from turtle import *
from math import *
import time

class Dancer(Turtle):

    def _init_(self,x,y):
        Turtle.__init__(self)
        self.shapesize(2,2,1)
        self.pu()
        self.speed(0)

    def arcright(self,r,a):
        theta = 2*pi*(360-2*a)/360
        c=2*r*sin(theta/2)
        self.rt(a)
        self.fd(c)

    def arcleft(self,r,a):
        theta = 2*pi*(360-2*a)/360
        c=2*r*sin(theta/2)
        self.lt(a)
        self.fd(c)

def main():
    t1 = Dancer(0,0)
    t2 = t1.clone()
    t3 = t1.clone()
    t4 = t1.clone()
    t5 = t1.clone()
    t6 = t1.clone()
    ts = t1.getscreen()
    ts.bgcolor("black")
    d = ts.delay(1)
#
    t1.ht()
    t2.ht()
    t3.ht()
    t4.ht()
    t5.ht()
    t6.ht()
#
    t1.color("red")
    t2.color("green")
    t3.color("blue")
    t4.color("magenta")
    t5.color("cyan")
    t6.color("yellow")
#
    t1.goto(440,0)
    t2.goto(520,0)
    t3.goto(600,0)
    t4.goto(680,0)
    t5.goto(760,0)
    t6.goto(840,0)
#
    t1.seth(180)
    t2.seth(180)
```

```
    t3.seth(180)
    t4.seth(180)
    t5.seth(180)
    t6.seth(180)
#
    t1.st()
    t2.st()
    t3.st()
    t4.st()
    t5.st()
    t6.st()
#
# SEQUENCE
#
# 1. All enter single-file stage left.
#
    for i in range(160):
        t1.fd(4)
        t2.fd(4)
        t3.fd(4)
        t4.fd(4)
        t5.fd(4)
        t6.fd(4)
    time.sleep(0.5)
#
# 2. All turn left 90.
#
    for i in range(30):
        t1.lt(3)
        t2.lt(3)
        t3.lt(3)
        t4.lt(3)
        t5.lt(3)
        t6.lt(3)
    time.sleep(0.5)
#
# 3. Circle outward 90, t1-t3 right, t4-t6 left.
#
    for i in range(30):
        t1.arcright(40,3)
        t2.arcright(80,3)
        t3.arcright(120,3)
        t4.arcleft(120,3)
        t5.arcleft(80,3)
        t6.arcleft(40,3)
    time.sleep(0.5)
#
# 4. All about face, 180.
#
    ts.tracer(1,10)
    for i in range(60):
        t1.lt(3)
        t2.lt(3)
        t3.lt(3)
        t4.rt(3)
        t5.rt(3)
        t6.rt(3)
    time.sleep(0.5)
#
# 5. All move towards each other.
#
    for i in range(5):
        t1.fd(2)
        t2.fd(2)
        t3.fd(2)
        t4.fd(2)
        t5.fd(2)
        t6.fd(2)
#
# 6. Turn back inward 90, t1-t3 left, t4-t6 right.
```

```
#
   for i in range(30):
      t1.arcleft(40,3)
      t2.arcleft(80,3)
      t3.arcleft(120,3)
      t4.arcright(120,3)
      t5.arcright(80,3)
      t6.arcright(40,3)
   t1.goto(-200,0)
   t2.goto(-120,0)
   t3.goto(-40,0)
   t4.goto(40,0)
   t5.goto(120,0)
   t6.goto(200,0)
   time.sleep(0.5)
#
# 7. All turn left 90, facing stage right.
#
#   ts.tracer(1,10)
   for i in range(30):
      t1.lt(3)
      t2.lt(3)
      t3.lt(3)
      t4.lt(3)
      t5.lt(3)
      t6.lt(3)
#
# 8. t1-t3 about face right.
#
   for i in range(60):
      t1.rt(3)
      t2.rt(3)
      t3.rt(3)
   time.sleep(0.5)
#
# 9. Gather at center.
#
   for i in range(40):
      t3.fd(1)
      t4.fd(1)
   for i in range(120):
      t2.fd(1)
      t5.fd(1)
   for i in range(200):
      t1.fd(1)
      t6.fd(1)
   t1.goto(0,0)
   t2.goto(0,0)
   t3.goto(0,0)
   t4.goto(0,0)
   t5.goto(0,0)
   t6.goto(0,0)
   time.sleep(0.5)
#
# LOOPING
#
# 10. Turn to initial dance headings
#
   t1.seth(30)
   t2.seth(90)
   t3.seth(150)
   t4.seth(210)
   t5.seth(270)
   t6.seth(330)
#
# 11. Loop (3 repetitions)
#
   for i in range(3):
#      11a. Move out from center.
      for j in range(100):
```

```
            t1.fd(2)
            t2.fd(2)
            t3.fd(2)
            t4.fd(2)
            t5.fd(2)
            t6.fd(2)
        # 11b. Turn right 90.
        for j in range(30):
            t1.rt(3)
            t2.rt(3)
            t3.rt(3)
            t4.rt(3)
            t5.rt(3)
            t6.rt(3)
        # 11c. Advance 60, 1/6 circle.
        for j in range(20):
            t1.arcright(100,3)
            t2.arcright(100,3)
            t3.arcright(100,3)
            t4.arcright(100,3)
            t5.arcright(100,3)
            t6.arcright(100,3)
        # 11d. Turn right 90 back toward center.
        for j in range(30):
            t1.rt(3)
            t2.rt(3)
            t3.rt(3)
            t4.rt(3)
            t5.rt(3)
            t6.rt(3)
        # 11e. Move back into center
        for j in range(100):
            t1.fd(2)
            t2.fd(2)
            t3.fd(2)
            t4.fd(2)
            t5.fd(2)
            t6.fd(2)
        t1.goto(0,0)
        t2.goto(0,0)
        t3.goto(0,0)
        t4.goto(0,0)
        t5.goto(0,0)
        t6.goto(0,0)
        # 11f. Face back out 180 from center.
        for j in range(60):
            t1.rt(3)
            t2.rt(3)
            t3.rt(3)
            t4.rt(3)
            t5.rt(3)
            t6.rt(3)
#
# End loop
#
# 12. Turn to outward headings.
#
    t1.seth(180)
    t2.seth(180)
    t3.seth(180)
    t4.seth(0)
    t5.seth(0)
    t6.seth(0)
#
# 13. Move to new starting positions
#
    for i in range(200):
        t1.fd(1)
        t6.fd(1)
    for i in range(120):
```

```
        t2.fd(1)
        t5.fd(1)
    for i in range(40):
        t3.fd(1)
        t4.fd(1)
    t1.goto(-200,0)
    t2.goto(-120,0)
    t3.goto(-40,0)
    t4.goto(40,0)
    t5.goto(120,0)
    t6.goto(200,0)
    time.sleep(0.5)
#
# CONDITIONAL
#
# 14. All turn left 90.
#
    for i in range(30):
        t1.lt(3)
        t2.lt(3)
        t3.lt(3)
        t4.lt(3)
        t5.lt(3)
        t6.lt(3)
    time.sleep(0.5)
#
# 15. Circle clockwise.
#
    for i in range(30):
        t1.arcleft(100,3)
        t2.arcleft(60,3)
        t3.arcleft(20,3)
        t4.arcleft(20,3)
        t5.arcleft(60,3)
        t6.arcleft(100,3)
    time.sleep(0.5)
#
# 16. All face forward 90.
#
    for i in range(30):
        t1.rt(3)
        t2.rt(3)
        t3.rt(3)
        t4.lt(3)
        t5.lt(3)
        t6.lt(3)
    time.sleep(0.5)
#
# 17. t1 becomes "traffic cop."
#
    for i in range(60):
        t1.rt(3)
    t1.shape("arrow")
    time.sleep(0.5)
    for i in range(90):
        t1.lt(1)
    time.sleep(0.5)
#
# 18. Turn right 90.
#
    for i in range(30):
        t2.rt(3)
        t3.rt(3)
        t4.lt(3)
        t5.lt(3)
        t6.lt(3)
    time.sleep(0.5)
#
# 19. Circle clockwise 60.
#
```

```
    for i in range(20):
        t2.arcright(60,3)
        t3.arcright(20,3)
        t4.arcright(20,3)
        t5.arcright(60,3)
        t6.arcright(100,3)
    time.sleep(0.5)
    t1.color("black")
#
# 20. Wait for traffic signal.
#
    t1.seth(0)
    time.sleep(1)
    t1.color("red")
#
# 21. Turn right 180.
#
    for i in range(60):
        t2.rt(3)
        t3.rt(3)
        t4.lt(3)
        t5.lt(3)
        t6.lt(3)
    time.sleep(0.5)
#
# 22. Circle left 60.
#
    for i in range(20):
        t2.arcleft(60,3)
        t3.arcleft(20,3)
        t4.arcleft(20,3)
        t5.arcleft(60,3)
        t6.arcleft(100,3)
    time.sleep(0.5)
    t1.color("black")
#
# 23. Wait for traffic signal.
#
    t1.seth(0)
    time.sleep(1)
    t1.color("red")
#
# 24. Continue counterclockwise 60.
#
    for i in range(20):
        t2.arcleft(60,3)
        t3.arcleft(20,3)
        t4.arcleft(20,3)
        t5.arcleft(60,3)
        t6.arcleft(100,3)
    time.sleep(0.5)
    t1.color("black")
#
# 25. Wait for traffic signal.
#
    t1.seth(180)
    time.sleep(1)
    t1.color("red")
#
# 26. Turn right 180.
#
    for i in range(60):
        t2.rt(3)
        t3.rt(3)
        t4.lt(3)
        t5.lt(3)
        t6.lt(3)
    time.sleep(0.5)
#
# 27. Circle right 60.
```

```
#
    for i in range(20):
        t2.arcright(60,3)
        t3.arcright(20,3)
        t4.arcright(20,3)
        t5.arcright(60,3)
        t6.arcright(100,3)
    time.sleep(0.5)
    t1.color("black")
#
# 28. t1 rejoins the ensemble.
#
    t1.seth(90)
    t1.shape("classic")
    t1.color("red")
    for i in range(90):
        t1.lt(1)
#
# 29. Circle right 90 back into starting line.
#
    for i in range(30):
        t1.arcright(100,3)
        t2.arcright(60,3)
        t3.arcright(20,3)
        t4.arcright(20,3)
        t5.arcright(60,3)
        t6.arcright(100,3)
    t1.goto(-200,0)
    t2.goto(-120,0)
    t3.goto(-40,0)
    t4.goto(40,0)
    t5.goto(120,0)
    t6.goto(200,0)
    time.sleep(0.5)
#
# COMBINED
#
# 30. set new heading.
#
    t1.seth(0)
    t2.seth(0)
    t3.seth(0)
    t4.seth(180)
    t5.seth(180)
    t6.seth(180)
    time.sleep(0.5)
#
# 31. Gather at center.
#
    for i in range(40):
        t3.fd(1)
        t4.fd(1)
    for i in range(120):
        t2.fd(1)
        t5.fd(1)
    for i in range(200):
        t1.fd(1)
        t6.fd(1)
    t1.goto(0,0)
    t2.goto(0,0)
    t3.goto(0,0)
    t4.goto(0,0)
    t5.goto(0,0)
    t6.goto(0,0)
    time.sleep(0.5)
#
# 32. Turn to initial dance headings
#
    t1.seth(30)
    t2.seth(90)
```

```
    t3.seth(150)
    t4.seth(210)
    t5.seth(270)
    t6.seth(330)
    time.sleep(0.5)
#
    t1.pd()
    t2.pd()
    t3.pd()
    t4.pd()
    t5.pd()
    t6.pd()
#
# 33. Move out from center.
#
    for j in range(40):
        t1.fd(2)
        t2.fd(2)
        t3.fd(2)
        t4.fd(2)
        t5.fd(2)
        t6.fd(2)
    time.sleep(0.5)
#
# 34. Turn left 90.
#
    for j in range(30):
        t1.lt(3)
        t2.lt(3)
        t3.lt(3)
        t4.lt(3)
        t5.lt(3)
        t6.lt(3)
    time.sleep(0.5)
#
# 35. Circle right 360.
#
    for i in range(120):
        t1.arcright(40,3)
        t2.arcright(40,3)
        t3.arcright(40,3)
        t4.arcright(40,3)
        t5.arcright(40,3)
        t6.arcright(40,3)
    time.sleep(0.5)
#
# 36. Turn right(90)
#
    for i in range(30):
        t1.rt(3)
        t2.rt(3)
        t3.rt(3)
        t4.rt(3)
        t5.rt(3)
        t6.rt(3)
    time.sleep(0.5)
#
# 37. Move out 80.
#
    for i in range(40):
        t1.fd(2)
        t2.fd(2)
        t3.fd(2)
        t4.fd(2)
        t5.fd(2)
        t6.fd(2)
    time.sleep(0.5)
#
# 38. Turn right 90
#
```

```
    for i in range(30):
        t1.rt(3)
        t2.rt(3)
        t3.rt(3)
        t4.rt(3)
        t5.rt(3)
        t6.rt(3)
    time.sleep(0.5)
#
# 39. Circle left 360.
#
    for i in range(120):
        t1.arcleft(20,3)
        t2.arcleft(20,3)
        t3.arcleft(20,3)
        t4.arcleft(20,3)
        t5.arcleft(20,3)
        t6.arcleft(20,3)
    time.sleep(0.5)
#
# 40. Turn right 90.
#
    for i in range(30):
        t1.rt(3)
        t2.rt(3)
        t3.rt(3)
        t4.rt(3)
        t5.rt(3)
        t6.rt(3)
    time.sleep(0.5)
#
# 41. Forward towards center.
#
    for i in range(40):
        t1.fd(2)
        t2.fd(2)
        t3.fd(2)
        t4.fd(2)
        t5.fd(2)
        t6.fd(2)
    time.sleep(0.5)
#
# 42. Turn right 90.
#
    for i in range(30):
        t1.rt(3)
        t2.rt(3)
        t3.rt(3)
        t4.rt(3)
        t5.rt(3)
        t6.rt(3)
    time.sleep(0.5)
#
# 43. Circle left 360.
#
    for i in range(120):
        t1.arcleft(40,3)
        t2.arcleft(40,3)
        t3.arcleft(40,3)
        t4.arcleft(40,3)
        t5.arcleft(40,3)
        t6.arcleft(40,3)
    time.sleep(0.5)
#
# 44. Turn left 90.
#
    for i in range(30):
        t1.lt(3)
        t2.lt(3)
        t3.lt(3)
```

```
        t4.lt(3)
        t5.lt(3)
        t6.lt(3)
    time.sleep(0.5)
#
# 45. Forward toward center.
#
    for i in range(20):
        t1.fd(2)
        t2.fd(2)
        t3.fd(2)
        t4.fd(2)
        t5.fd(2)
        t6.fd(2)
    time.sleep(0.5)
#
# 46. Turn right 90.
#
    for i in range(30):
        t1.rt(3)
        t2.rt(3)
        t3.rt(3)
        t4.rt(3)
        t5.rt(3)
        t6.rt(3)
    time.sleep(0.5)
#
# 47. Circle left 360.
#
    for i in range(120):
        t1.arcleft(20,3)
        t2.arcleft(20,3)
        t3.arcleft(20,3)
        t4.arcleft(20,3)
        t5.arcleft(20,3)
        t6.arcleft(20,3)
    time.sleep(0.5)
#
# 48. Turn left 90.
#
    for i in range(30):
        t1.lt(3)
        t2.lt(3)
        t3.lt(3)
        t4.lt(3)
        t5.lt(3)
        t6.lt(3)
    time.sleep(0.5)
#
# 49. Forward into center.
#
    for i in range(20):
        t1.fd(2)
        t2.fd(2)
        t3.fd(2)
        t4.fd(2)
        t5.fd(2)
        t6.fd(2)
    time.sleep(0.5)
    t1.goto(0,0)
    t2.goto(0,0)
    t3.goto(0,0)
    t4.goto(0,0)
    t5.goto(0,0)
    t6.goto(0,0)
#
# 50. Turn to exit heading.
#
    t1.seth(180)
    t2.seth(180)
```

```
    t3.seth(180)
    t4.seth(180)
    t5.seth(180)
    t6.seth(180)
    time.sleep(0.5)
#
# 51. Exit single-file in order stage right.
#
    for i in range(40):
        t1.pu()
        t1.fd(2)
    for i in range(40):
        t2.pu()
        t1.fd(2)
        t2.fd(2)
    for i in range(40):
        t3.pu()
        t1.fd(2)
        t2.fd(2)
        t3.fd(2)
    for i in range(40):
        t4.pu()
        t1.fd(2)
        t2.fd(2)
        t3.fd(2)
        t4.fd(2)
    for i in range(40):
        t5.pu()
        t1.fd(2)
        t2.fd(2)
        t3.fd(2)
        t4.fd(2)
        t5.fd(2)
    for i in range(600):
        t6.pu()
        t1.fd(2)
        t2.fd(2)
        t3.fd(2)
        t4.fd(2)
        t5.fd(2)
        t6.fd(2)
#
if __name__=="__main__":
    msg=main()
    print("DONE!")
    mainloop()
```

Kalizuko, "You need to understand variables, lists to change the elements, arithmetic operators, such as +,-,*,/, more advanced operators, such as %, and there are more, for example, if statements, if true, do this, if not, do something else, loops, which are useful for moving content, functions, animating one job only, turtle to do your animations, and to troubleshoot you need to understand errors."

Kalizuko works on his laptop and drinks water while Dramata examines the codes and watches the videos and tries the codes posted on the website listed at the beginning and the end.

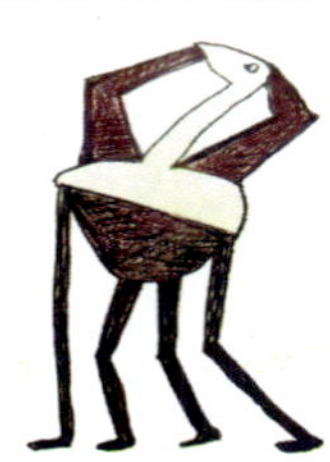

Segas: has legs for walking and legs for dancing.

Dramata, "I can code!"

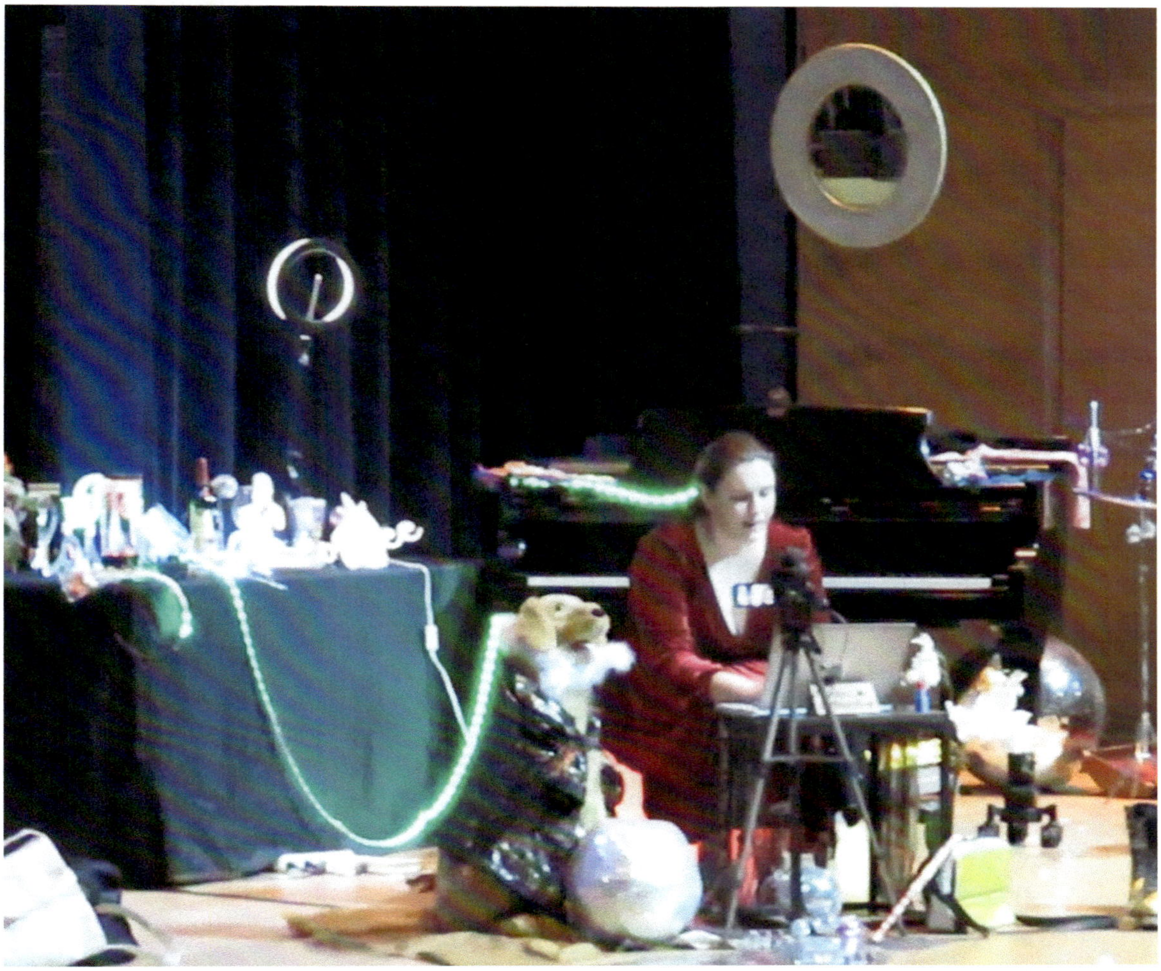

"I can code!" – a scene from a theater performance: Dramata sings the computer code

Kalizuko, "Yep, Yup, Yes!"
Dramata, "Would you describe your own challenges for me, please?"
Kalizuko, "That would be the conditions. I never know what to expect."
Dramata, "That should be easy, just dance a tango with the two partners."

`Three dancers dressed in black dance while alternating partners.`

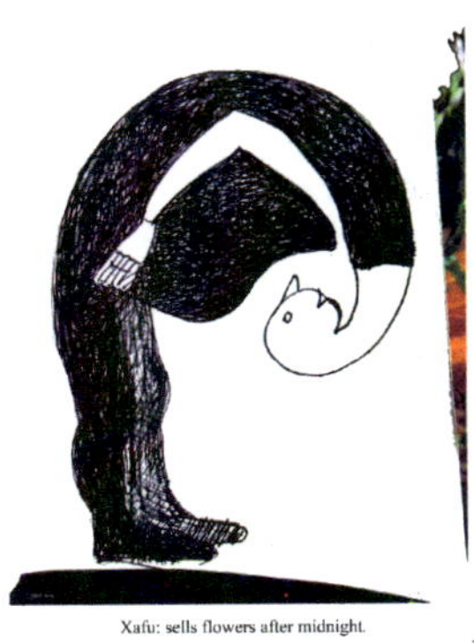

Xafu: sells flowers after midnight.

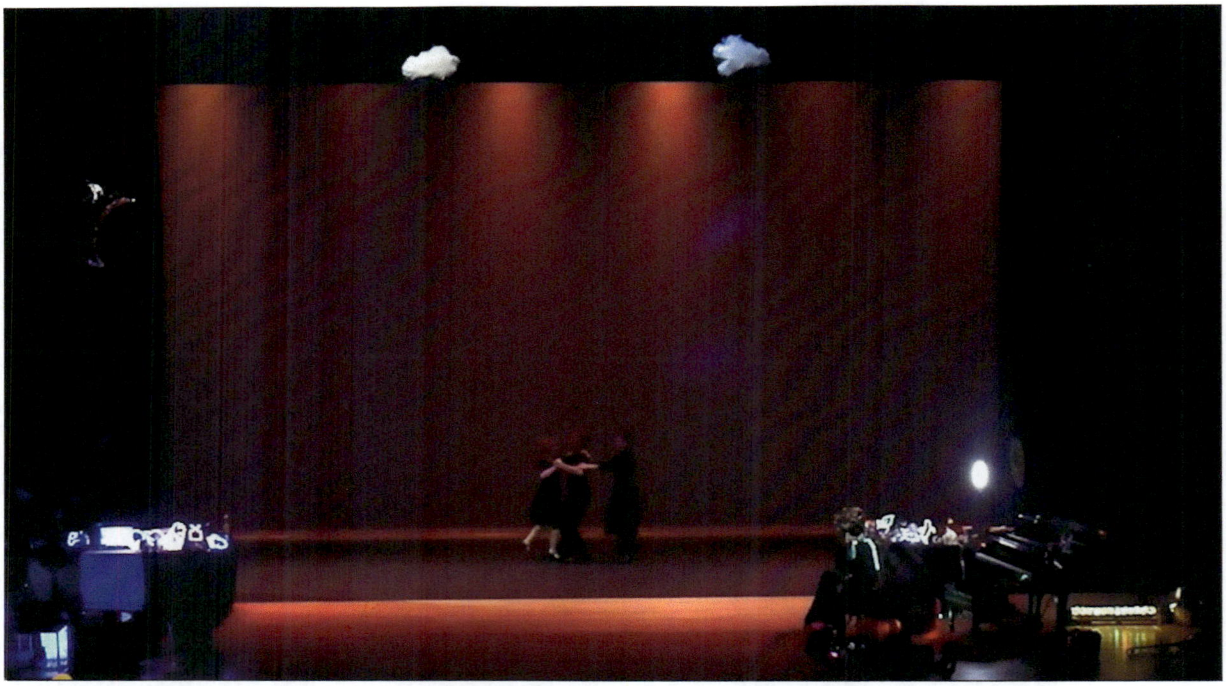

The partners - a scene from a theater performance

Kalizuko, "I can dance!"
Dramata, "I feel ready to become a coder"
Kalizuko, "I feel ready for a dancing career"
Dramata + *Kalizuko,* "Let's do both!"

Dramata sings to a computer code projected on the screen, while Kalizuko dances with a large stuffed bear.

Dramata, "You made me curious!"
Kalizuko, "You made me dance!"

The End.

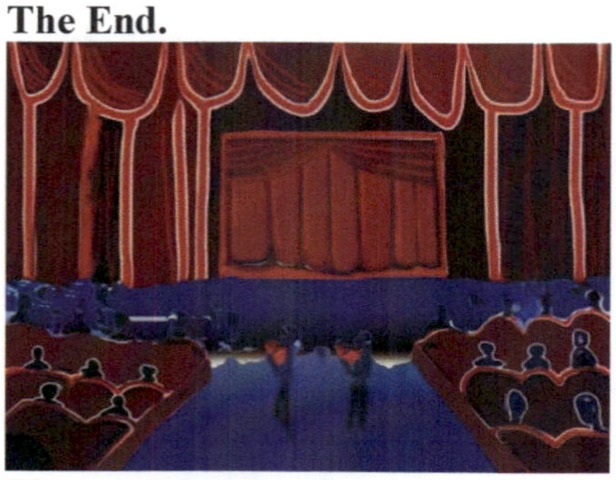

The curtain goes down (or closes).

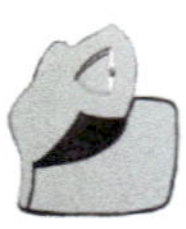

Hadura: always waits for her followers.

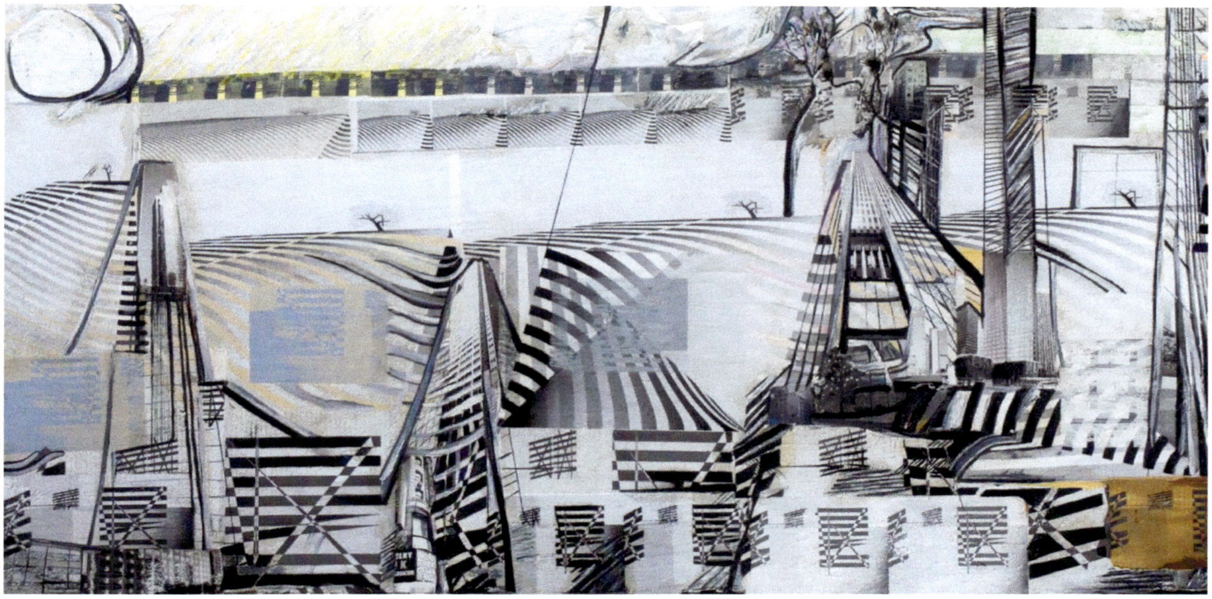

Commuters' Tunes. © 2025 by Anna Ursyn

CHAPTER 5
CODE DANCE

JOY TO ANGER, CELEBRATION TO PROTEST

INTRODUCTION

We enjoy happiness, celebrating, and participating in a joyful crowd. Group emotions affect us. When we hear about injustice, things that should not be happening or are harmful, we join in to protest to sustain and protect the values and habits we believe in.

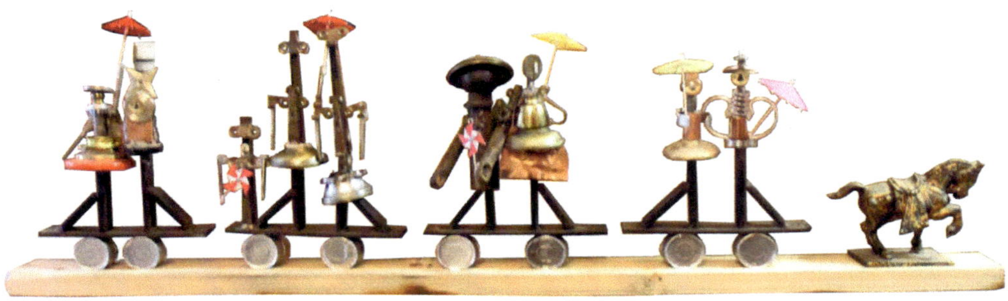

A Parade (Joy). © 2025 by Anna Ursyn

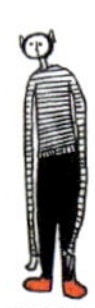

Rynago: makes wide turns.

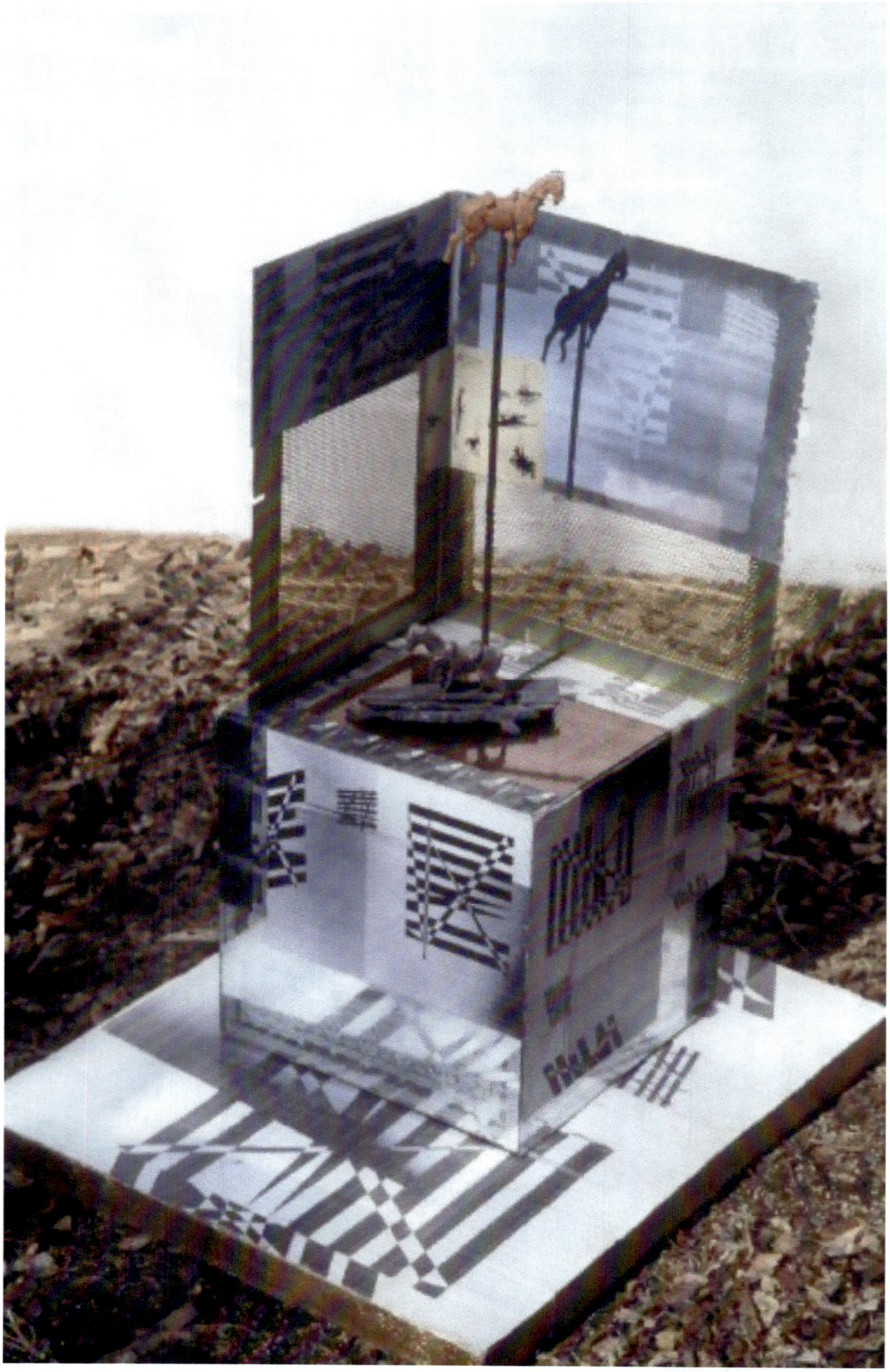

Hero Horse. © 2025 by Anna Ursyn

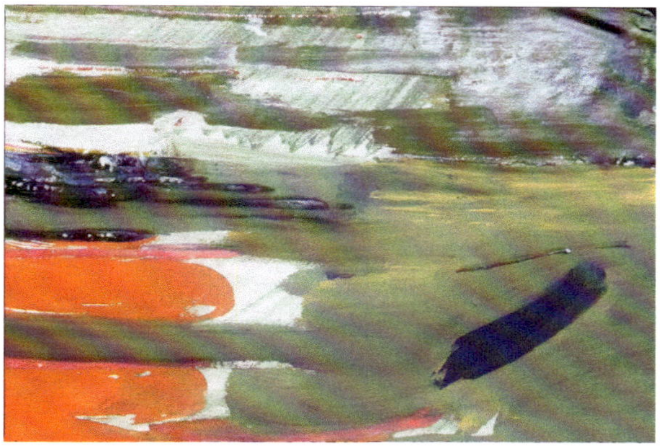

A Cloudy Day. © 2025 by Anna Ursyn

SYNOPSIS

Kalizuko meets with Dramata in person this time. They join a joyful parade without knowing much about its members and their reasons. An unexpected event triggers the participants to protest. While participating, they practice their newly acquired skills.

SUMMARY

A sudden shift can happen even to a large group. We participate in what we believe in.

CAST OF CHARACTERS

Kalizuko, a coder, now a dancer
Dramata, a dancer, now a coder
A Parader (celebrator)
A Protester
A Mob of Paraders
A Mob of Protesters

STAGE DIRECTIONS

A joyful parade involves many colorful, happy characters, while the protest shows the angry ones. The projections of the street are alternated with the codes. The music illustrates the two moods.

`The action happens beyond time and place.`

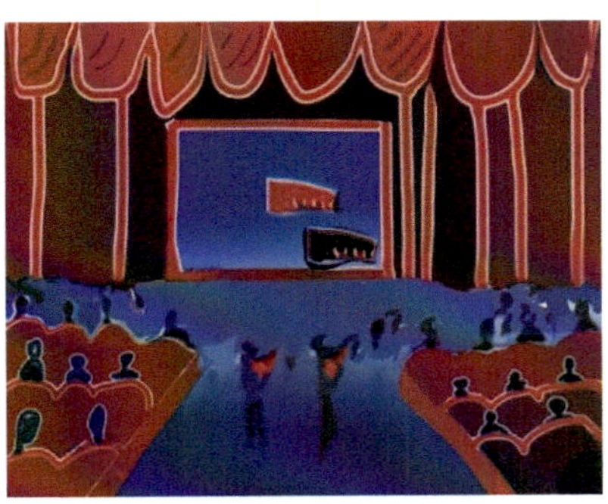

The Curtain goes up (or opens).

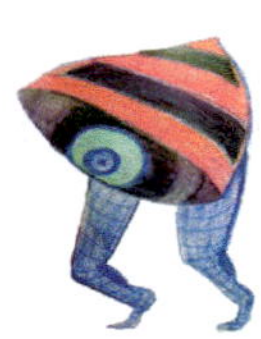

Lugatar: knows well where to go.

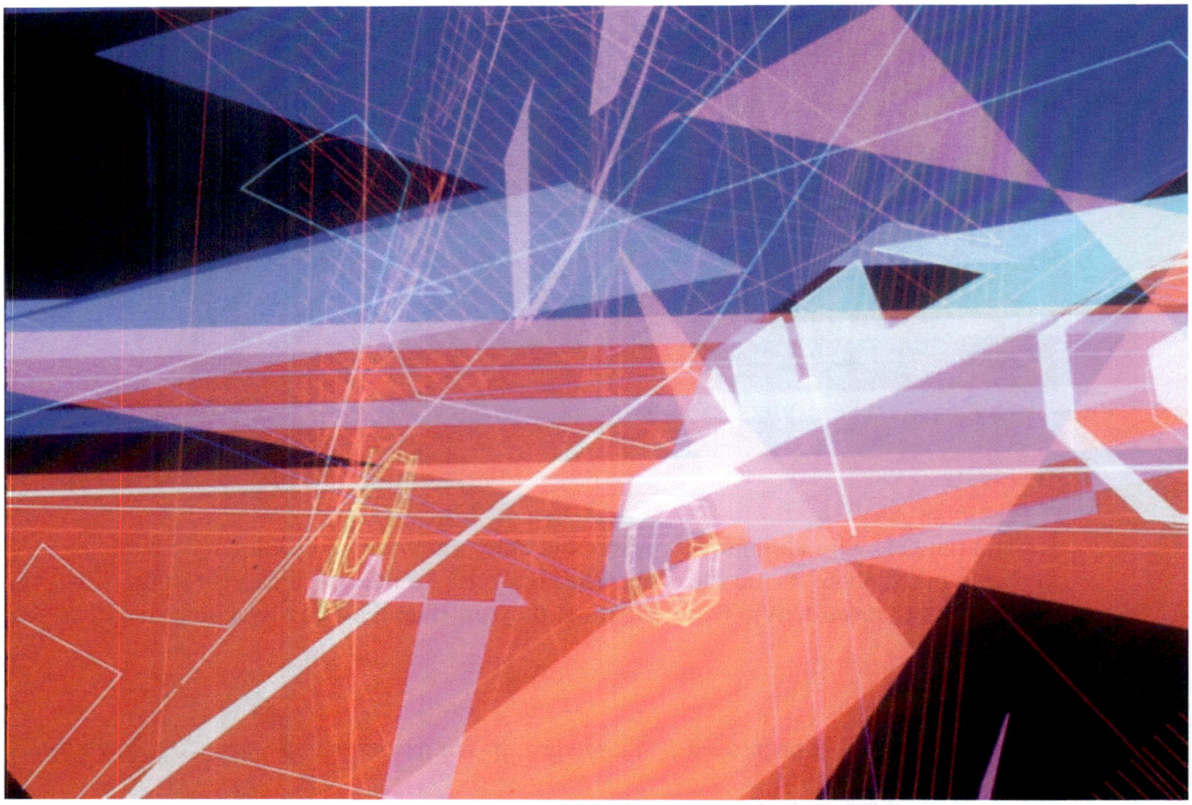

The Thirties I. Outputs from computer codes.© 2025 by Anna Ursyn.

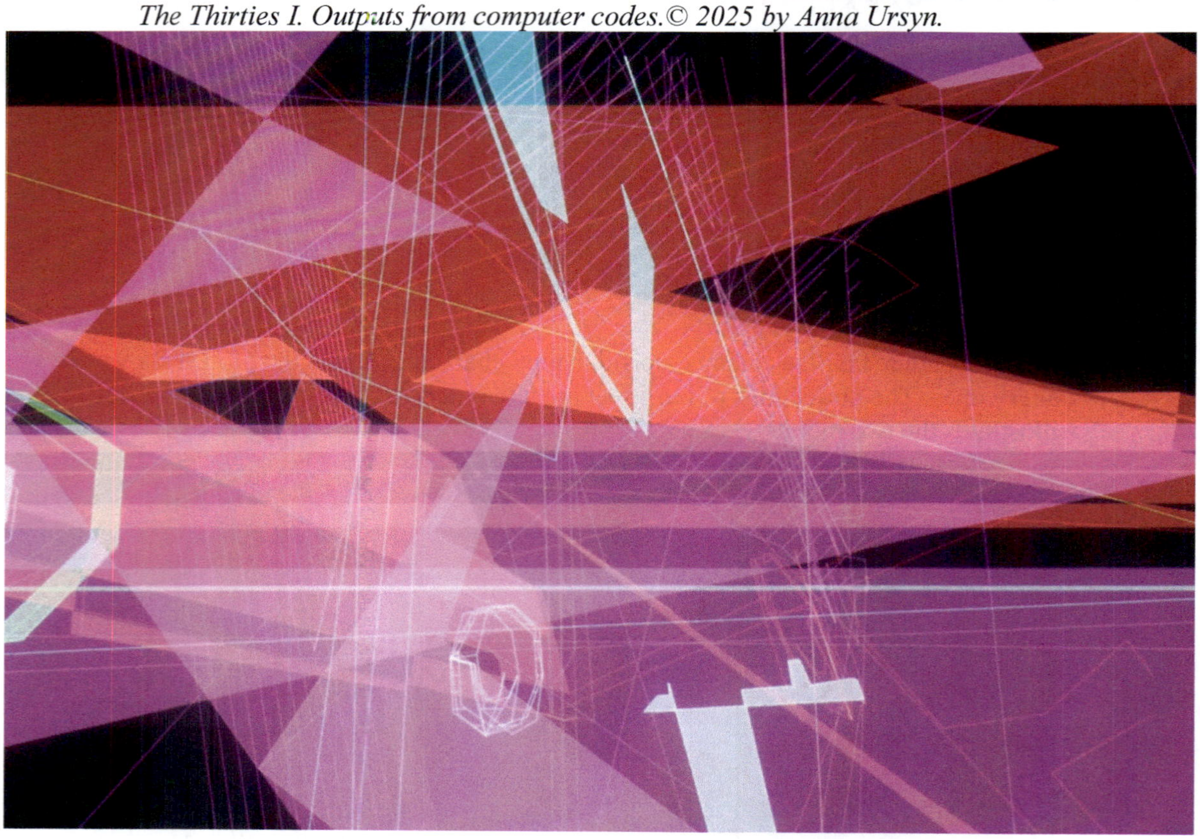

The Thirties II. Outputs from computer codes.© 2025 by Anna Ursyn

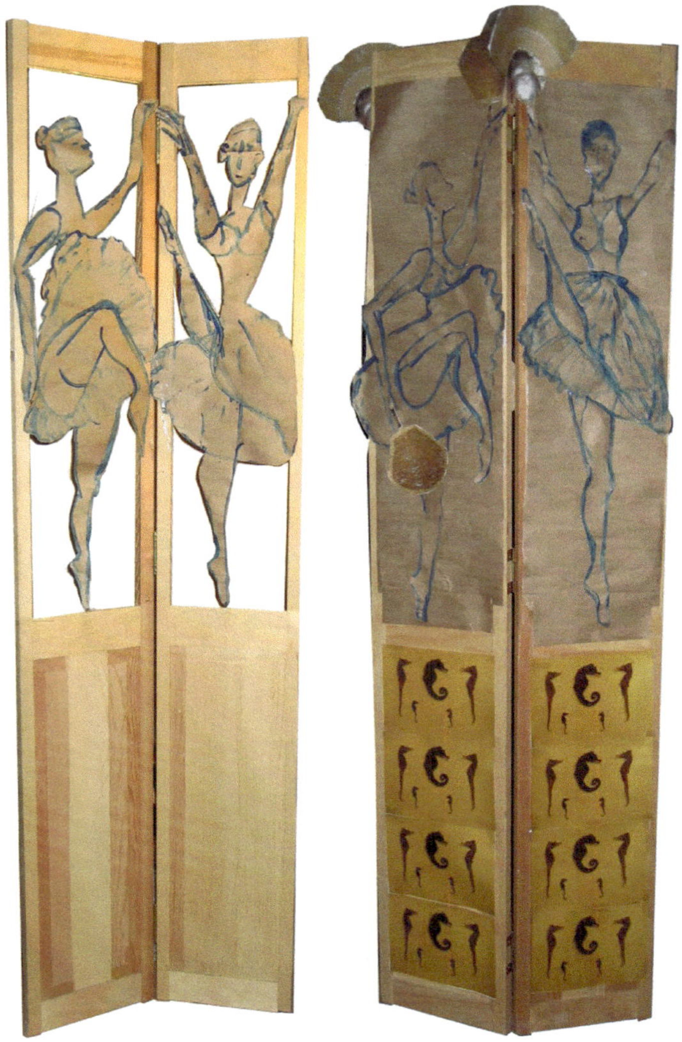

The Dancers, © 2025 by Anna Ursyn

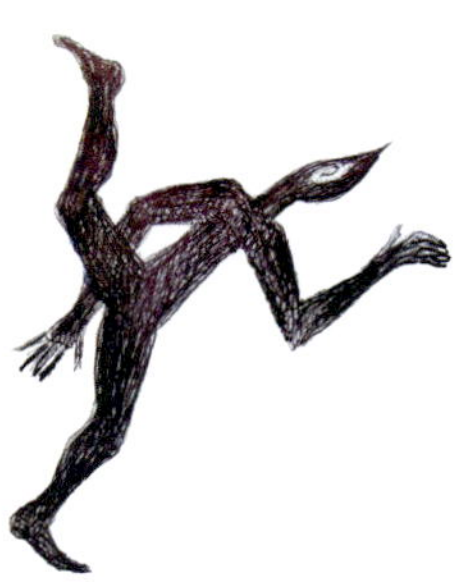

Lorkagon: makes the least amount of steps within a certain distance.

Lorkagon: makes the least number of steps in a certain distance.

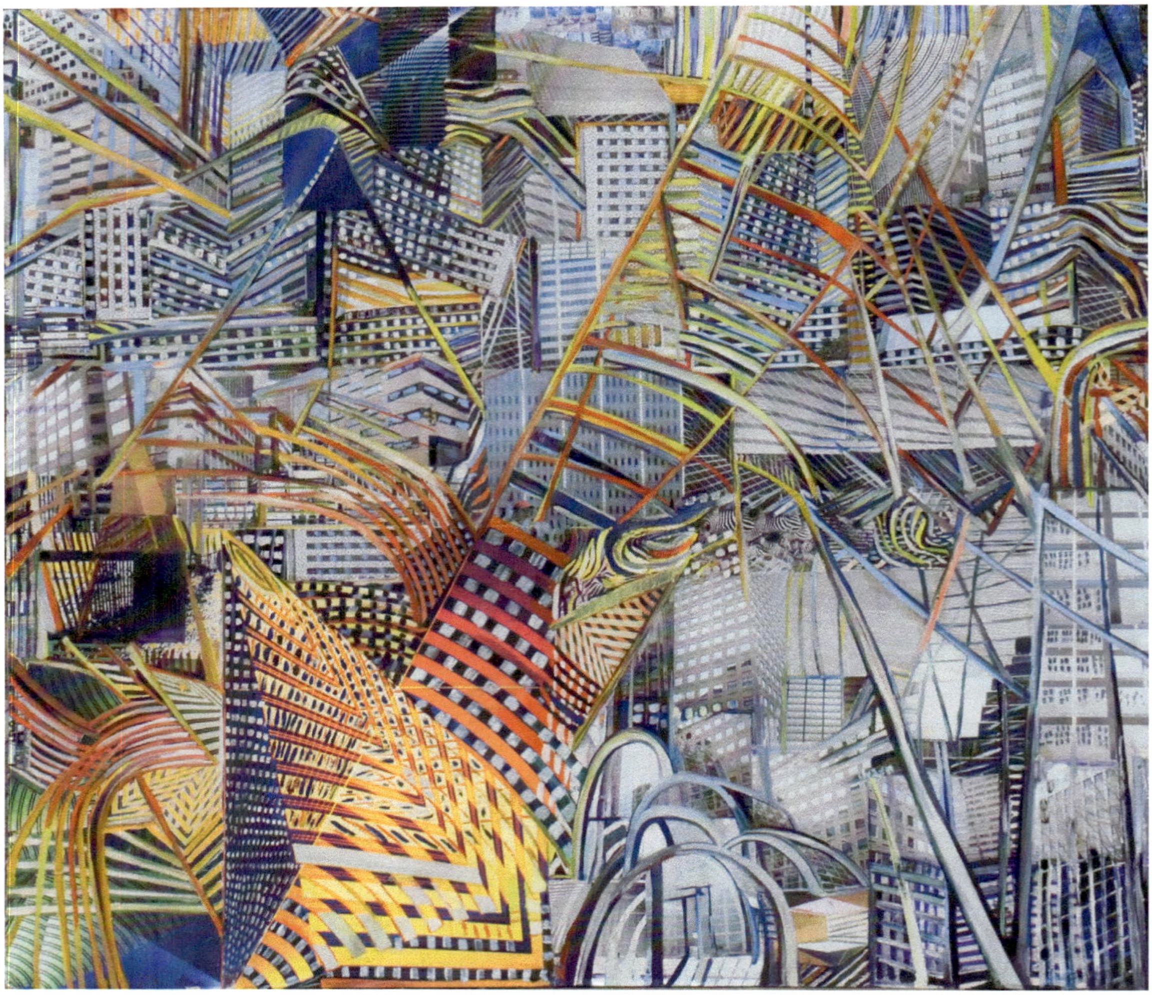

Ideas & Dogmas. © *2025 by Anna Ursyn*

THE SCRIPT FOR CODE DANCE

Dramata, "What a day for a joyful parade. Everybody dressed up according to their professions, I gather?"
Kalizuko, "What's the occasion?"
Dramata, "No idea, but it's good to master our newly acquired skills since you came from that far."
Kalizuko, "We can even dance in here! How do you know when your dance looks good?"
Dramata, "If it feels good, it looks good."

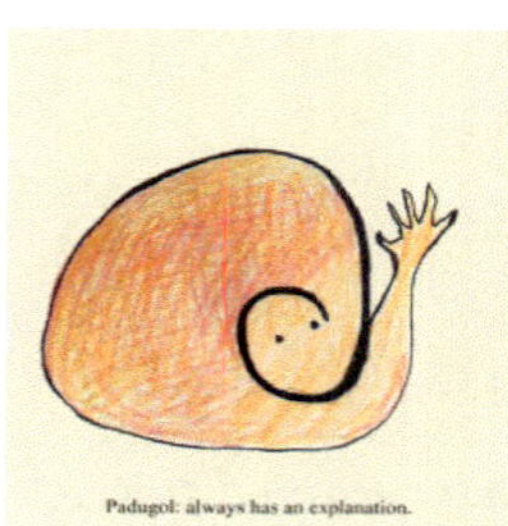

Padugol: always has an explanation.

Dramata, "All while coding. Since the parade seems joyful, let's do the waltz!"

```
import pygame
from pygame.locals import *

# Initialize Pygame
pygame.init()

# Set up the window
width, height = 640, 480
screen = pygame.display.set_mode((width, height))
clock = pygame.time.Clock()

# Define colors
WHITE = (255, 255, 255)
RED = (255, 0, 0)

# Define constants for the dancers
DANCER_WIDTH = 50
DANCER_HEIGHT = 50
DANCER_SPEED = 5
```

```
dancer images
l_image = pygame.image.load("dancer1.png")
2_image = pygame.image.load("dancer2.png")

itial positions of the dancers
l_x, dancer1_y = width // 4, height // 2
2_x, dancer2_y = width // 4 * 3, height // 2

e loop
 = True
unning:
vent in pygame.event.get():
event.type == QUIT:
running = False

ear the screen
n.fill(WHITE)

ove the dancers
er1_x += DANCER_SPEED
er2_x -= DANCER_SPEED

ap the dancers around the screen
ncer1_x > width:
ncer1_x = -DANCER_WIDTH
ncer2_x < -DANCER_WIDTH:
ncer2_x = width

aw the dancers
n.blit(dancer1_image, (dancer1_x, dancer1_y))
n.blit(dancer2_image, (dancer2_x, dancer2_y))

date the display
me.display.flip()

ntrol the frame rate
k.tick(30)

he game
e.quit()
```

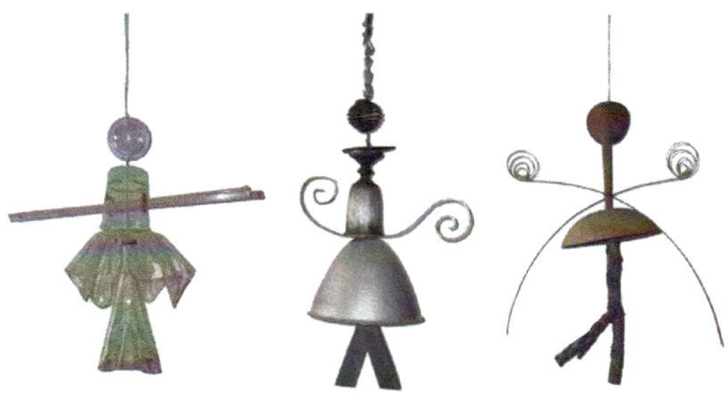

Three Prima ballerinas, © 2024 by A. Ursyn

Kalizuko, " In this code, you used the Pygame library to create a window and handle the game loop. This program initializes it, sets up the window, defines colors, and loads dancer images in their initial positions. The game loop continuously moves the dancers in opposite directions to return when they reach the edges. Finally, it updates the display and controls the frame rate. For the code to work, you must provide your dancing photos (you.png and me.png), so let's have someone take pictures of us while we dance."

`Dramata asks one of the participants to take a picture of them dancing.`

Kalizuko, "The waltz is a traditional ballroom dance characterized by its smooth, flowing movements and graceful turns. Let's follow the basic sequence of the waltz: for a starting position, we face each other, with my right hand holding your left hand, my left hand resting on your back, and your right hand resting on my left shoulder."
Dramata, "Good start! The primary step in the waltz is a box step or box pattern, with four movements performed in a square. The partners move counterclockwise around the space."
Kalizuko continues, "We take four steps: forward, side, close, and side, returning to the starting position."
Dramata, "Nice. Now, the turns add elegance and variety to the dance, which is performed in different directions and styles, depending on personal preference."
Kalizuko, "Common turns in dancing include the natural turn to the right, reverse turn to the left, and underarm turn where the partner turns under a raised arm to change directions.
Dramata, "Oh, very sorry, Sir…."
The waltz incorporates an astonishing rise-and-fall action, adding a sense of lightness and elegance to the dance. We lower as we move forward and rise on our toes as we move backward."
A Parader "Watch your step!"
Kalizuko, "We can practice intricate footwork, additional turns, embellishments, and patterns at different tempos, ranging from slow and romantic to faster and livelier. The key is maintaining a smooth and flowing motion, connecting with you as we gracefully flow across the parade."
Kalizuko," So sorry, Madam…"

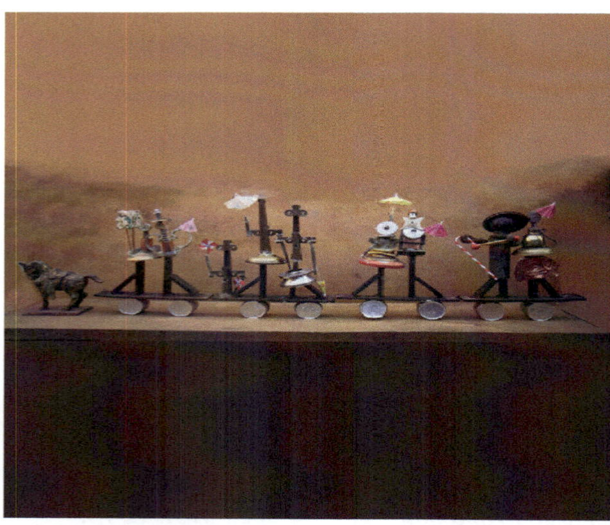
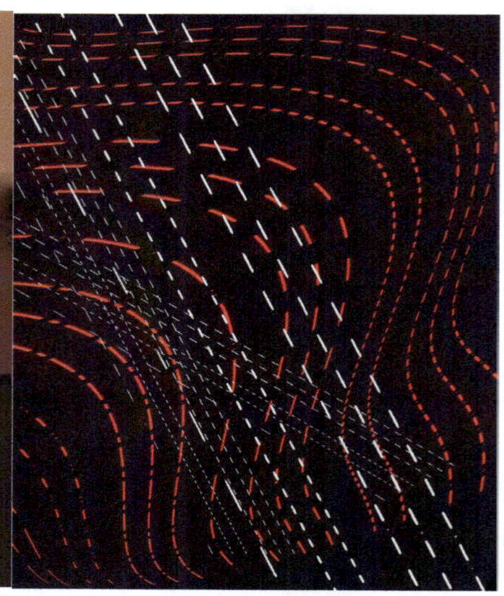

A Parade (Joy). *Joyful and Angry Steps.* © *2025 by Anna Ursyn*

`A loud sound can be heard, and people around them turn angry.`

Dramata, in a scared voice, "What happened?"

Kalizuko, “No clue, but they are all furious. Shell, we leave?”
Dramata, “Nope, we’ll do the angry dance.”
Kalizuko, “An anger dance is a form of Latihan. It removes angry energy from a body.”

```
import time

def anger_dance(duration):
    print("The Anger Dance begins!")

    for i in range(duration):
        print("Step", i+1, ": Stomp your feet!")
        time.sleep(0.5) # Pause for half a second

        print("Step", i+1, ": Clench your fists!")
        time.sleep(0.5) # Pause for half a second

        print("Step", i+1, ": Shout out loud!")
        time.sleep(0.5) # Pause for half a second

        print("Step", i+1, ": Release your anger!")
        time.sleep(1) # Pause for one second

    print("The Anger Dance is over!")

# Example usage
anger_dance(5) # Run the anger dance for 5 steps
```

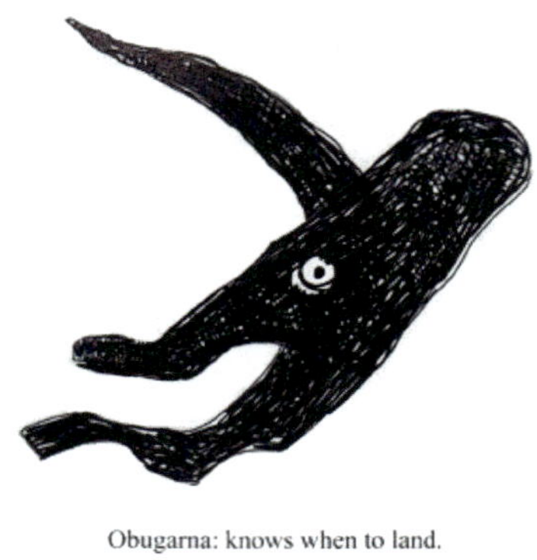

Obugarna: always knows when to land.

Dramata, “In this code, the anger_dance() function takes a duration parameter, which represents the number of iterations the dance goes through. Each step consists of stomping feet, clenching fists, shouting loudly, and releasing anger. The function uses the time.sleep() method to pause the program for a specified time between each step.”
Kalizuko, “An angry dance can help release suppressed emotions. Here are steps to incorporate into your routine: stand tall, stomp and march, punch and jab, use sharp movements, keep your fists clenched, use isolations, jump, incorporate floorwork, turn and spin, and use the space around you. Adapt these steps to your own style and comfort level, and prioritize safety. Add personal touches to express your anger even stronger. Remember to adapt these steps to your own style and comfort level. You must prioritize your safety and avoid any movements that may cause injury. Add your personal touches and improvise to express your anger stronger.”

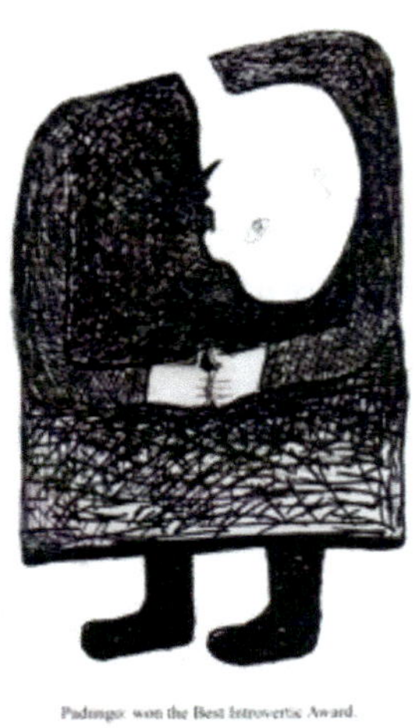

Padungo: won the Best Introvertic Award.

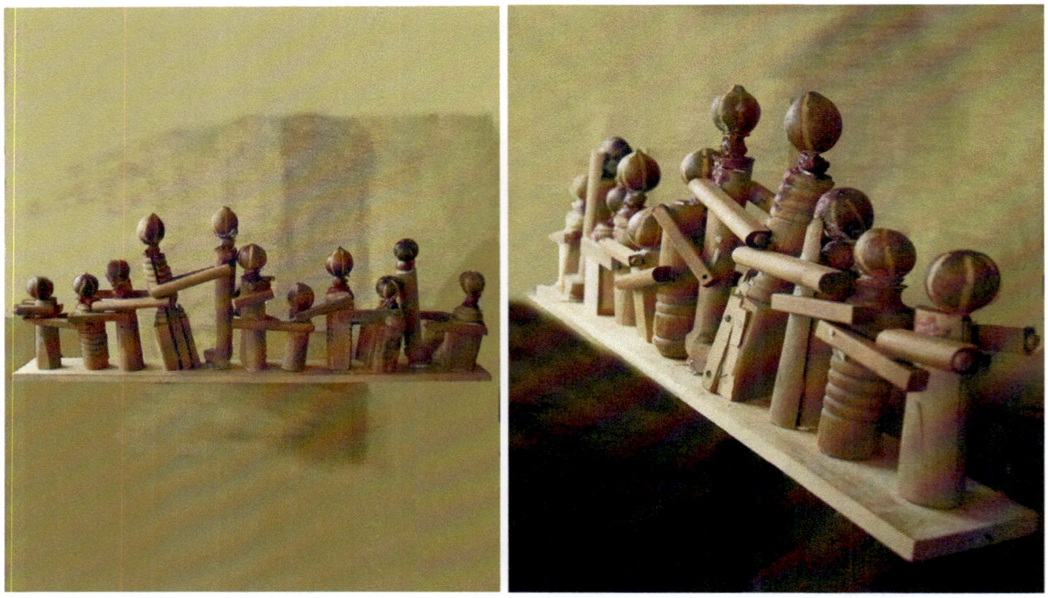

A Protest (Anger). © *2025 by Anna Ursyn*

A Protester, "What is your name, please?"
Kalizuko, "Why? It's Kalizuko."
The angry mob of Protesters, "Kalizuko for the CIO! Kalizuko for the CIO! Kalizuko for the CIO!
Kalizuko for the CIO!"

The End

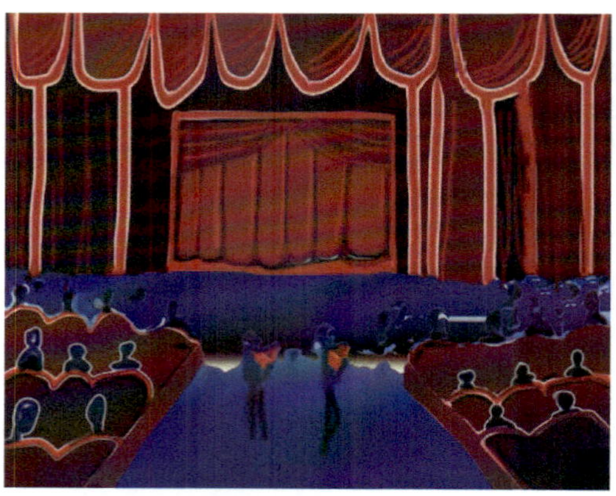

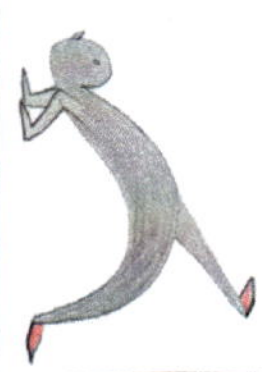

Samtua: always in balance.

The Curtain goes down (or closes)

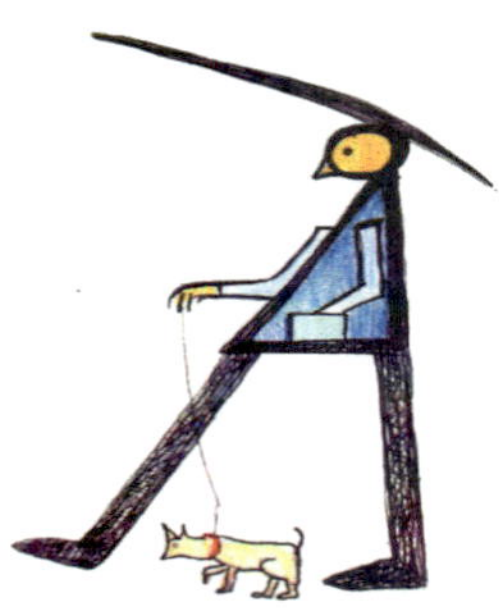

Pikutun: can adjust his tempo.

Pikutun: can adjust his tempo.

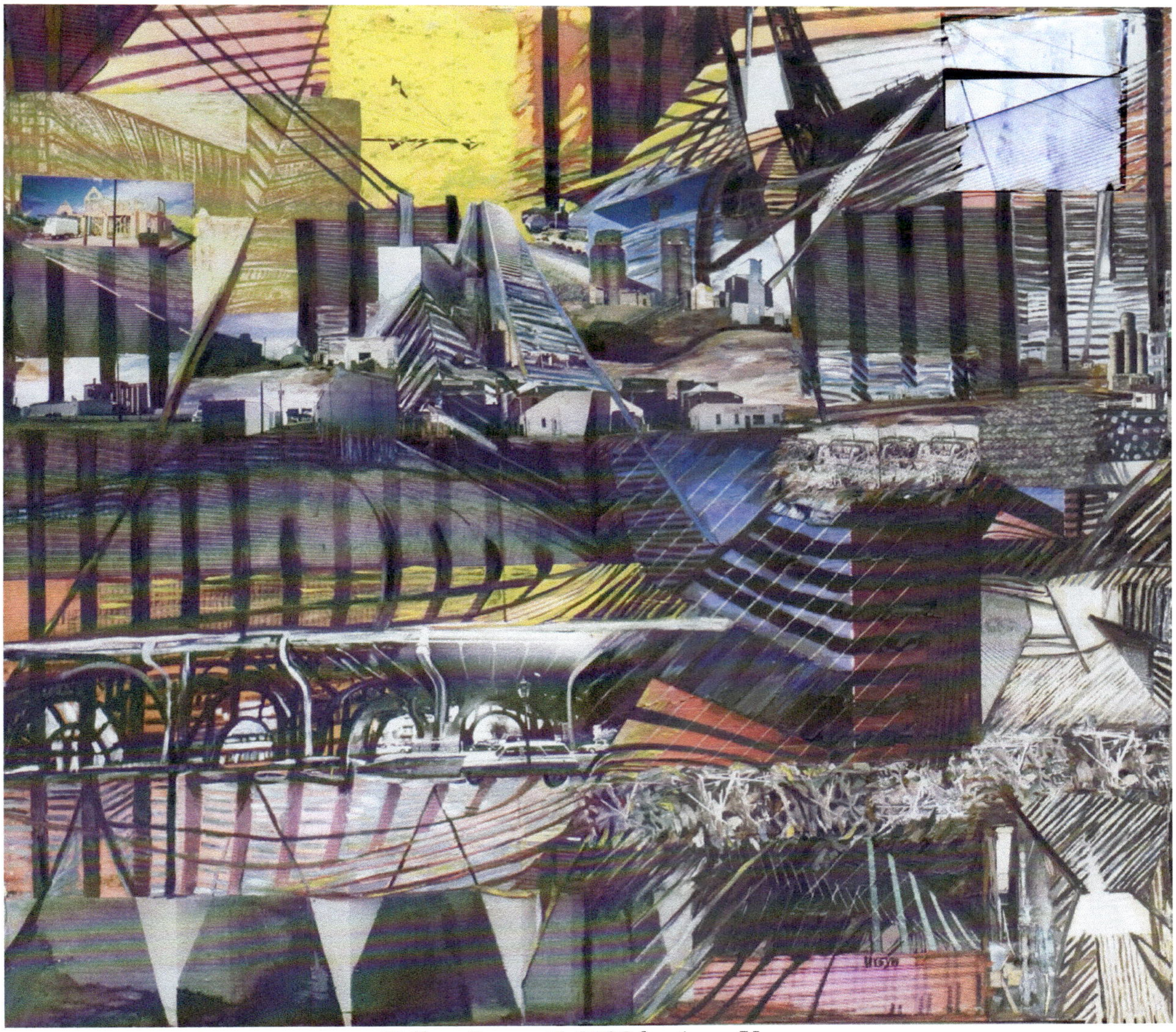

Yellow Pages. © 2025 by Anna Ursyn

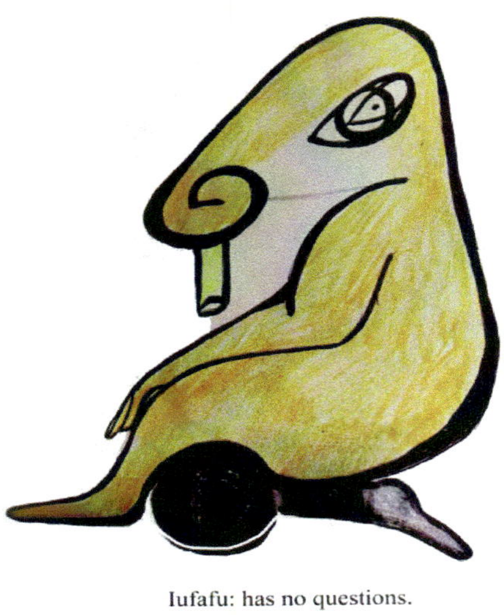

Lulafu: never has questions.

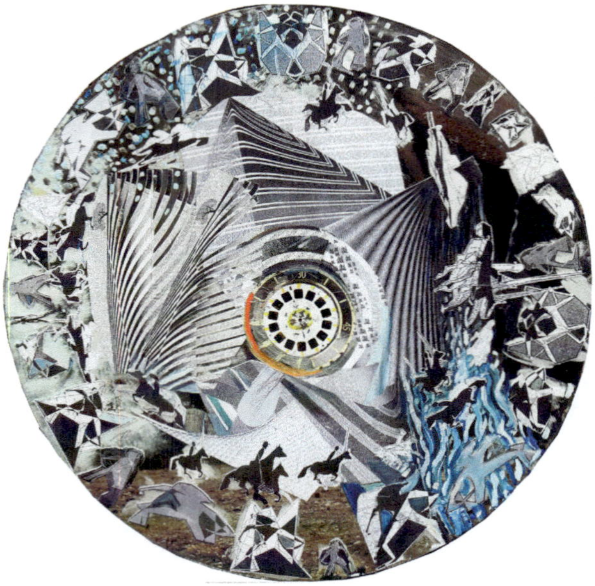

Manhole Cover. © 2025 by Anna Ursyn.

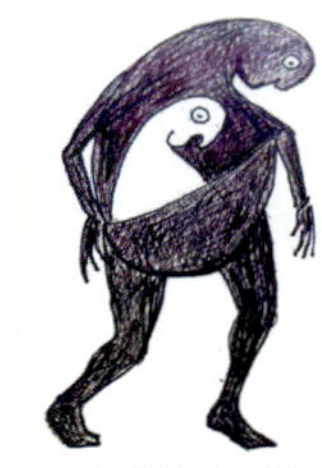

Nuvago: never feels lost

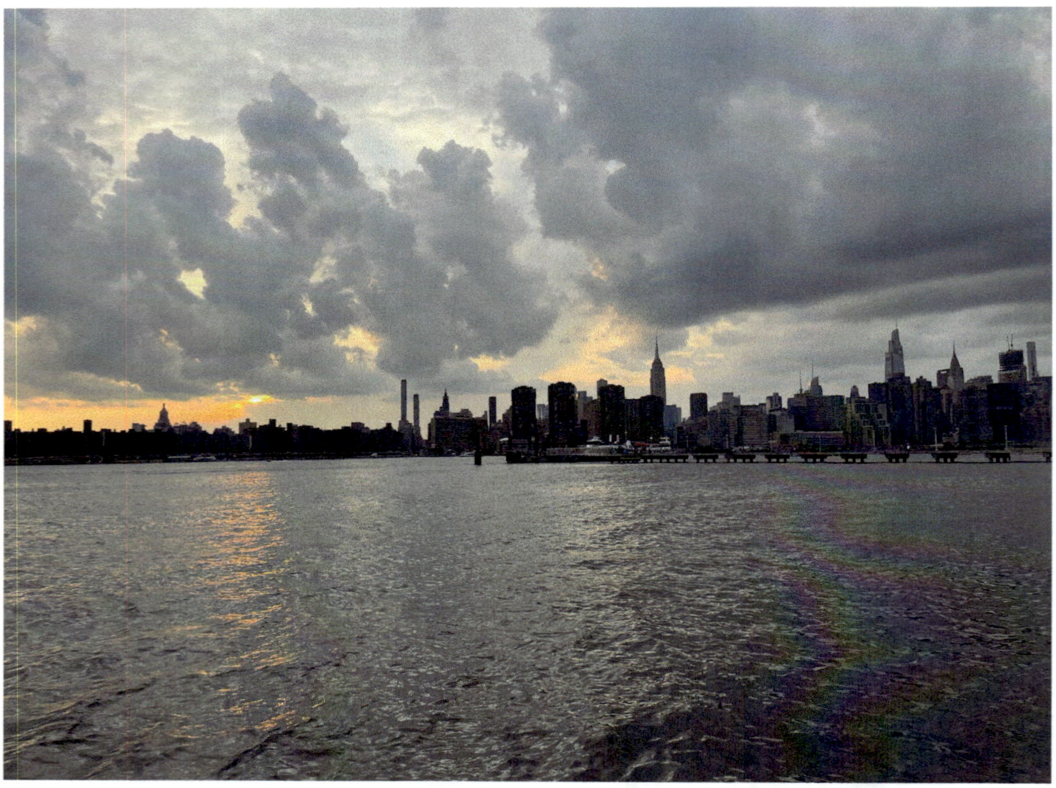

Cityscape. © 2025 by Matylda Czarnecka

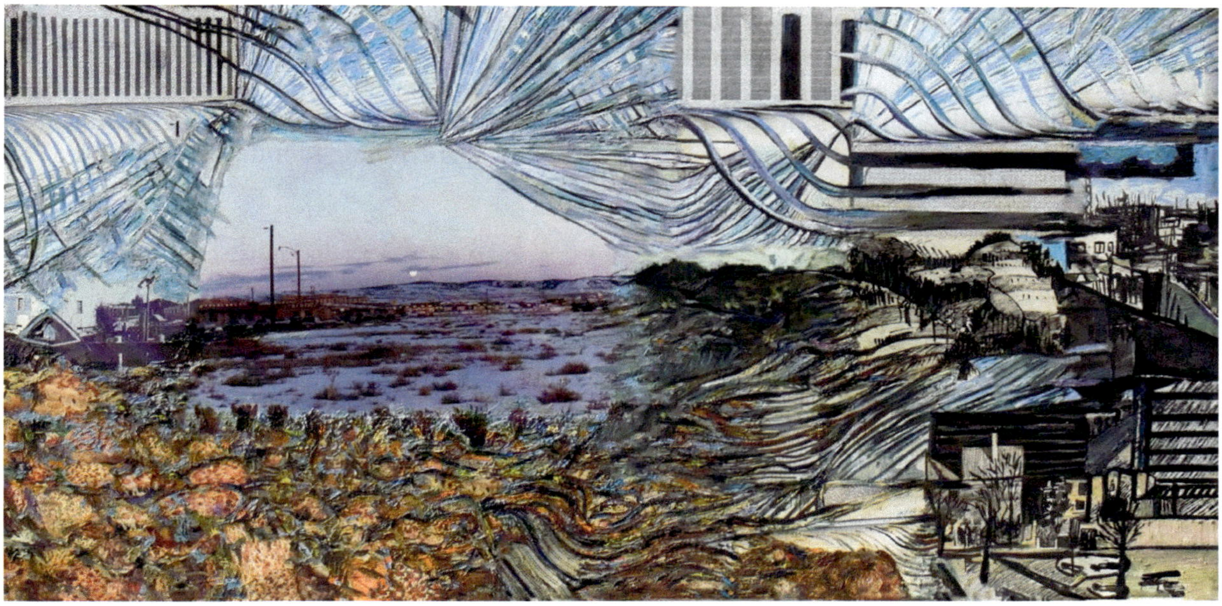

Change of Matter. © 2025 by Anna Ursyn.

CHAPTER 6

PROJECTS FOLLOWING DANCE CODE SCRIPT

Knowledge-based learning projects take the form of visual materials, first animated as videos, and then used for the actual performance to explore the impact of technology on our daily lives. Art students contributed with their research, voices, characters, animation, and art to this enterprise. Students' contributions resulted in an artistic video based on this script, recorded to obtain time-based and graphical data. Faculty and students from UNC's Theater Arts and Dance Department danced and improvised.

DEVELOPING ANIMATED VIDEOS

Videos and performances resulting from this script explore metaphorical connections between the rules and restrictions in code writing and dancing. They involved multidisciplinary, international collaboration and included programmers, musicians, students, faculty, and staff, and students. An artistic exploration of coding allows producing a dance performance guided by the rules in codes as a way of learning through metaphors. To visualize ideas behind programming by applying connections students danced, sang, and worked on an explanatory video (please see the list of collaborators).

This set facilitates learning how to write programs by translating dance performances into computing codes and provides metaphors between the dance steps and steps in coding. The explanatory video by Irwin Stuart Smith shows a set of basic rules behind programming as visual trajectories. Six animated triangular turtles with assigned colors of red, green, blue, magenta, cyan, and yellow were used, like in RGB and CMYK models, with a black set as a background. They symbolized six animated characters. The dancers were to wear color-coded matching the moving triangles costumes. Dancers were improvising to music.

At some point, a red arrow became larger and followed the trajectories recorded on the video. Animated arrows followed the code in Python, with music to explain coding through the dance steps to be outputted as a work of art. This explanatory video was first used with one dancer improvising by following the triangles projected on the floor, which was later used as a projection above the six dancers performing for the same content during the performance.

The video with color-coded triangles was paired with a story and illustrated by dances, and the visual outcomes of the codes created for different dances: tango, waltz, cha-cha, foxtrot, swing, and salsa, followed by jazz-based composition.

STAGE PERFORMANCE

Those who contributed to this enterprise offered their research, codes, music, dances, and voices. The preparatory videos were recorded before, with Izzy Sajdak dancing while wearing color-coded triangle-shaped coverings. An animated preparatory video served as a guide to establish the timing in planning for the performance.

The explanatory video by Stuart Smith with overlayed footage of Mary Jo Lasky dancing by following the trajectories projected from the original explanatory video served as a backdrop for the performance when the dancers improvised. During the performance, this video was projected on the backdrop of the stage of the theater, showing generic statements of the codes that were visible on the giant screen behind the six dancers performing to the same video. In the performance and on the website, there is singing to the code on display.

The program consisted of three connected parts. First, the two characters sit on red and blue-lit cubes in front of their laptops, interacting online. At the same time, a well-lit, colorful space on the stage allows the dancers to perform major coding and dance-related ideas in front of projected codes, musical notations, visuals, and props. Traditional dances accompanied the codes and musical notations projected during the performance.

During the second part, the videos and graphics served as visual tools for learning programming: to carry information with explanatory power. The distinctive colors on their costumes matched the arrows in the video. The projected visual explanation of coding followed the six-color arrows resulting from the codes. Six dancers performed the dances: waltz, tango, foxtrot, cha-cha, salsa, and swing (specifically Lindy Hop for swing.) They were dressed in color-coded costumes (red, green, blue, magenta, cyan, and yellow. The color-coded programs that matched each dancer's leotard were projected at the end of the performance when each dancer was performing an improvised solo dance. Codes were projected for each dance matching the outfit of each student dancer. Hopefully, it would help the audience to focus on the movement of just one dance/code and one performer per each dance's theme. Also, adding color supported the visual quality of the whole spectacle. At the end of the performance, the whole group improvised.

For the third part of the spectacle, Mary Jo Lasky danced to a poem about a coder who taught a dancer about coding and the dancer teaching a coder about dancing to a poem recitation by Izzy Sajdak. Miscommunication occurs when topics from different disciplines are discussed. As a result, the dancer unintentionally breaks the rules in coding and the coder in dancing.

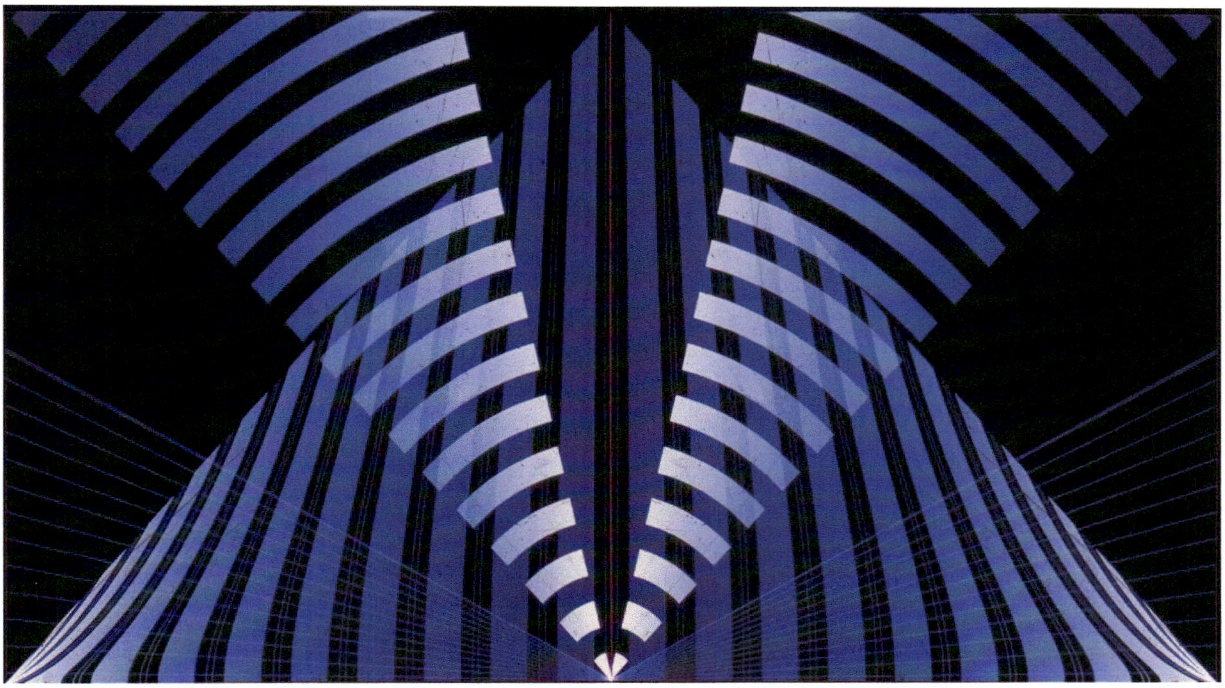

The Flow. The output from the computer code. © 2025 by Anna Ursyn.

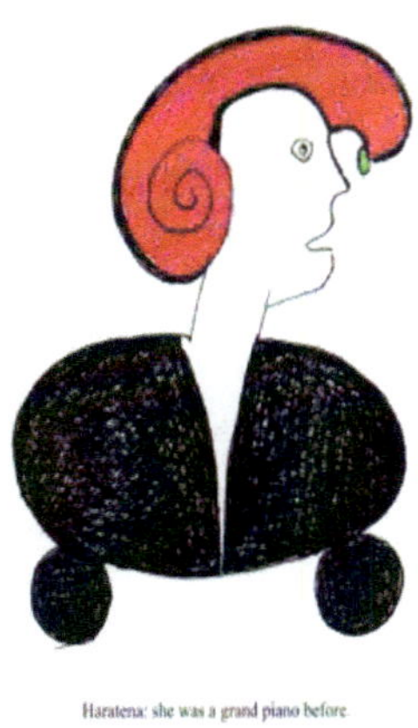

Haratena: she was a grand piano before.

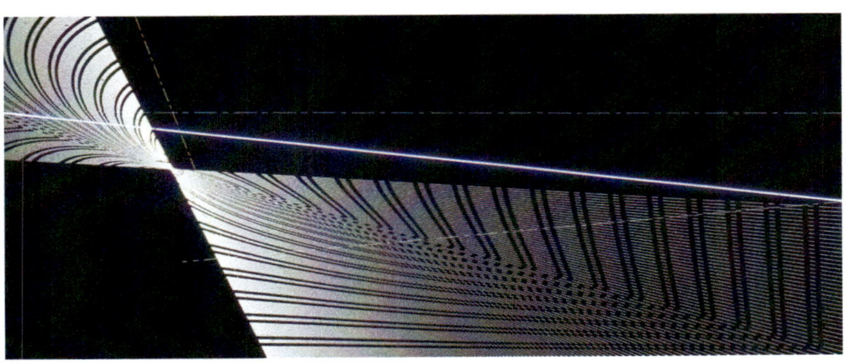

The New Swirl. The output from the computer code. © 2025 by Anna Ursyn

PROJECTS RESULTING FROM THE SCRIPT

DANCE CODE PERFORMANCE AT THE CAMPUS COMMONS PERFORMANCE HALL (CCPH), UNC - https://arts.unco.edu/campus-commons/:

https://www.youtube.com/watch?v=0YED8KkB7wE

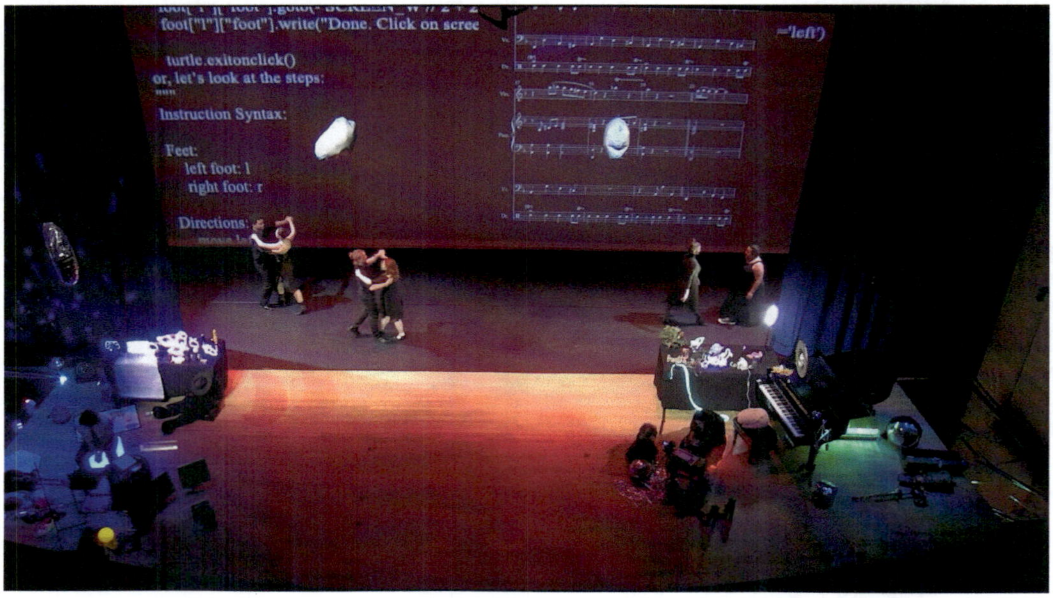

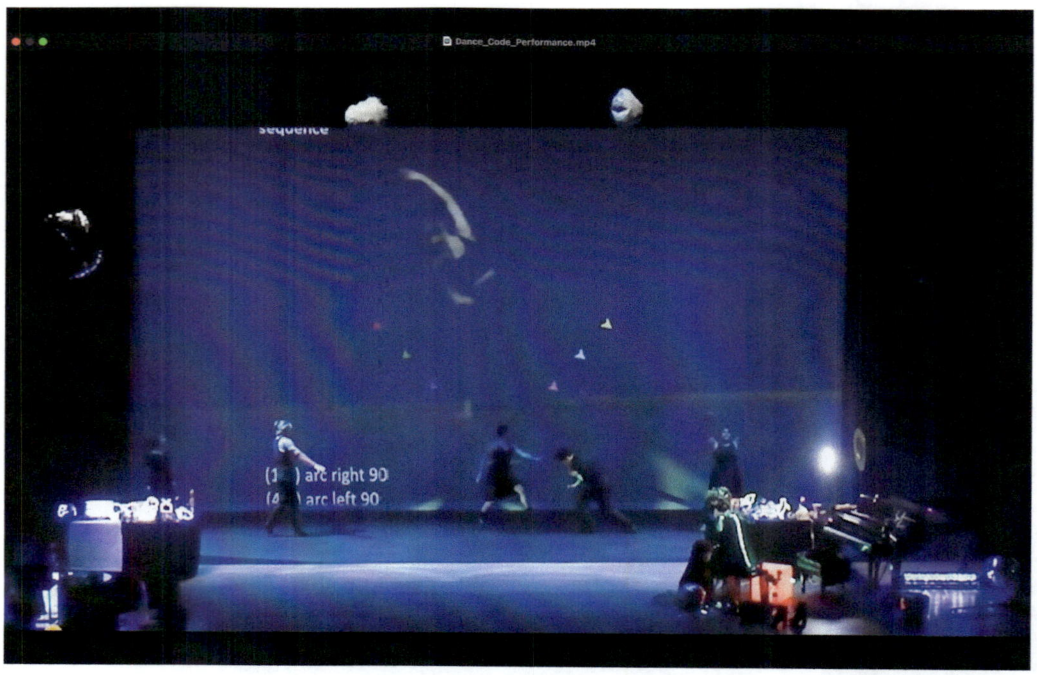

EXPLANATORY VIDEO BY STUART SMITH:

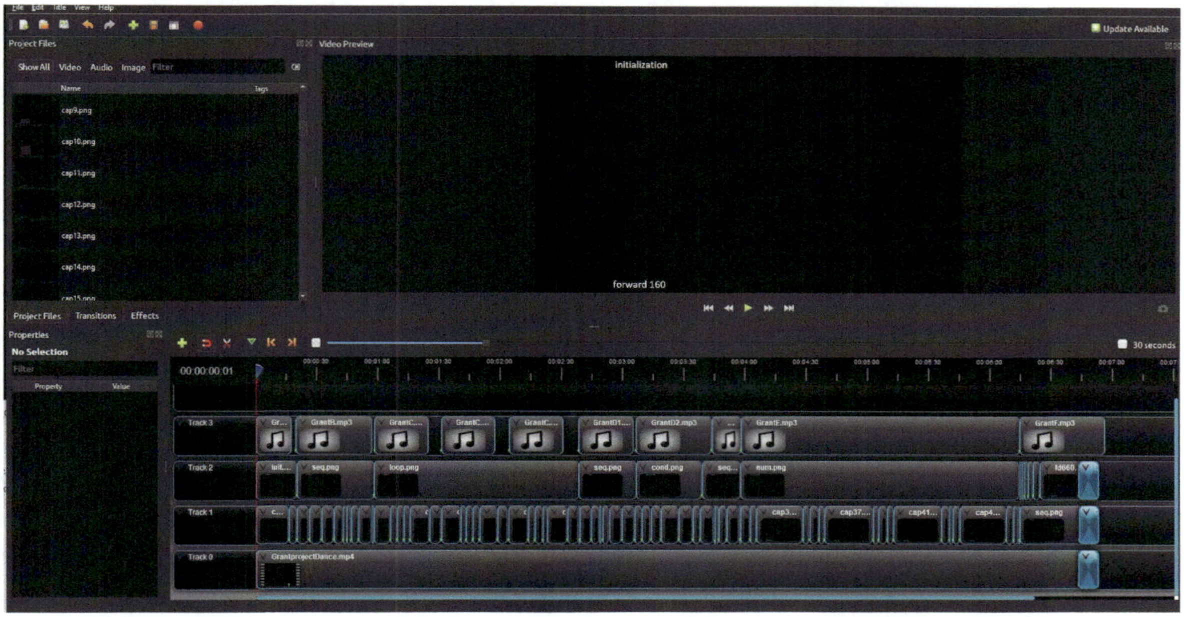

Initialization

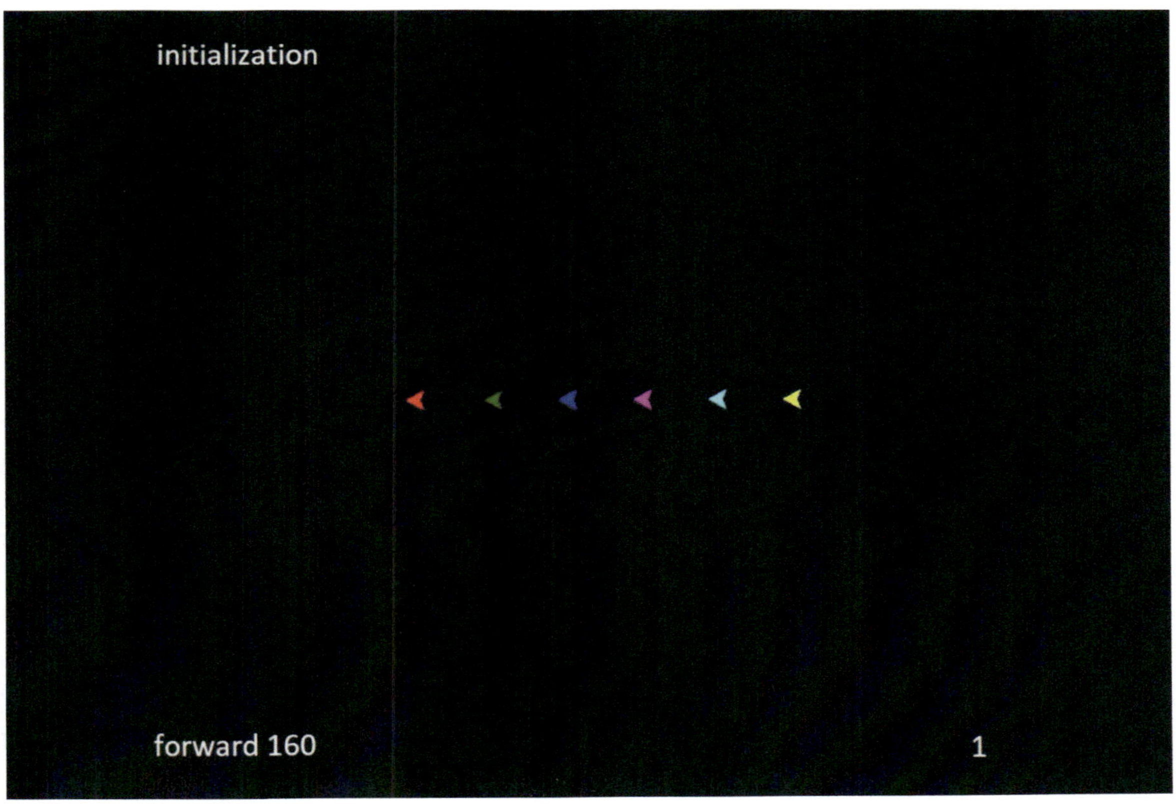

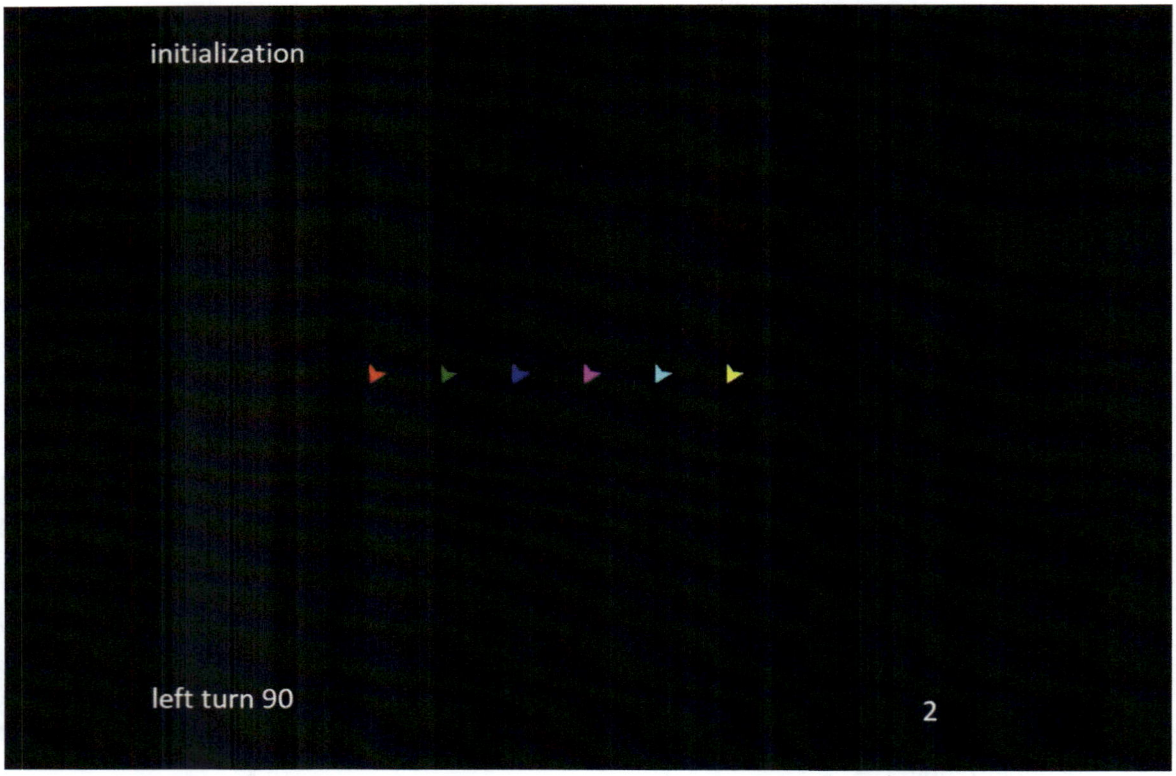
initialization
left turn 90
2

Sequence

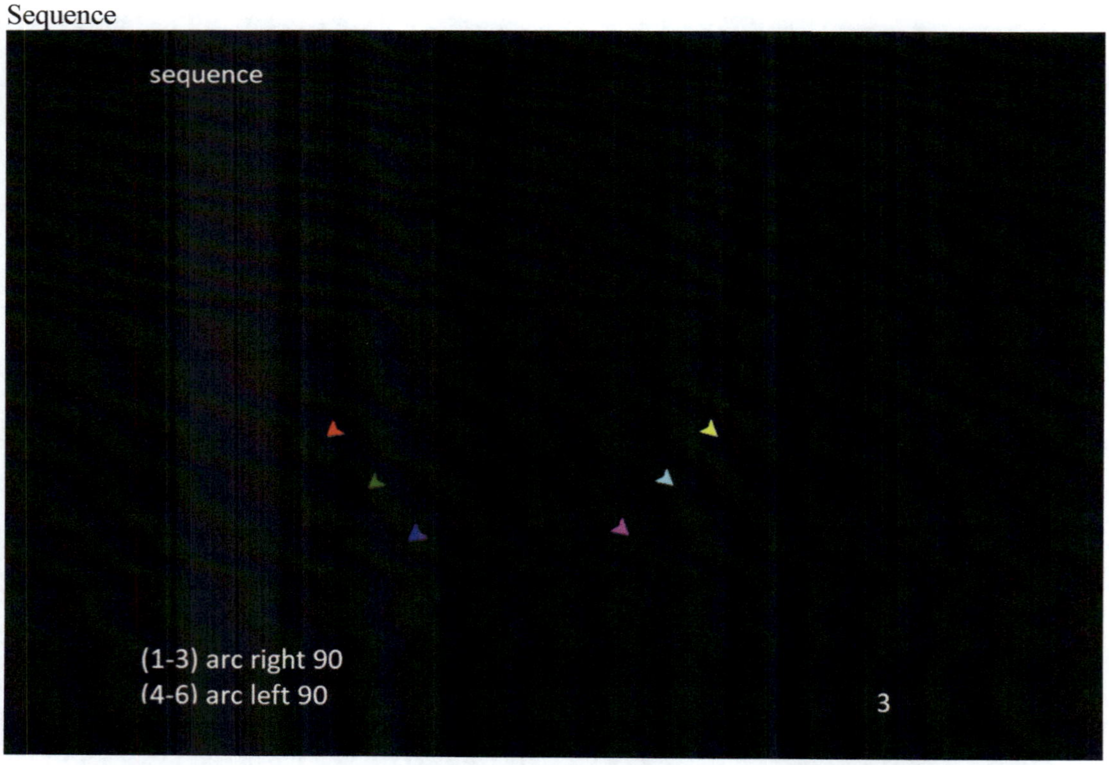
sequence
(1-3) arc right 90
(4-6) arc left 90
3

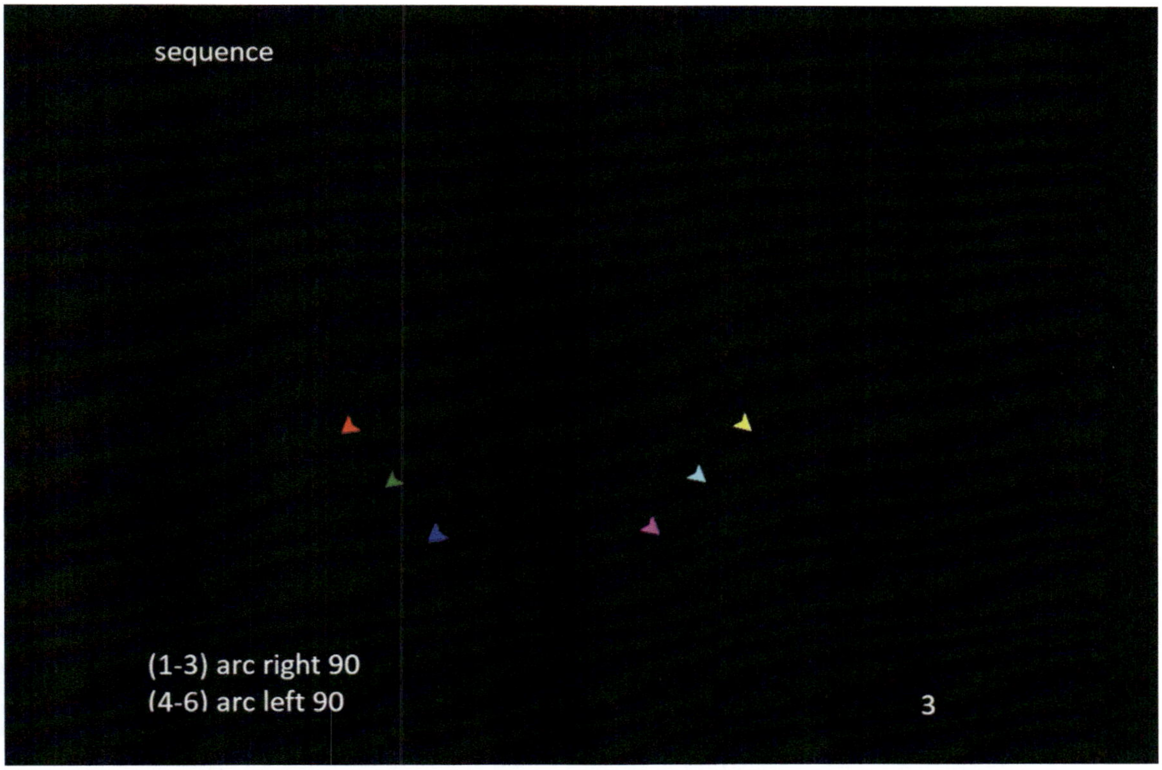
sequence
(1-3) arc right 90
(4-6) arc left 90
3

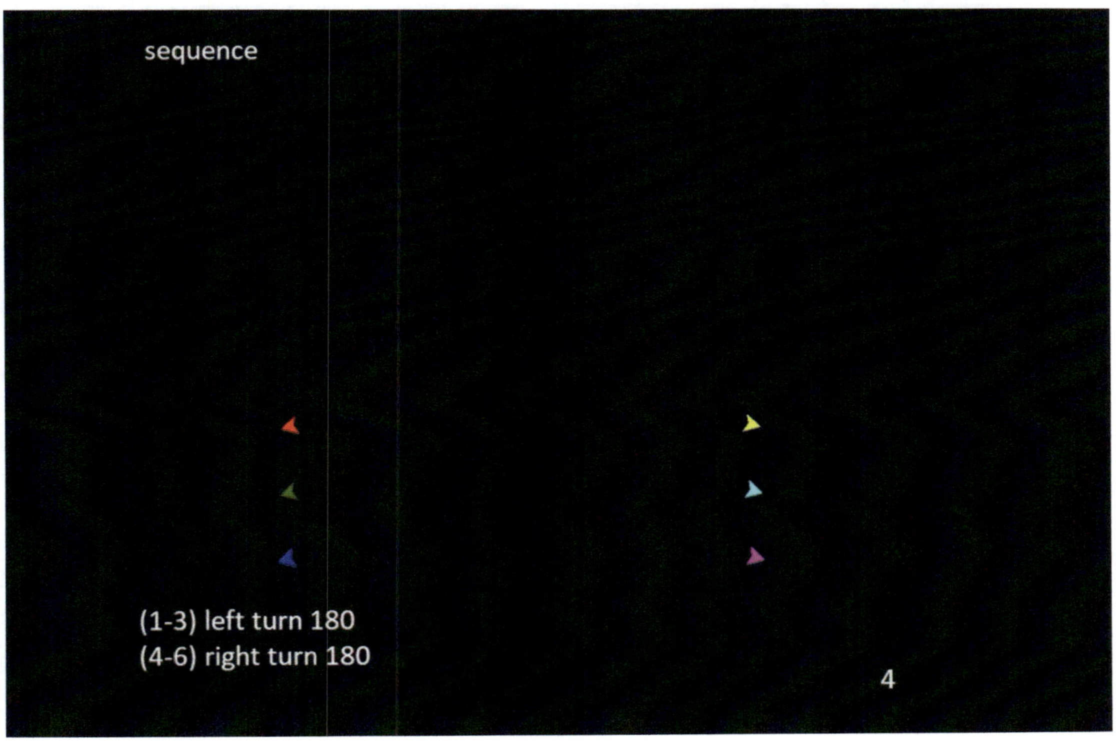
sequence
(1-3) left turn 180
(4-6) right turn 180
4

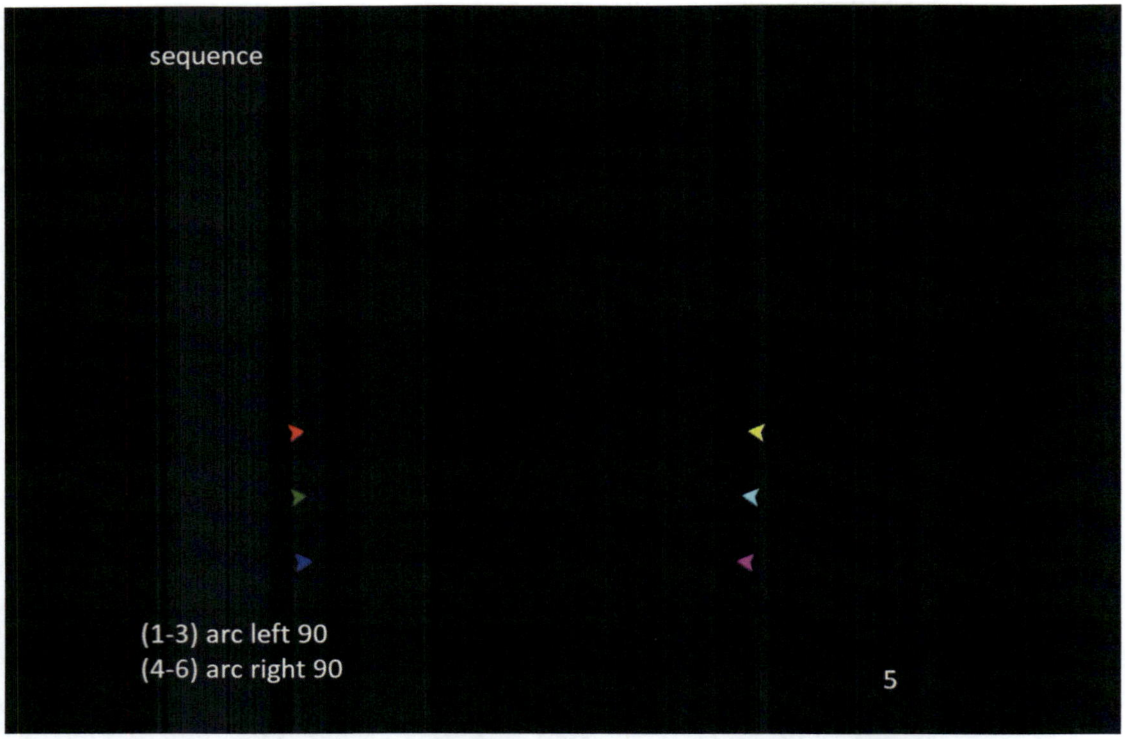
sequence
(1-3) arc left 90
(4-6) arc right 90
5

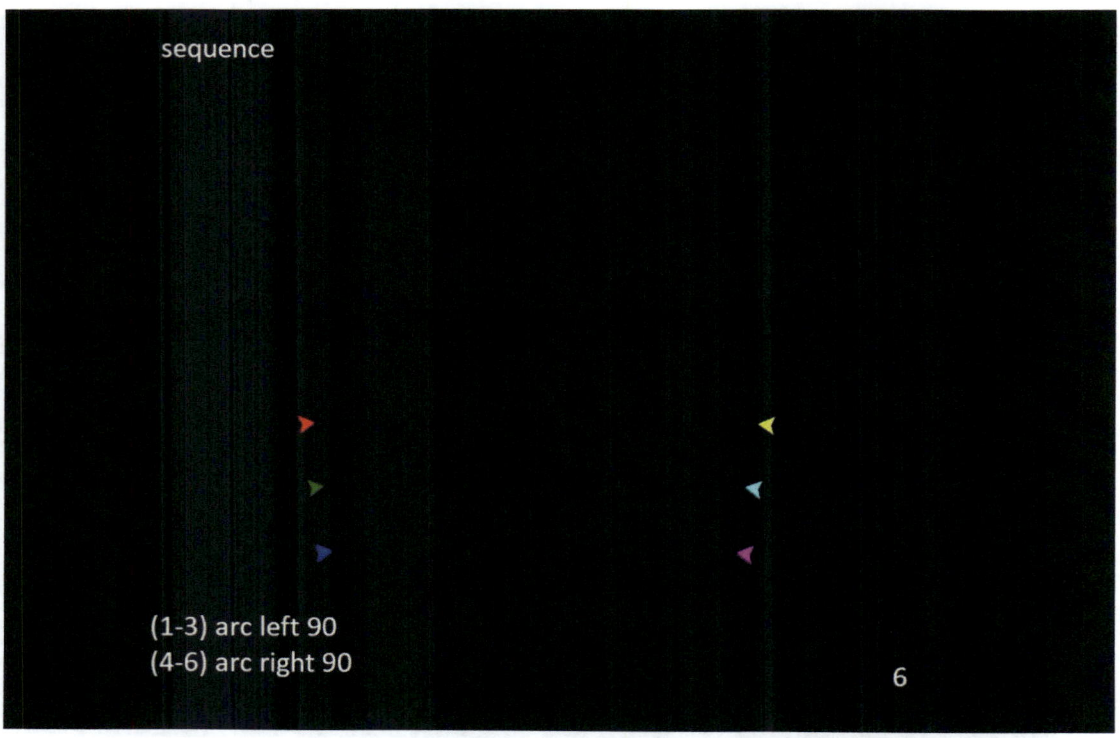
sequence
(1-3) arc left 90
(4-6) arc right 90
6

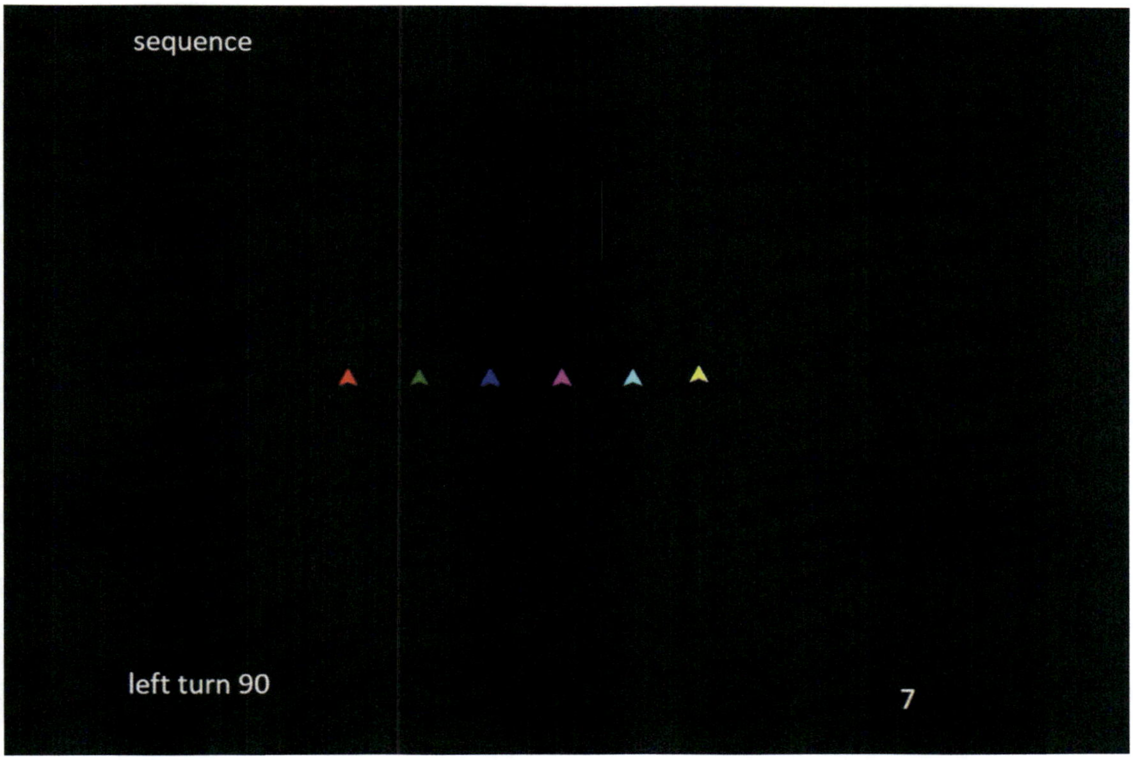
sequence
left turn 90
7

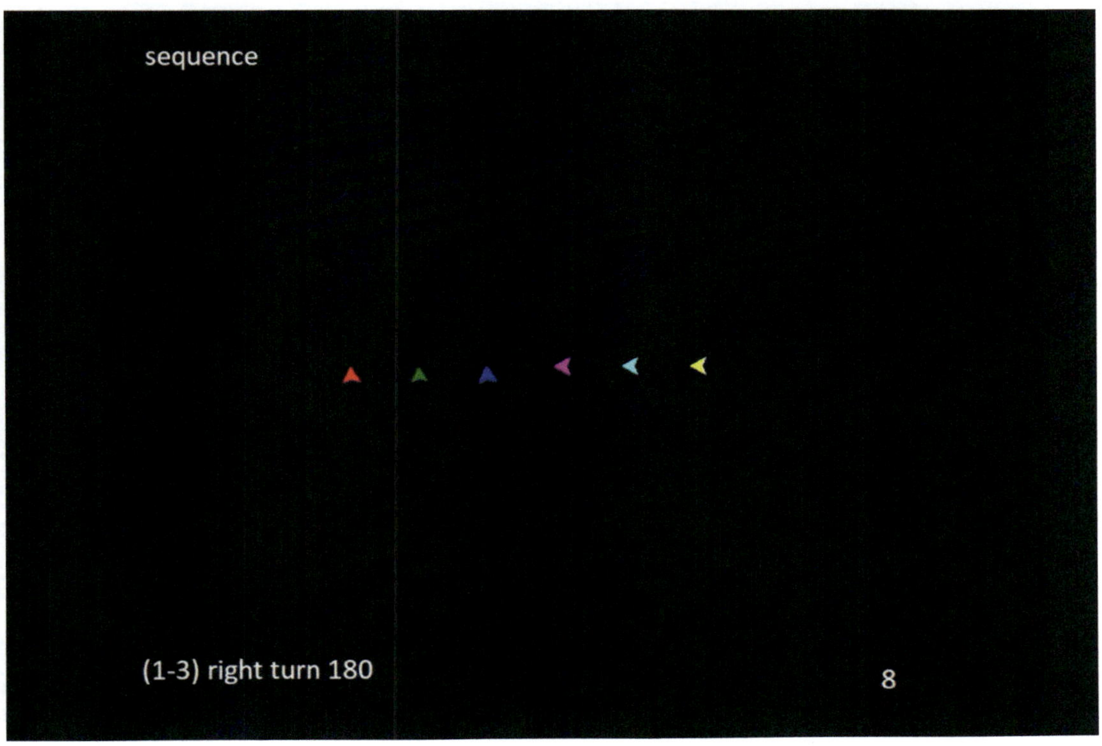
sequence
(1-3) right turn 180
8

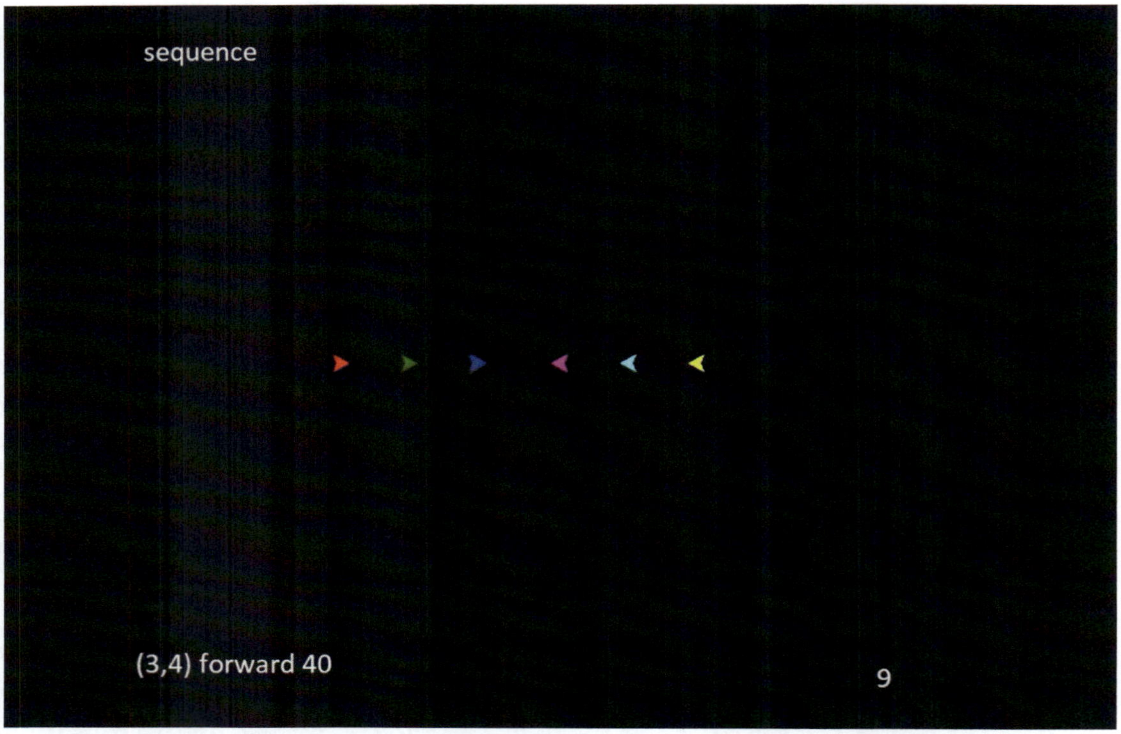
sequence
(3,4) forward 40
9

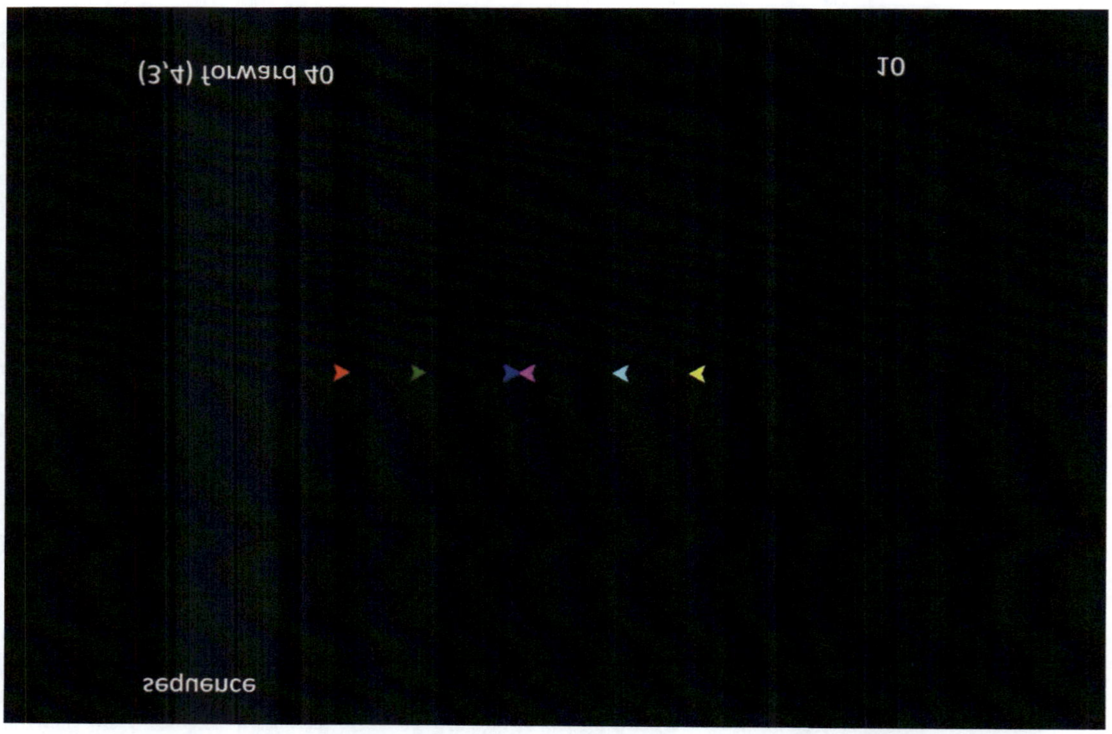
(3,4) forward 40
10
sequence

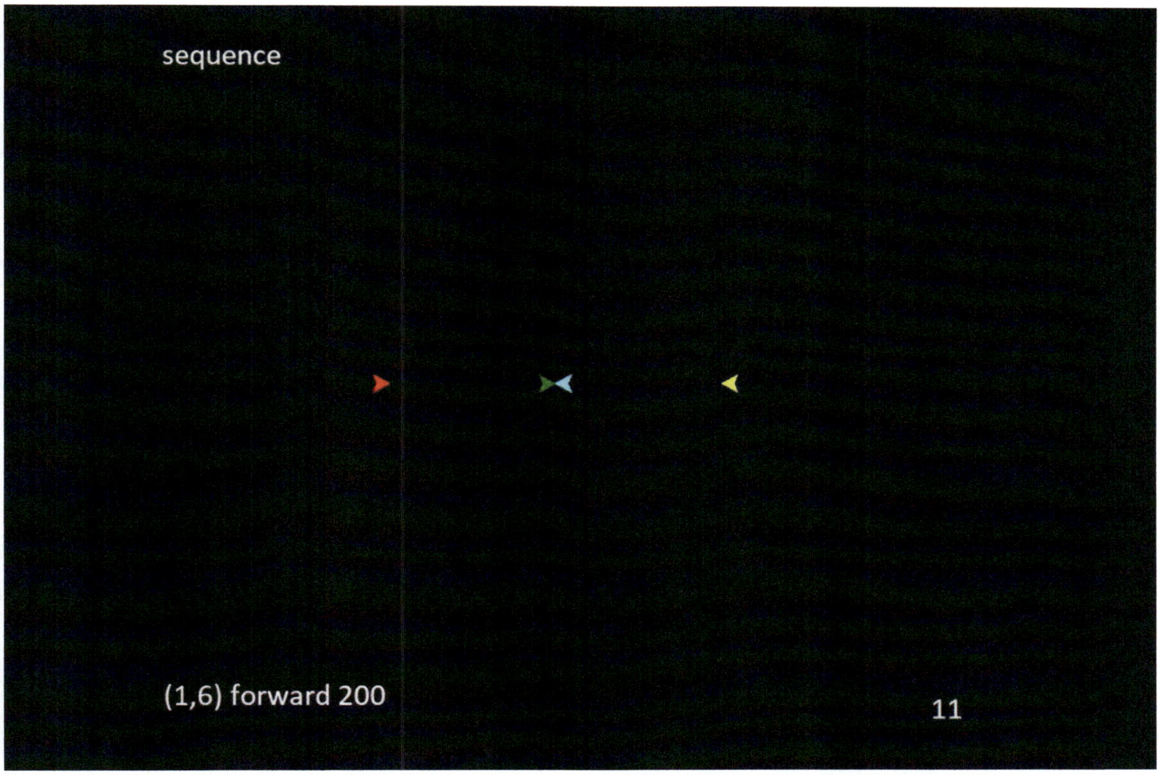
sequence
(1,6) forward 200
11

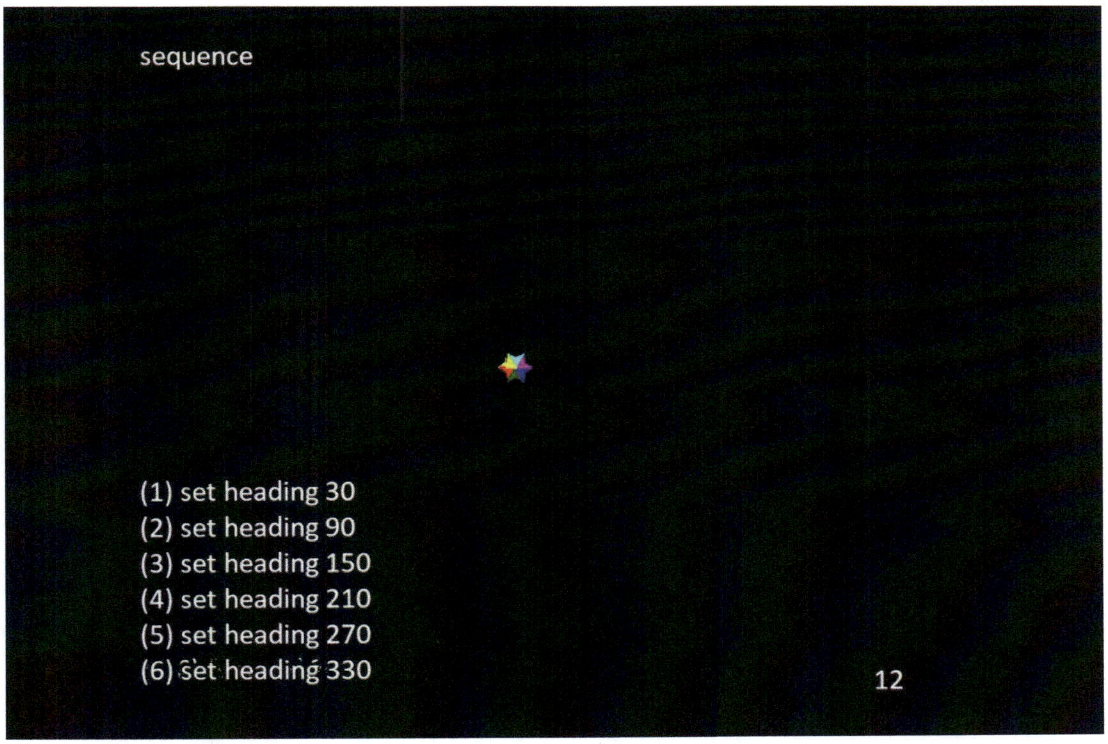
sequence
(1) set heading 30
(2) set heading 90
(3) set heading 150
(4) set heading 210
(5) set heading 270
(6) set heading 330
12

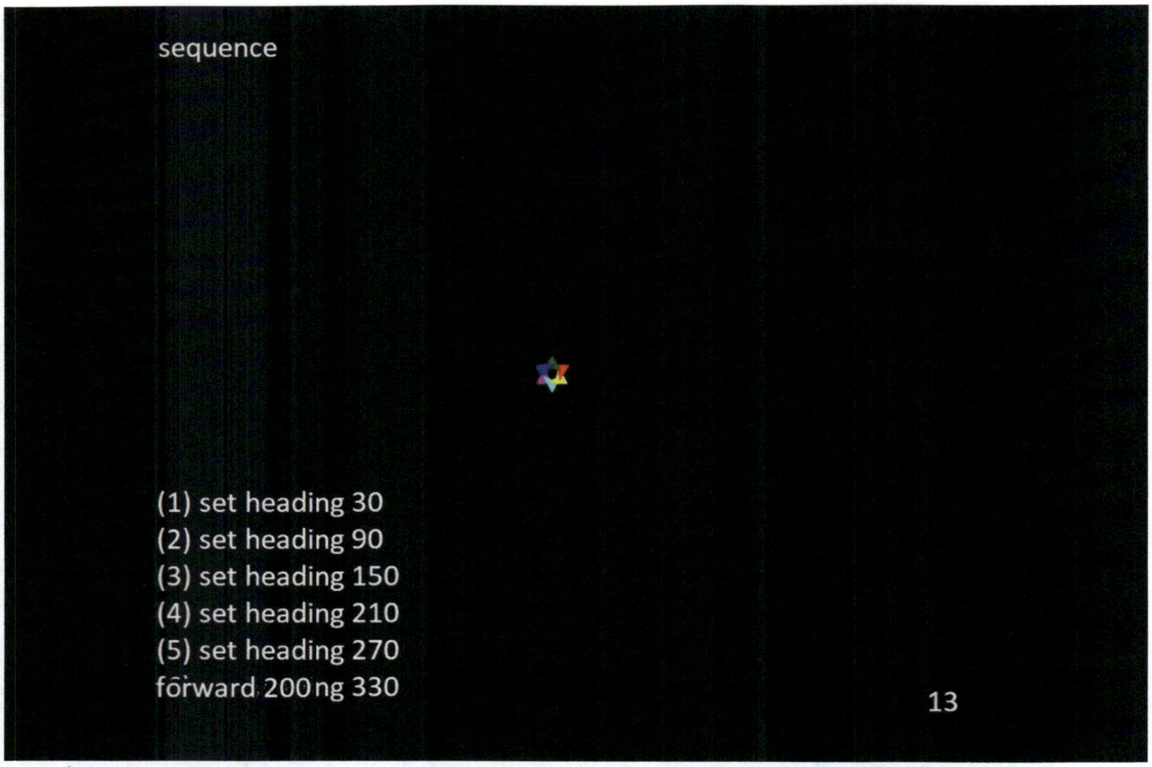
sequence
(1) set heading 30
(2) set heading 90
(3) set heading 150
(4) set heading 210
(5) set heading 270
forward 200ng 330
13

Looping

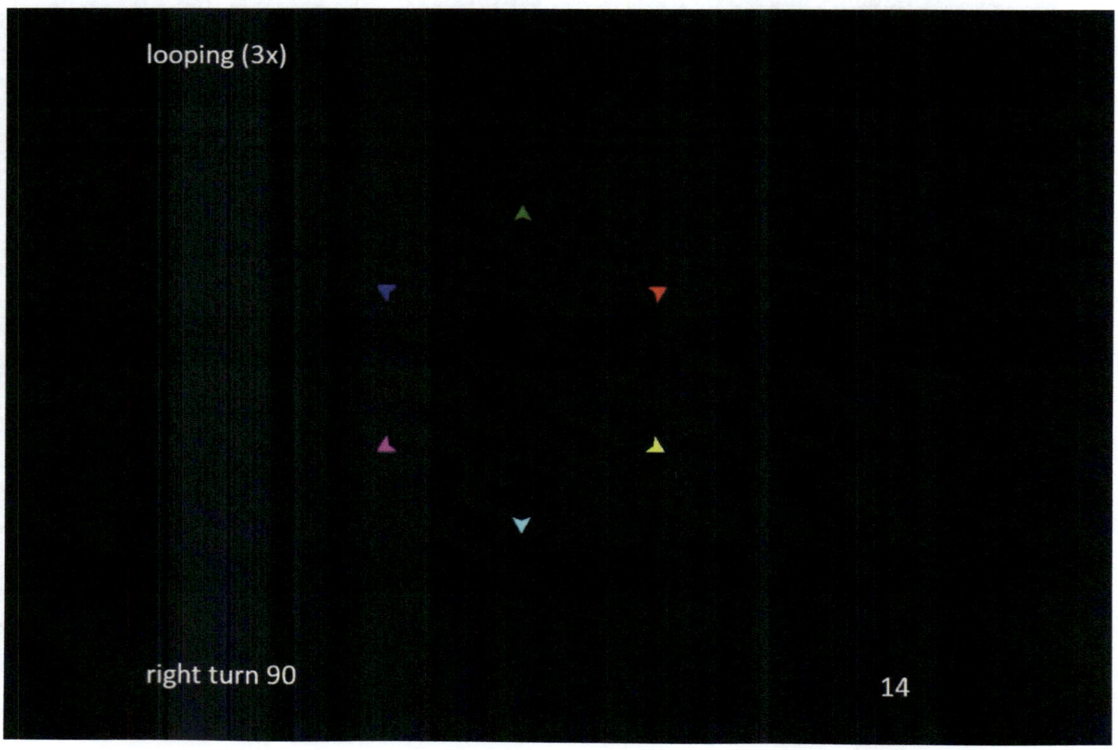
looping (3x)
right turn 90
14

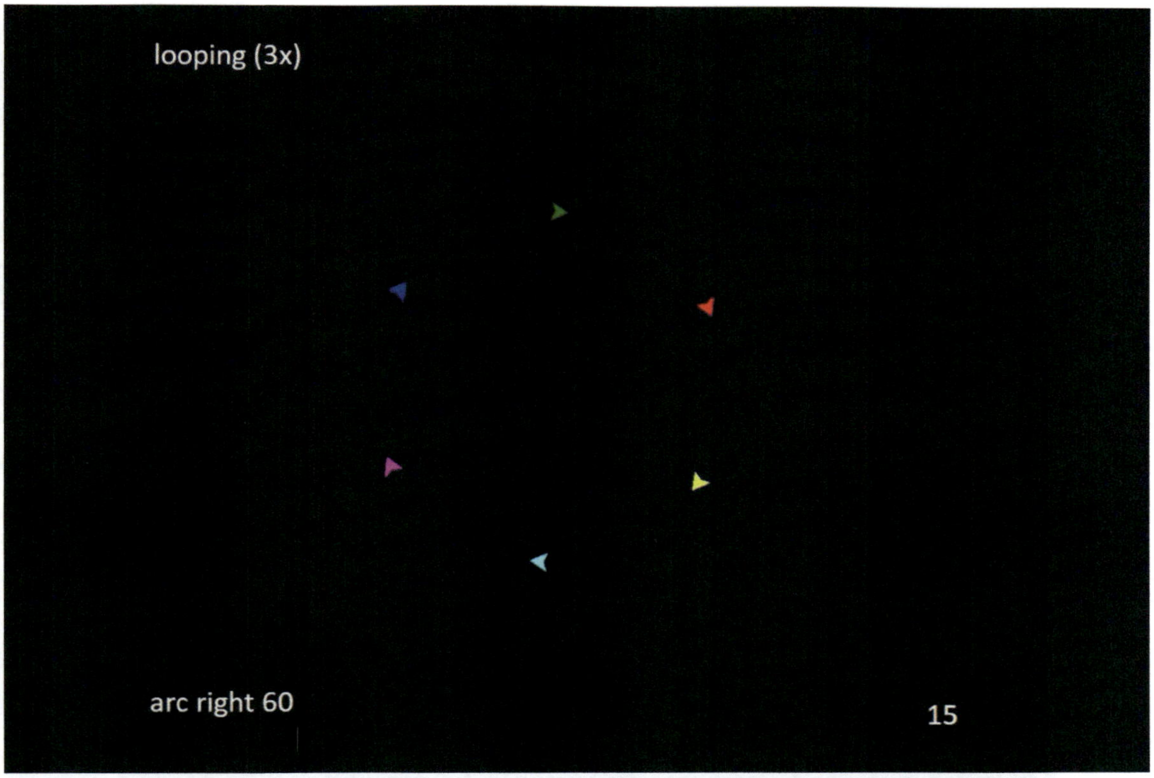
looping (3x)
arc right 60
15

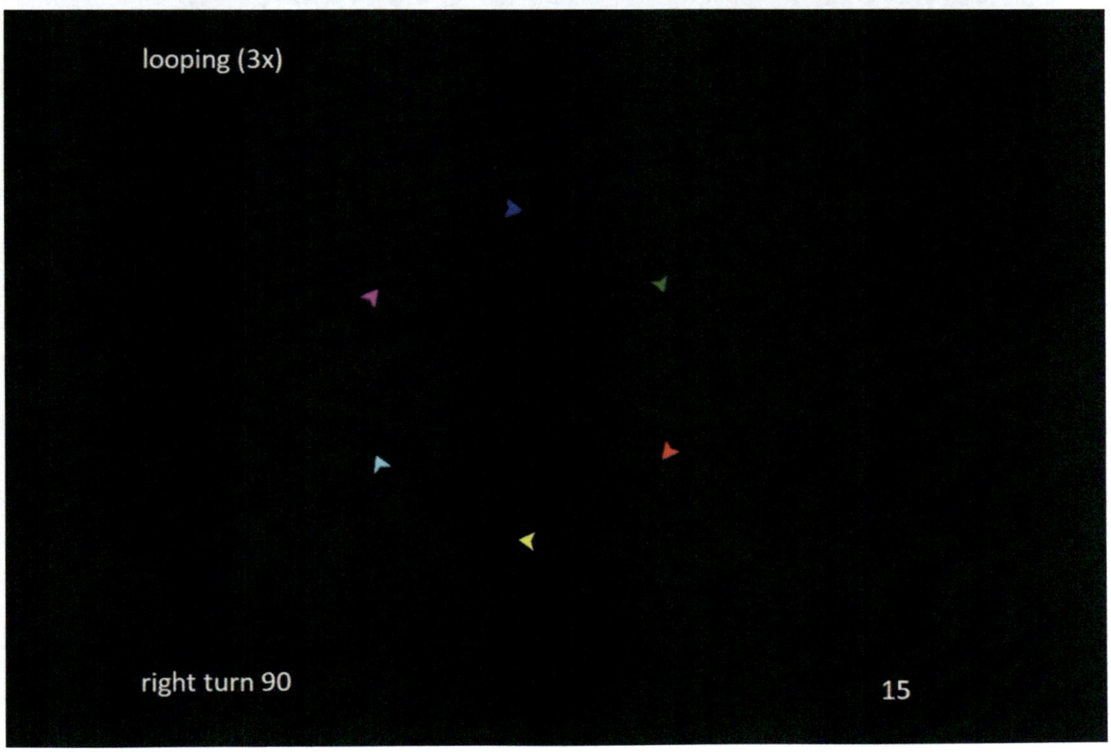
looping (3x)
right turn 90
15

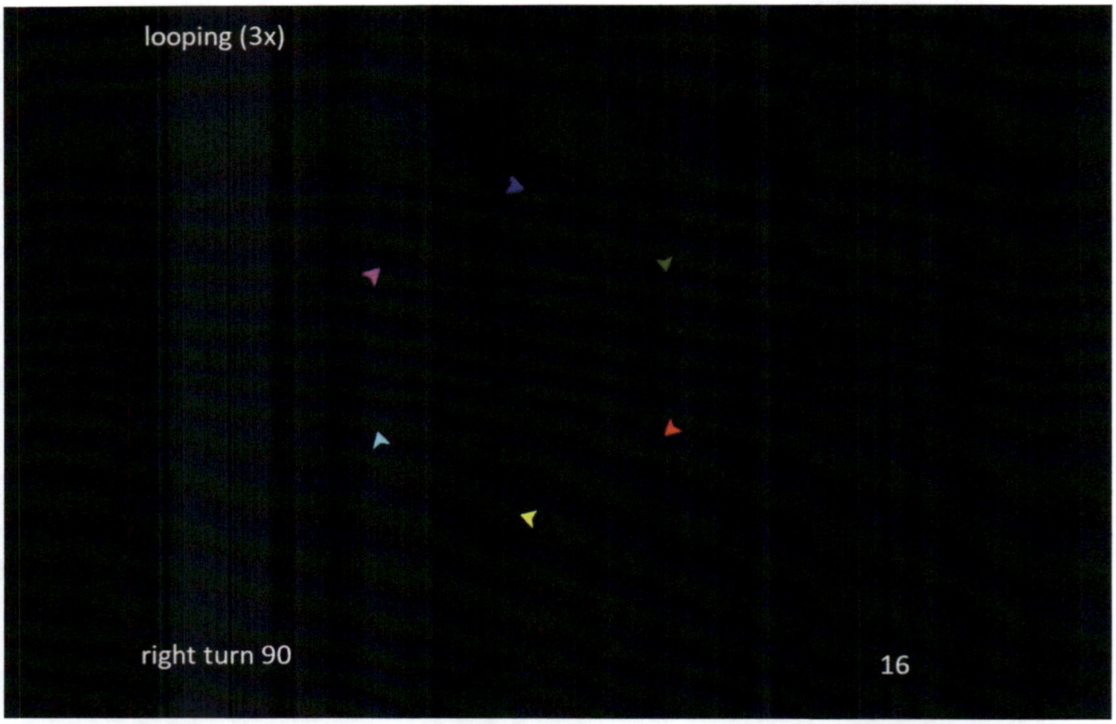
looping (3x)
right turn 90
16

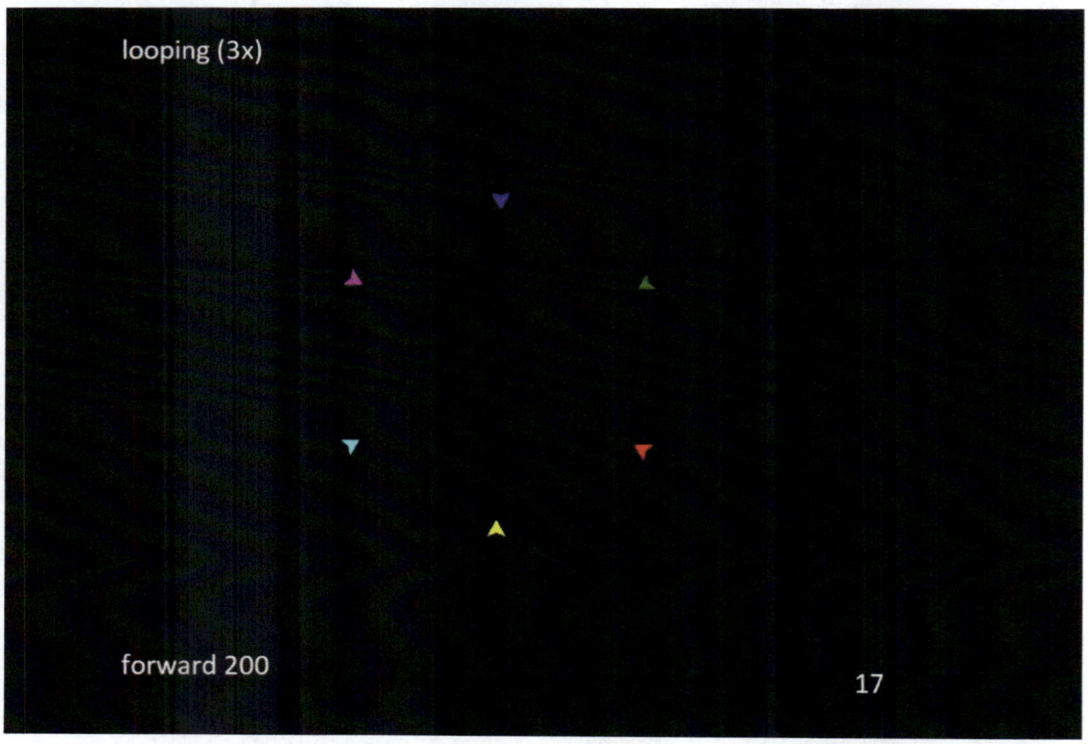
looping (3x)
forward 200
17

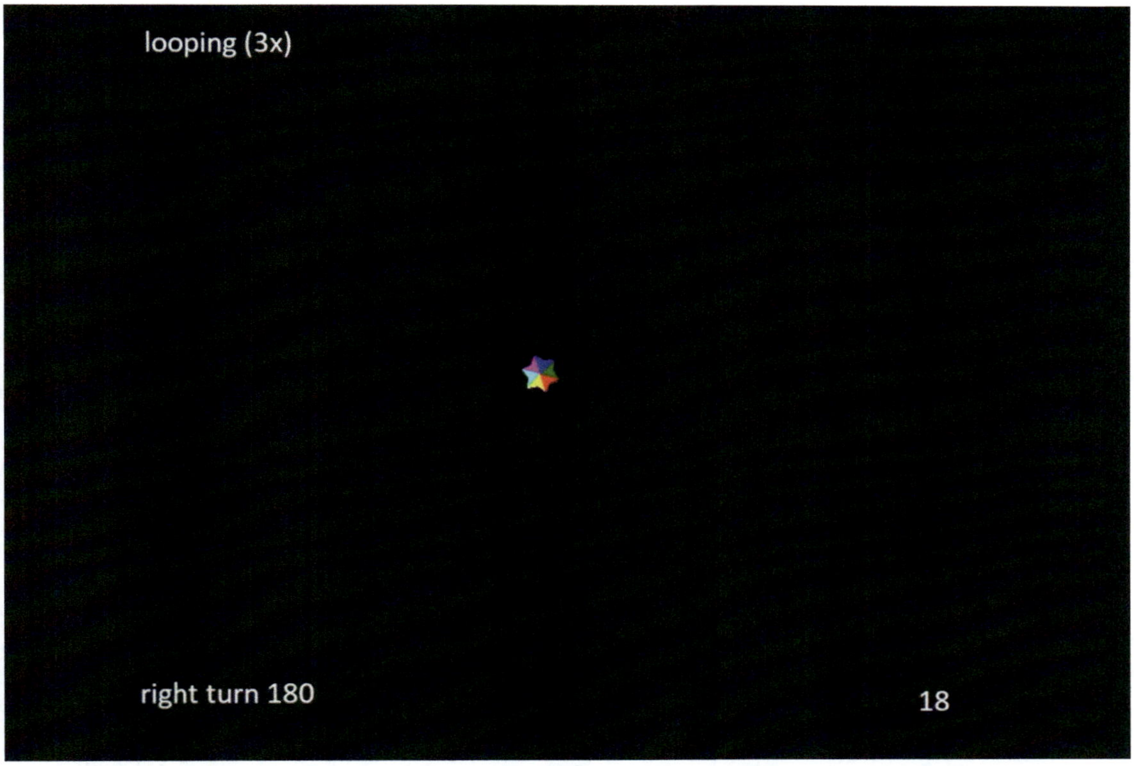
looping (3x)
right turn 180
18

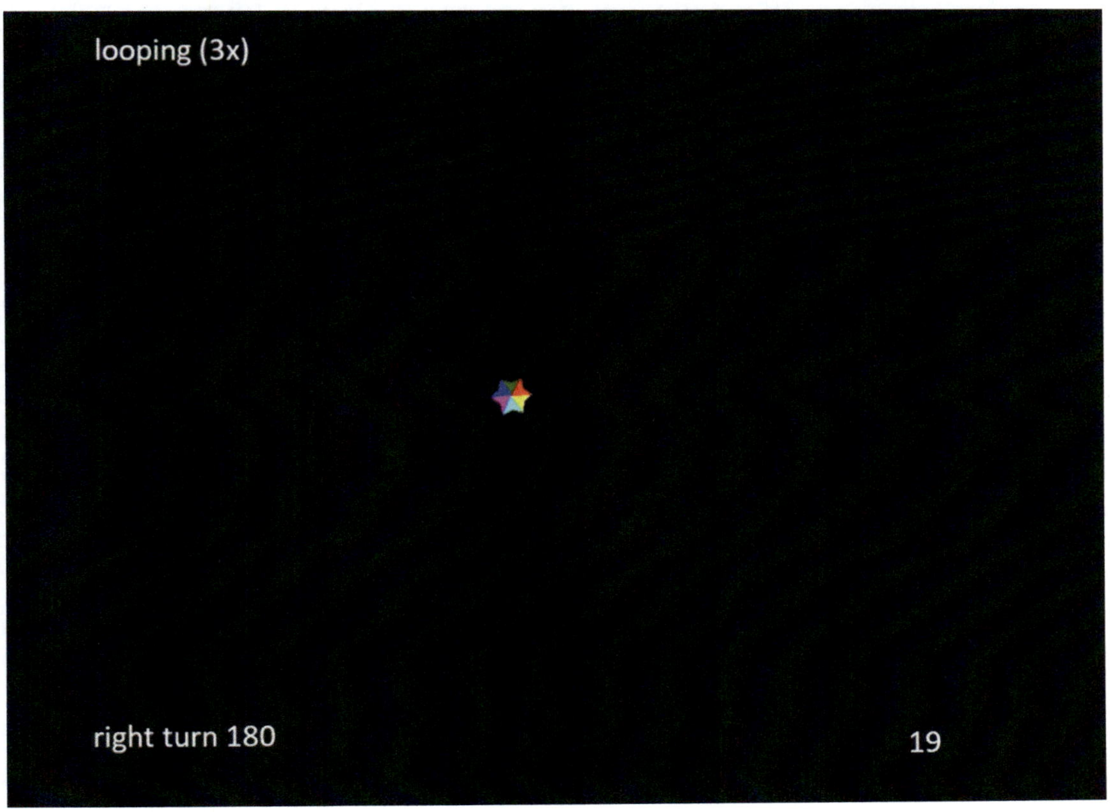
looping (3x)
right turn 180
19

Sequence

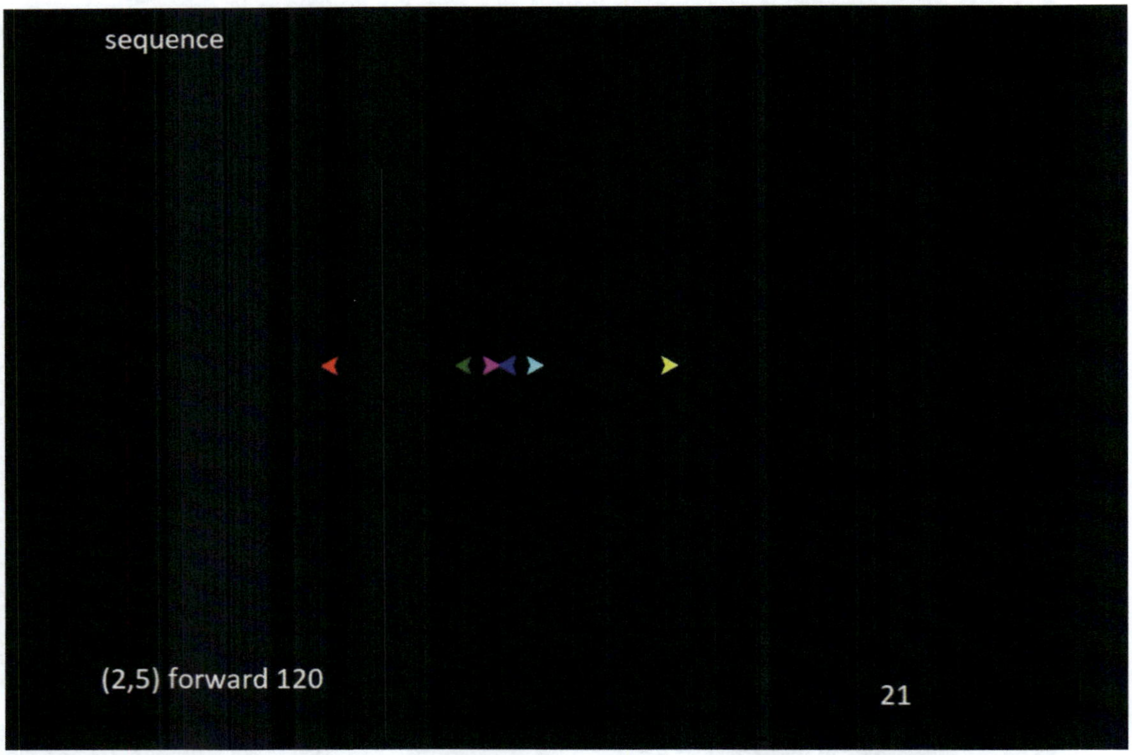

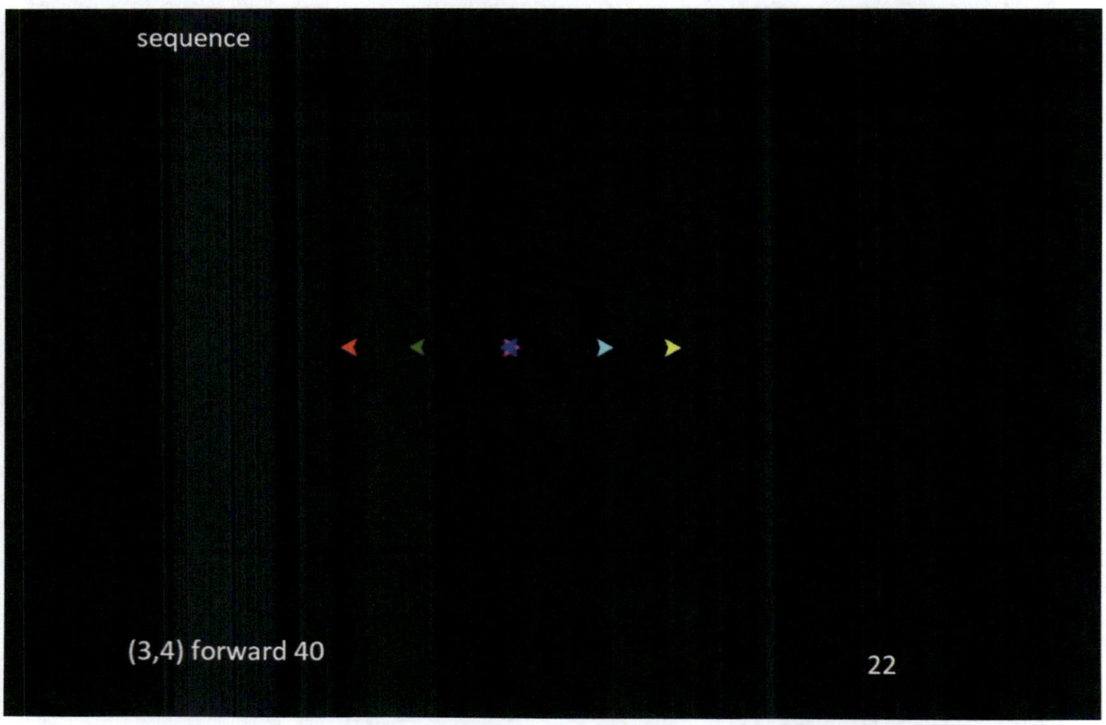

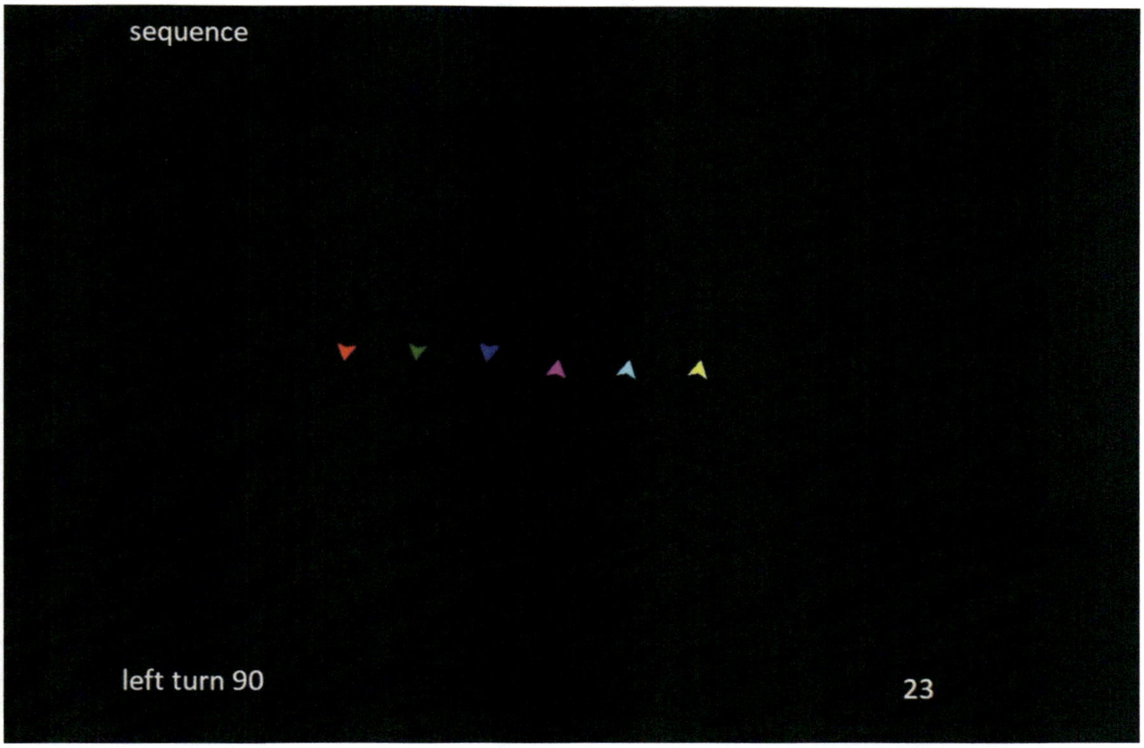
sequence
left turn 90
23

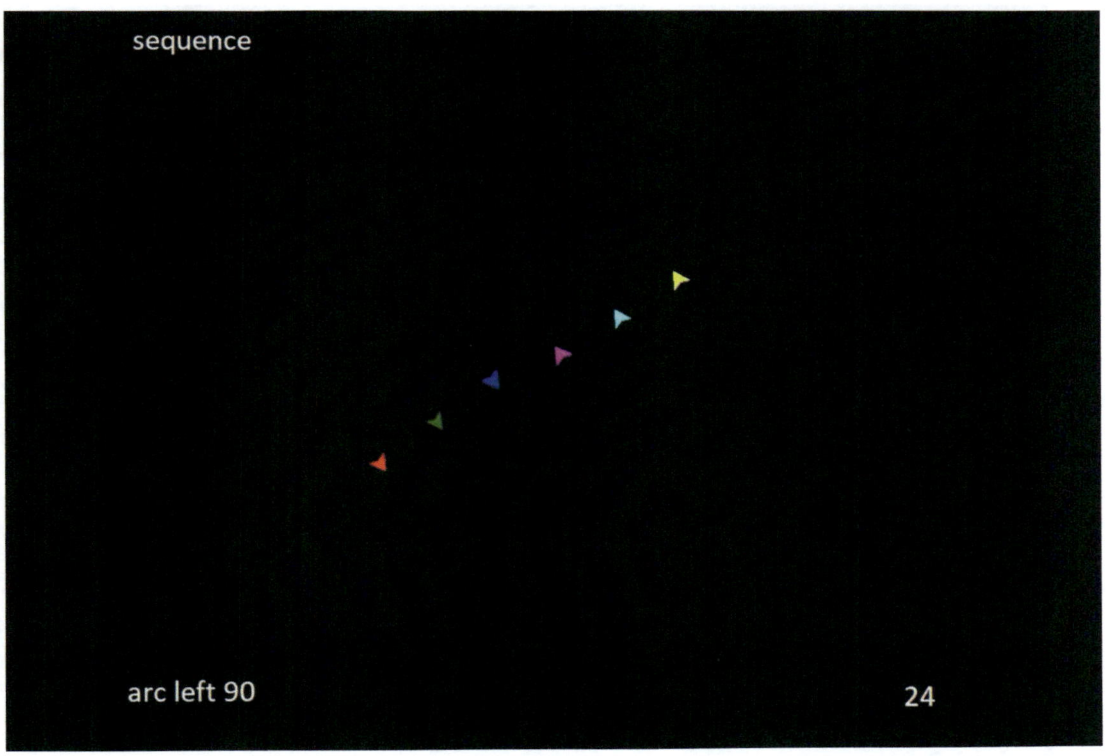
sequence
arc left 90
24

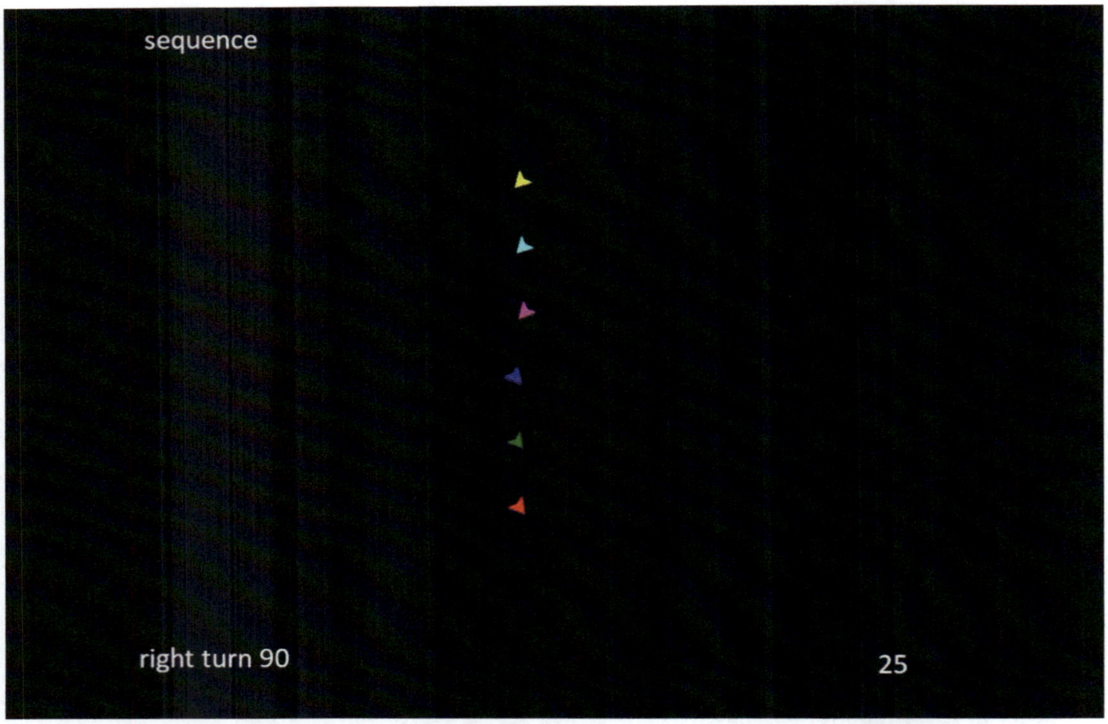
sequence
right turn 90
25

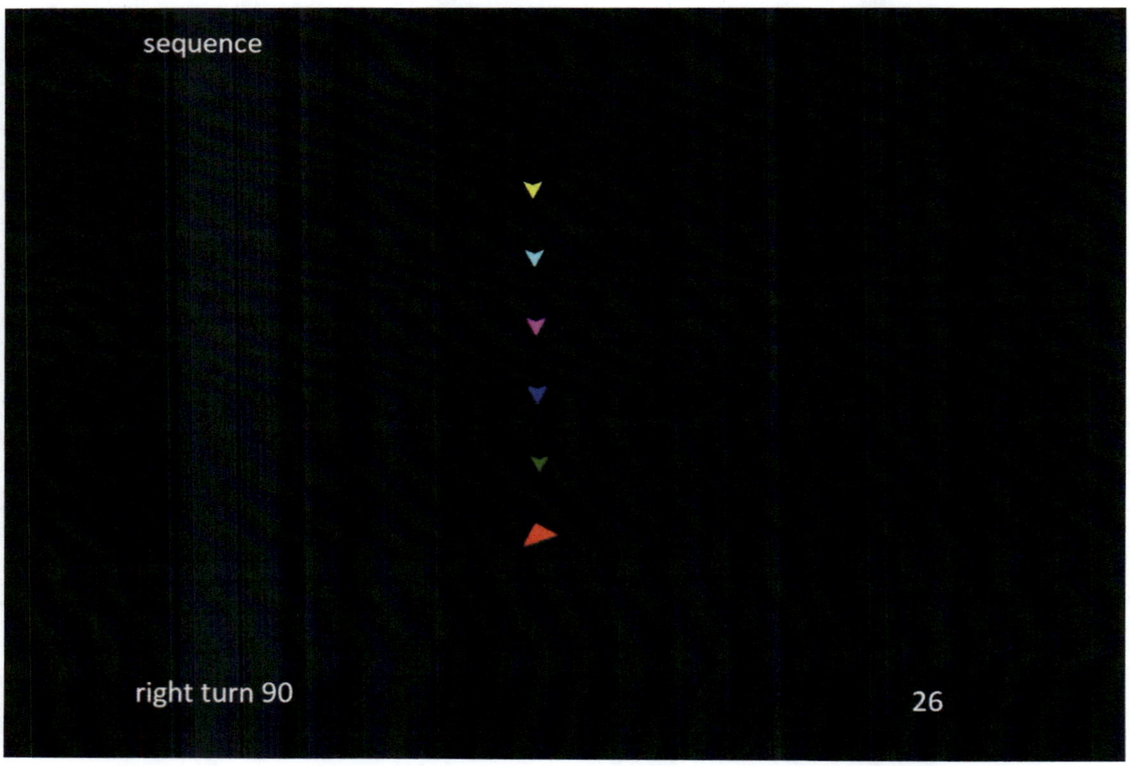
sequence
right turn 90
26

Conditional Execution

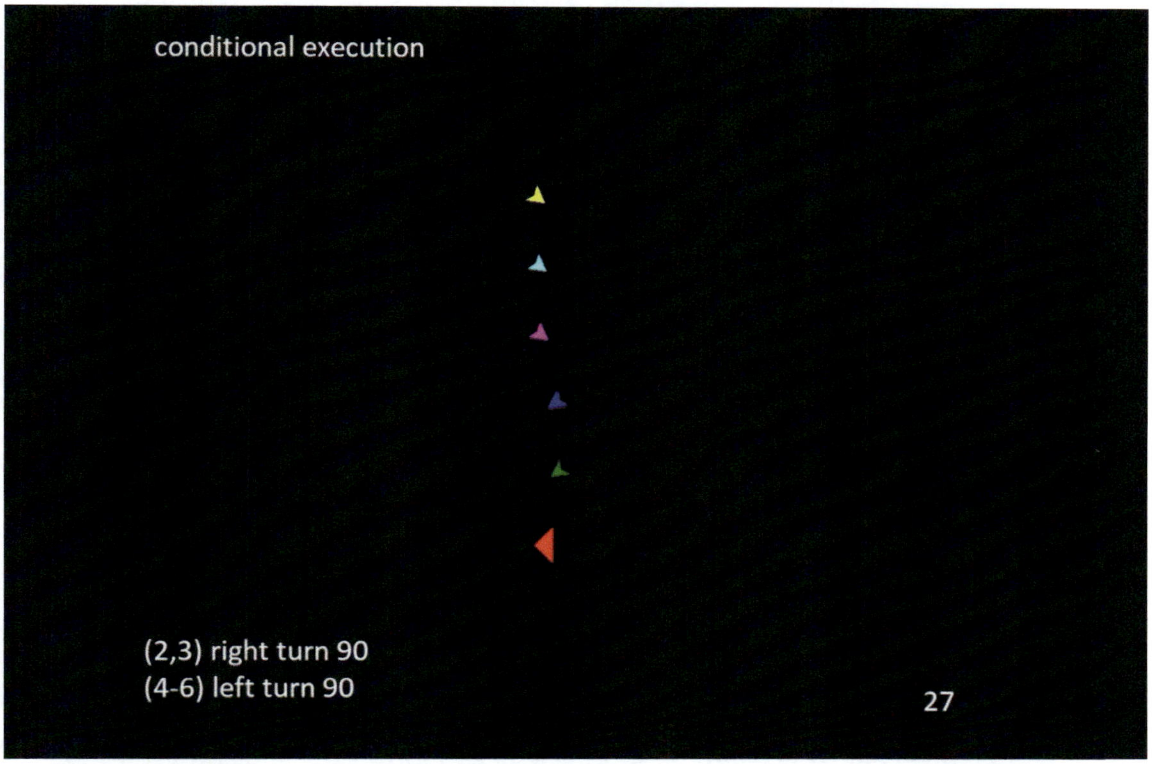

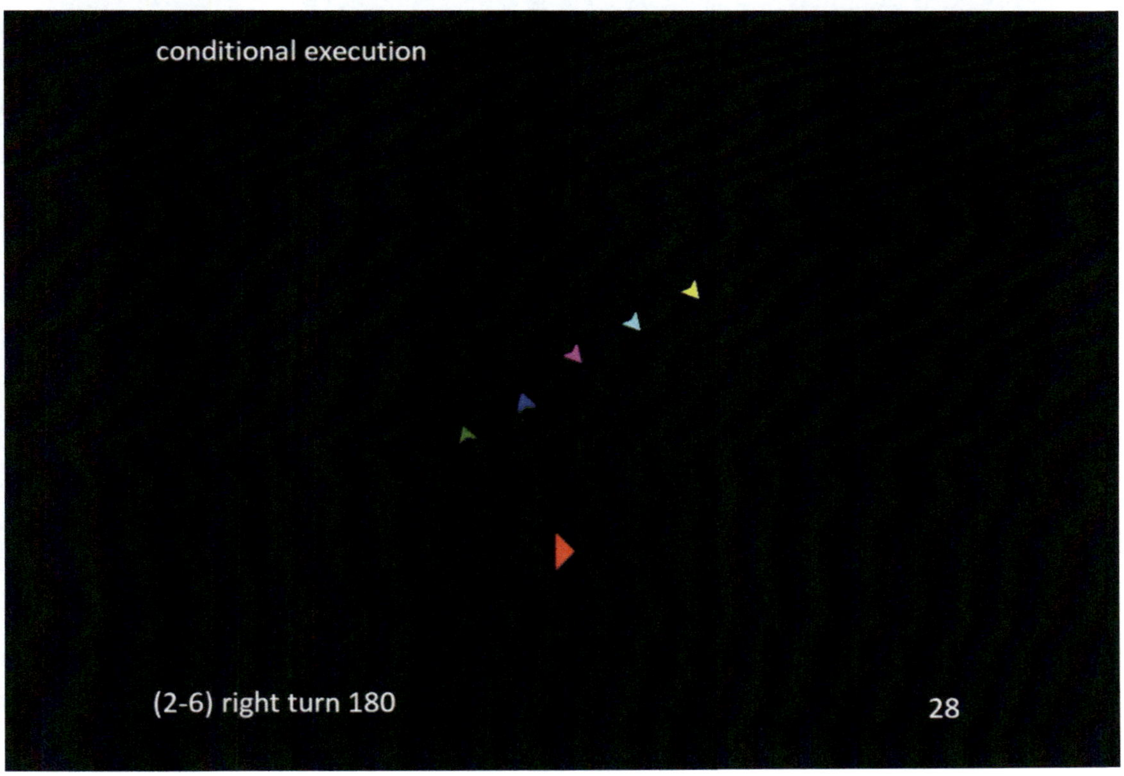

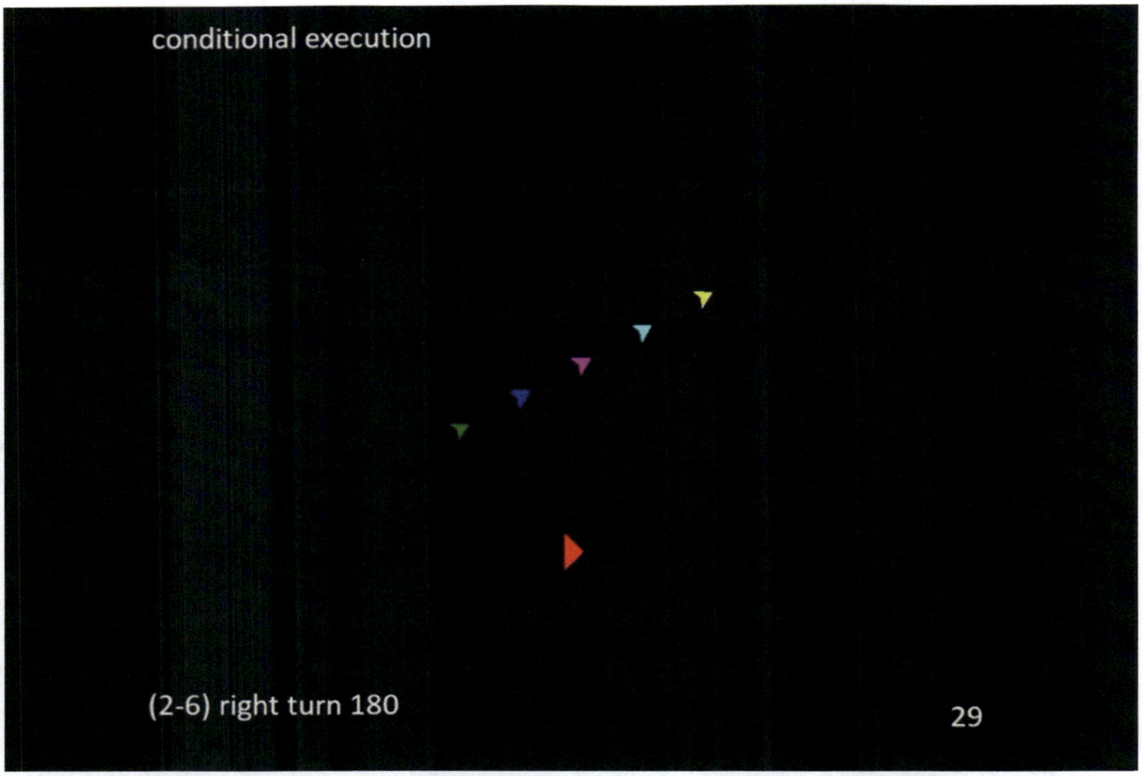
conditional execution
(2-6) right turn 180
29

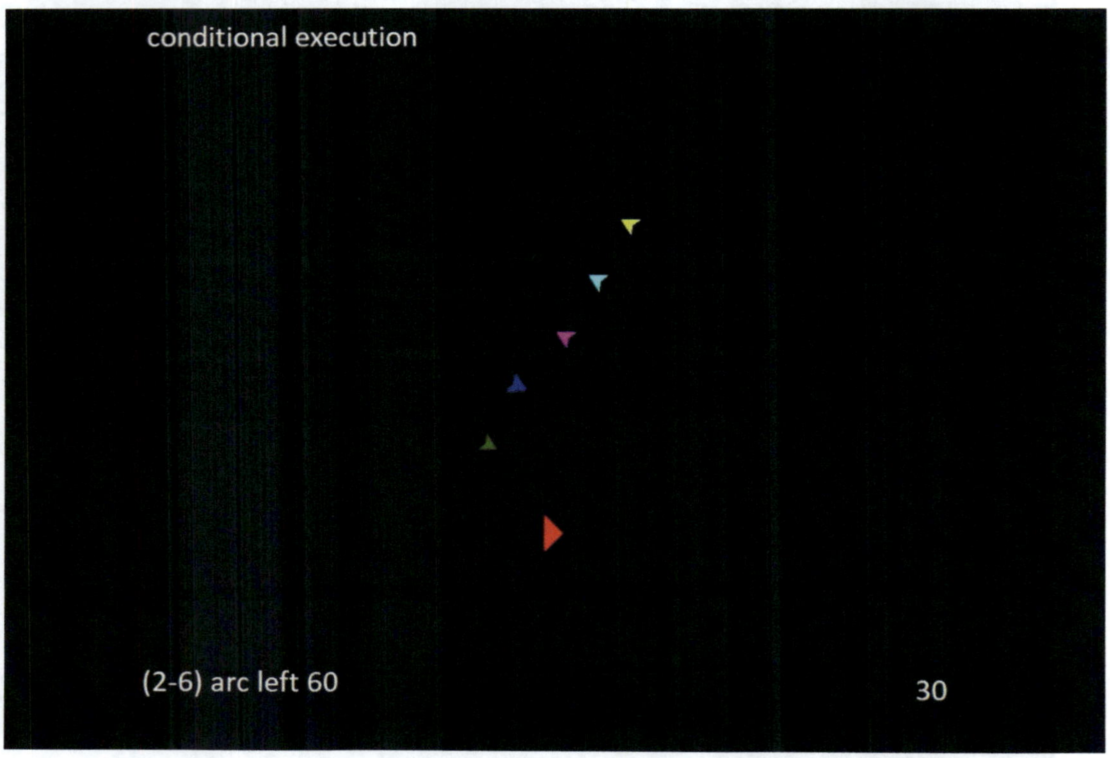
conditional execution
(2-6) arc left 60
30

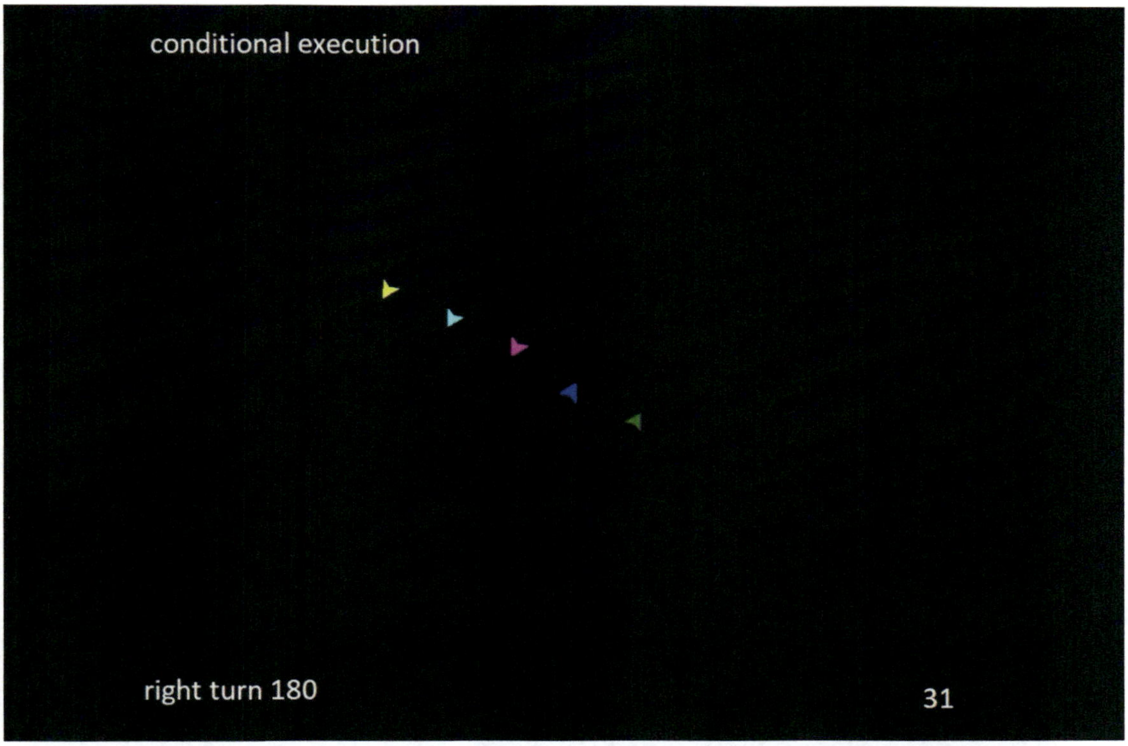
conditional execution
right turn 180
31

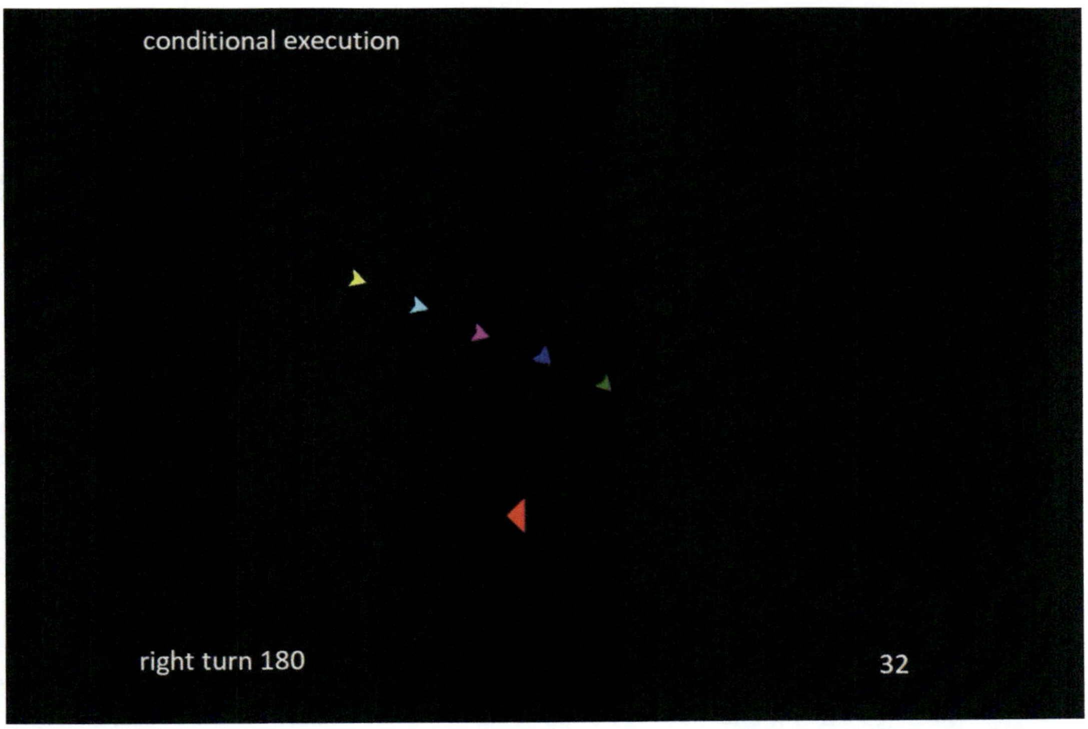
conditional execution
right turn 180
32

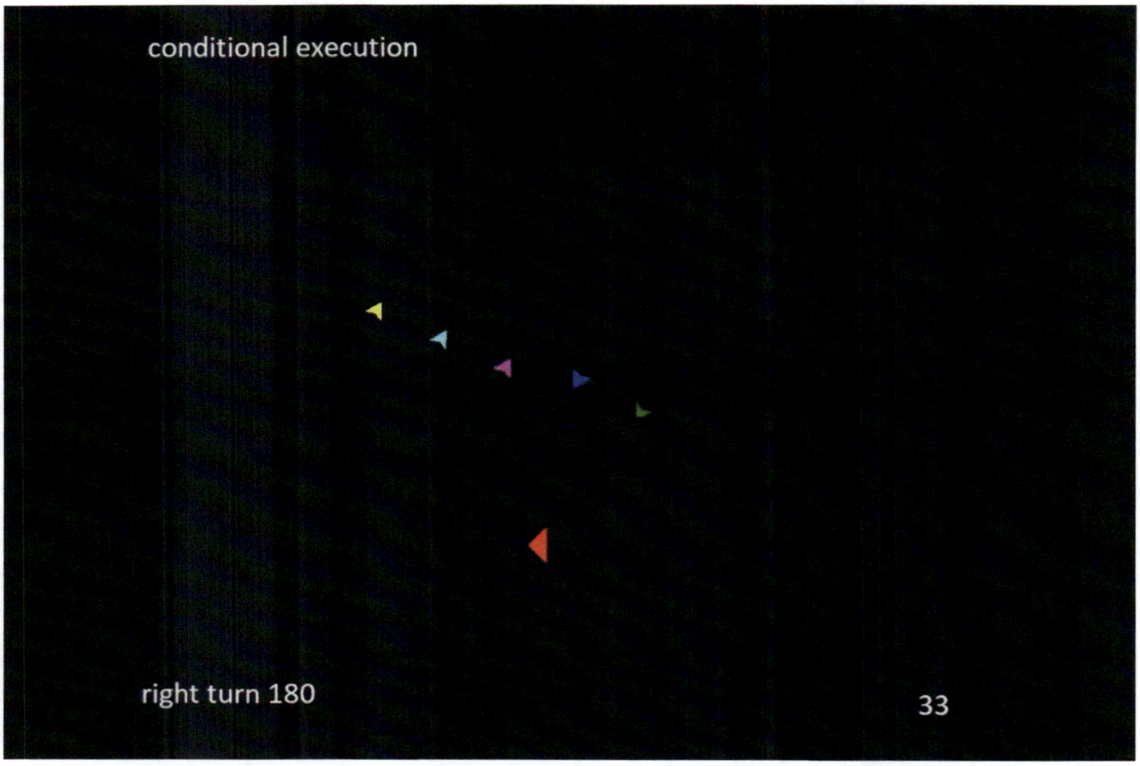
conditional execution
right turn 180
33

Sequence

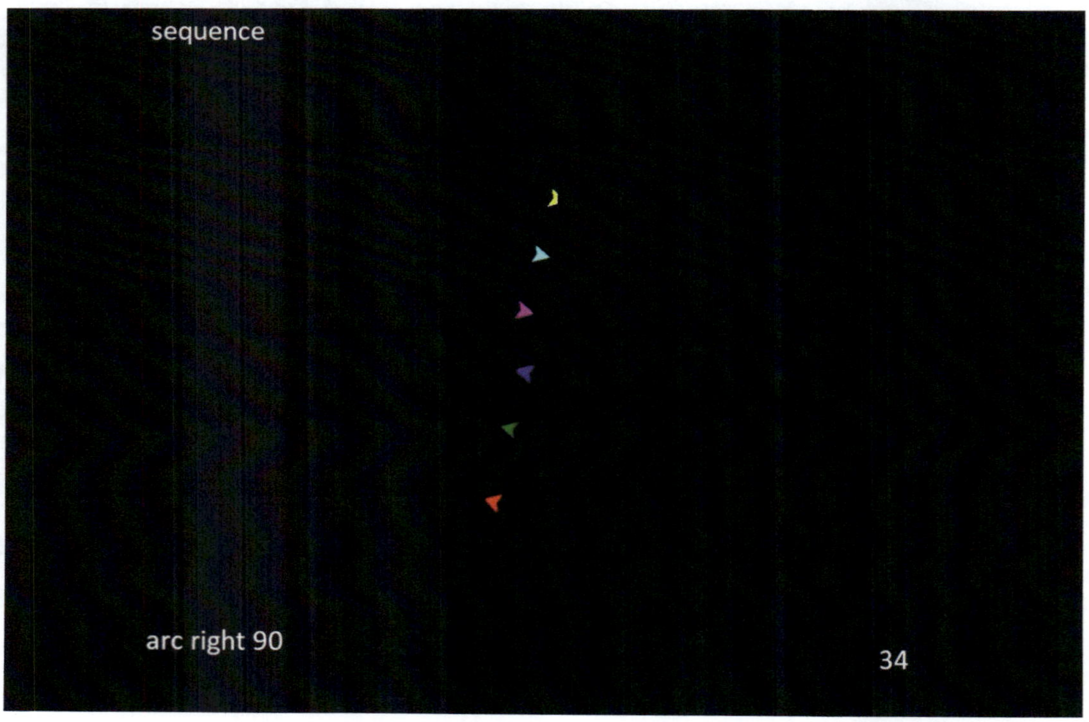
sequence
arc right 90
34

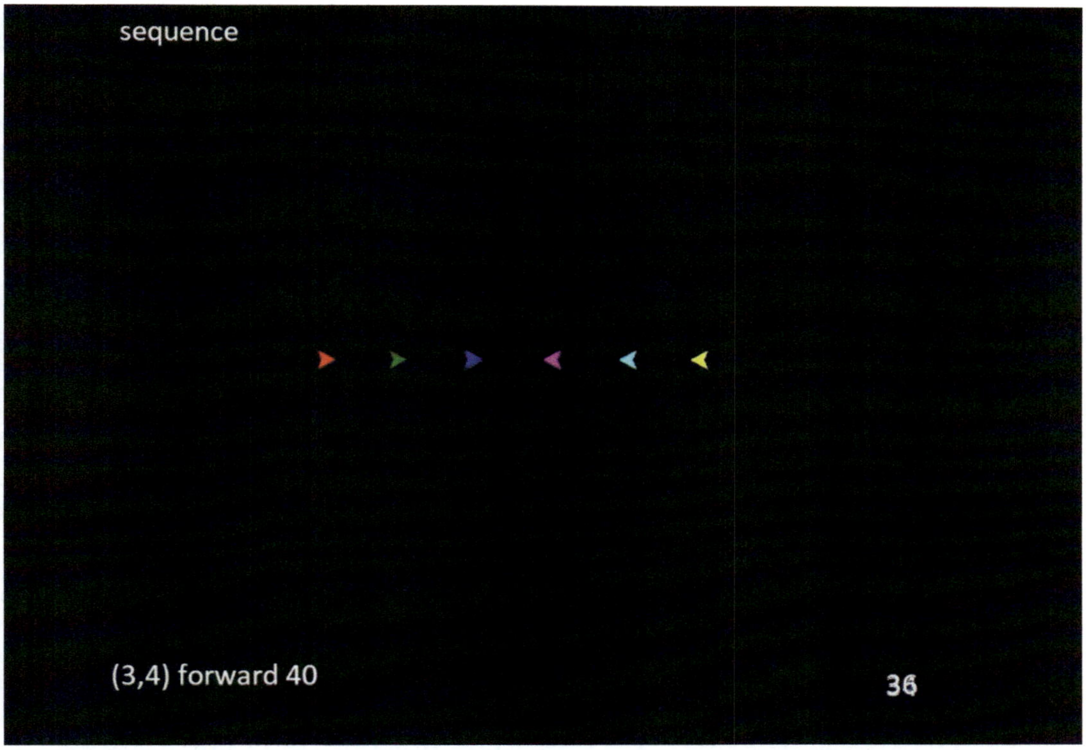
sequence
(3,4) forward 40
36

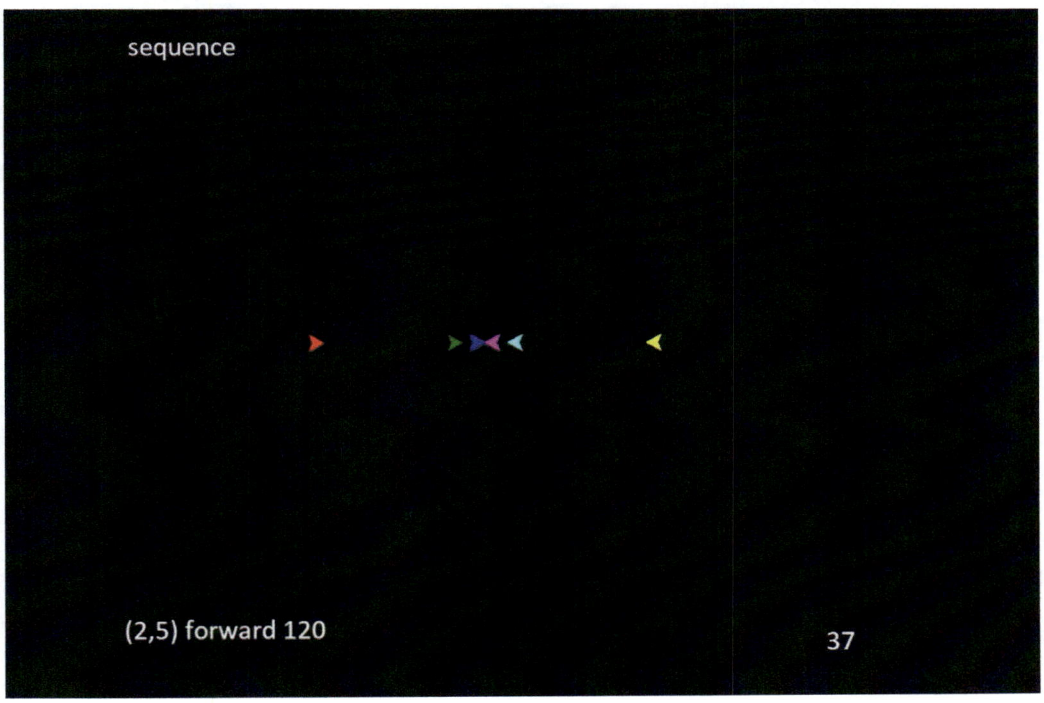
sequence
(2,5) forward 120
37

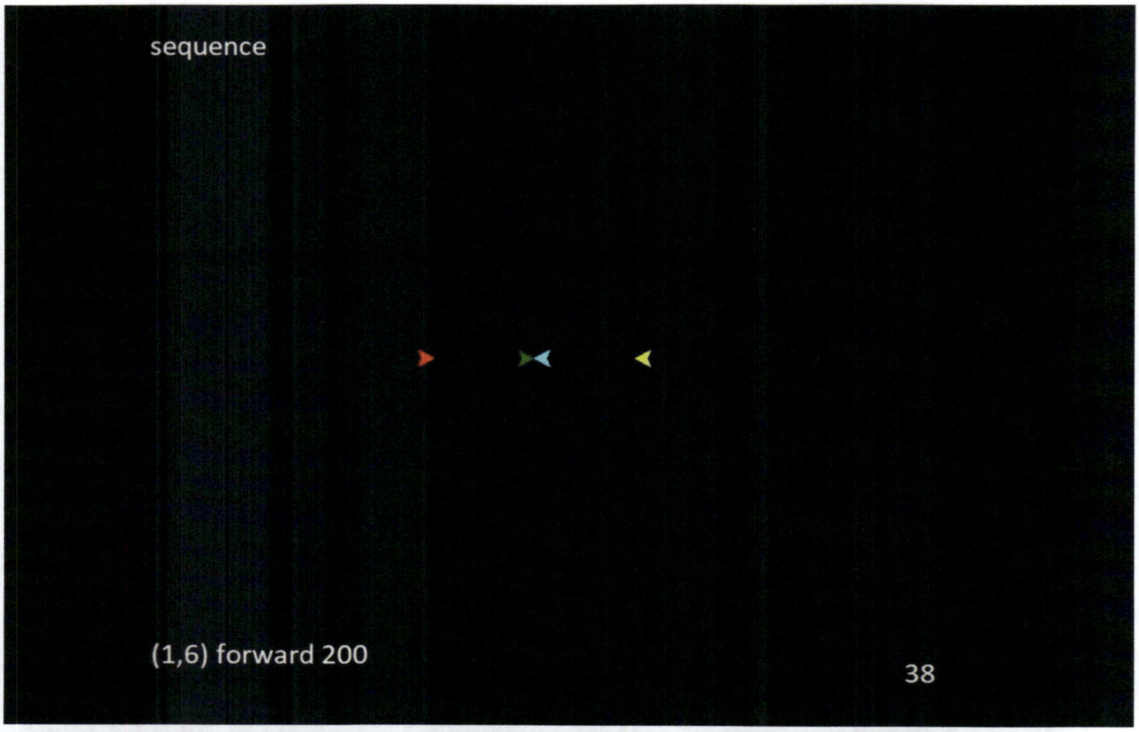
sequence
(1,6) forward 200
38

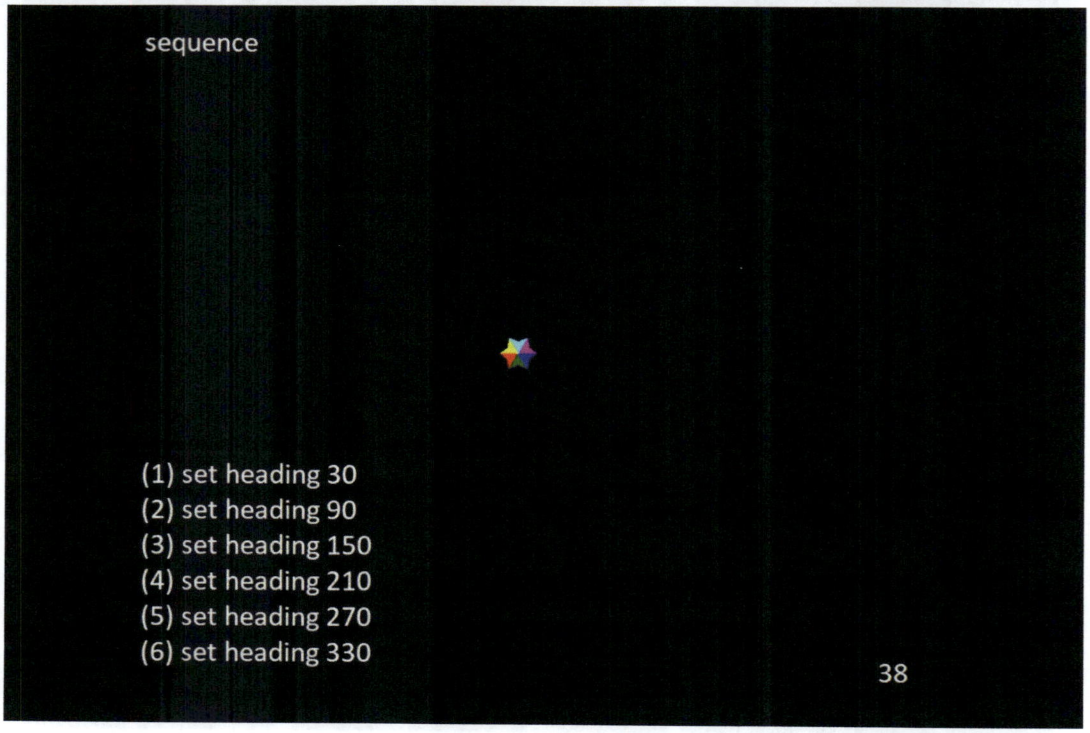
sequence
(1) set heading 30
(2) set heading 90
(3) set heading 150
(4) set heading 210
(5) set heading 270
(6) set heading 330
38

Numeric Computation

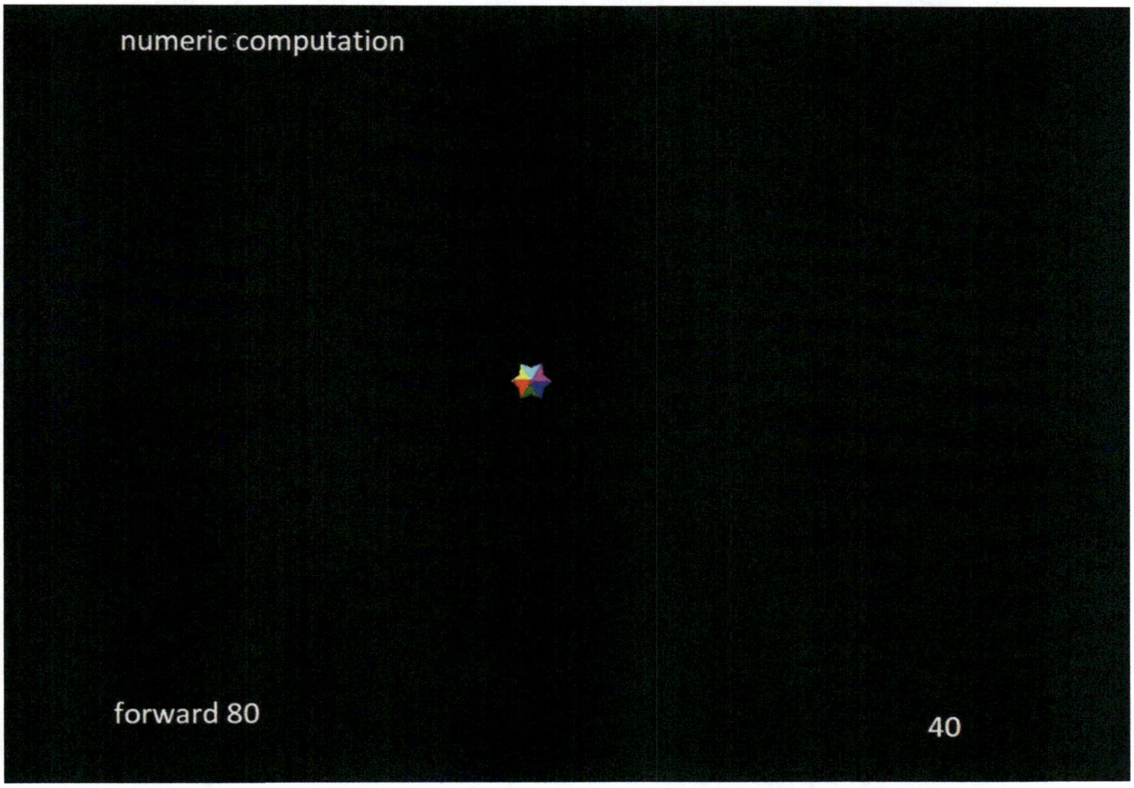

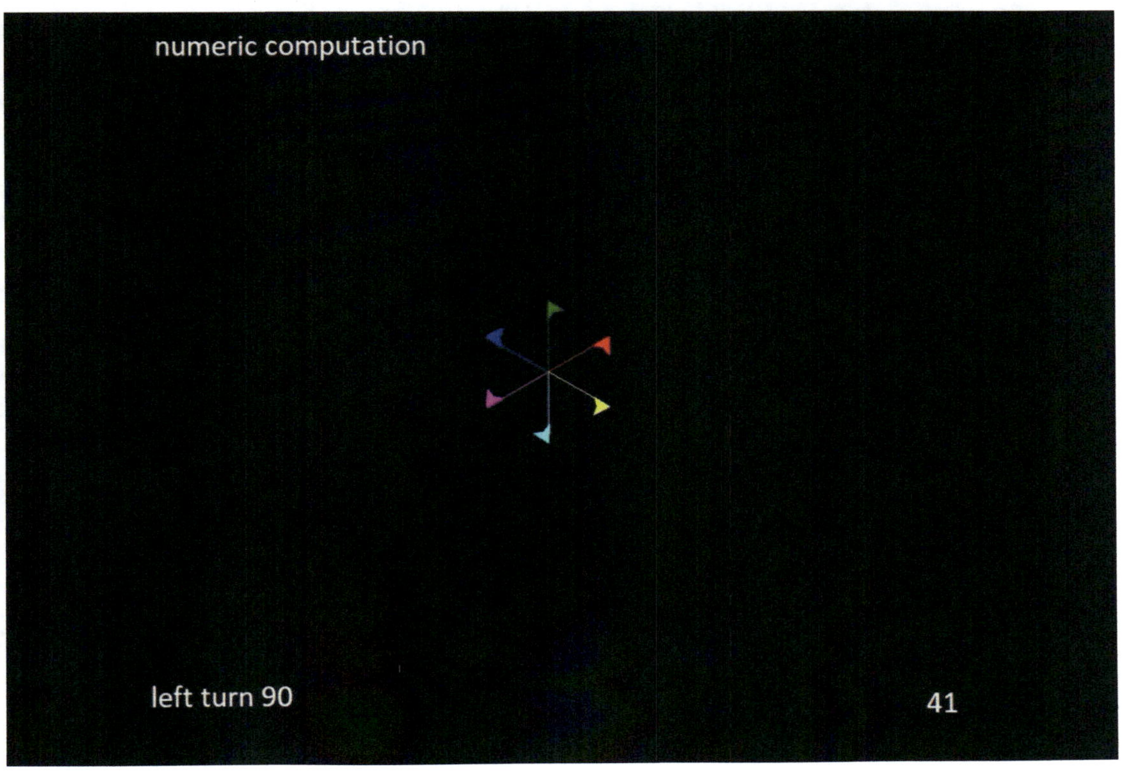

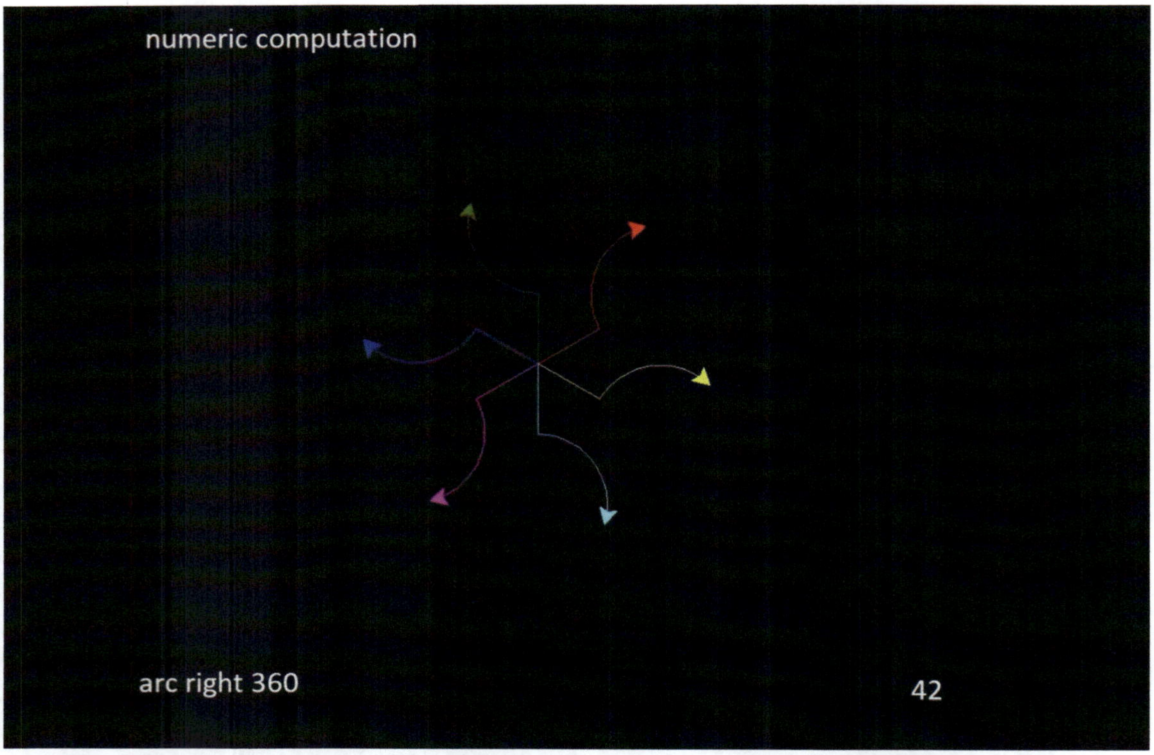
numeric computation
arc right 360
42

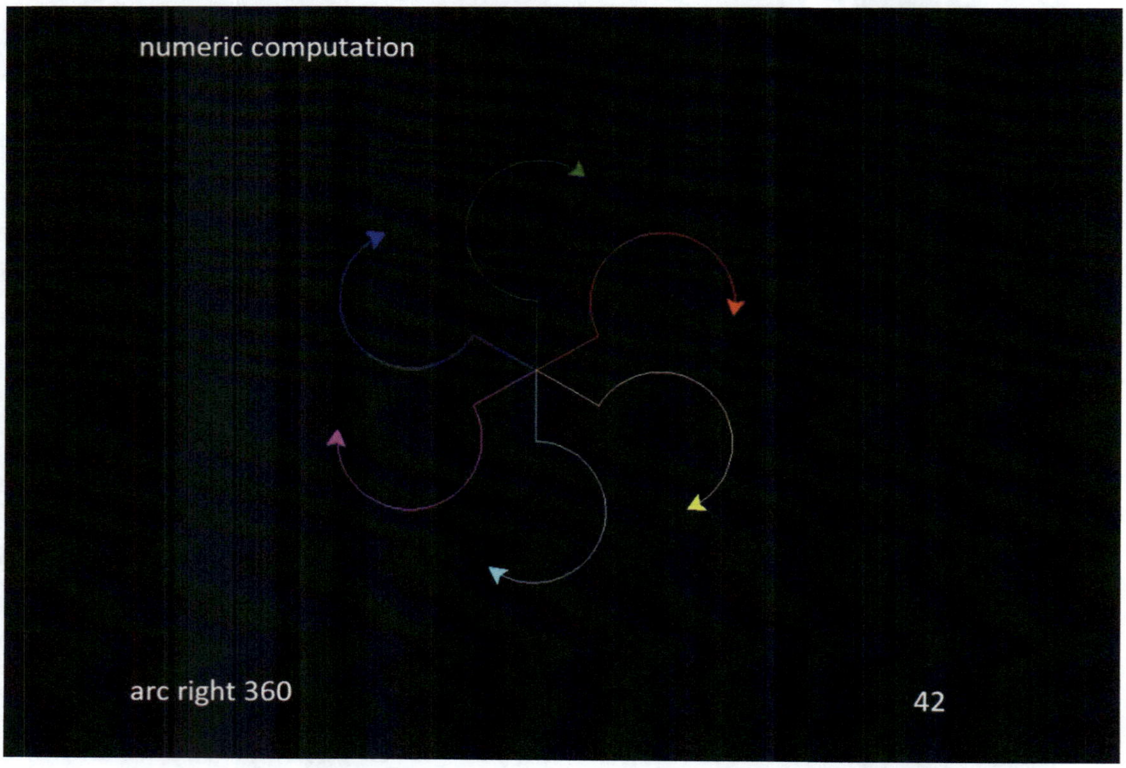
numeric computation
arc right 360
42

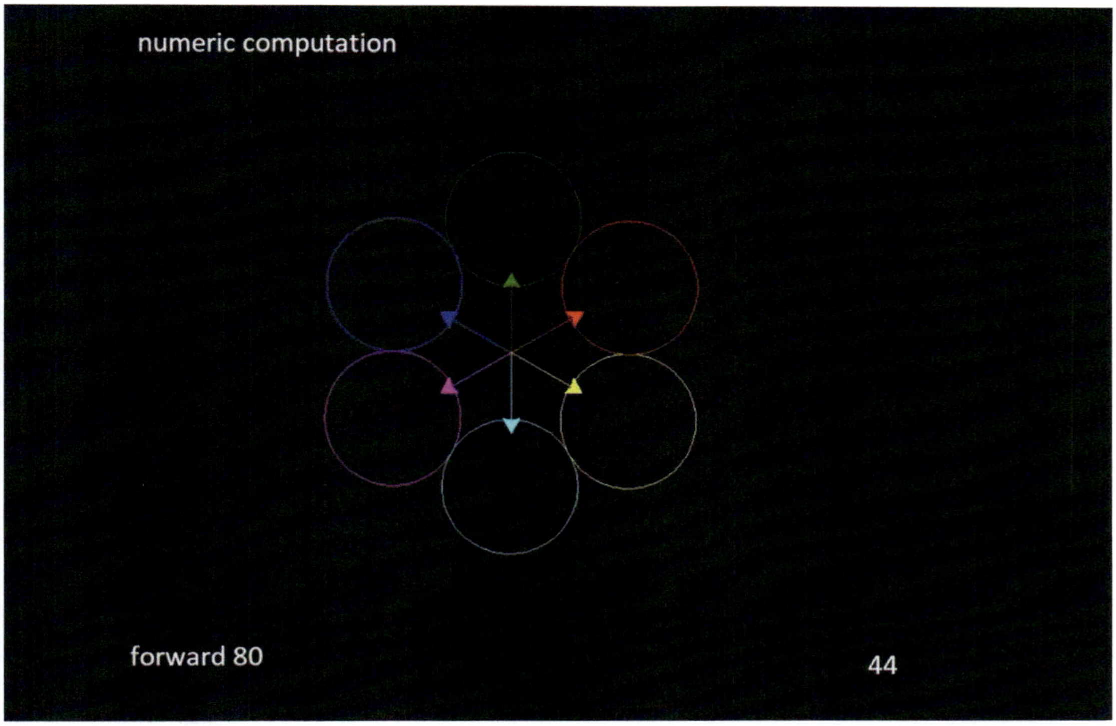
numeric computation
forward 80
44

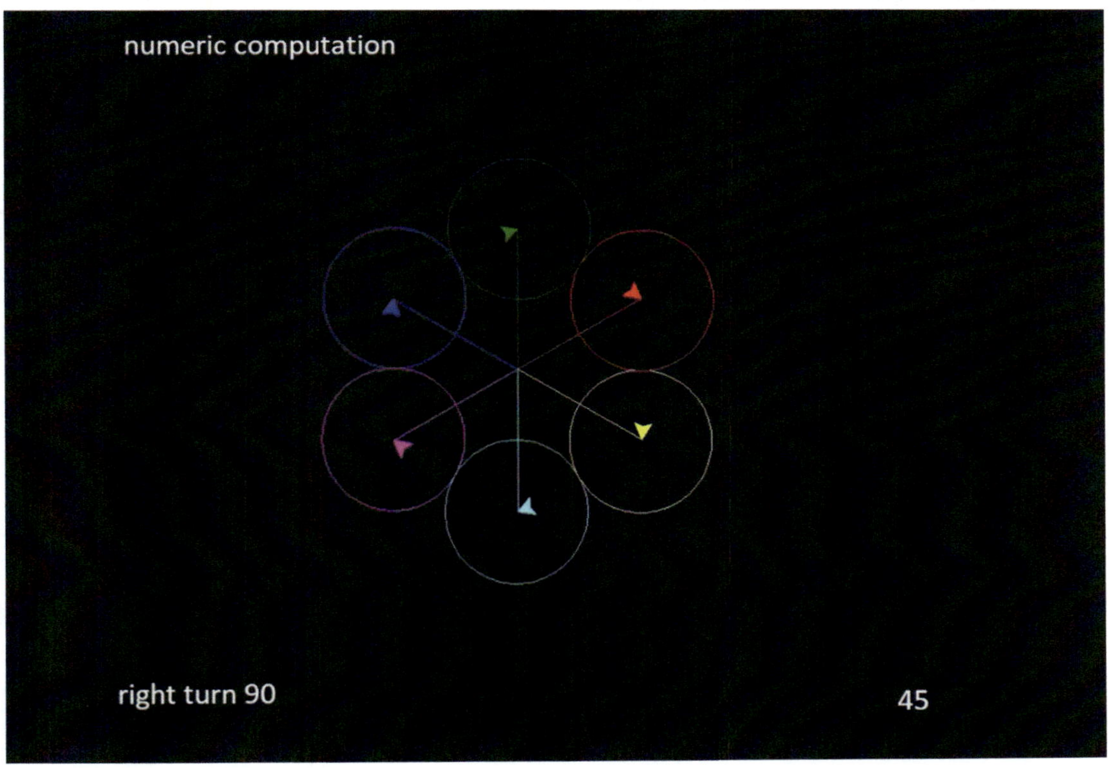
numeric computation
right turn 90
45

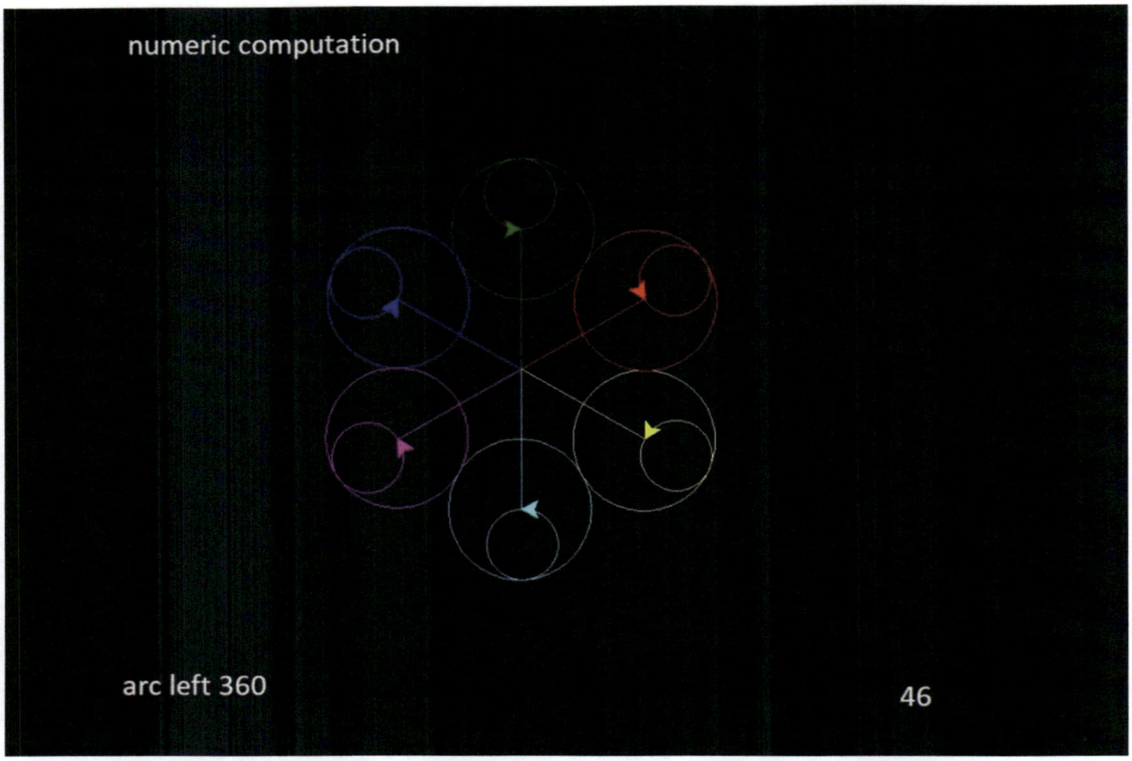
numeric computation
arc left 360
46

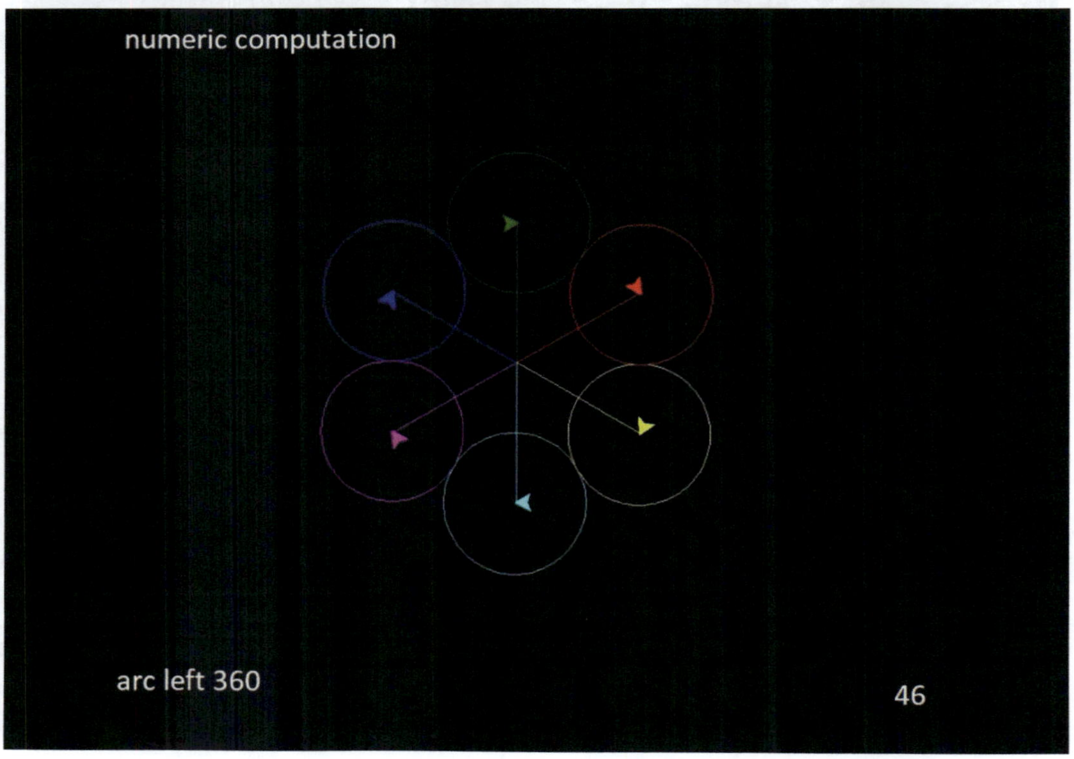
numeric computation
arc left 360
46

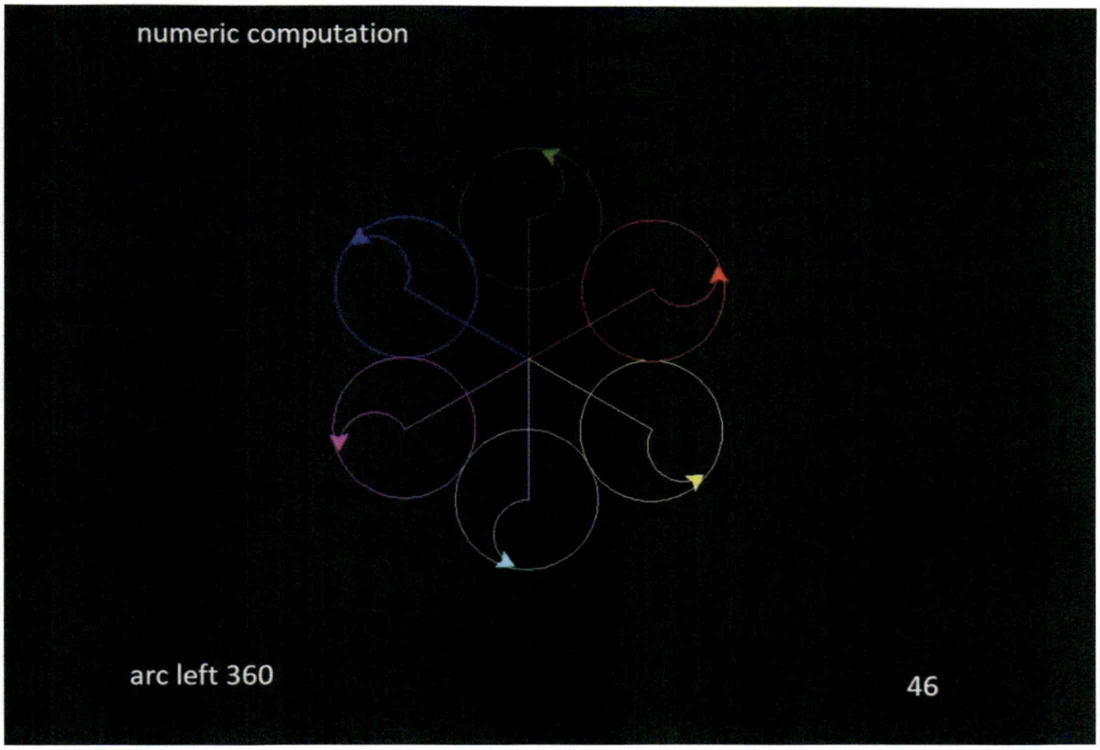
numeric computation
arc left 360
46

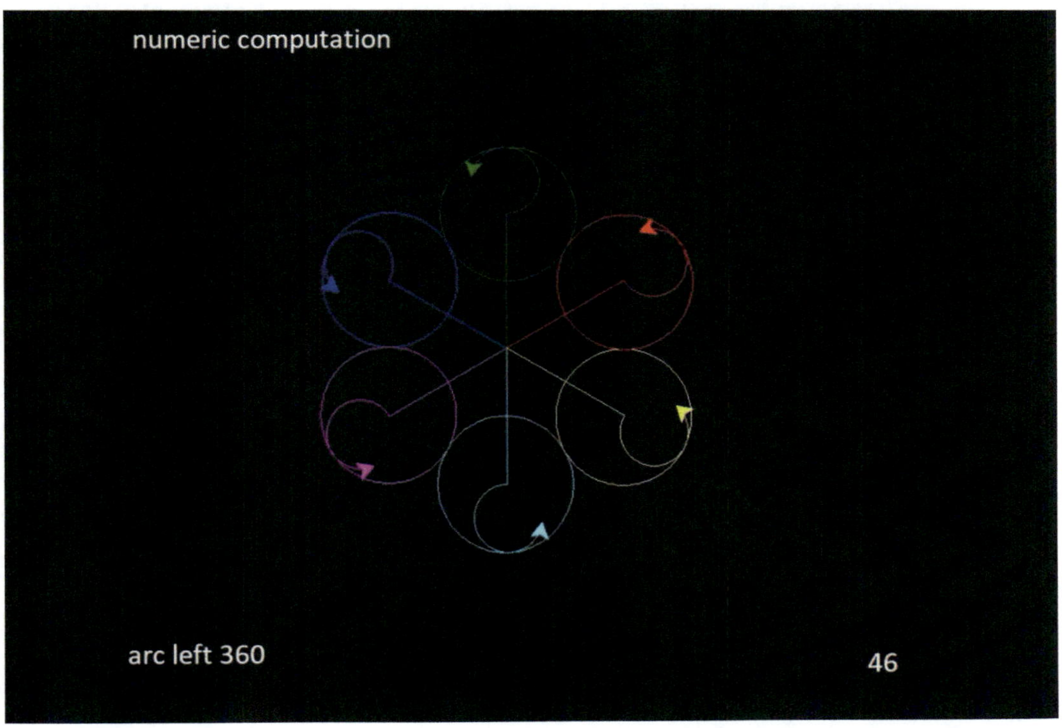
numeric computation
arc left 360
46

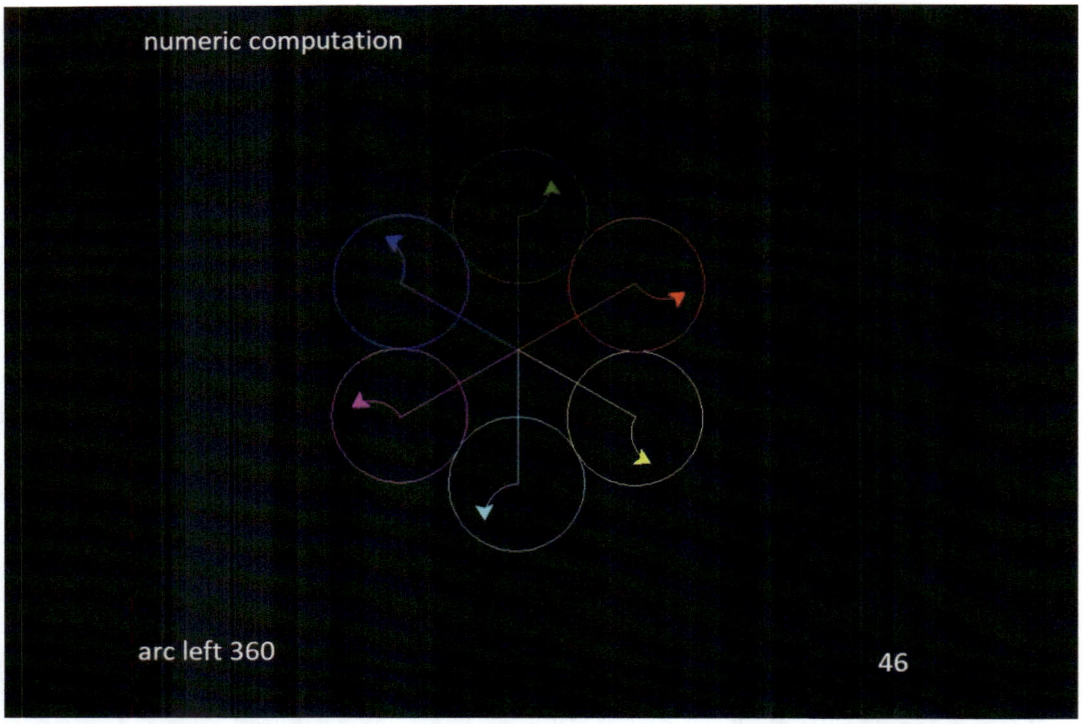
numeric computation
arc left 360
46

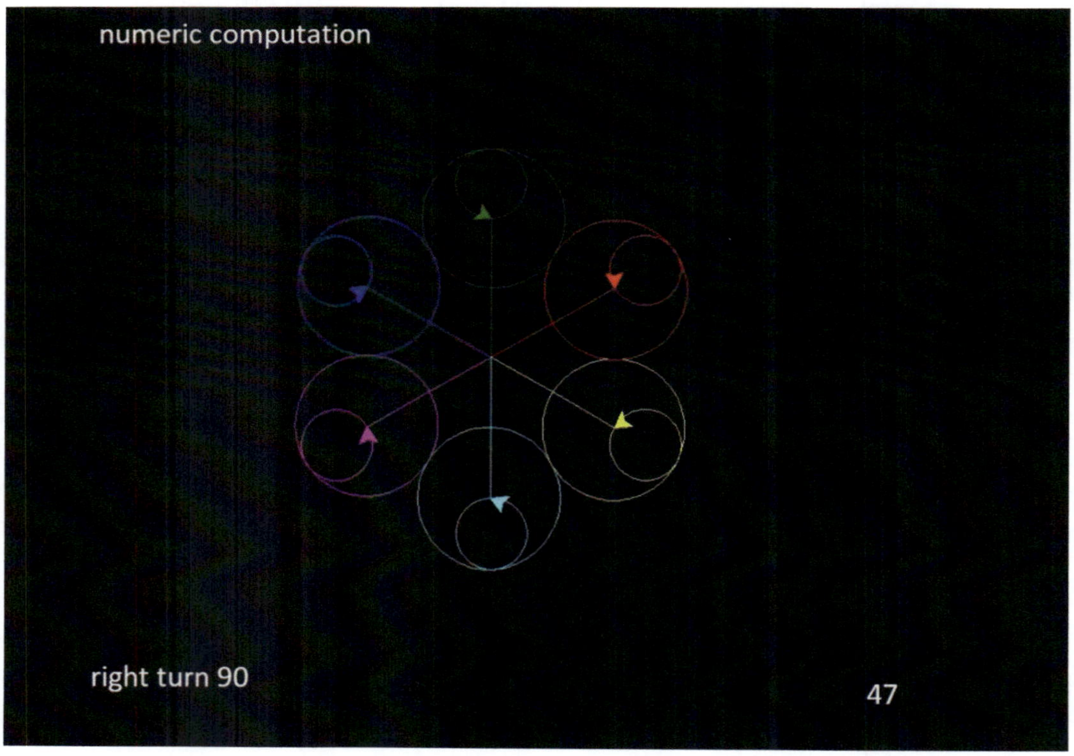
numeric computation
right turn 90
47

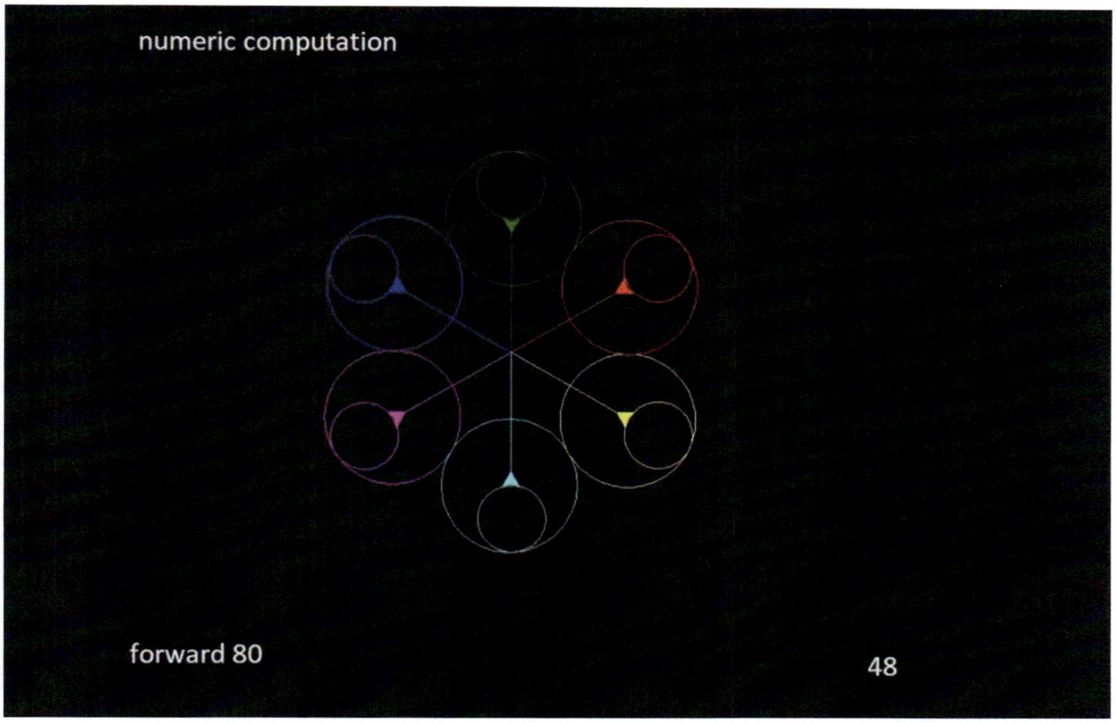
numeric computation
forward 80
48

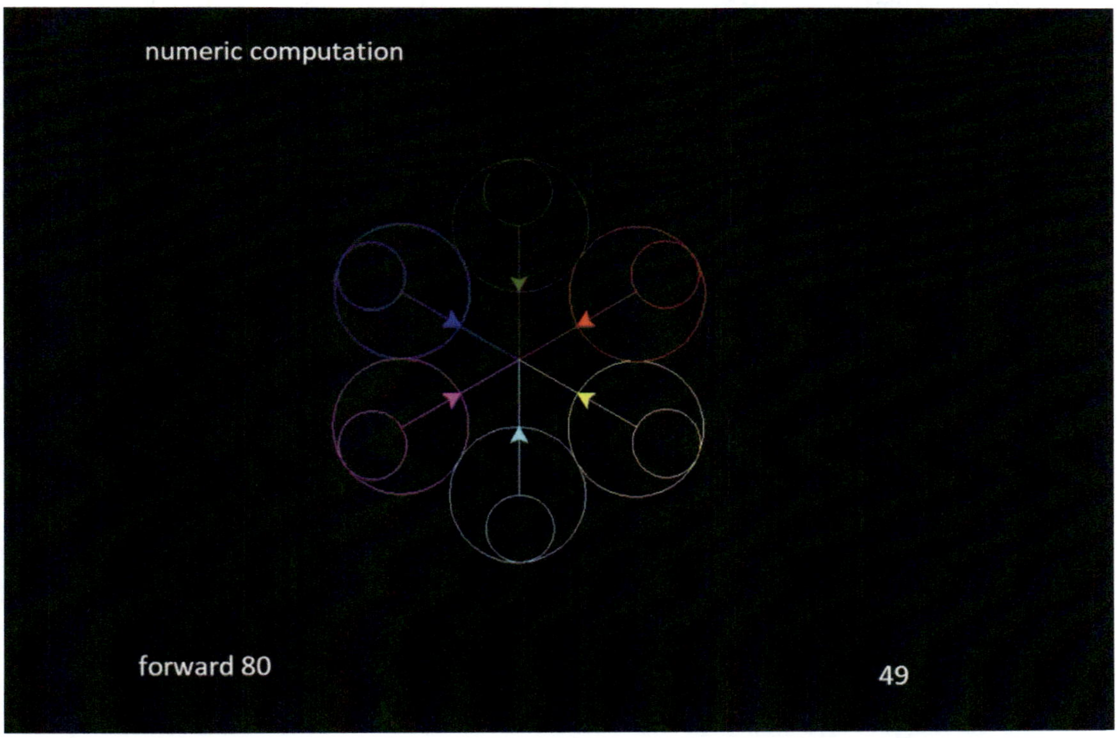
numeric computation
forward 80
49

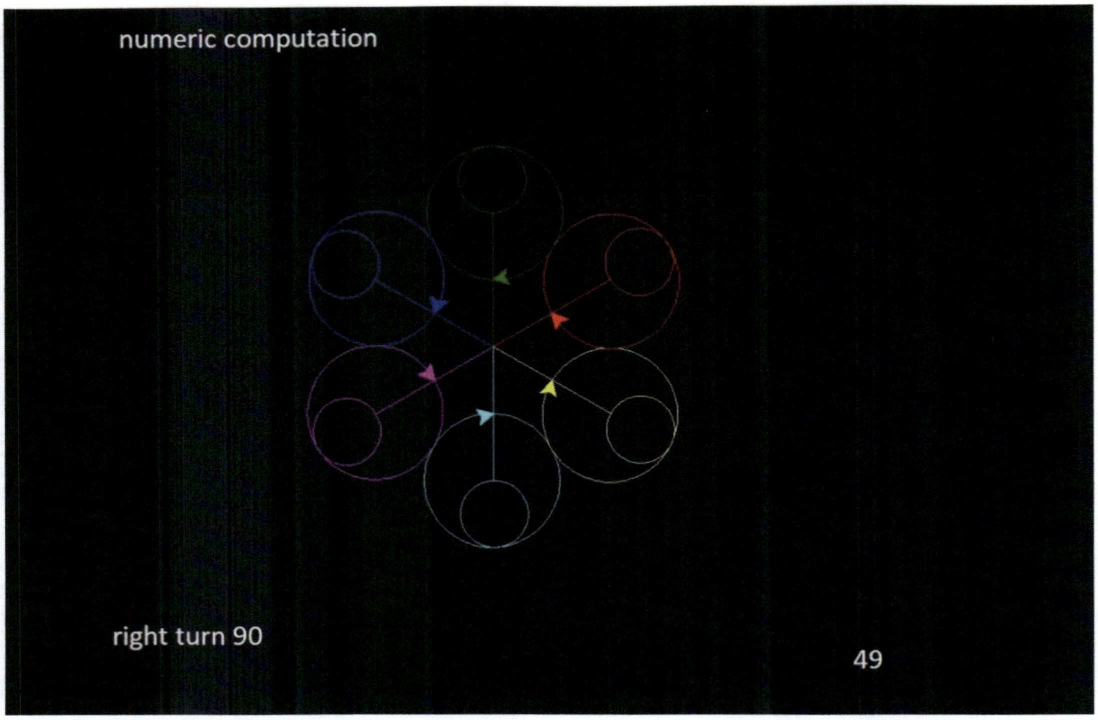
numeric computation
right turn 90
49

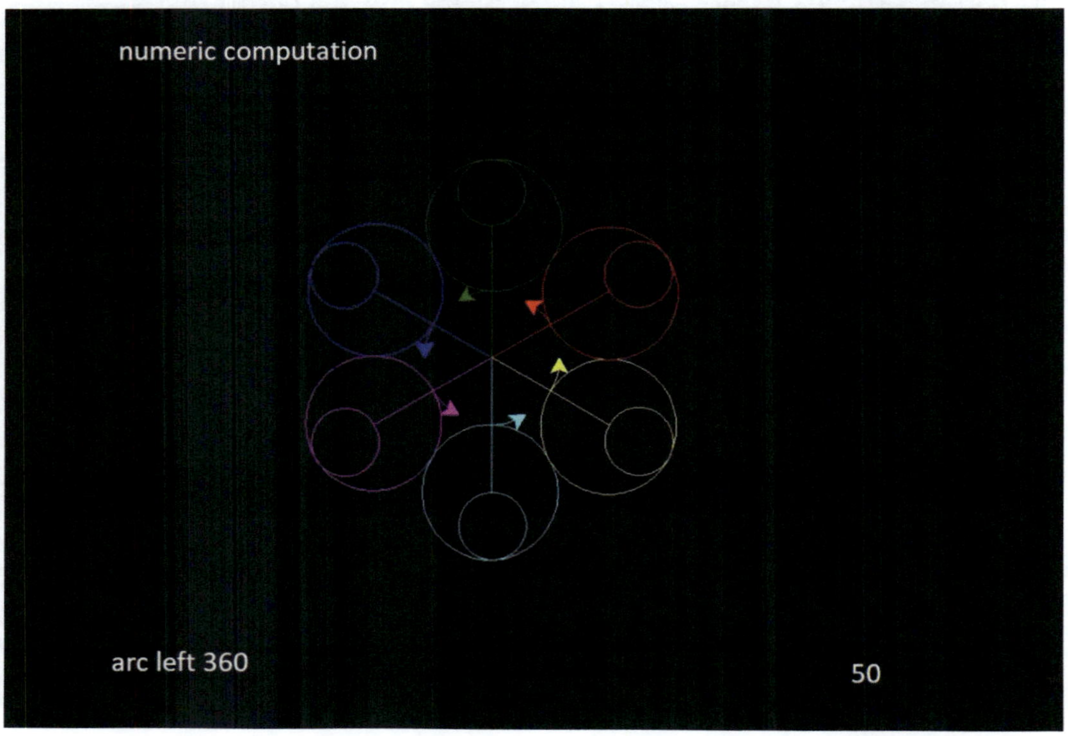
numeric computation
arc left 360
50

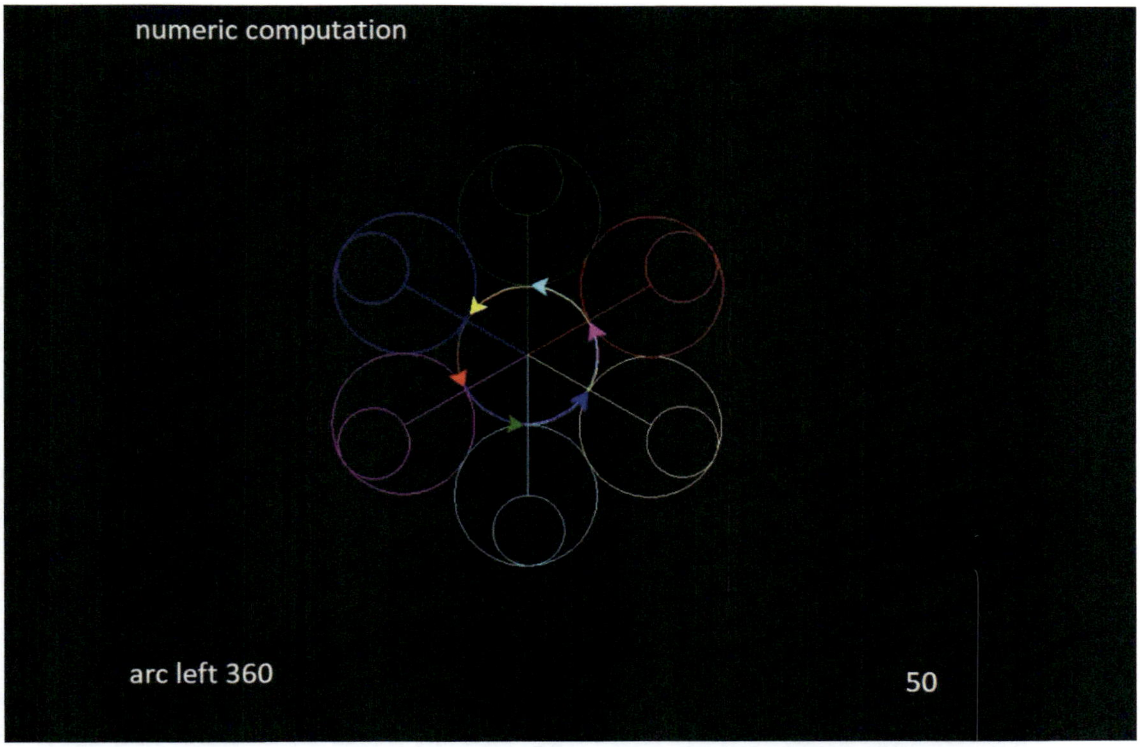
numeric computation
arc left 360
50

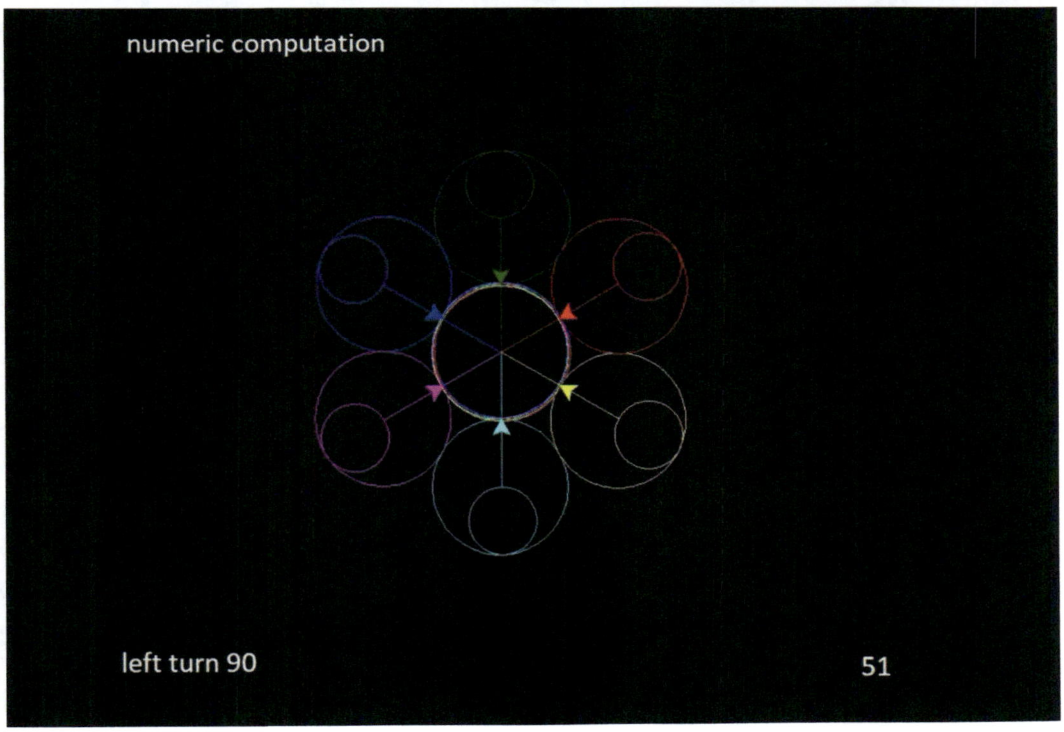
numeric computation
left turn 90
51

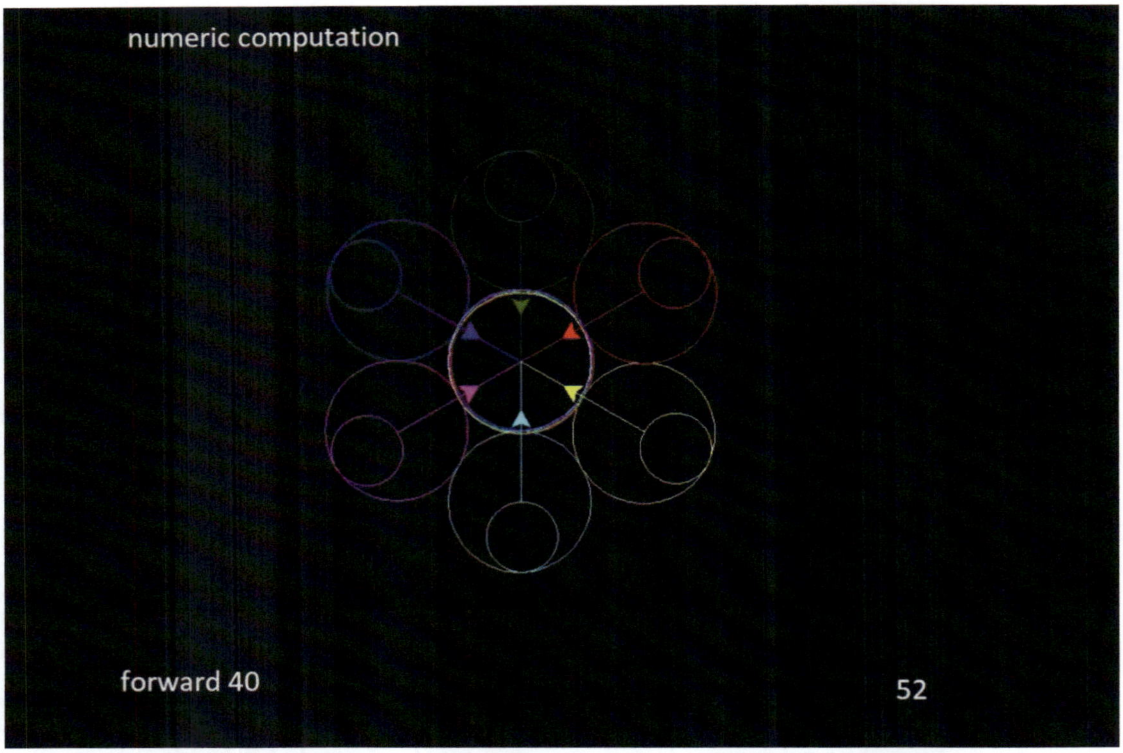
numeric computation
forward 40
52

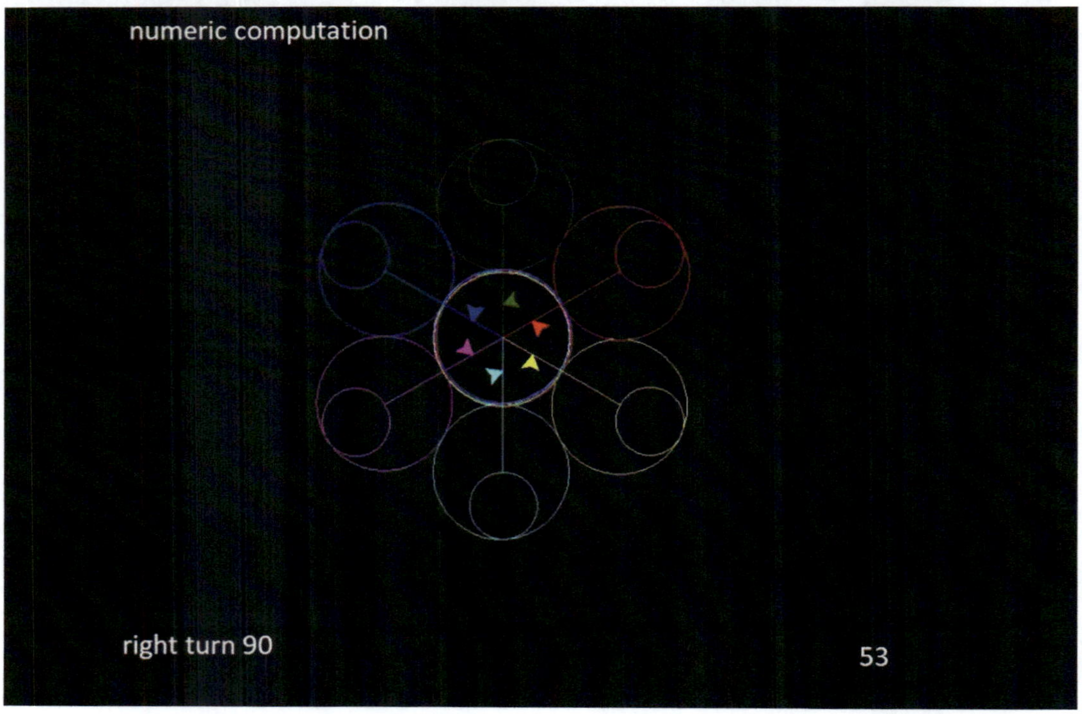
numeric computation
right turn 90
53

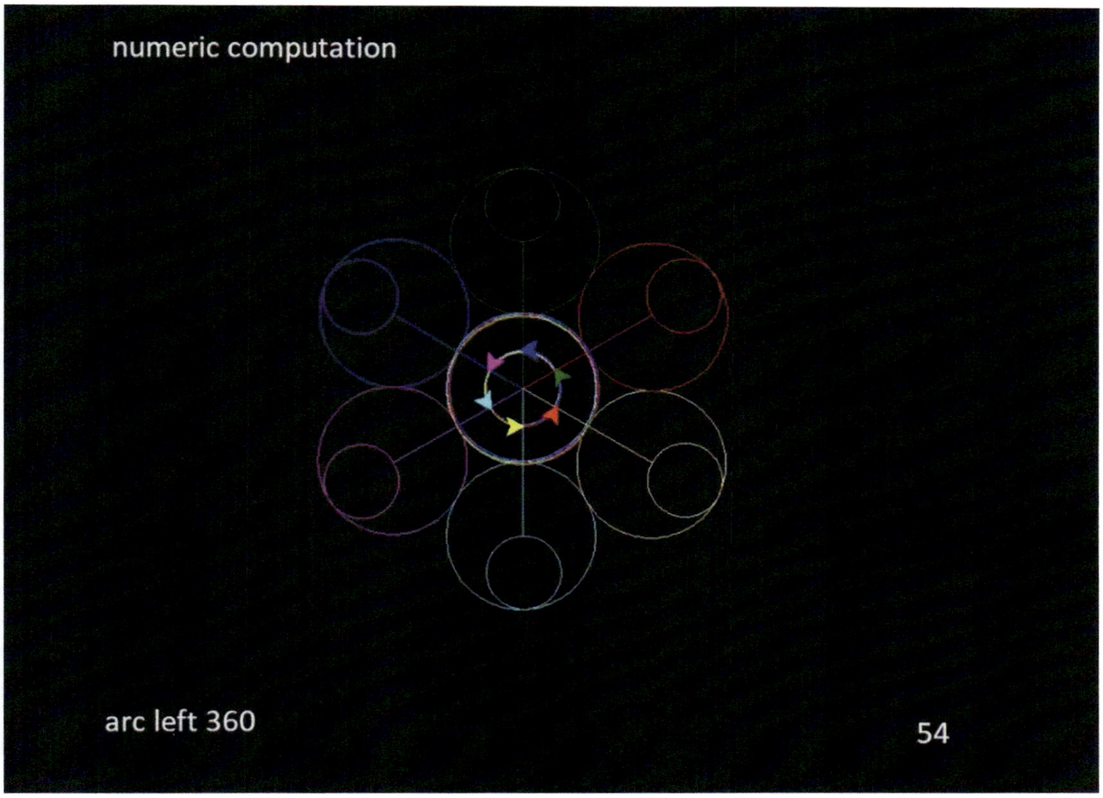
numeric computation
arc left 360
54

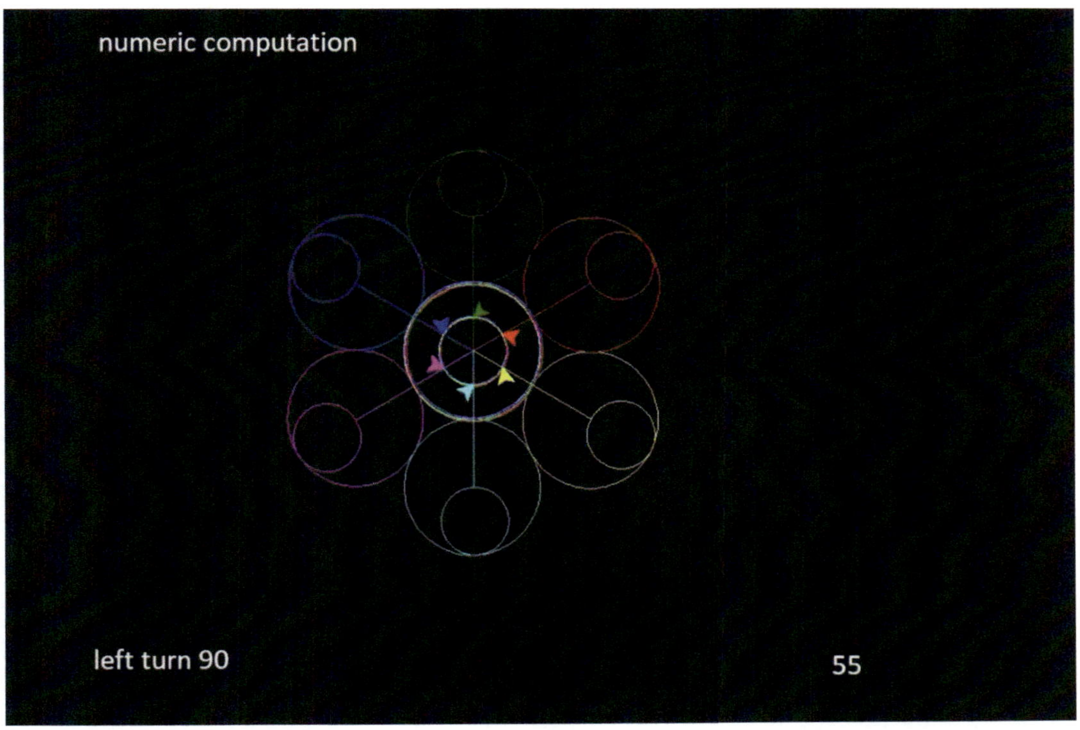
numeric computation
left turn 90
55

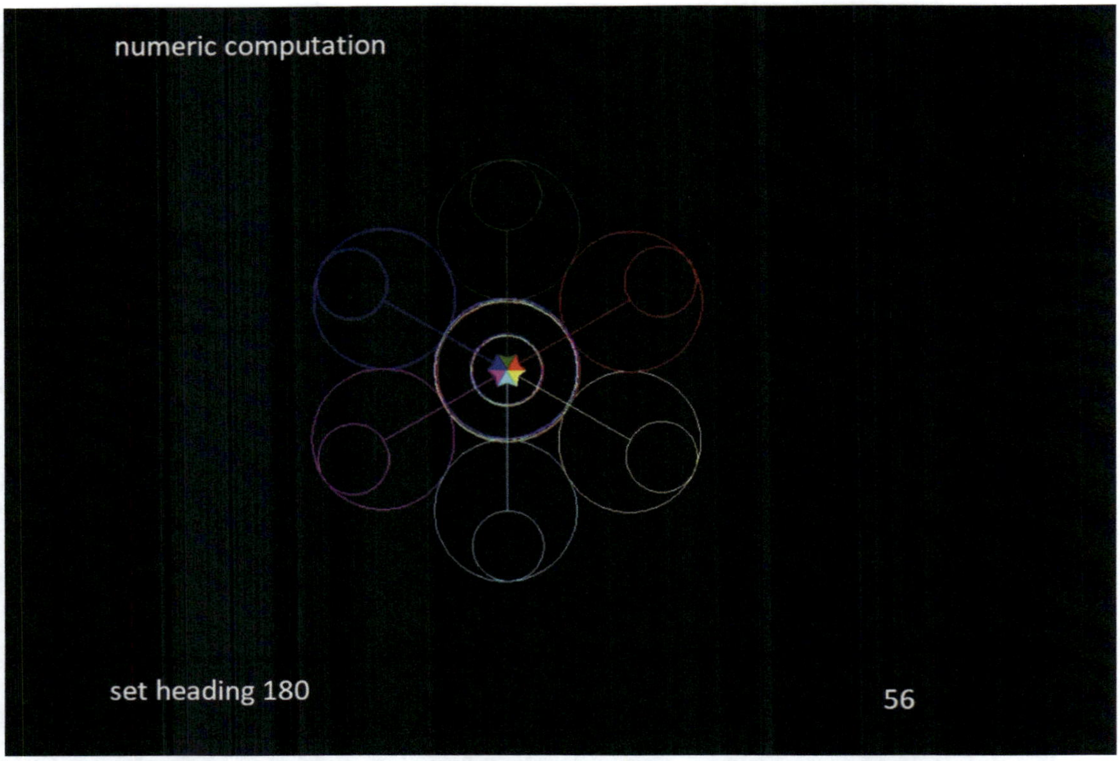
numeric computation
set heading 180
56

Sequence

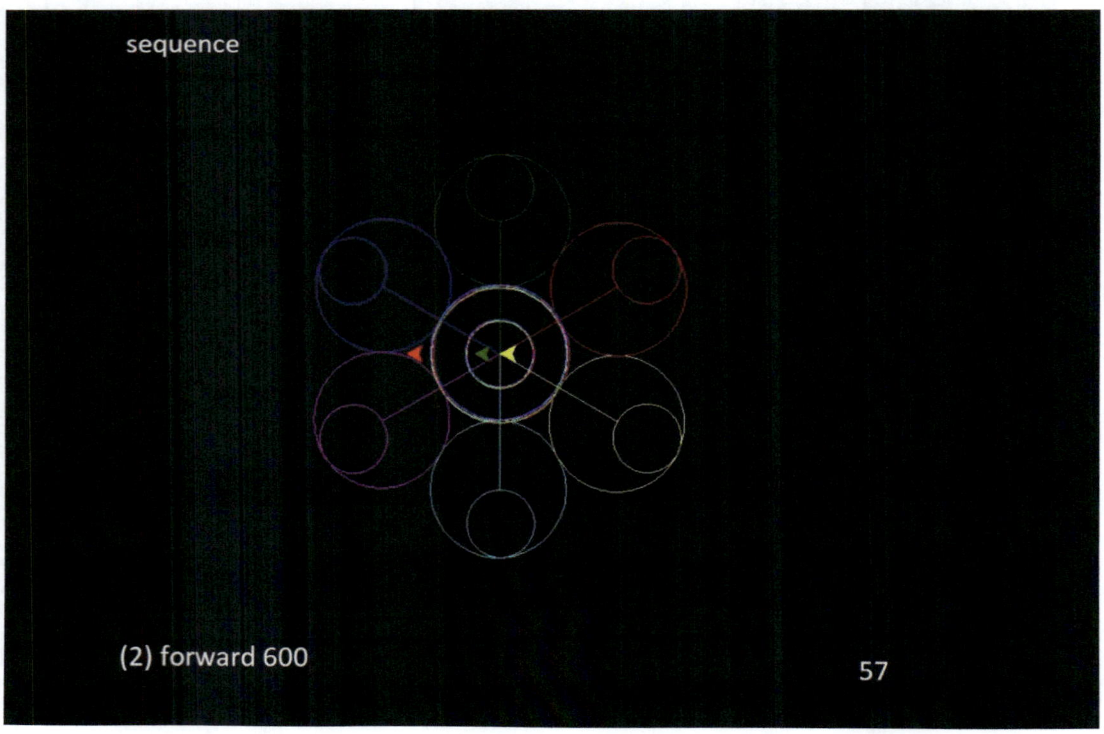
sequence
(2) forward 600
57

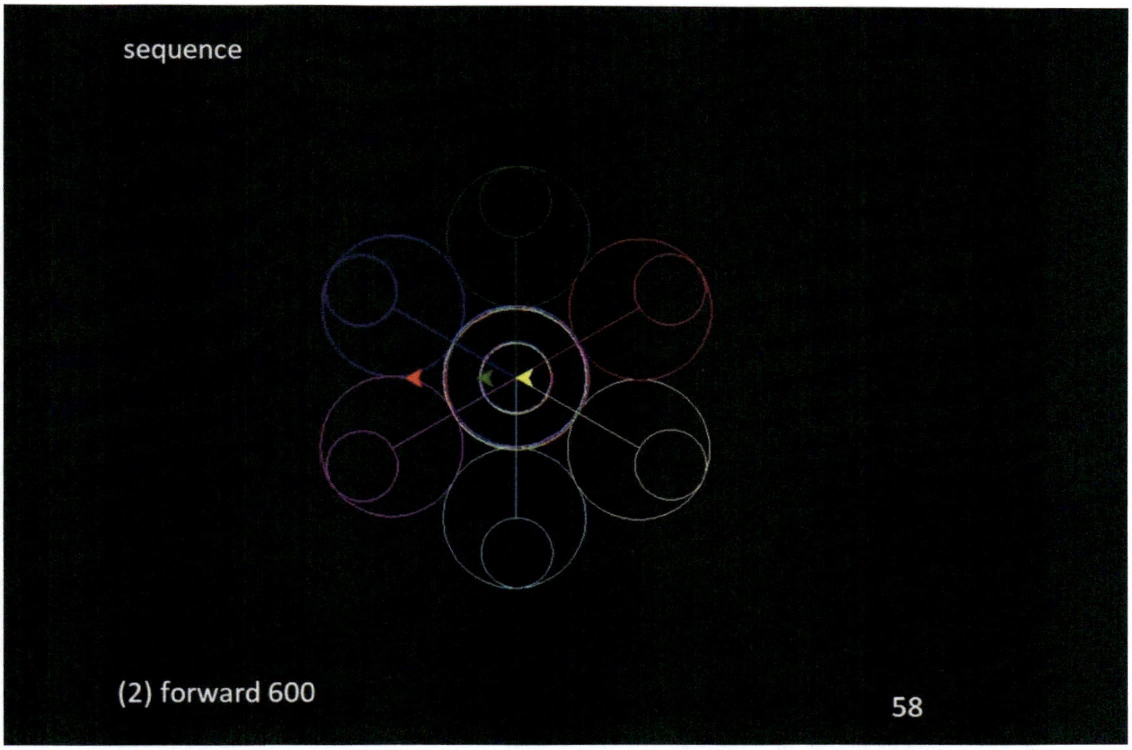
sequence
(2) forward 600
58

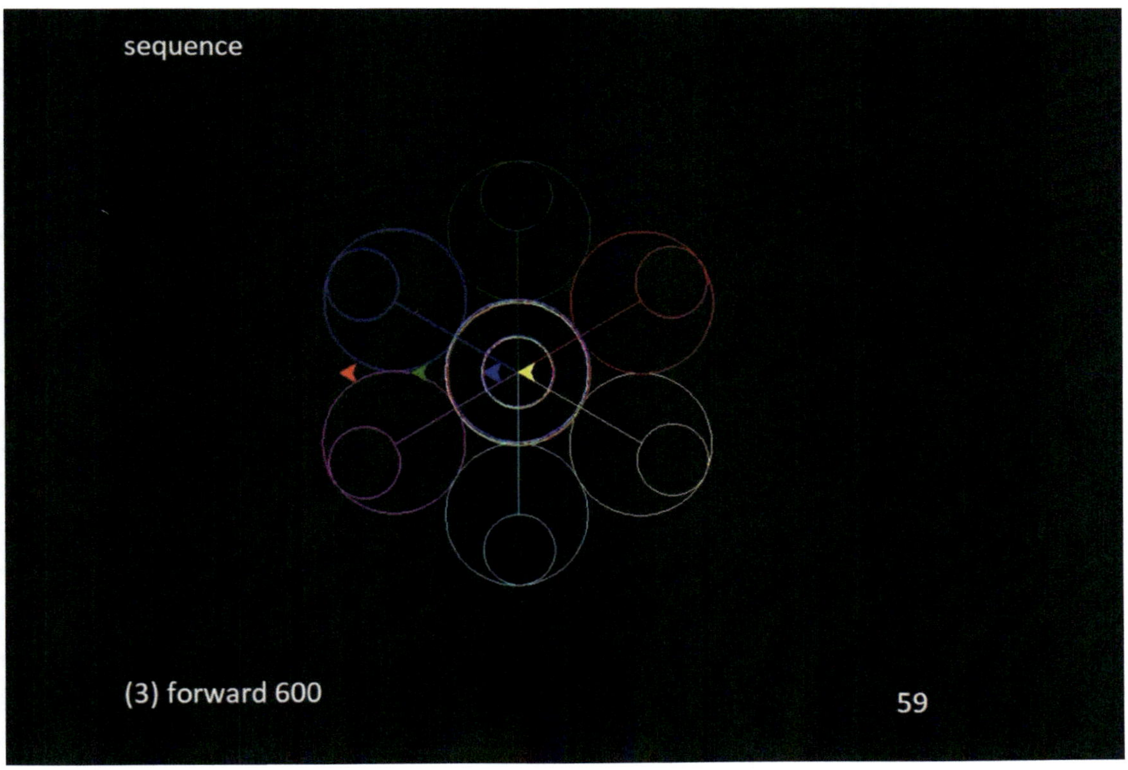
sequence
(3) forward 600
59

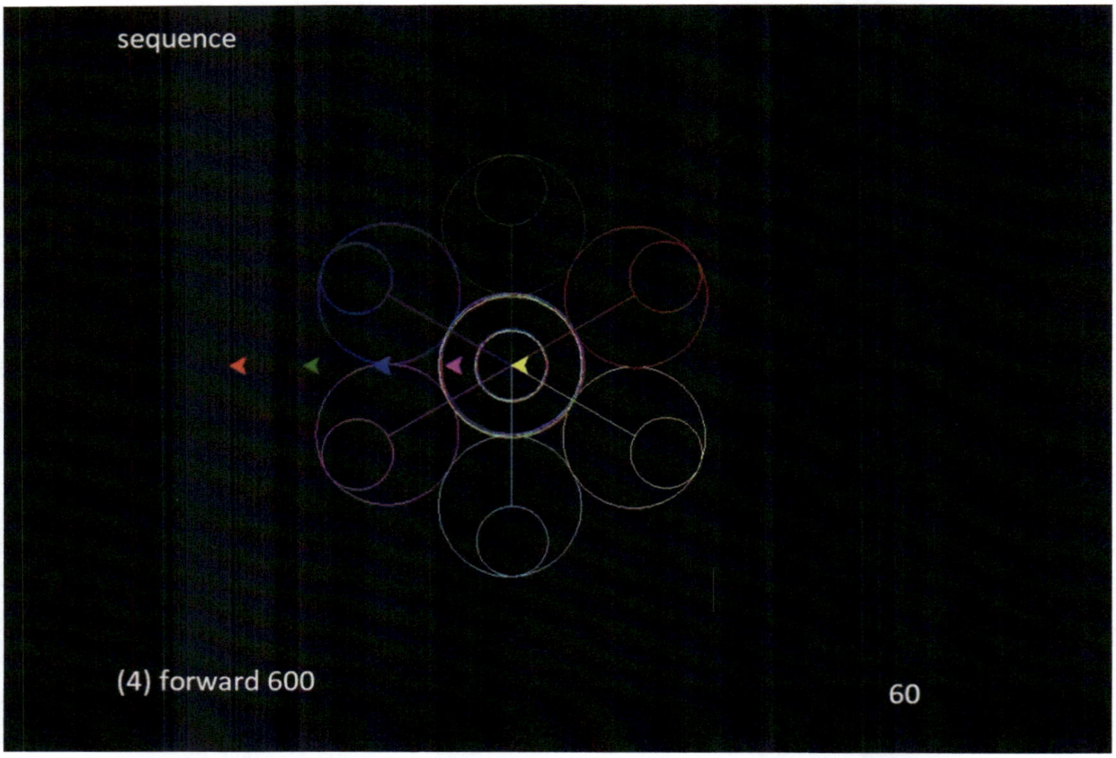
sequence
(4) forward 600
60

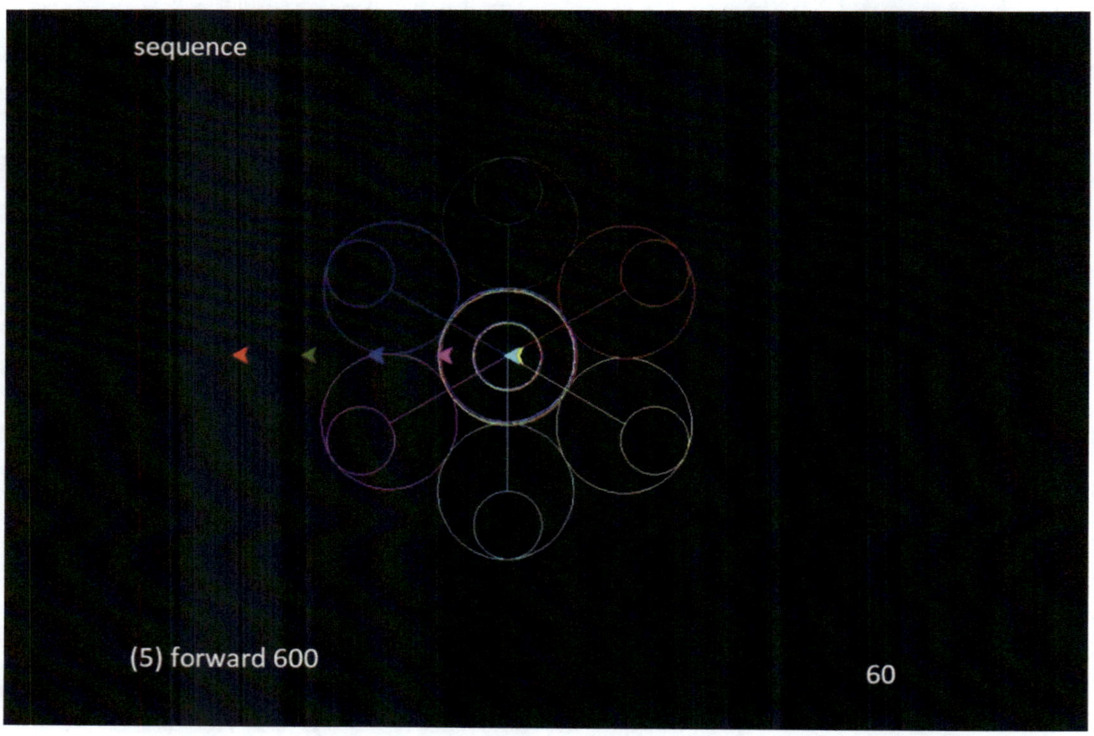
sequence
(5) forward 600
60

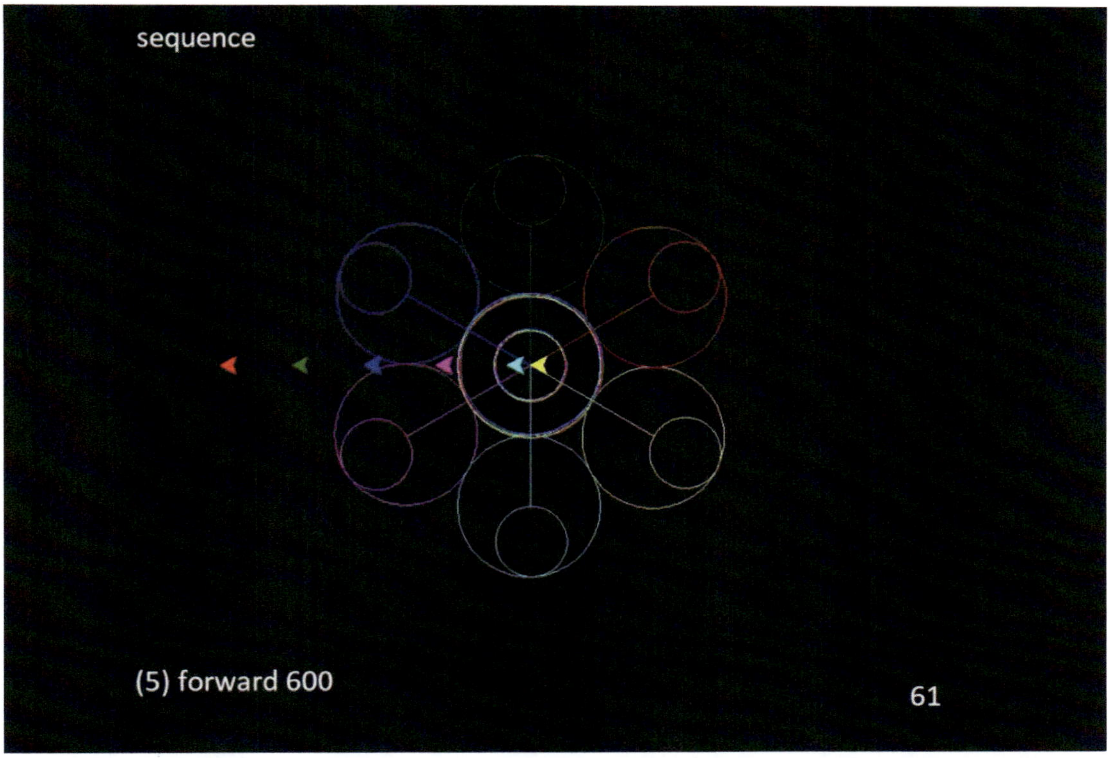
sequence
(5) forward 600
61

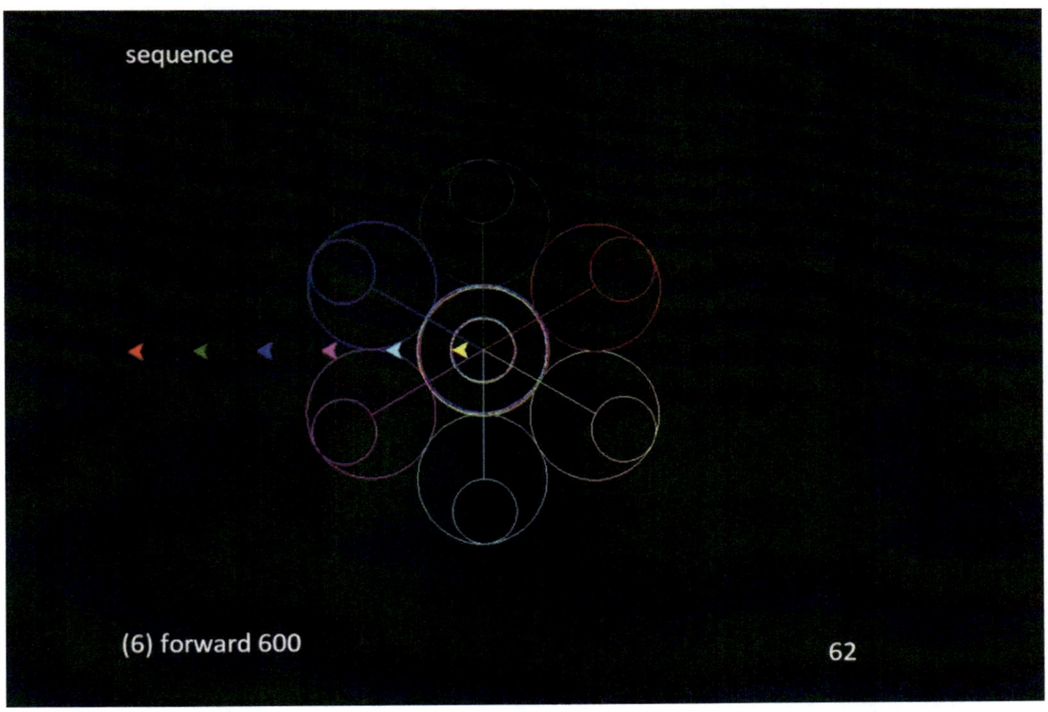
sequence
(6) forward 600
62

CODING THROUGH DANCE METAPHORS VIDEO:

Video created by Stuart Smith with Mary Jo Lasky dancing was used as a projection for the performance. Mary Jo was my student, an Art & Design major, and Dance minor at the University of Northern Colorado.

https://www.youtube.com/watch?v=pNRIIVzRHEE

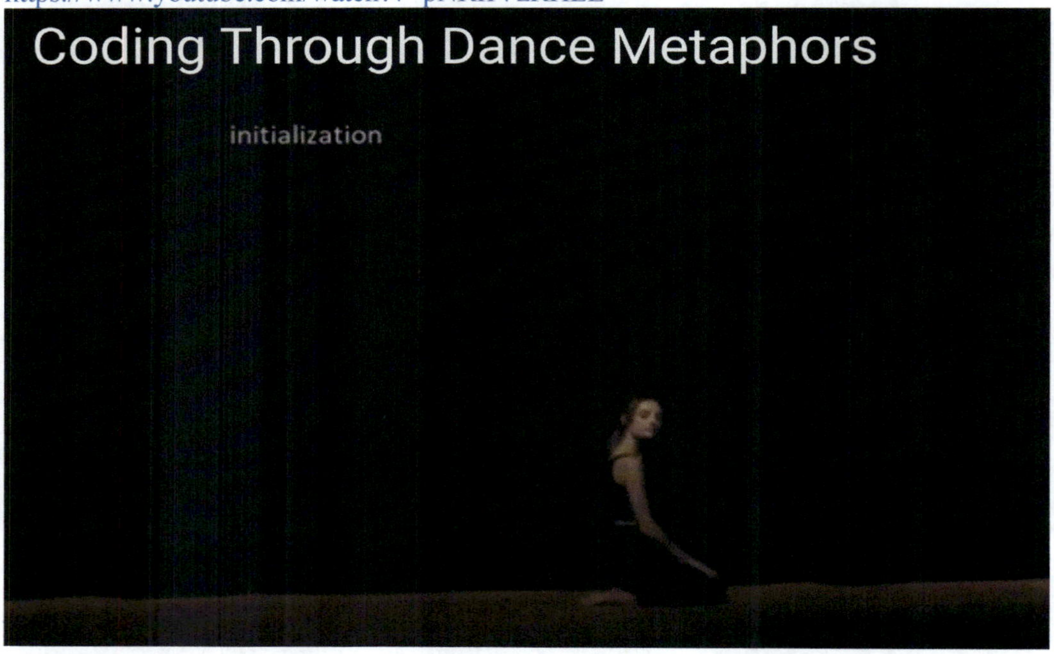

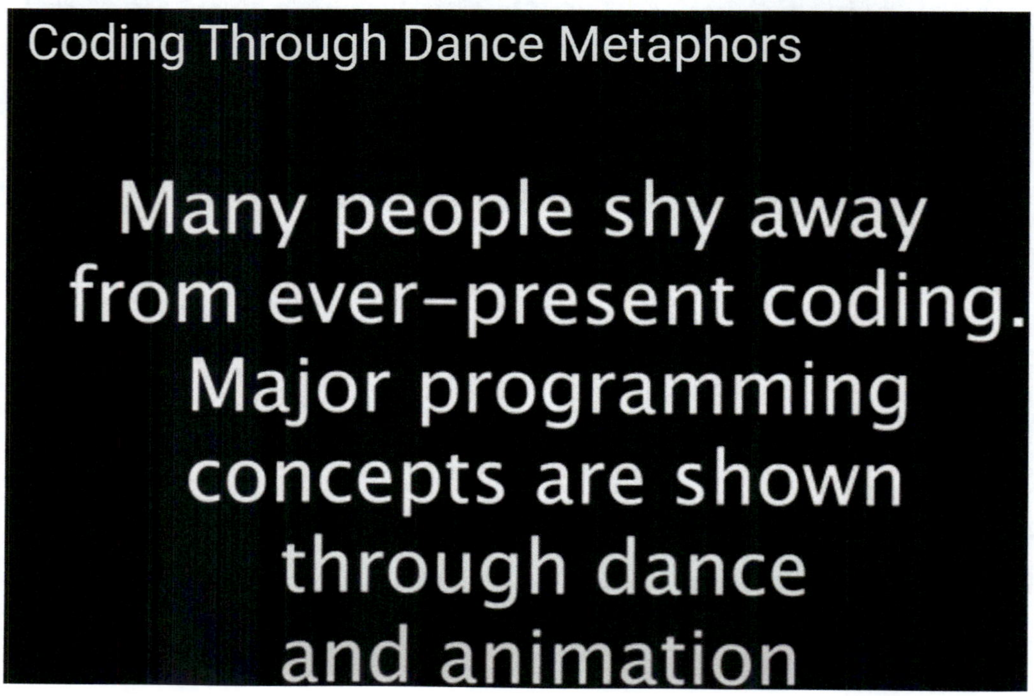

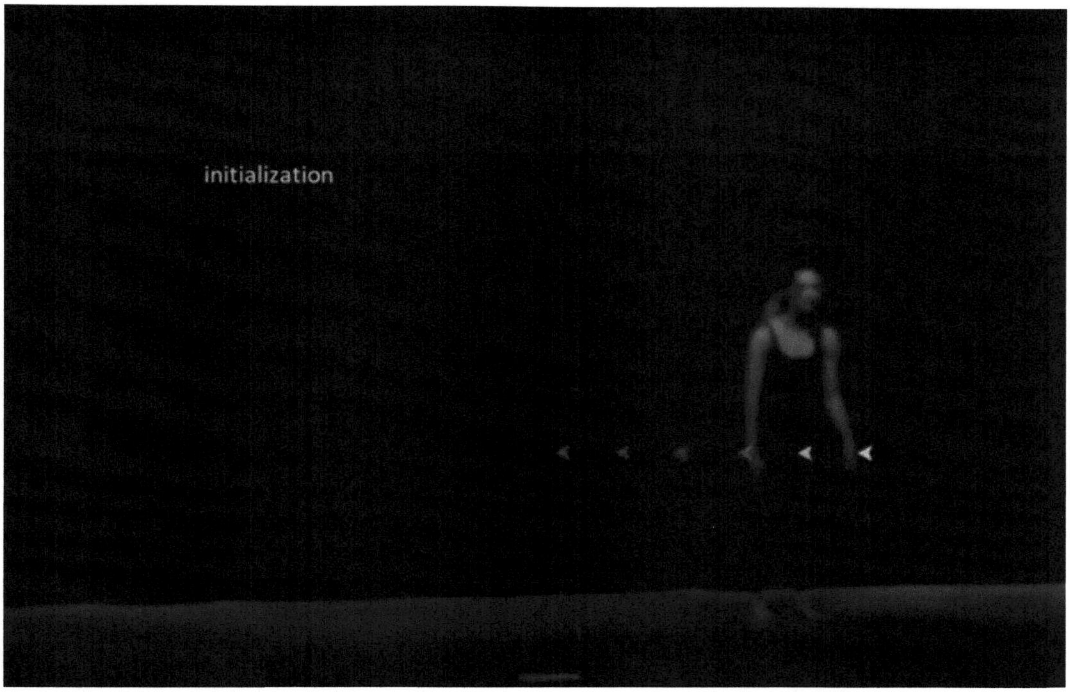
initialization

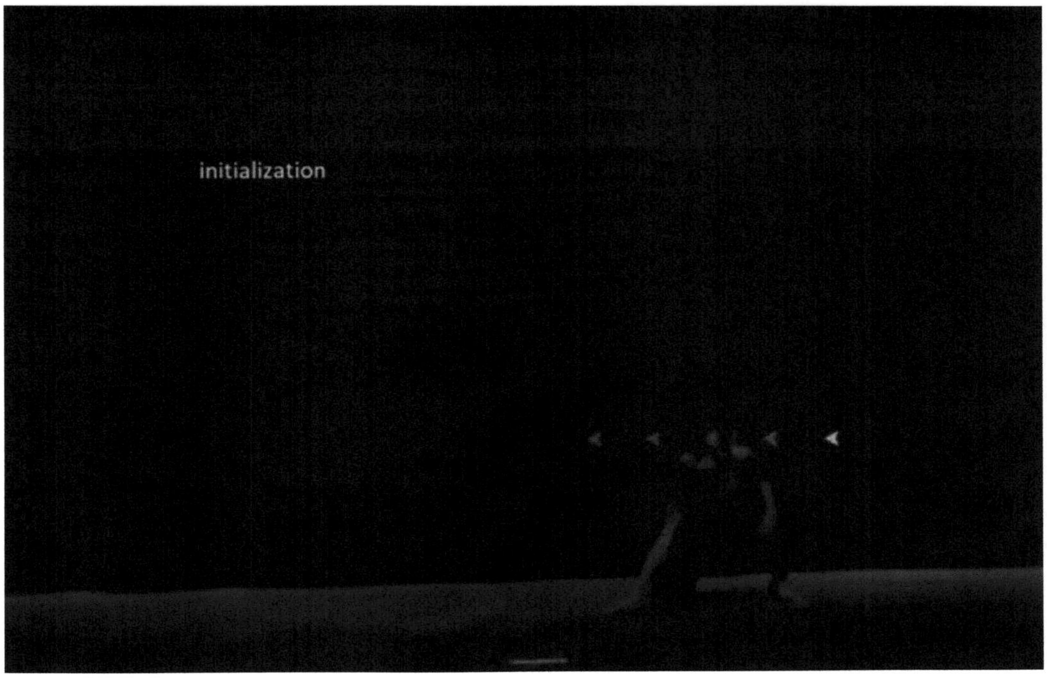
initialization

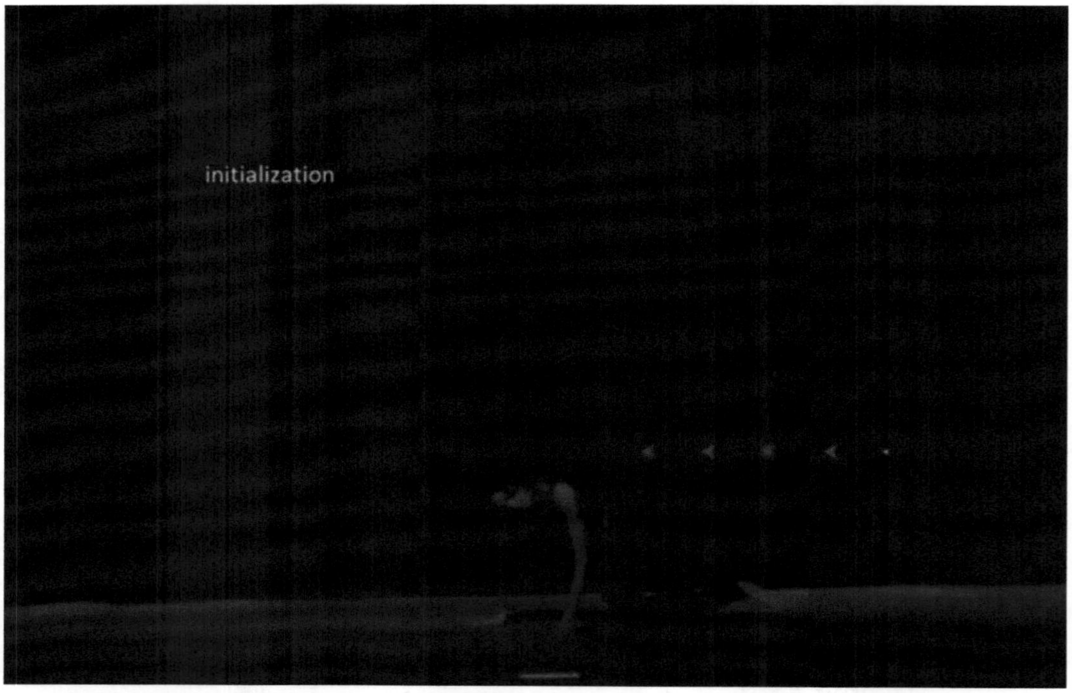
initialization

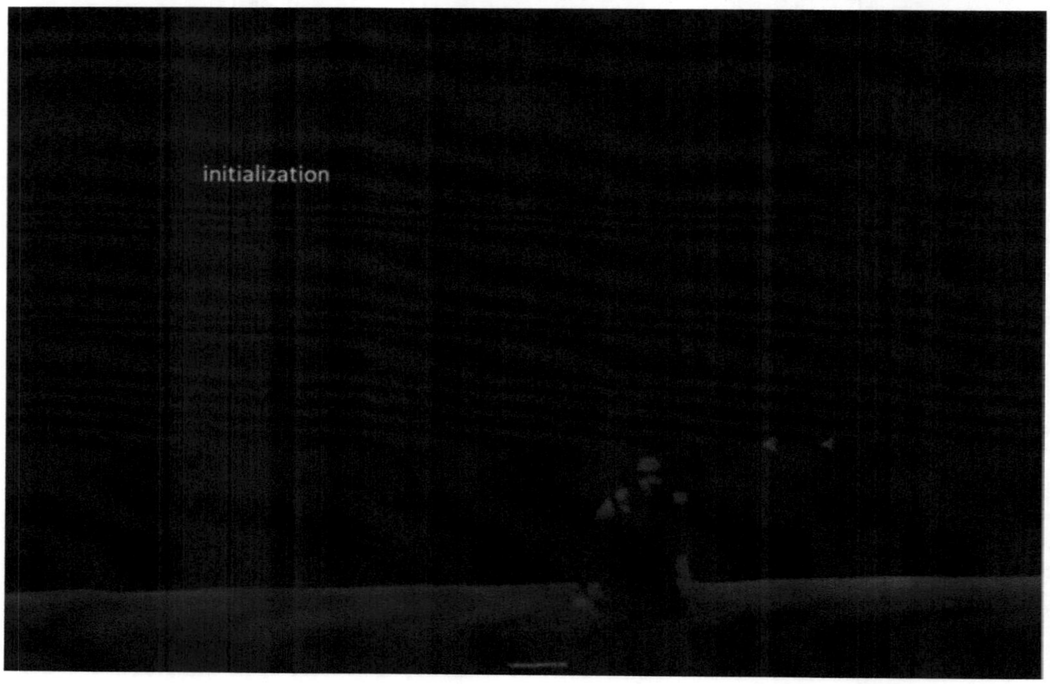
initialization

Conditional Execution

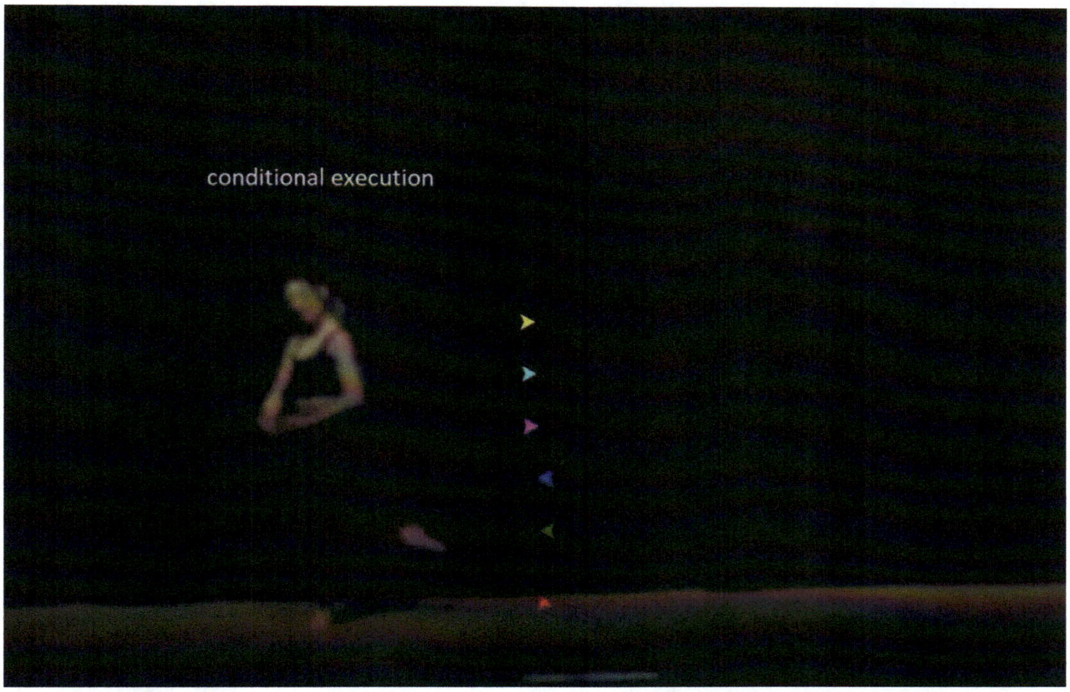

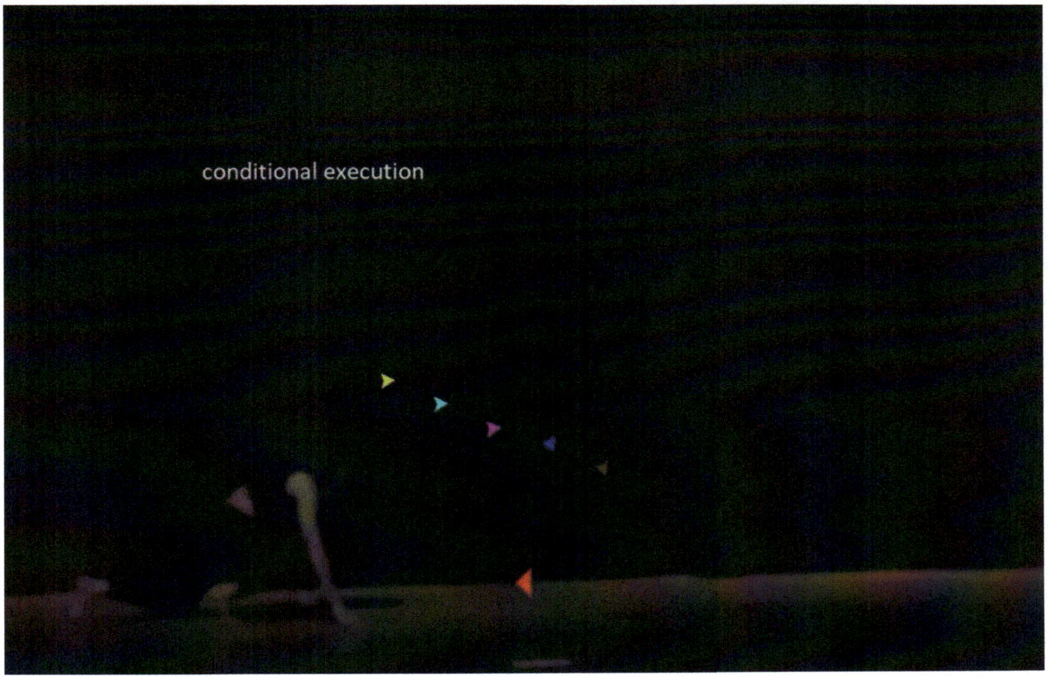

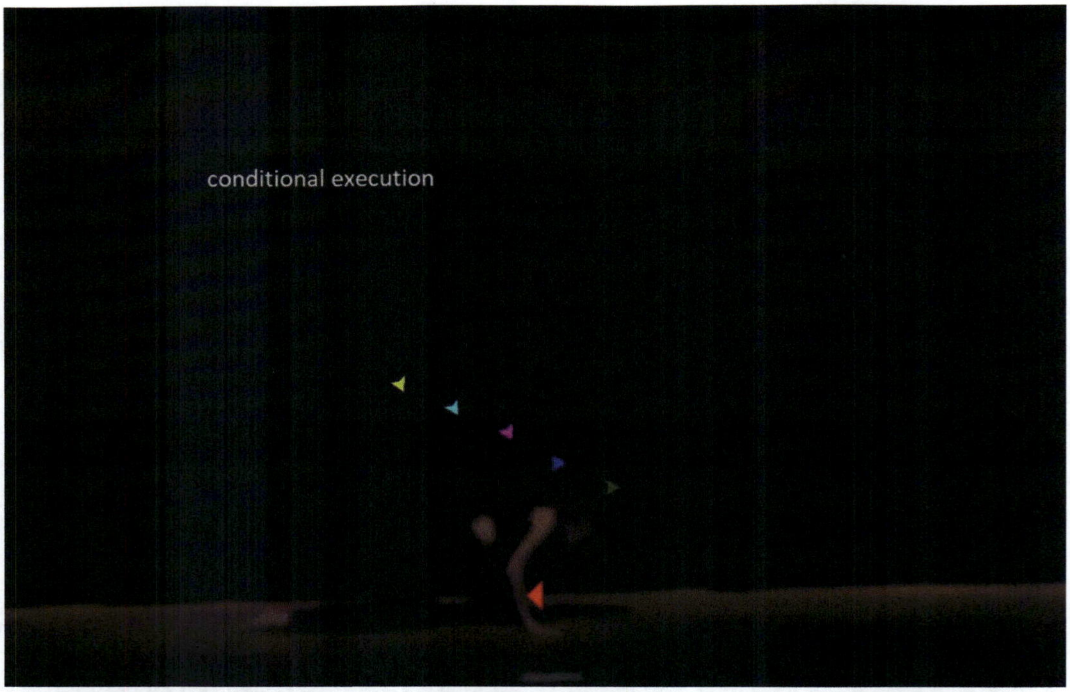
conditional execution

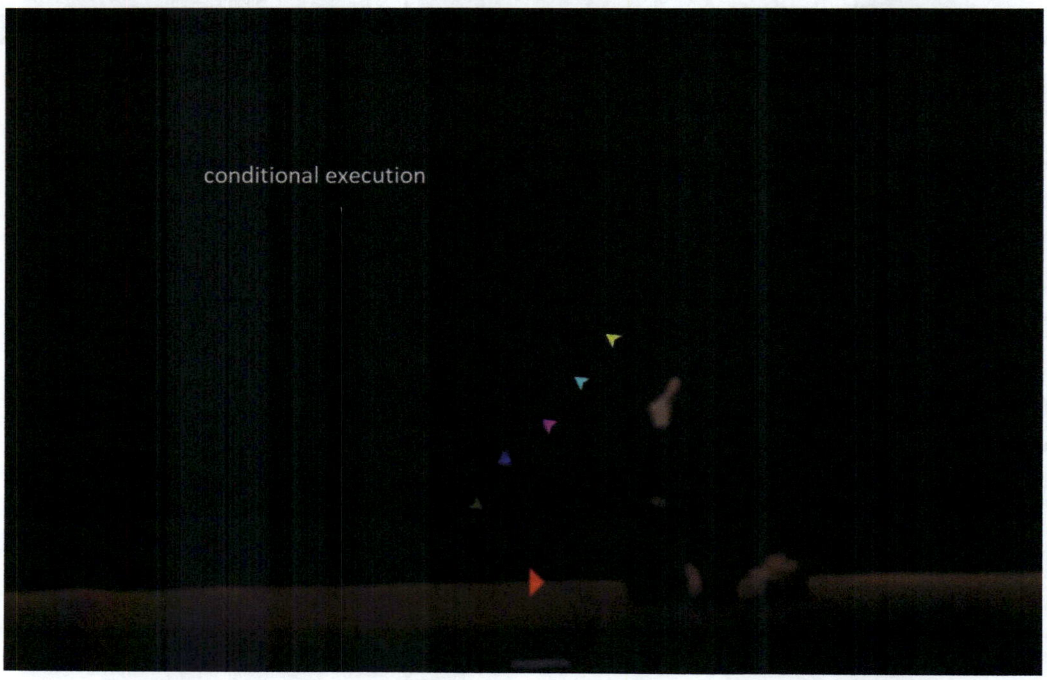
conditional execution

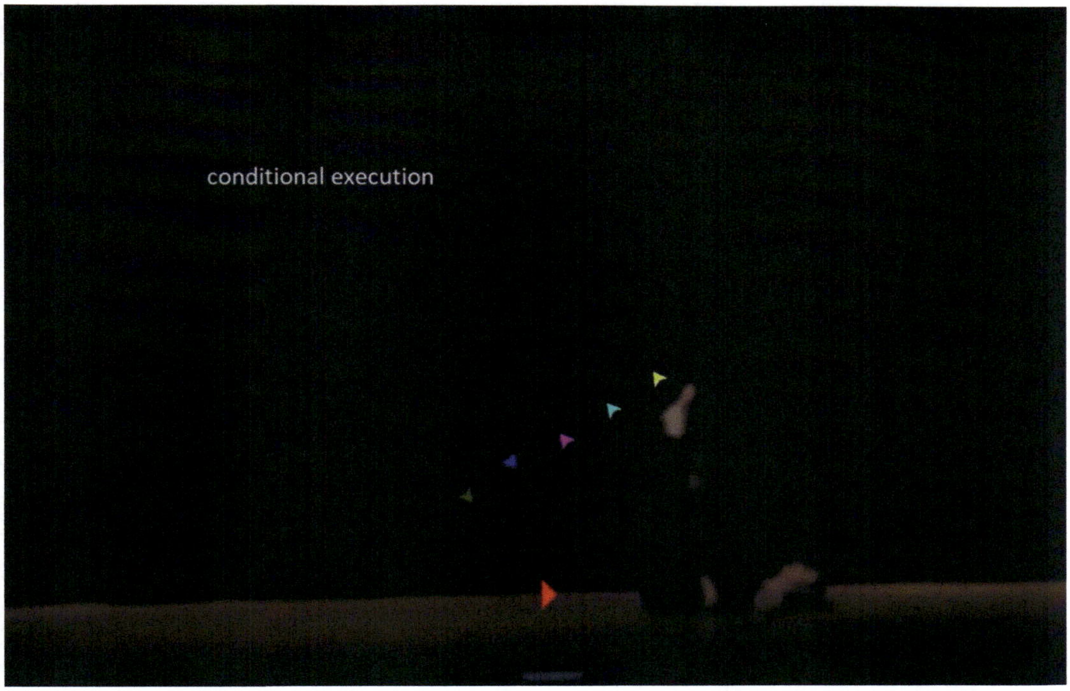
conditional execution

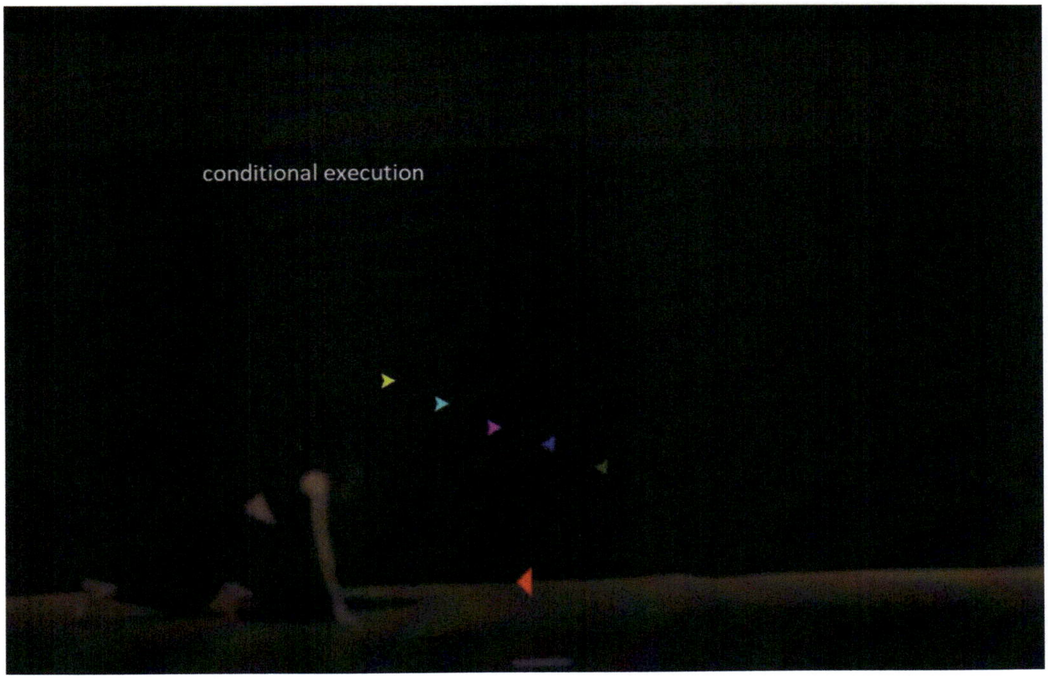
conditional execution

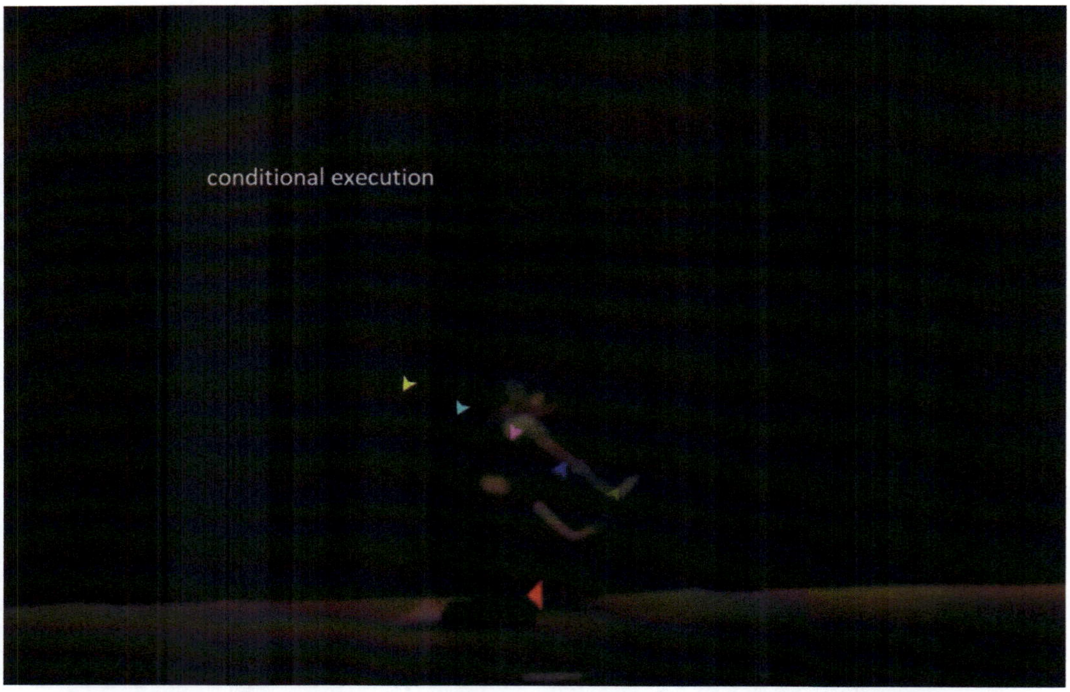
conditional execution

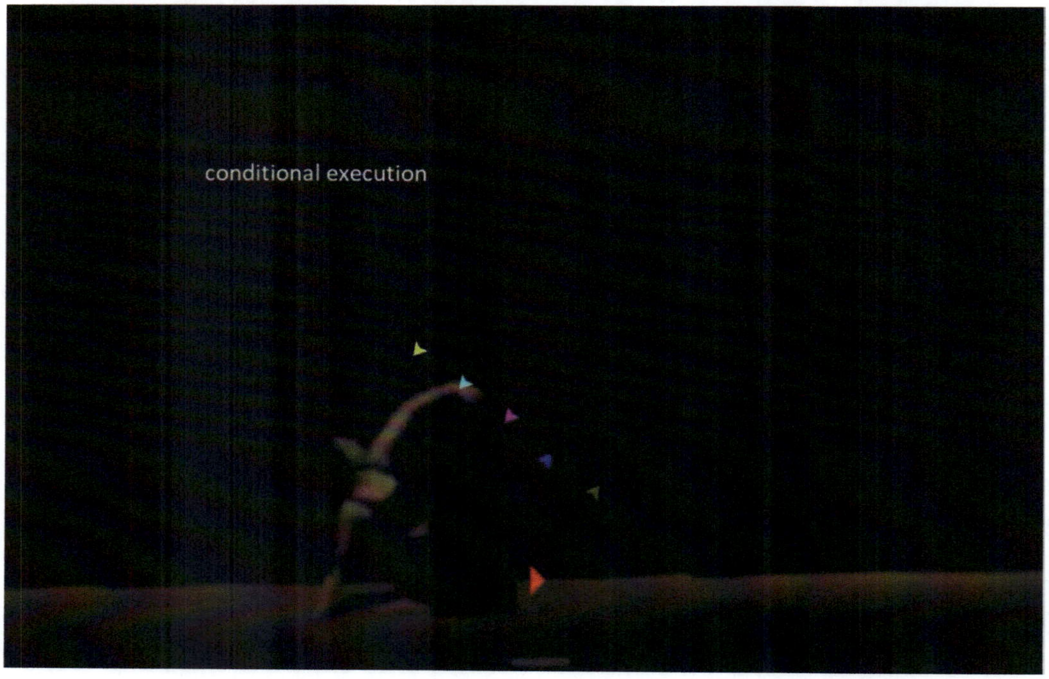
conditional execution

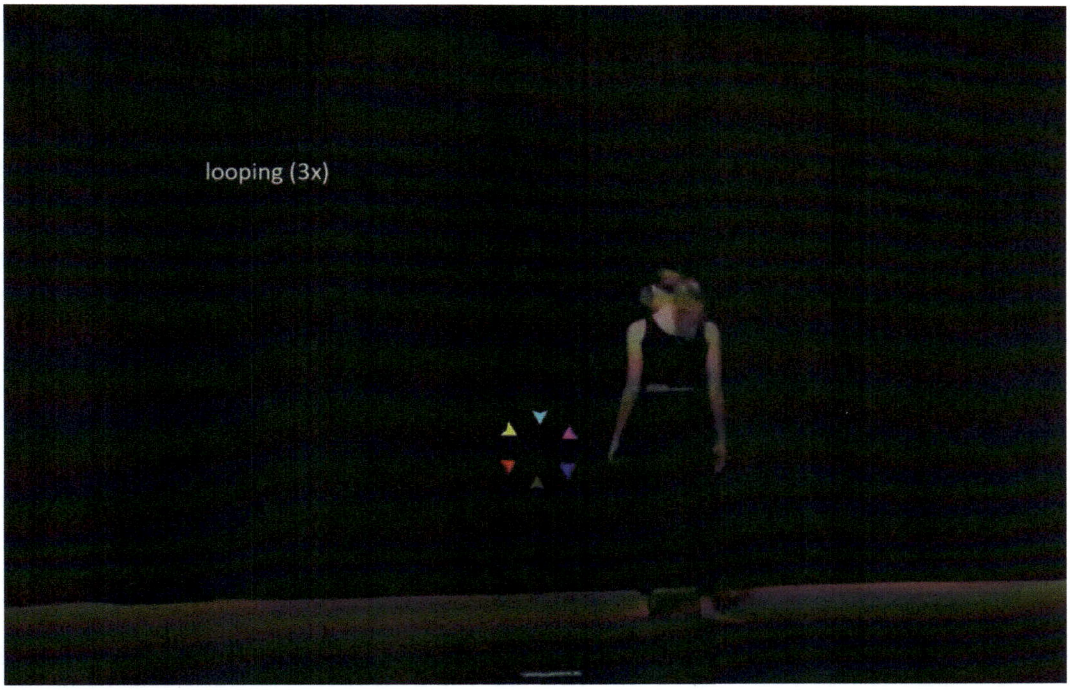
looping (3x)

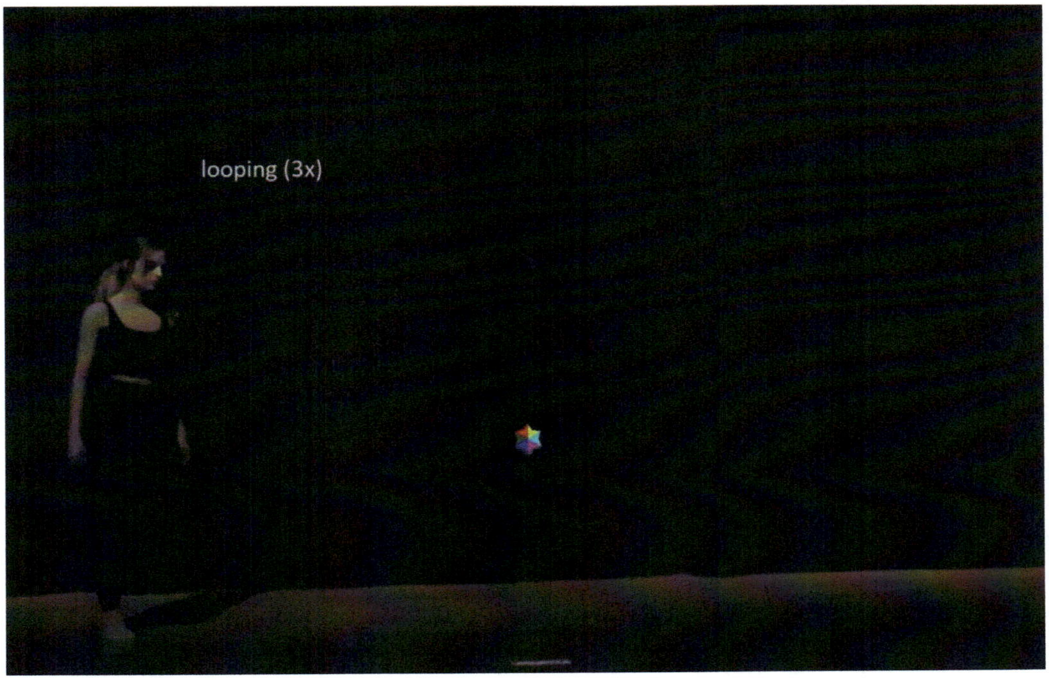
looping (3x)

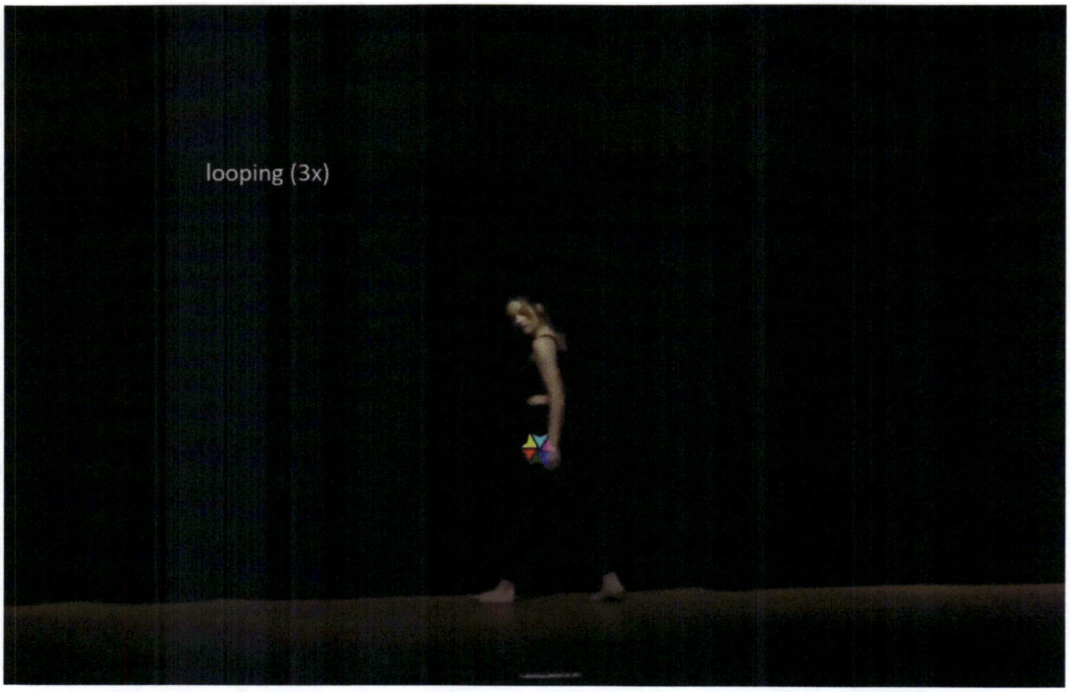
looping (3x)

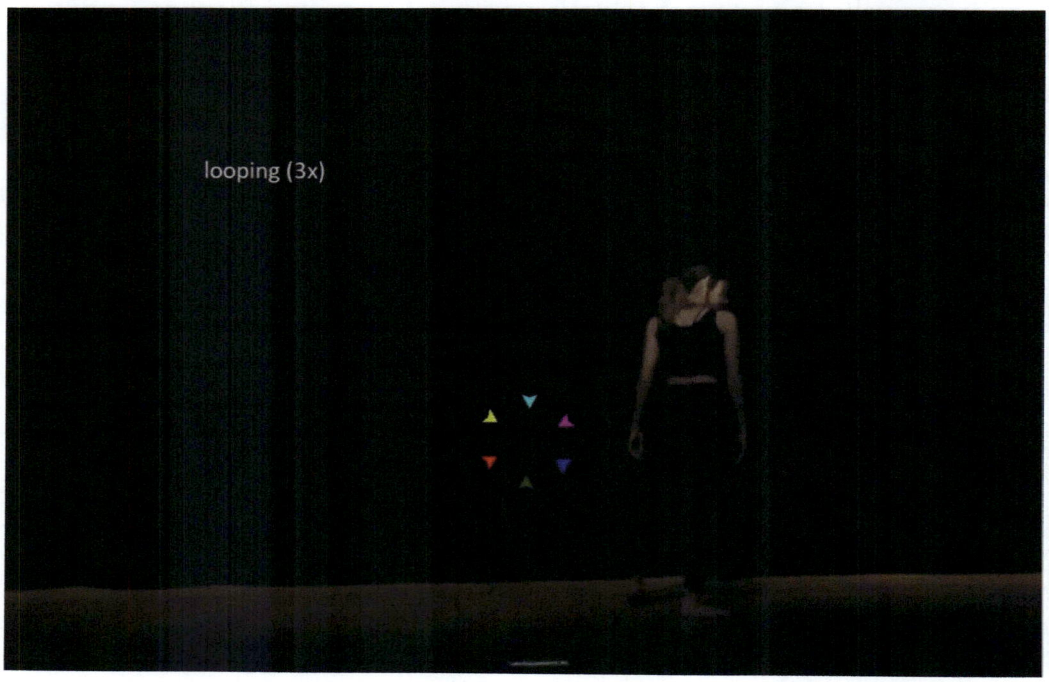
looping (3x)

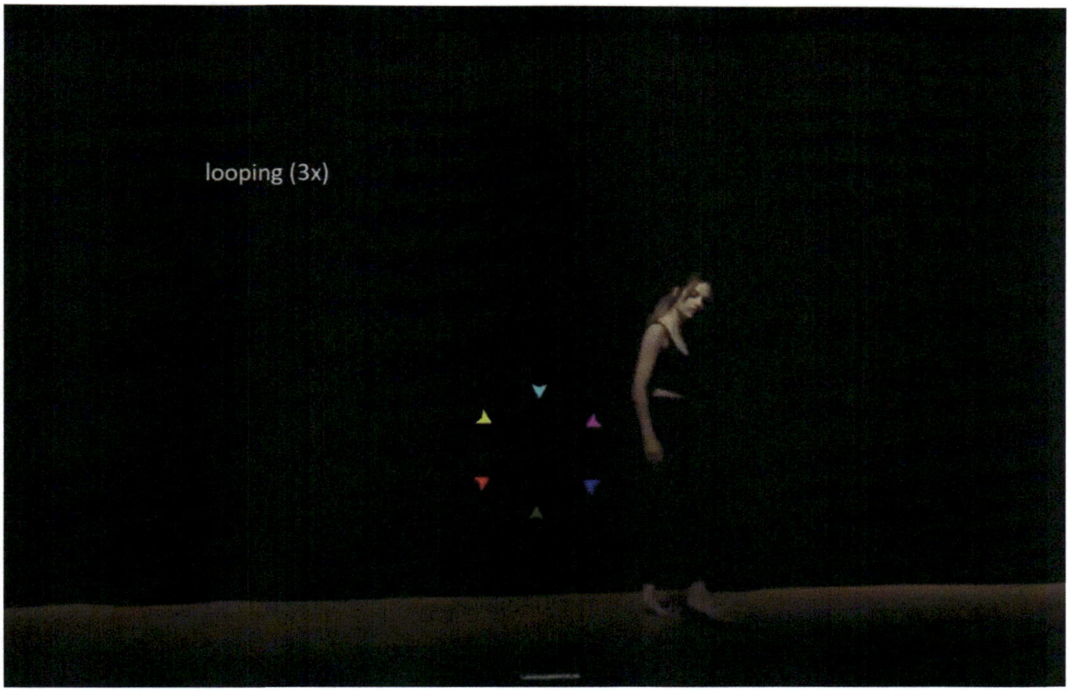
looping (3x)

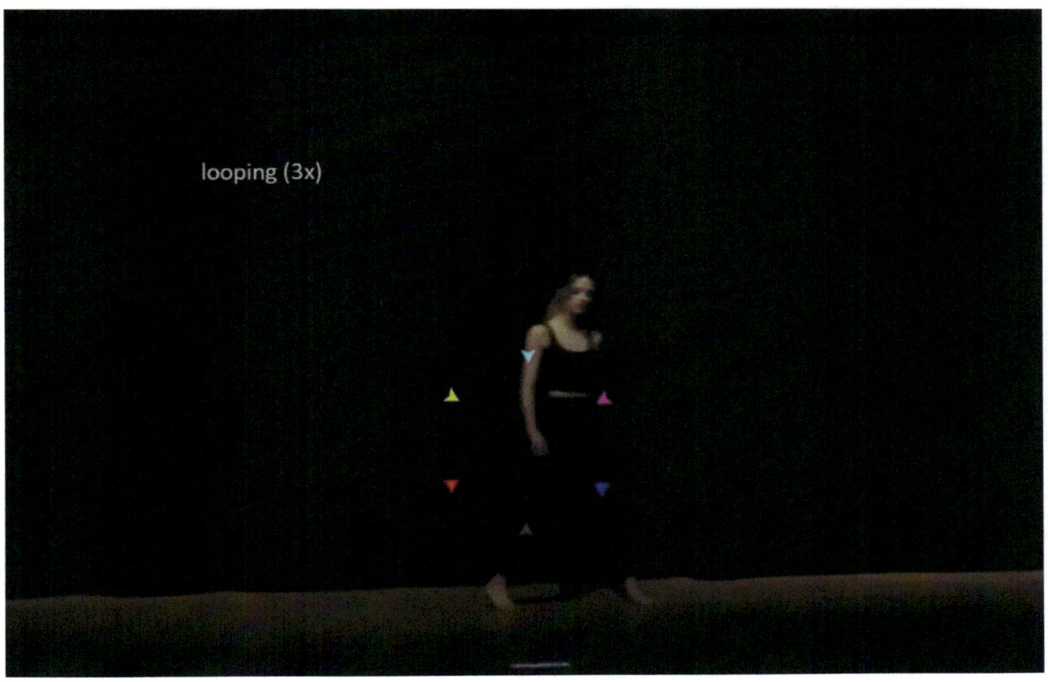
looping (3x)

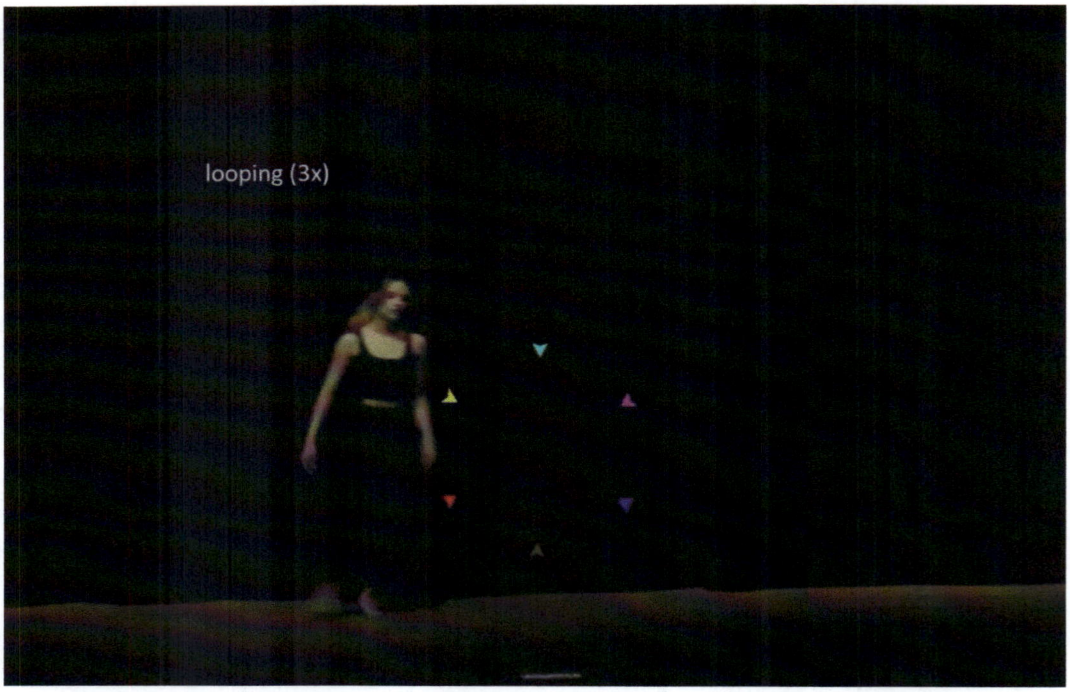
looping (3x)

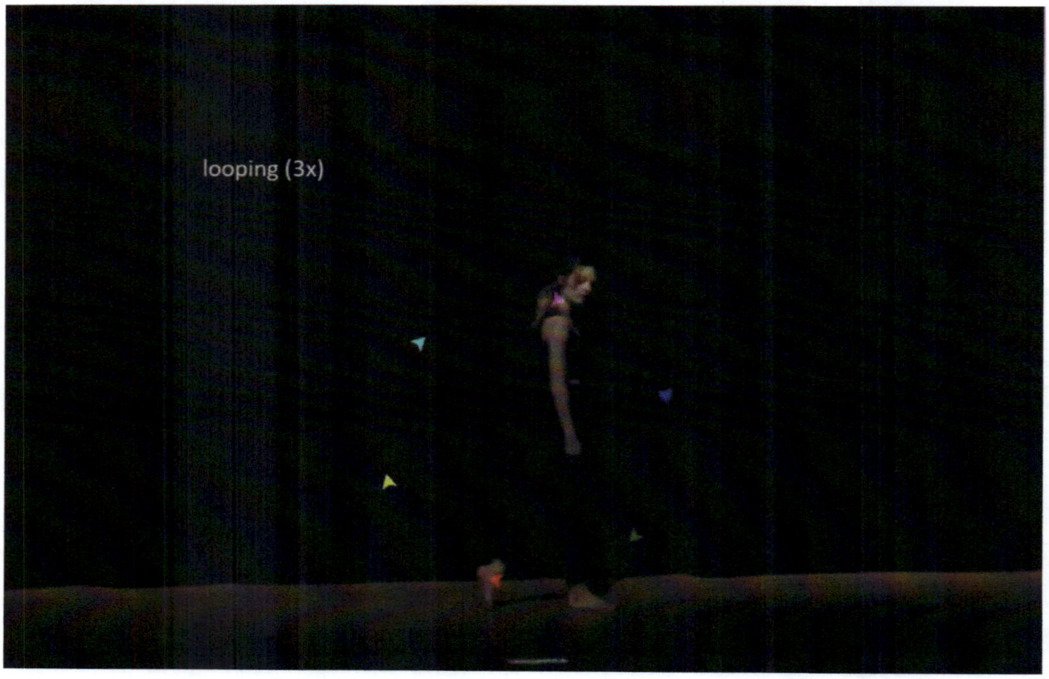
looping (3x)

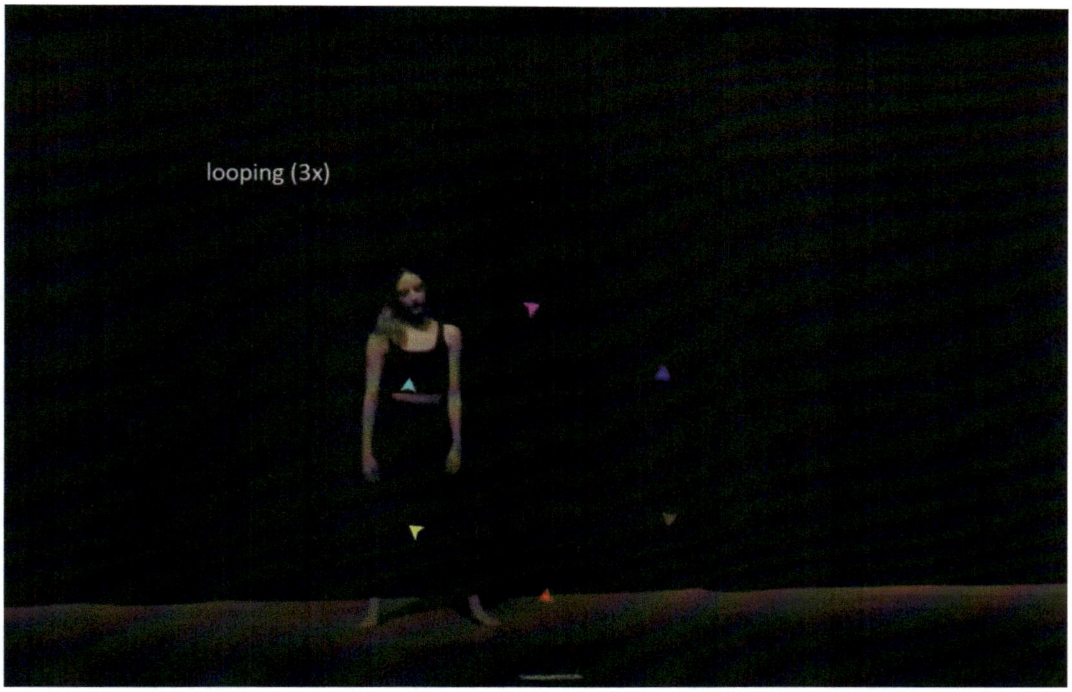
looping (3x)

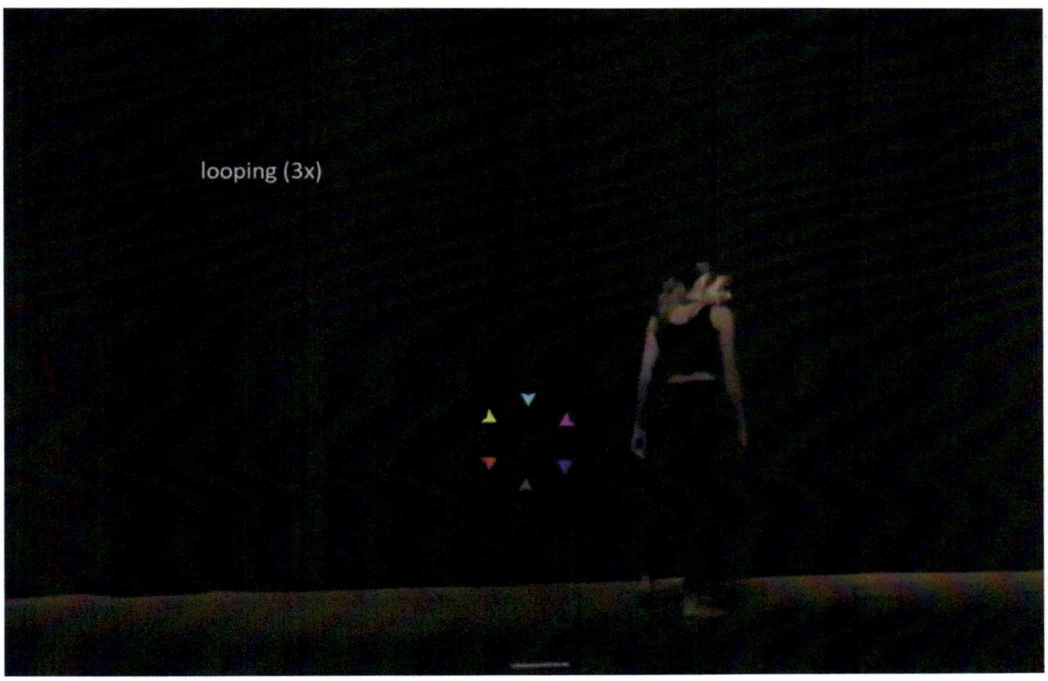
looping (3x)

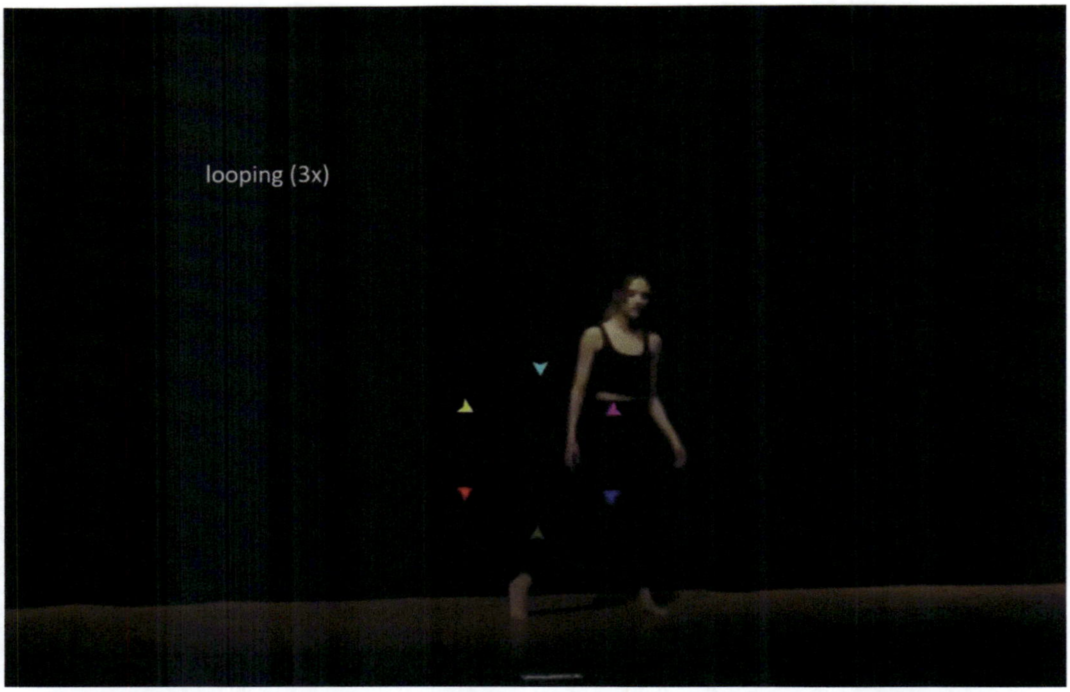
looping (3x)

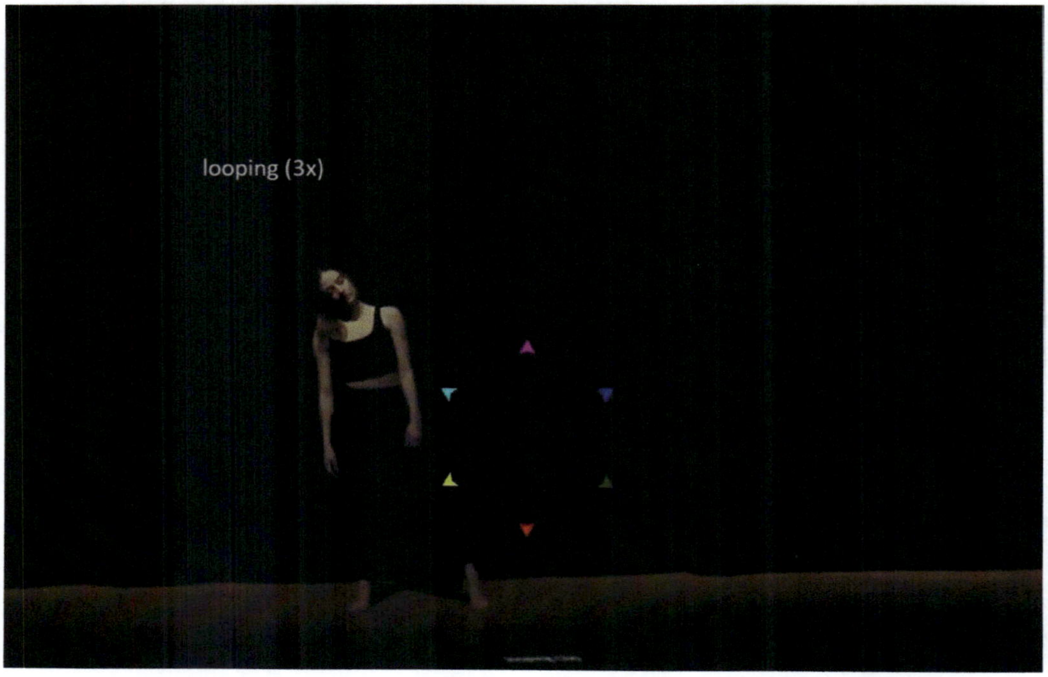
looping (3x)

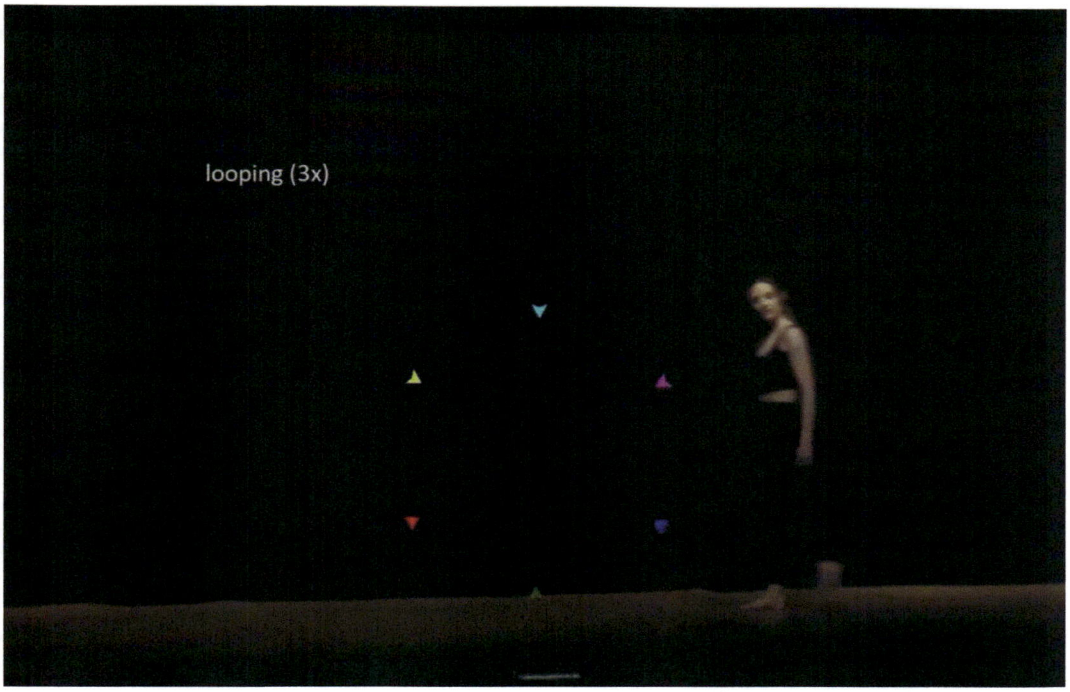
looping (3x)

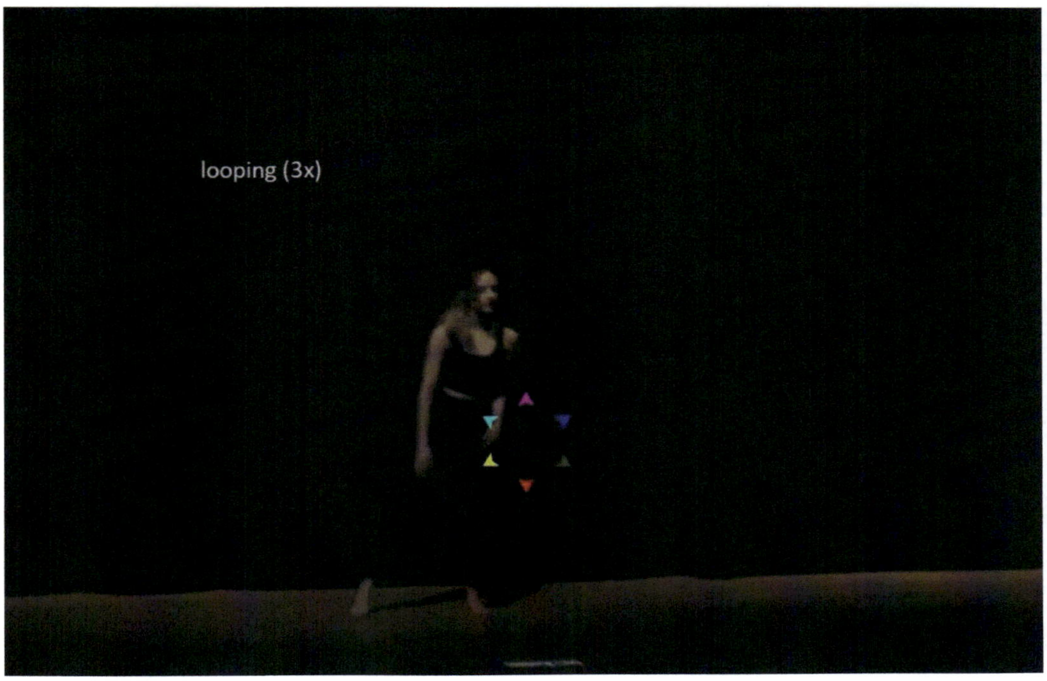
looping (3x)

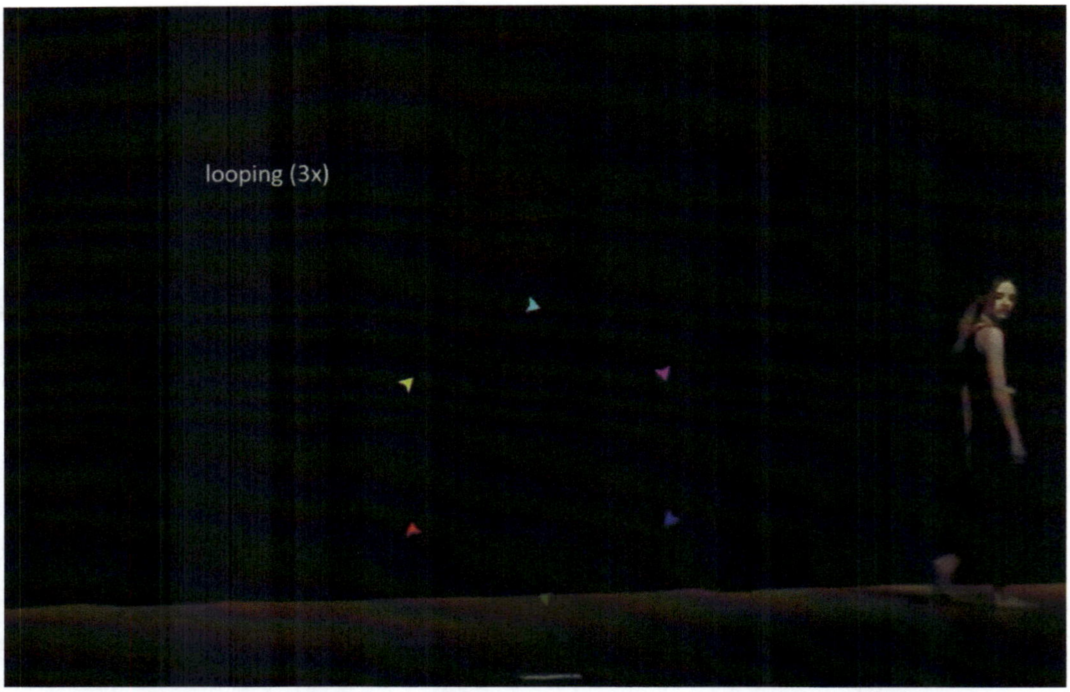
looping (3x)

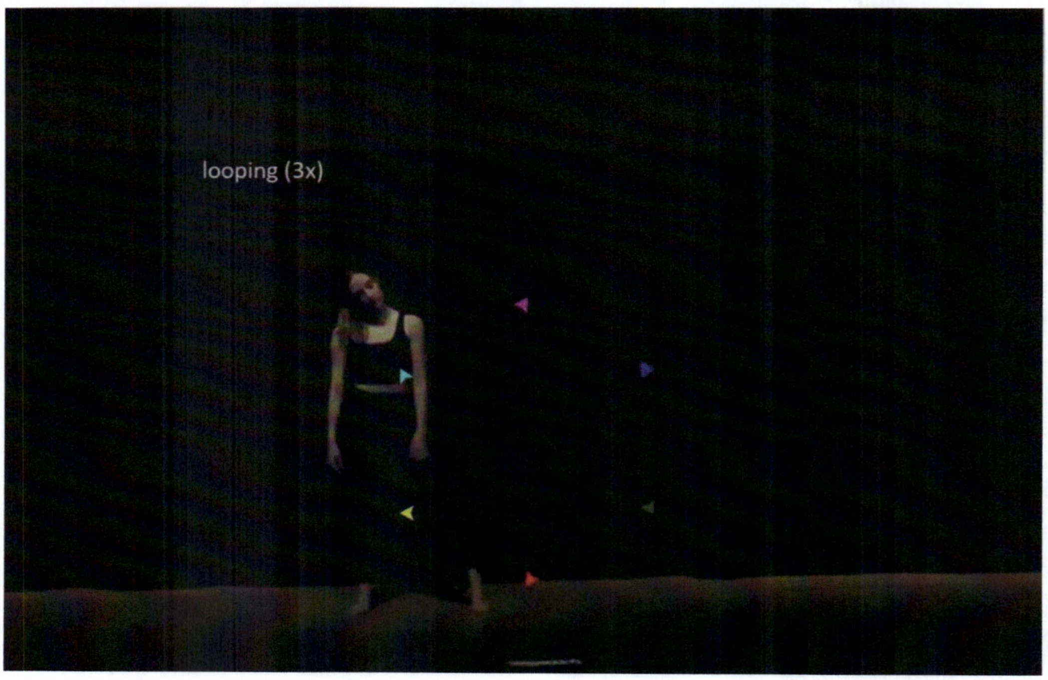
looping (3x)

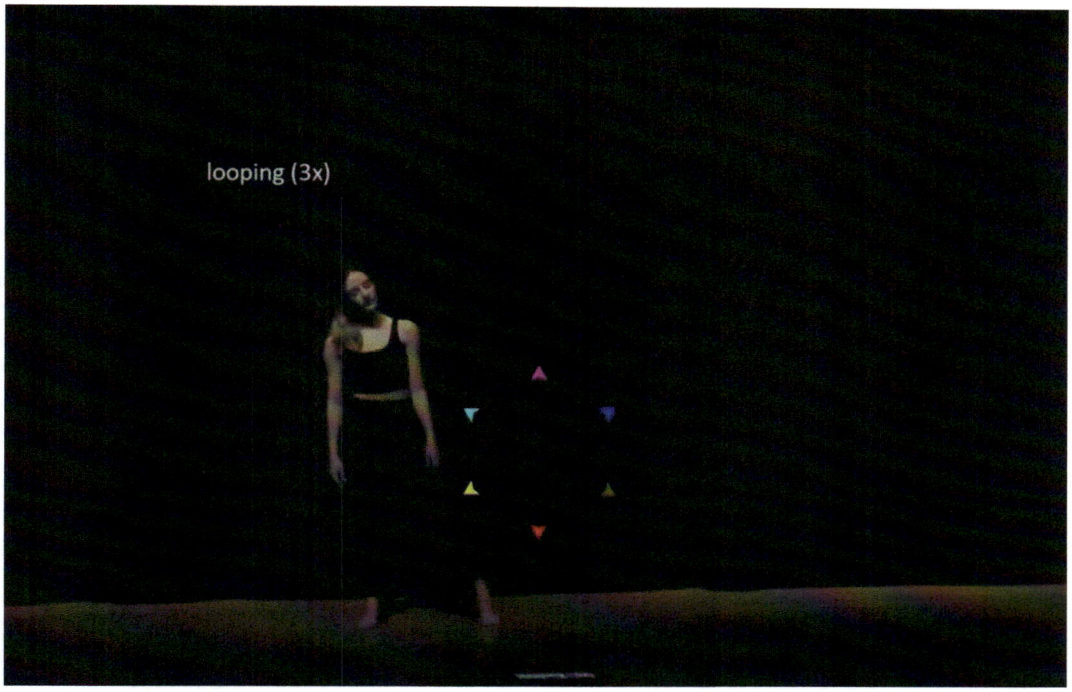
looping (3x)

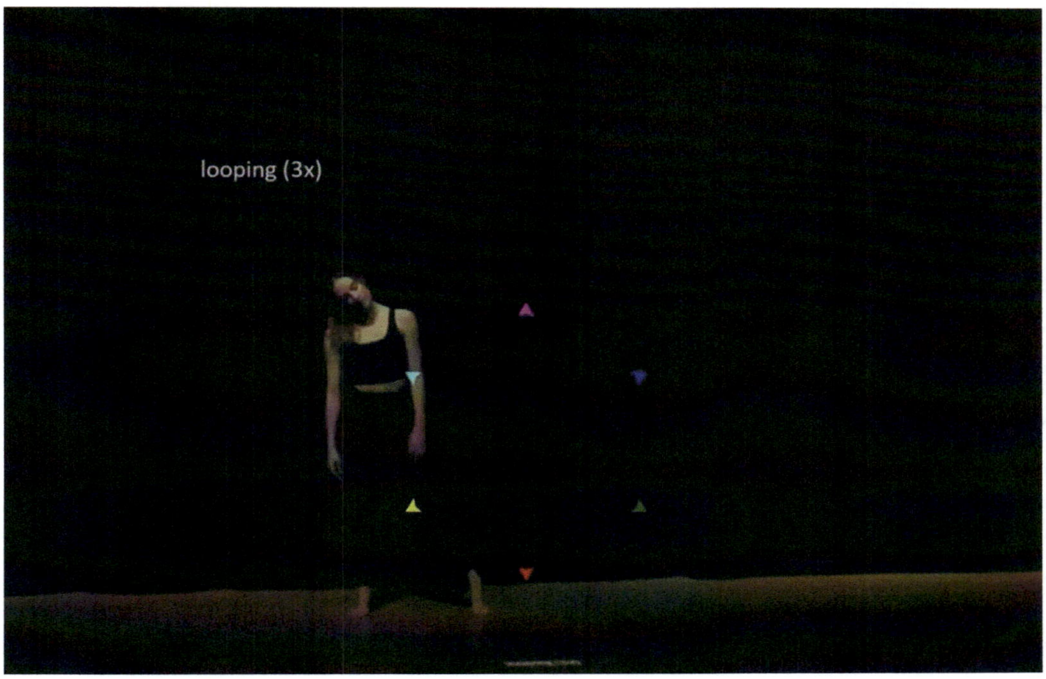
looping (3x)

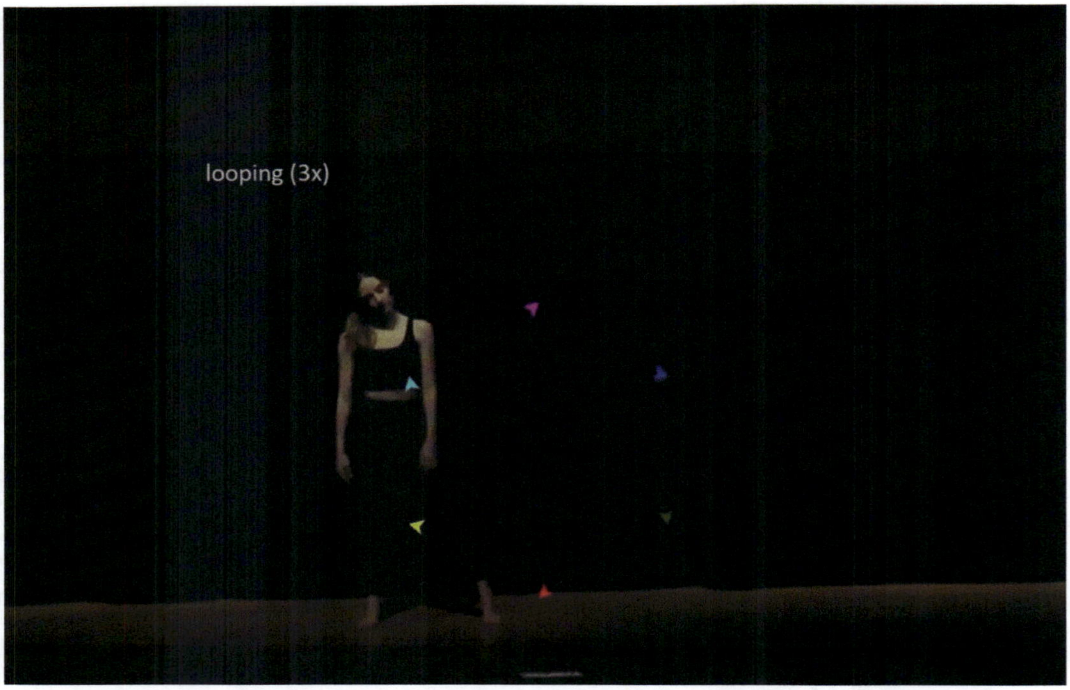
looping (3x)

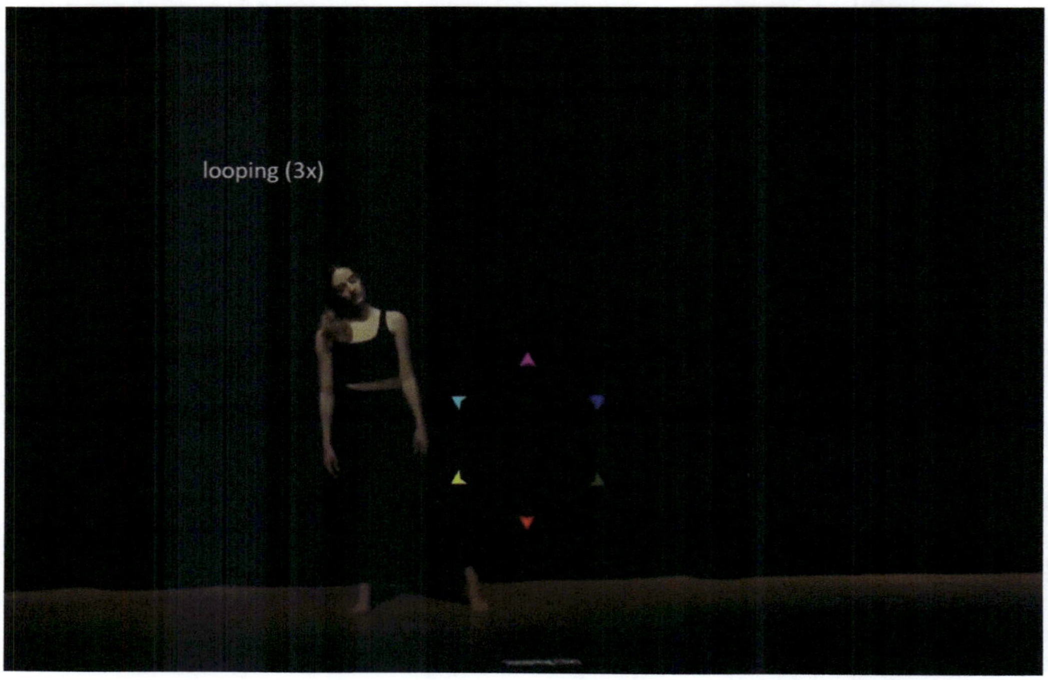
looping (3x)

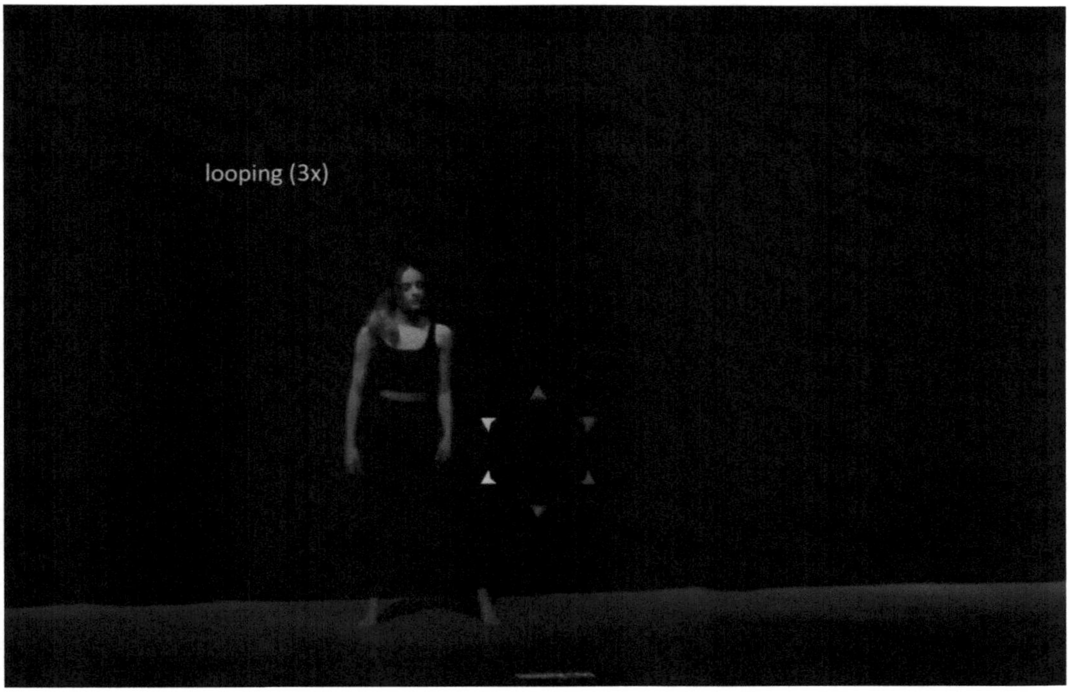
looping (3x)

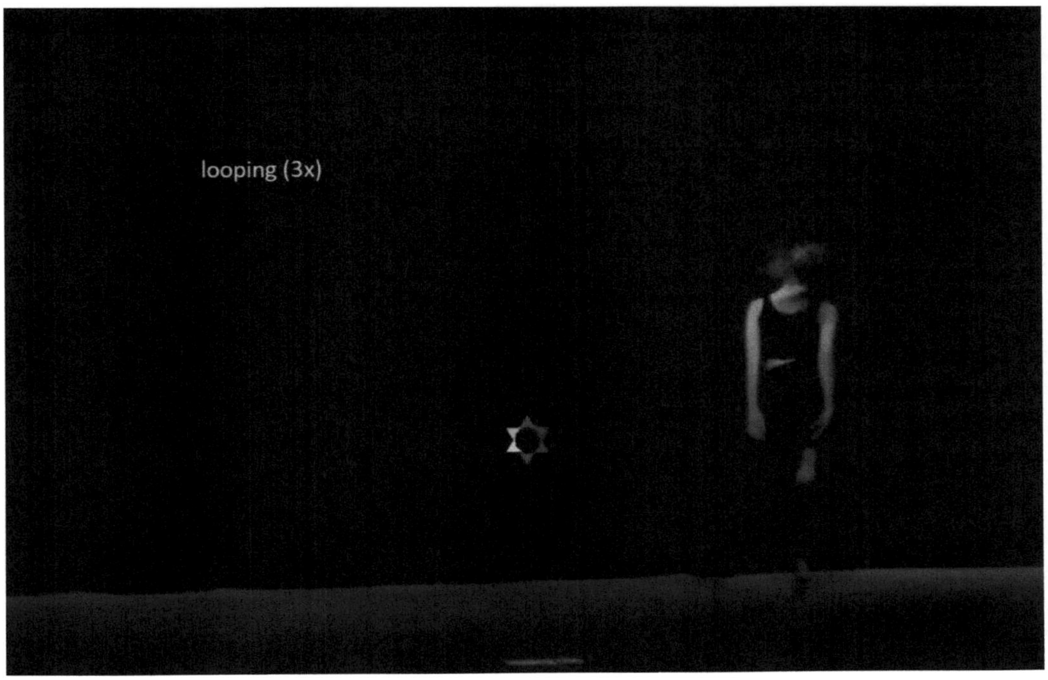
looping (3x)

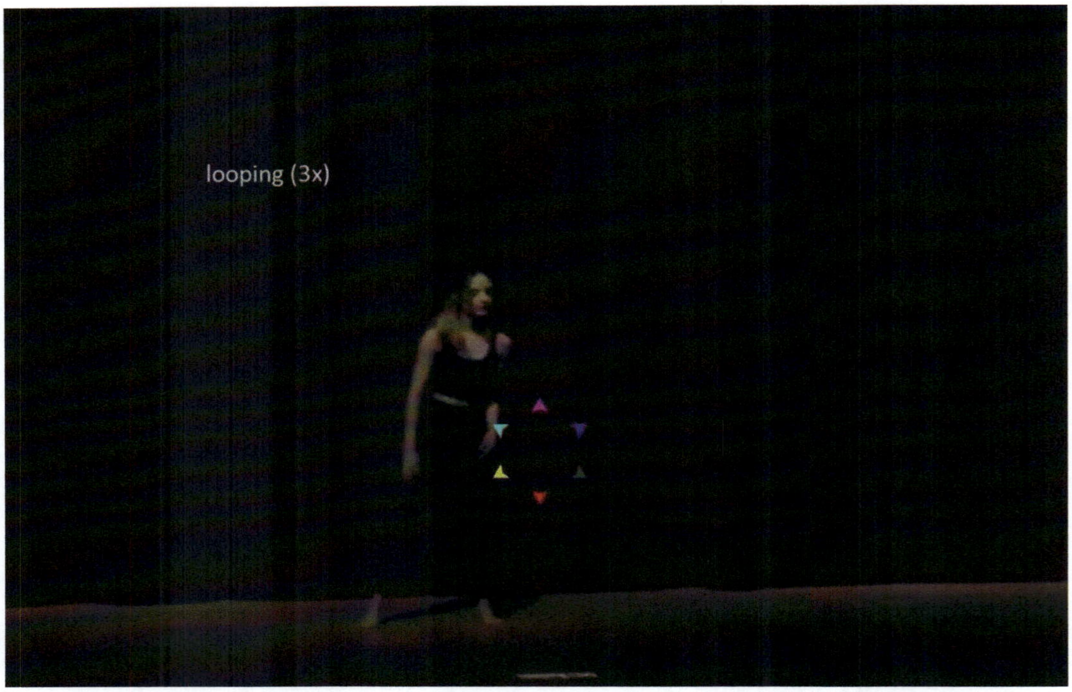
looping (3x)

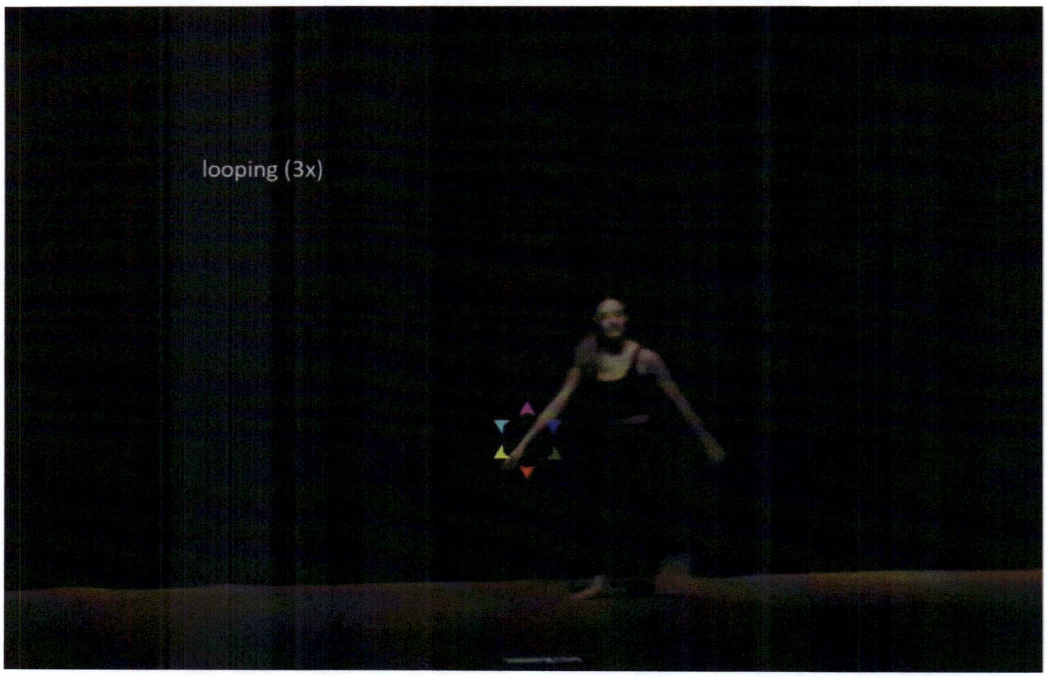
looping (3x)

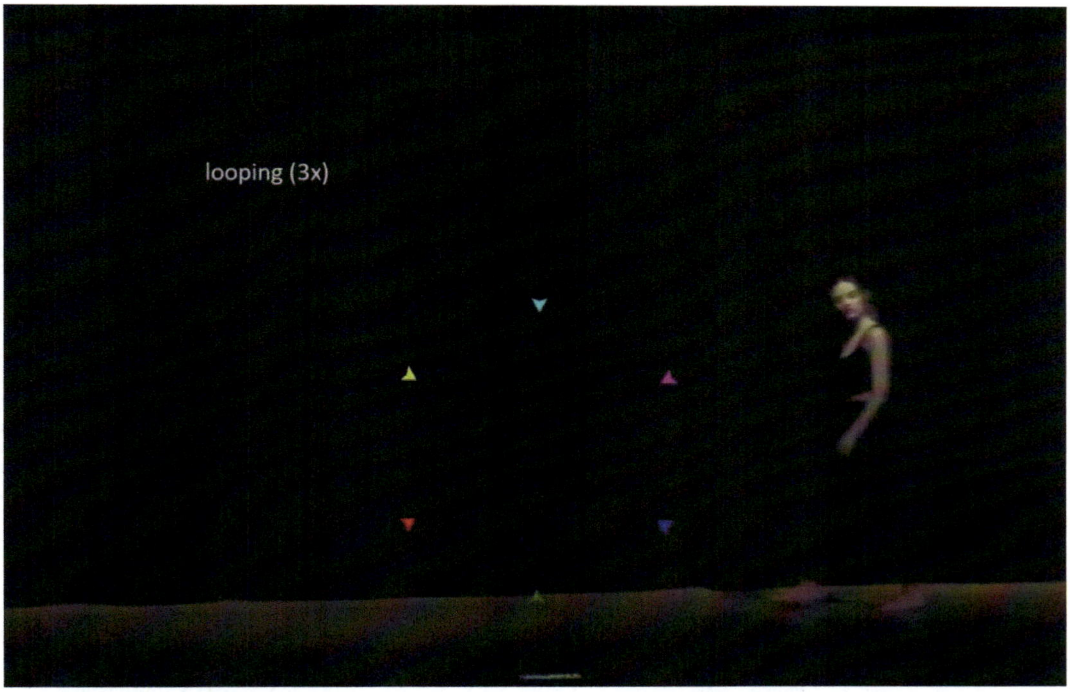
looping (3x)

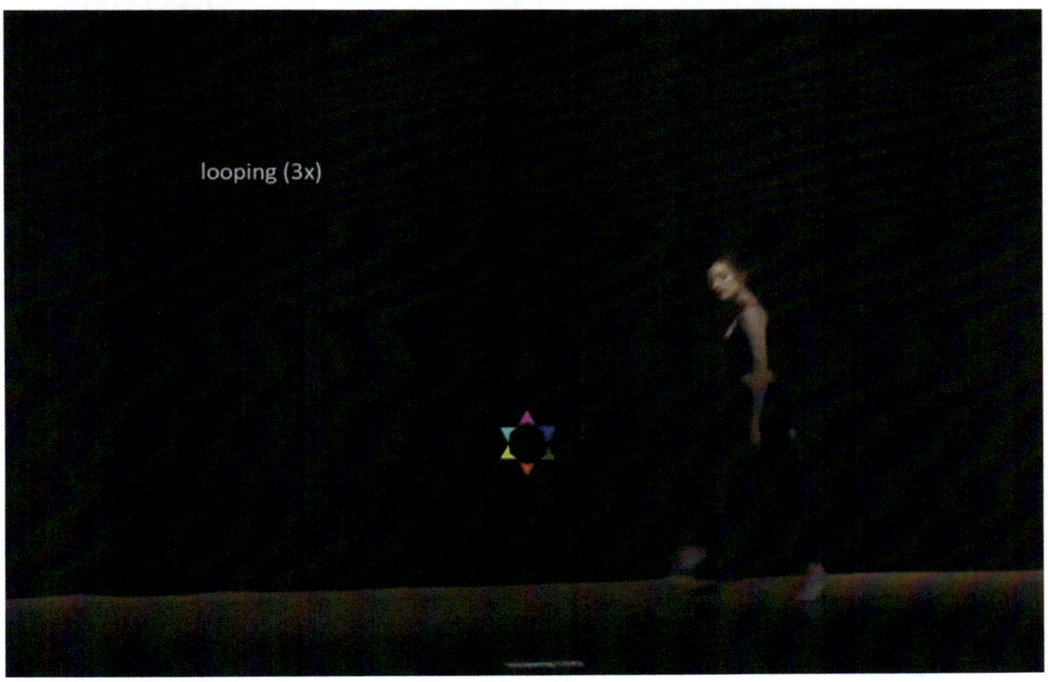
looping (3x)

Numeric computation

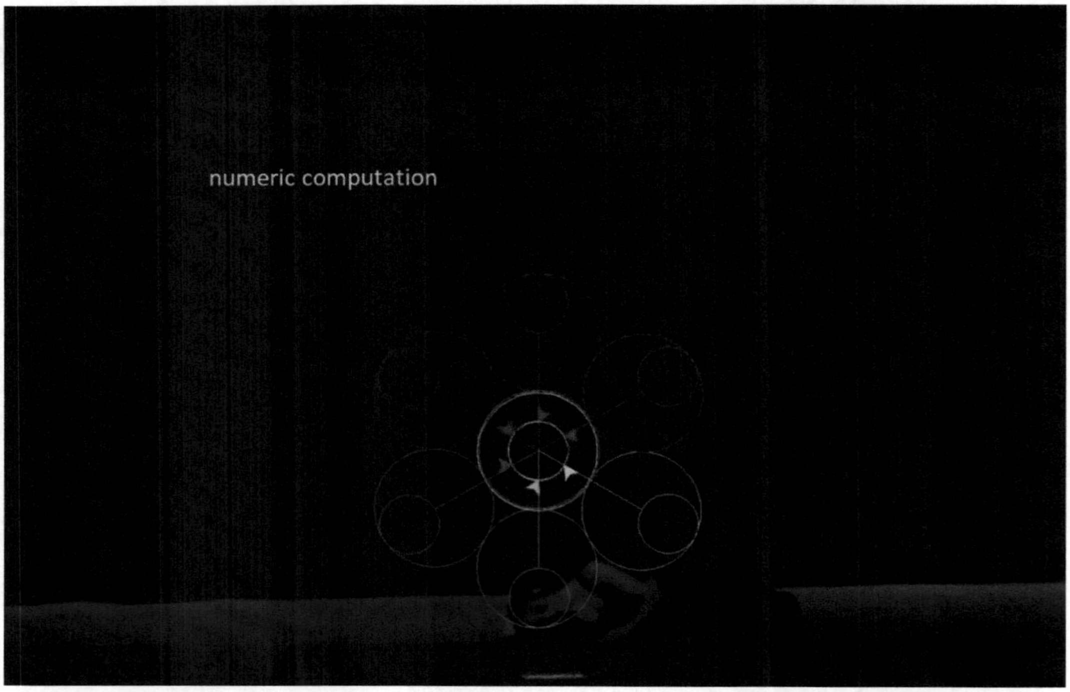

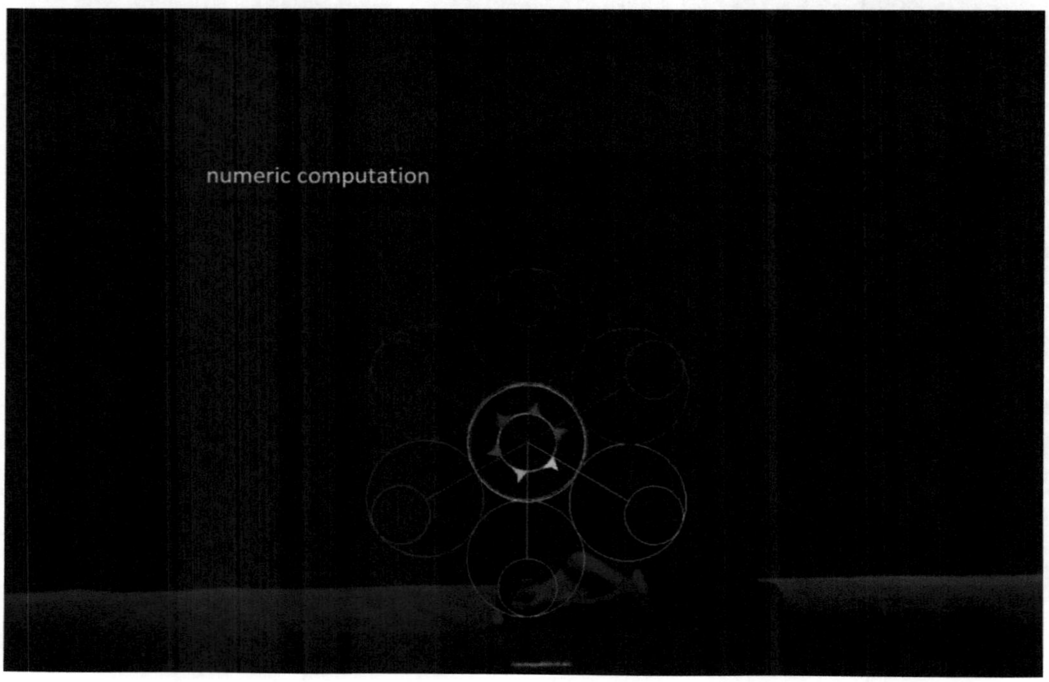

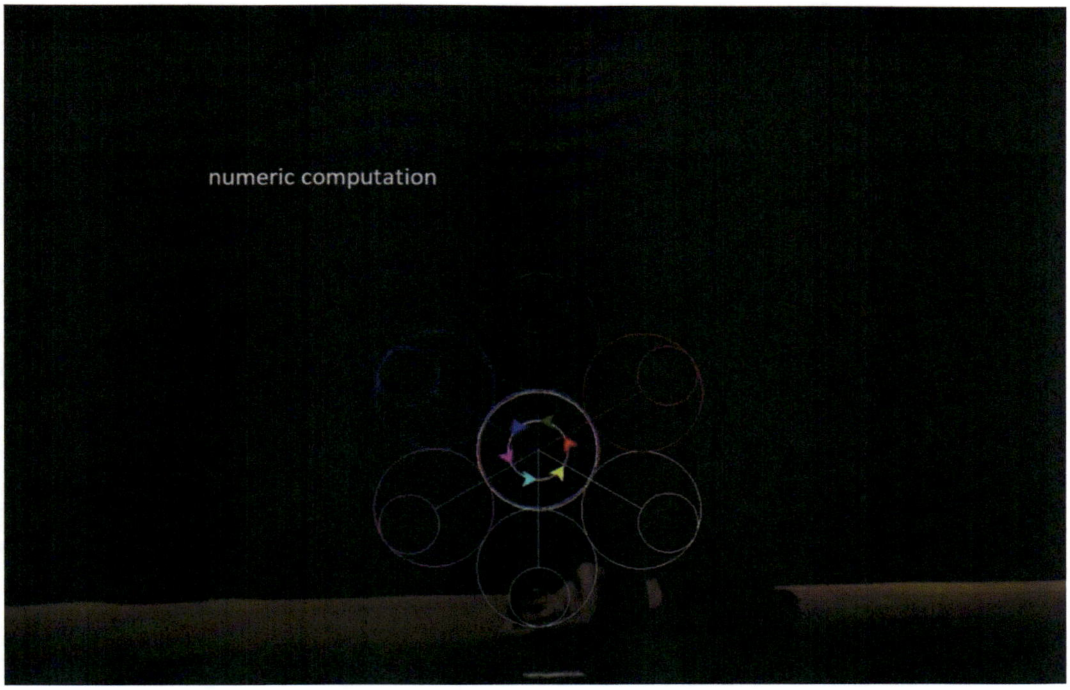
numeric computation

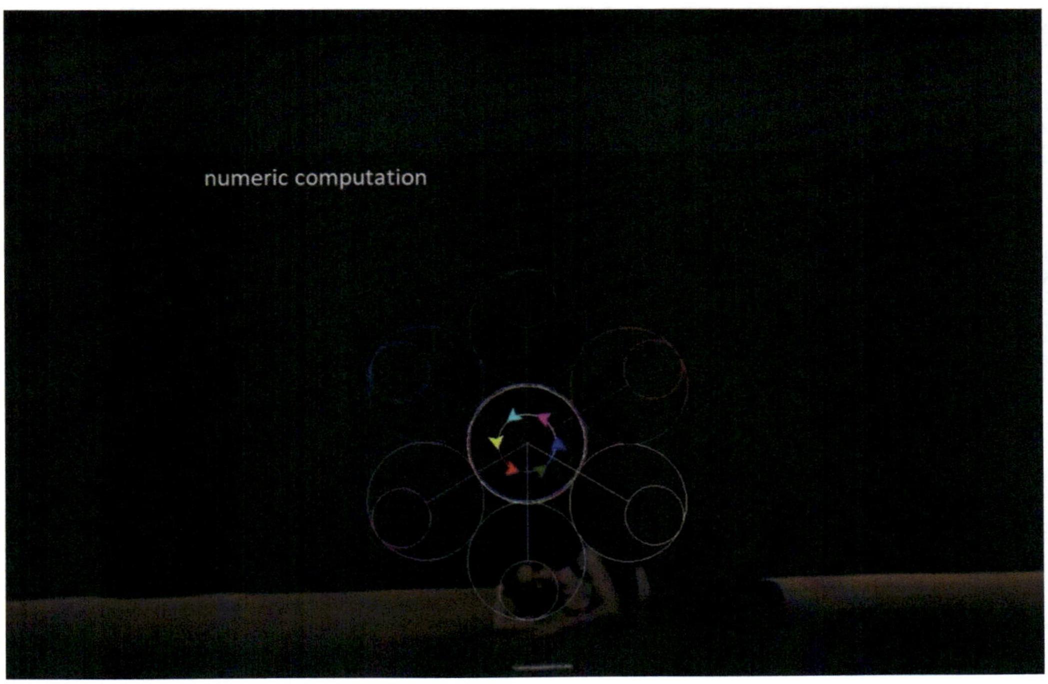
numeric computation

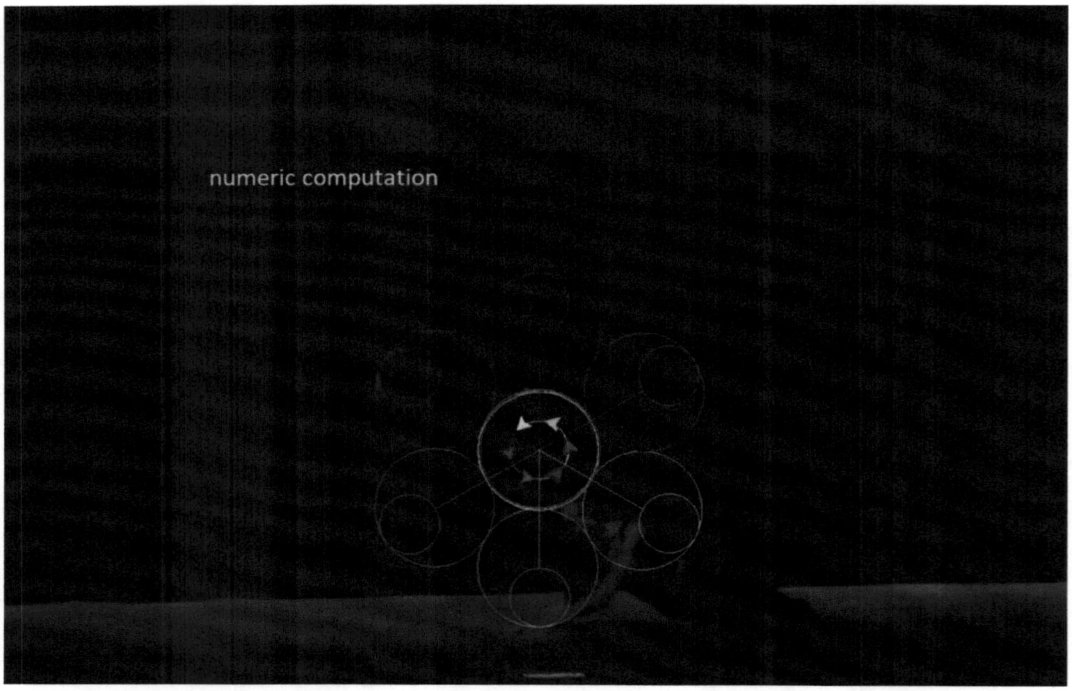
numeric computation

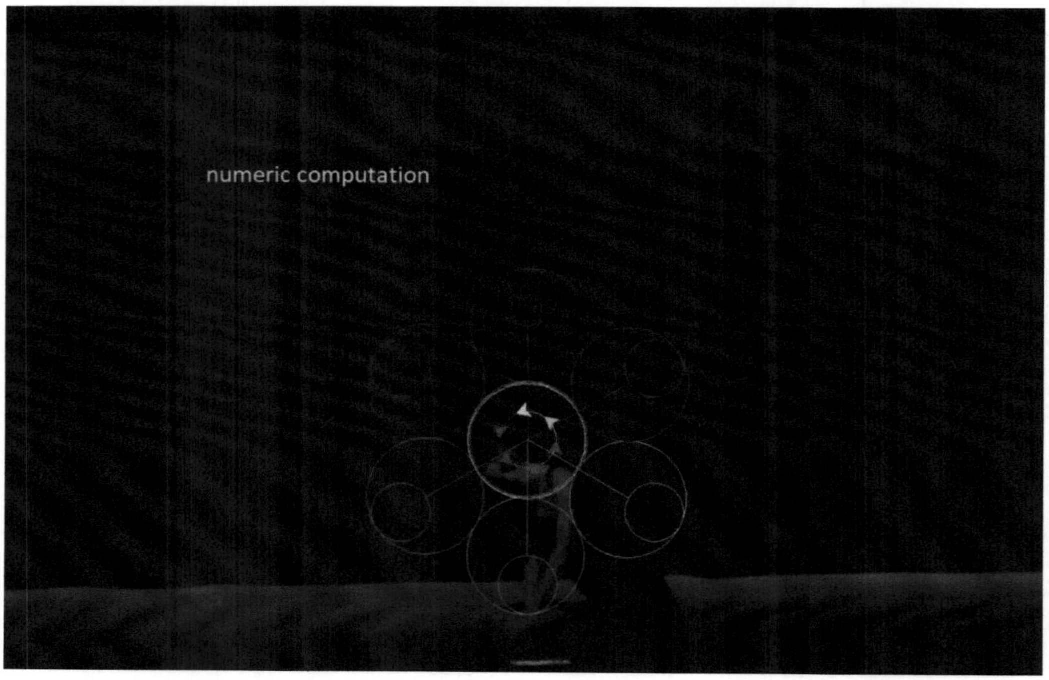
numeric computation

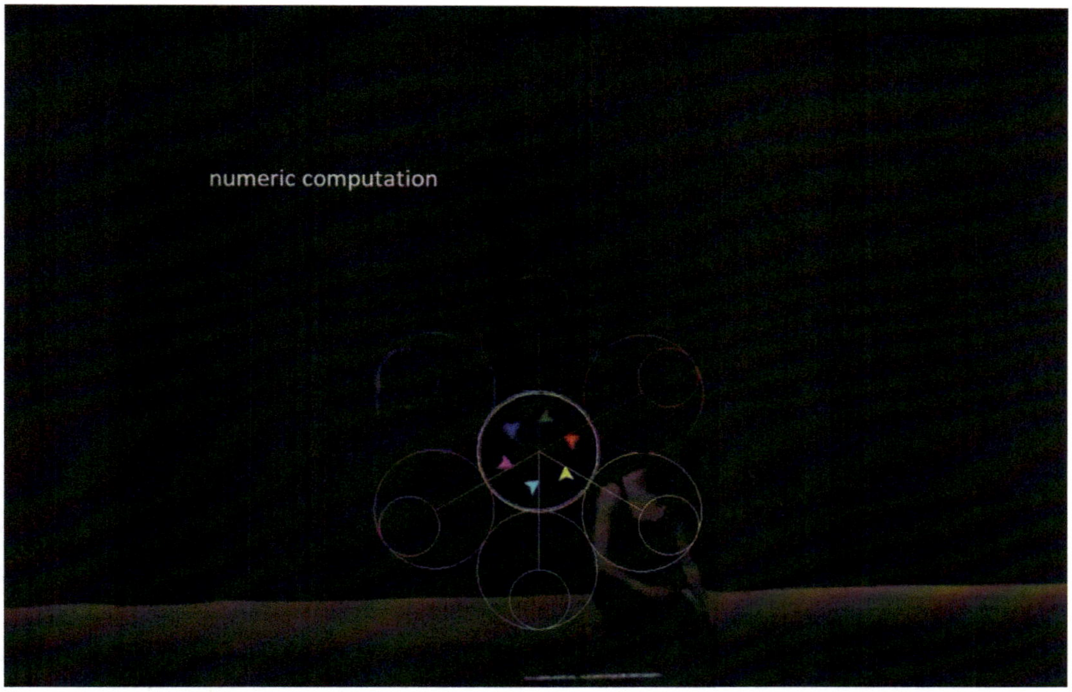
numeric computation

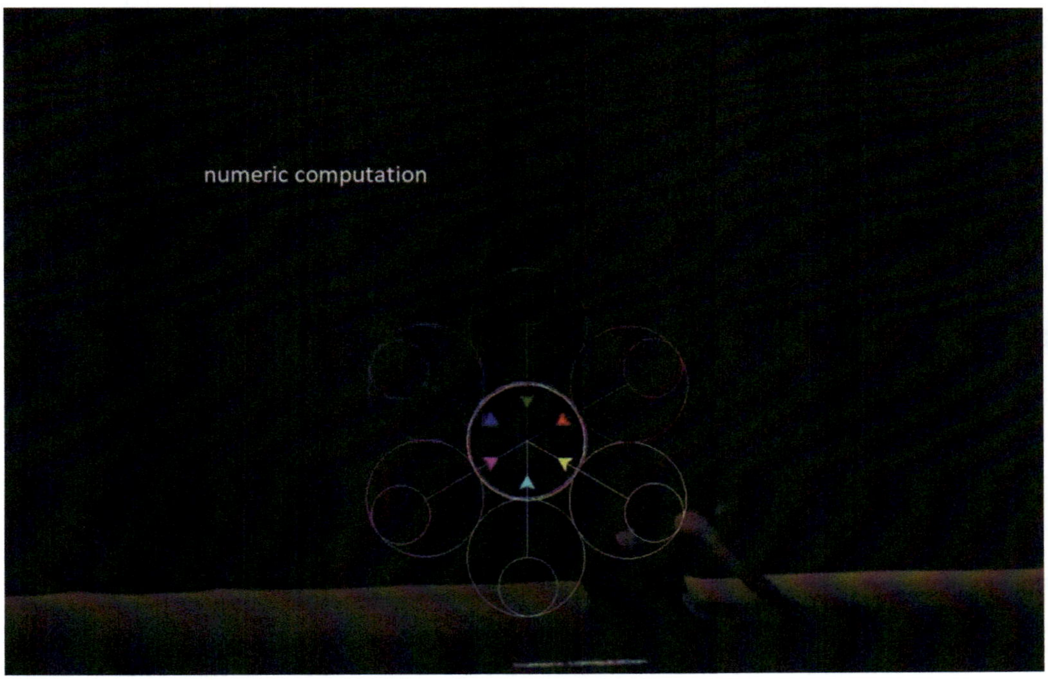
numeric computation

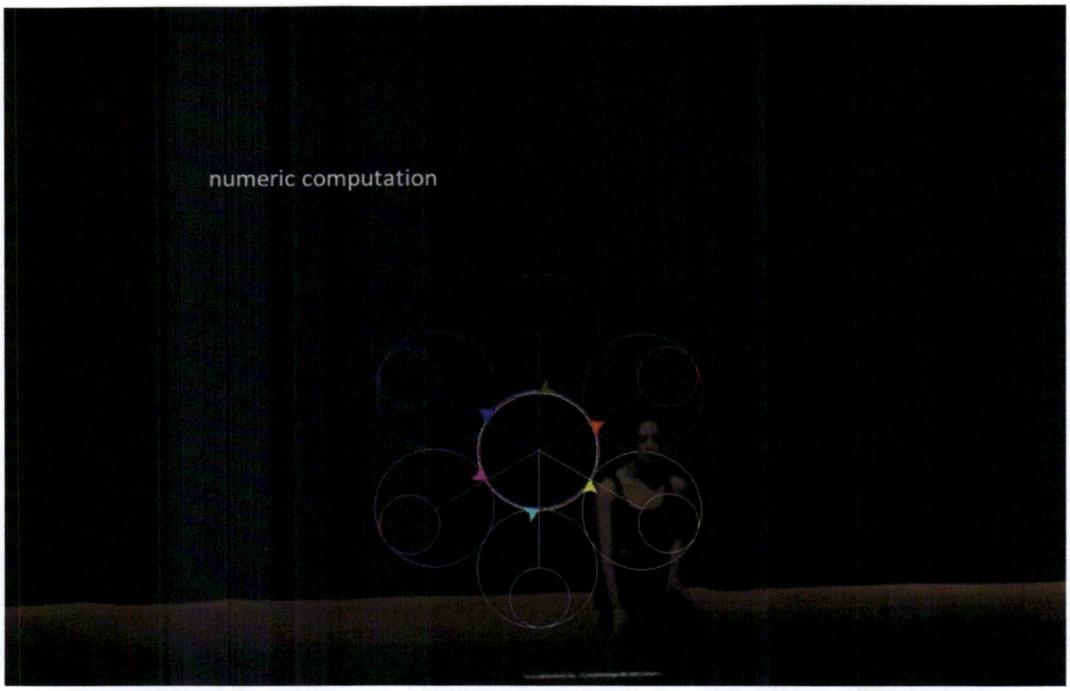
numeric computation

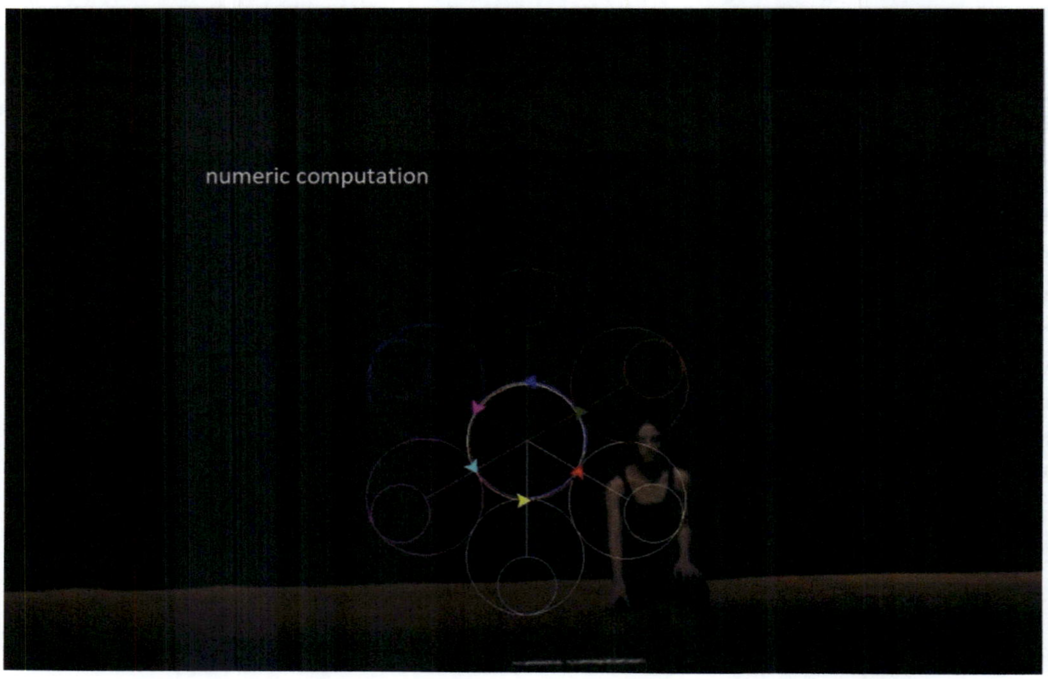
numeric computation

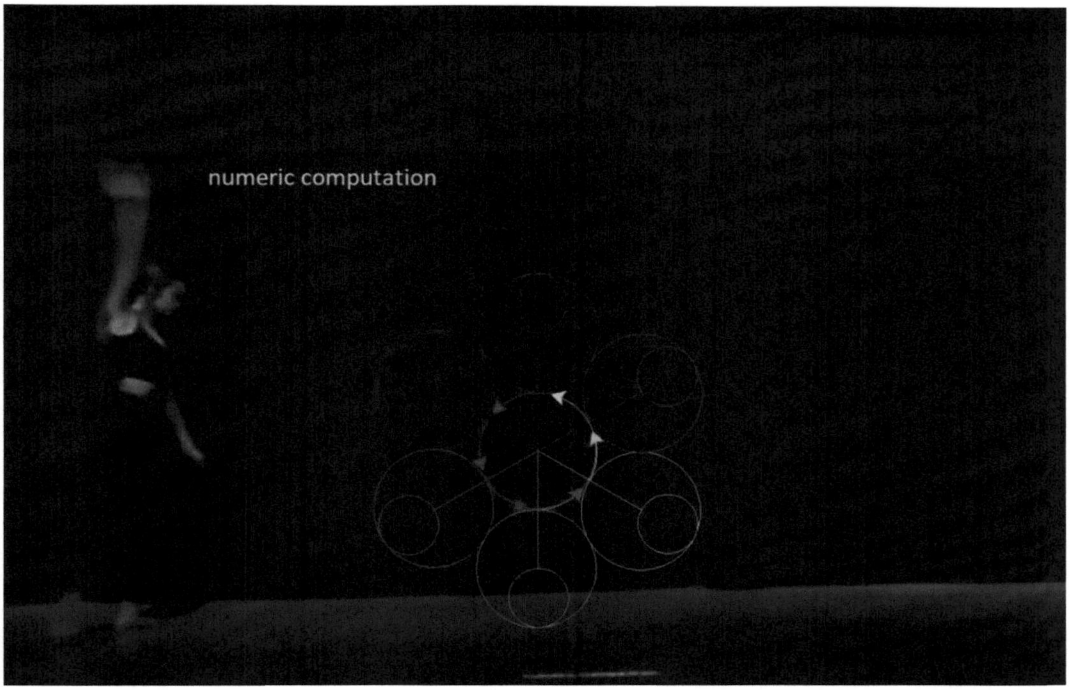
numeric computation

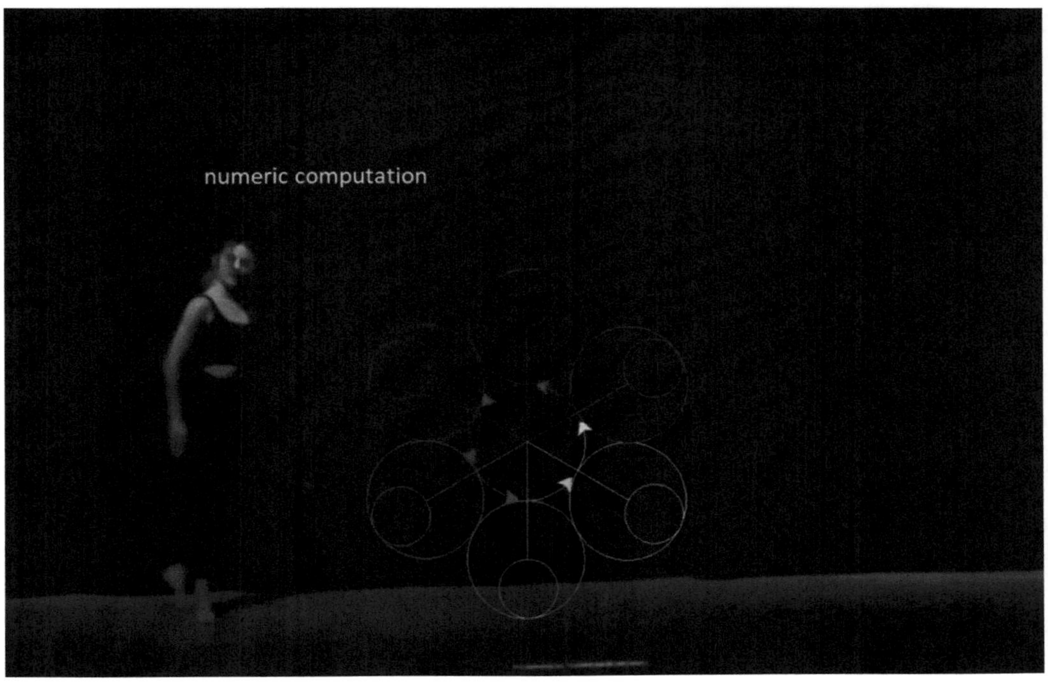
numeric computation

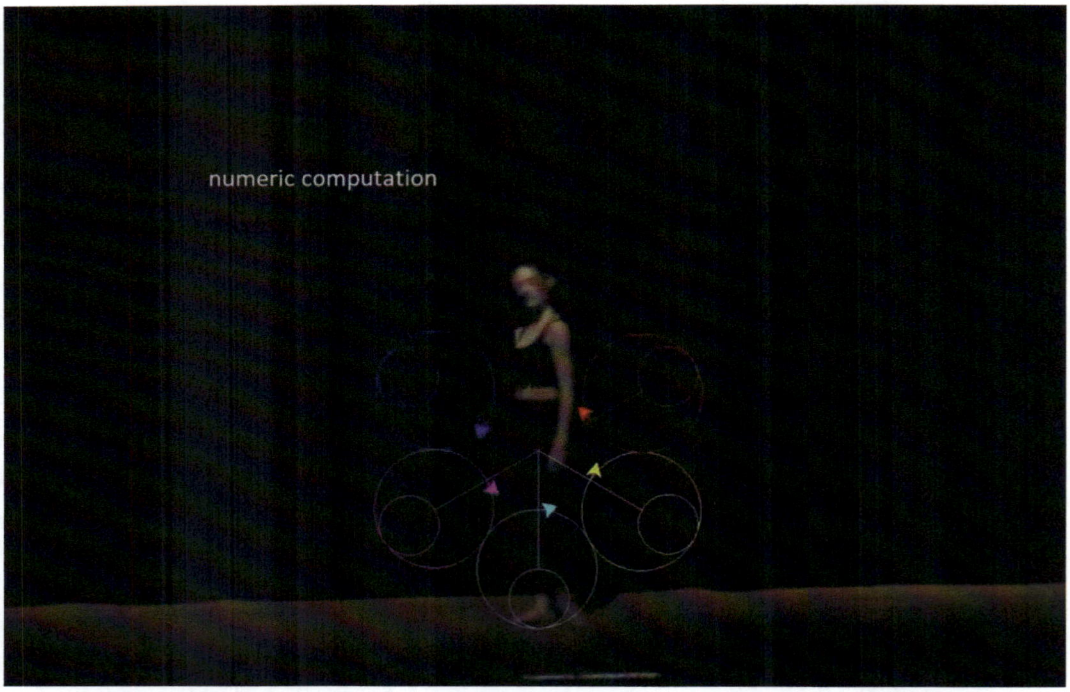
numeric computation

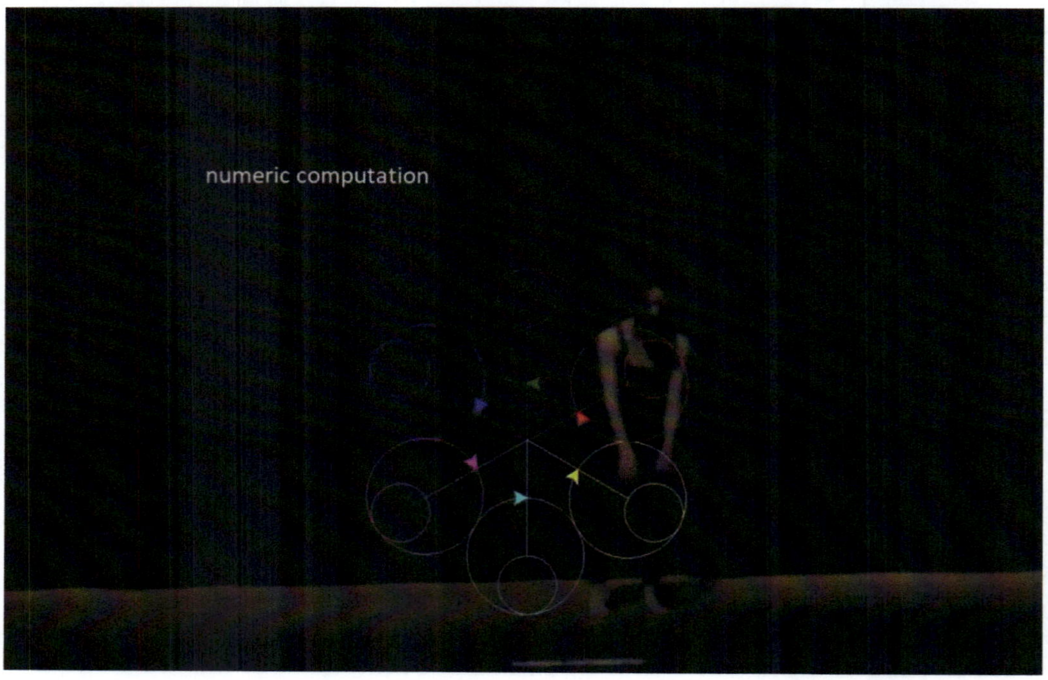
numeric computation

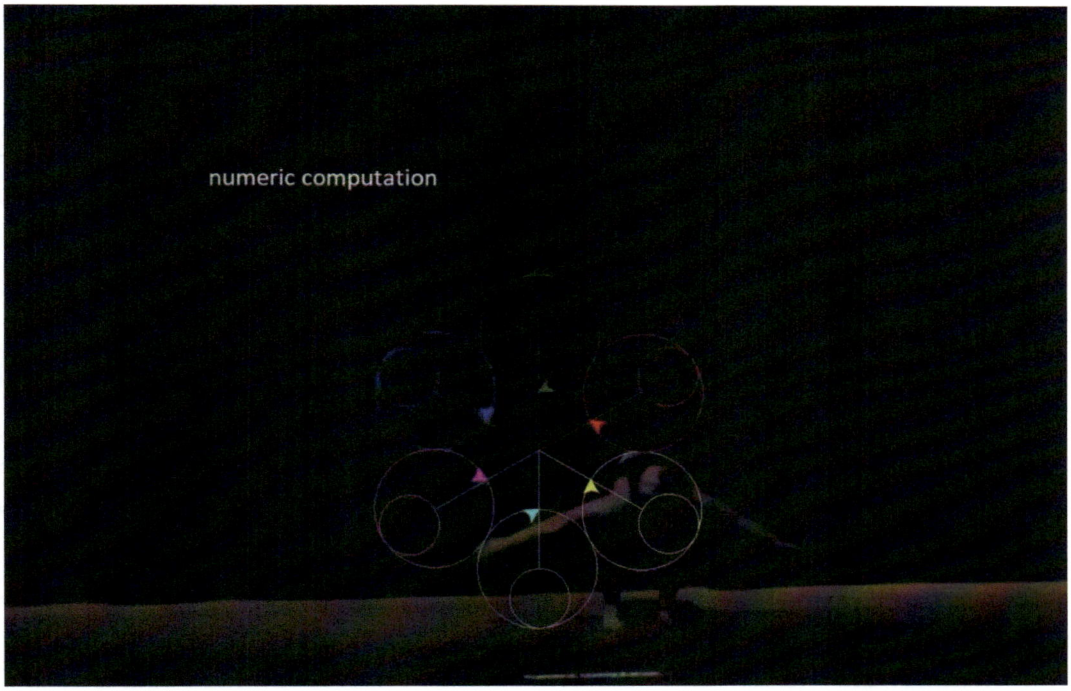
numeric computation

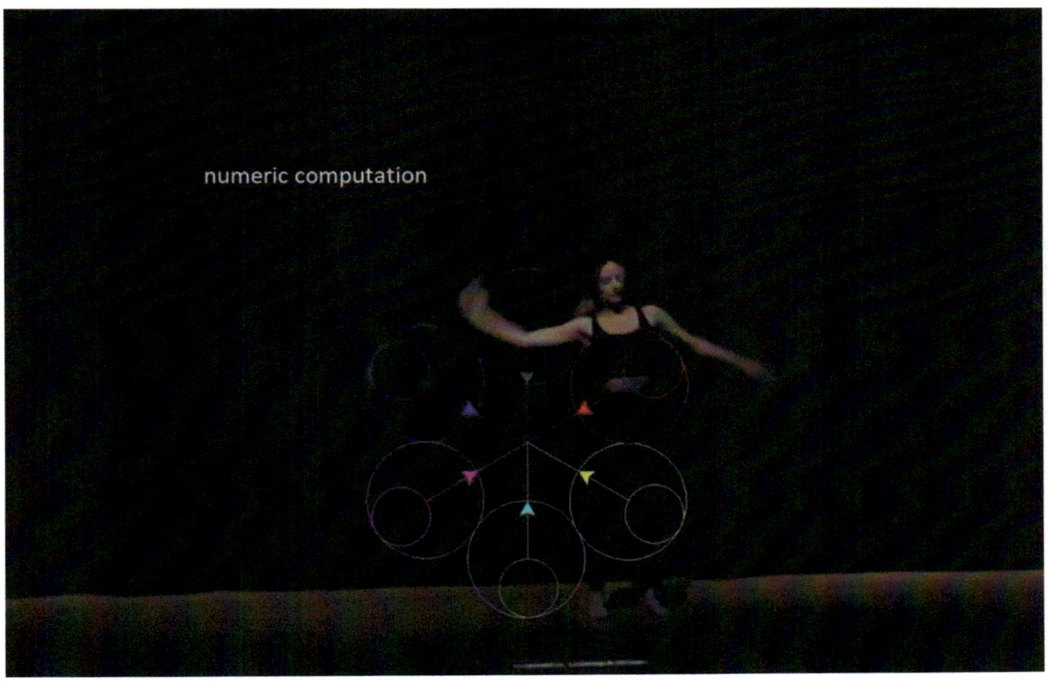
numeric computation

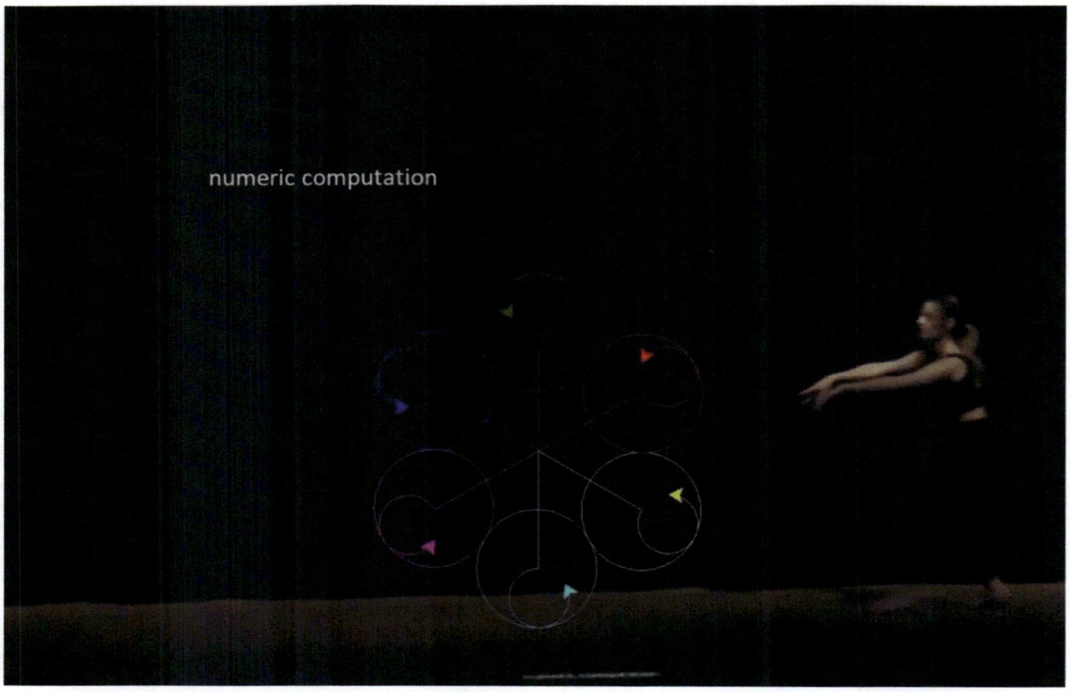
numeric computation

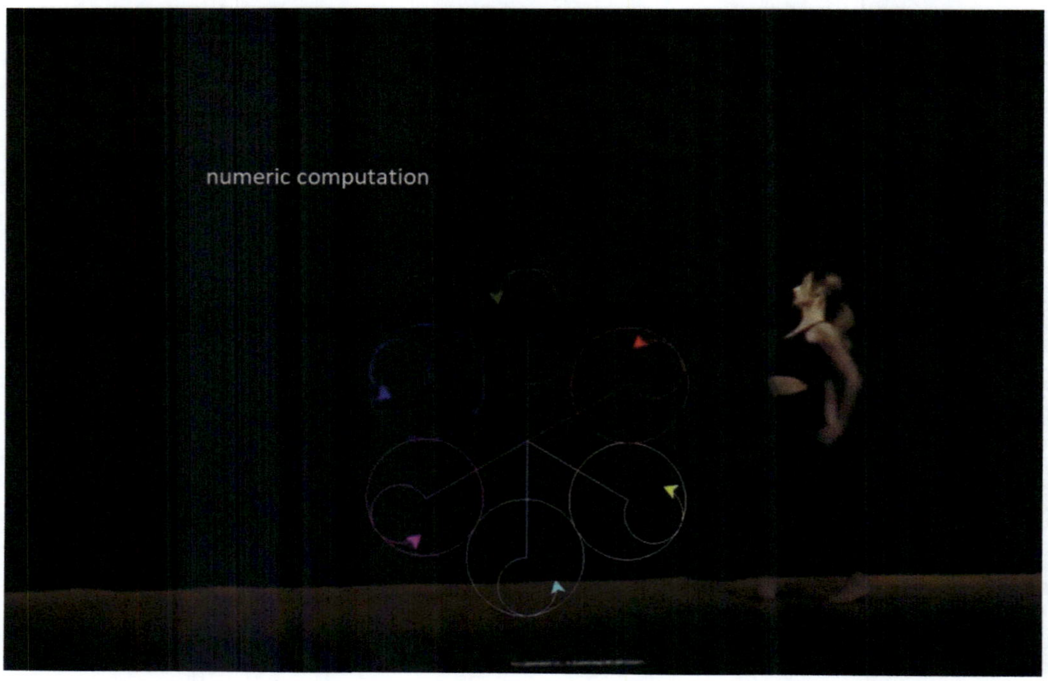
numeric computation

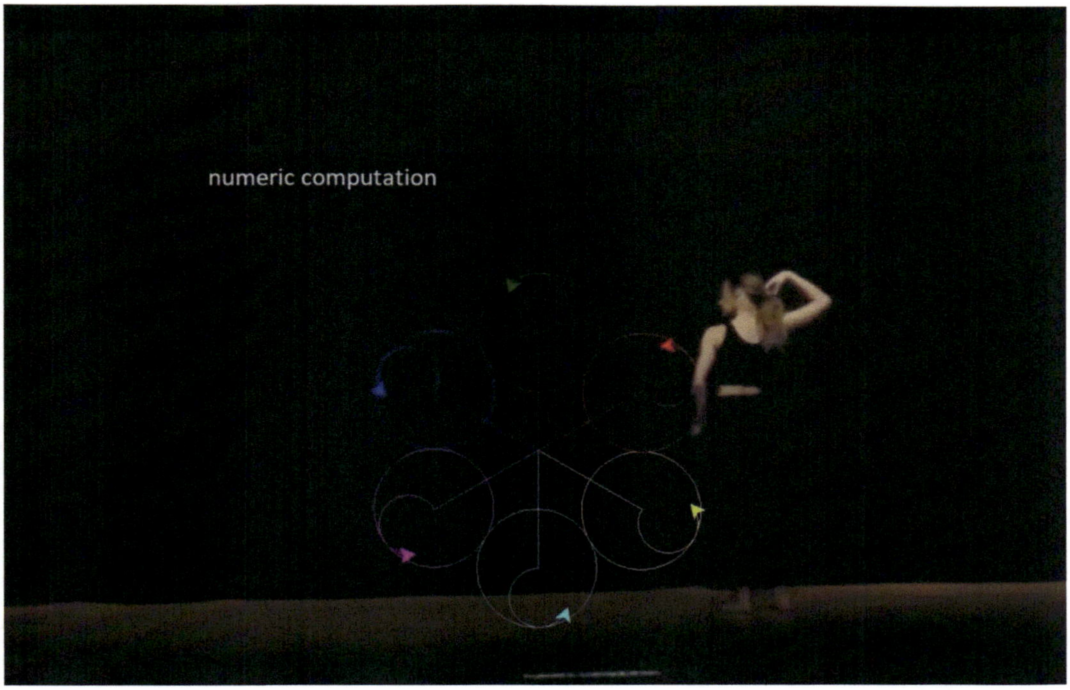
numeric computation

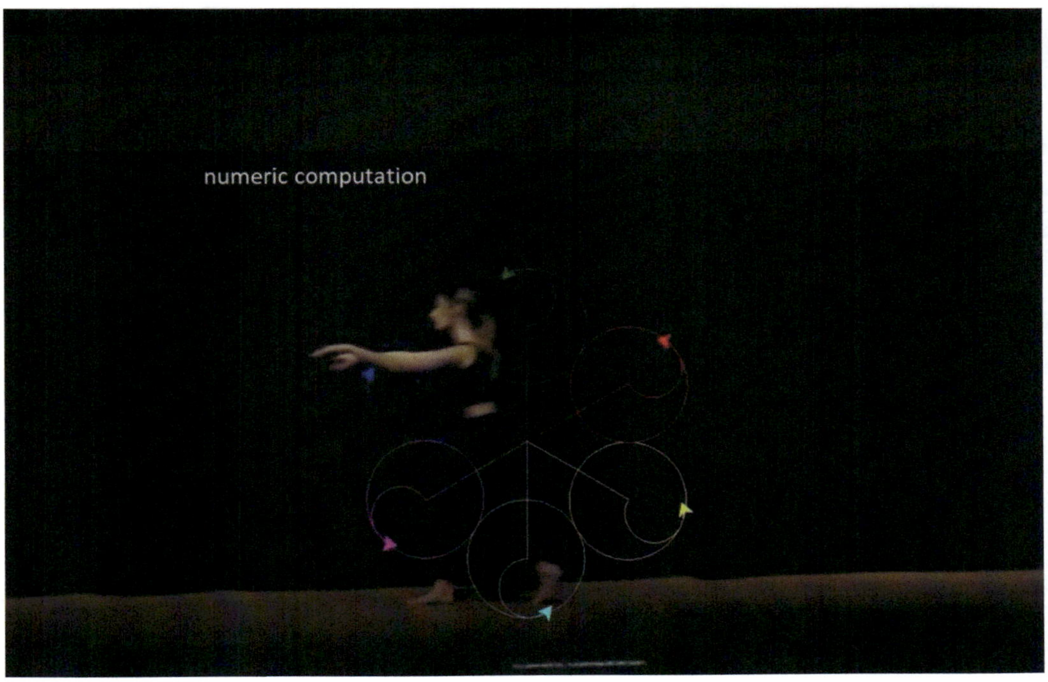
numeric computation

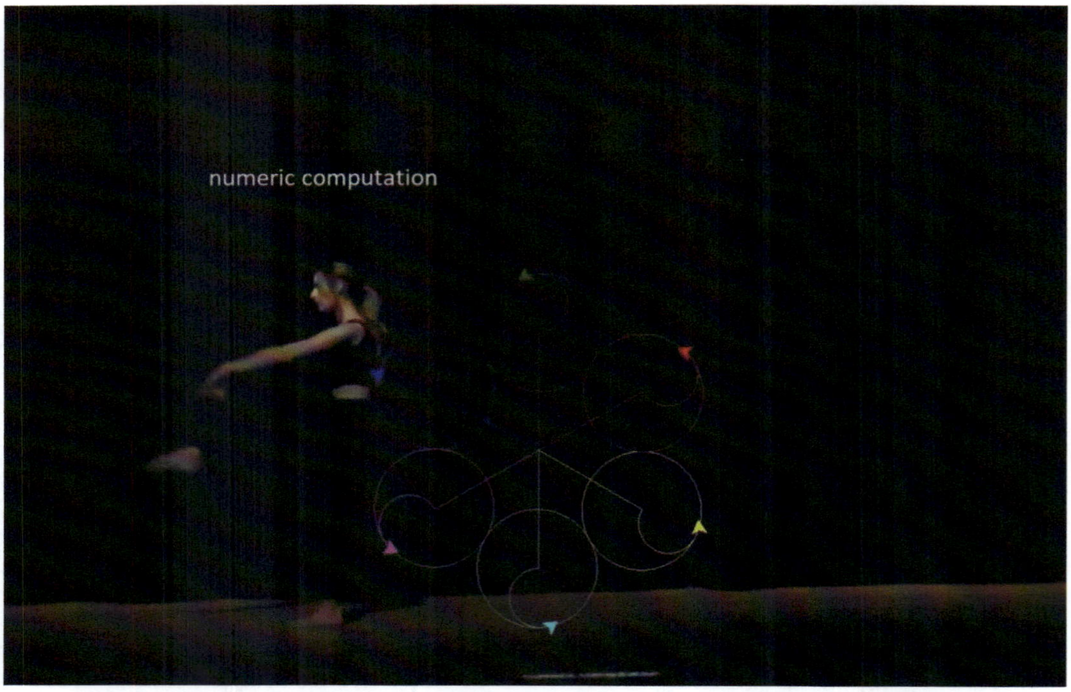
numeric computation

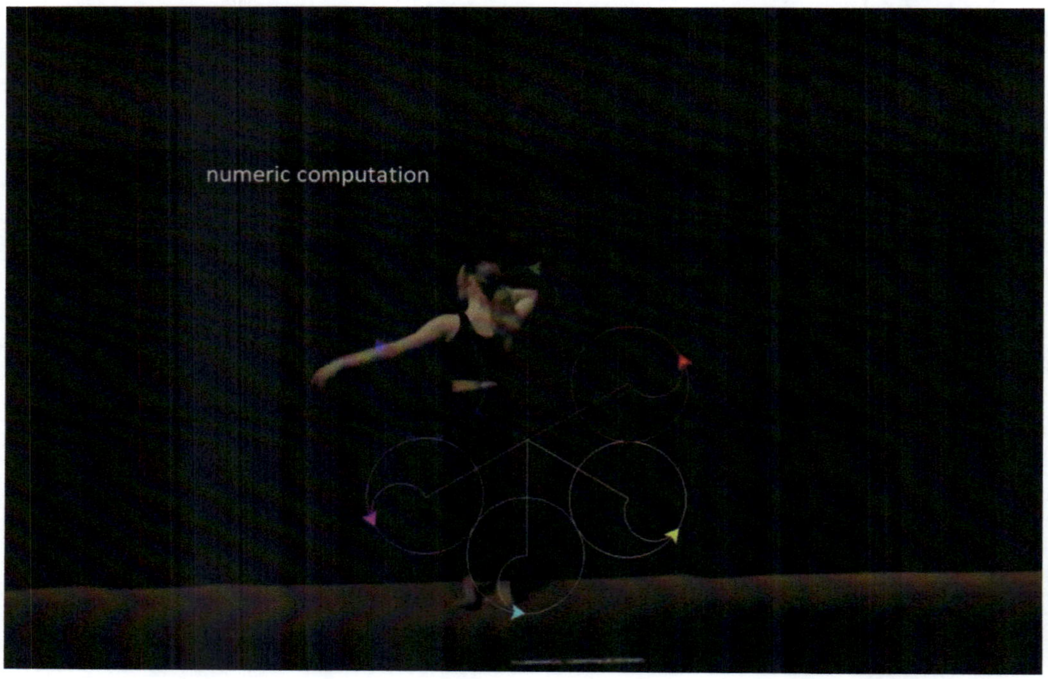
numeric computation

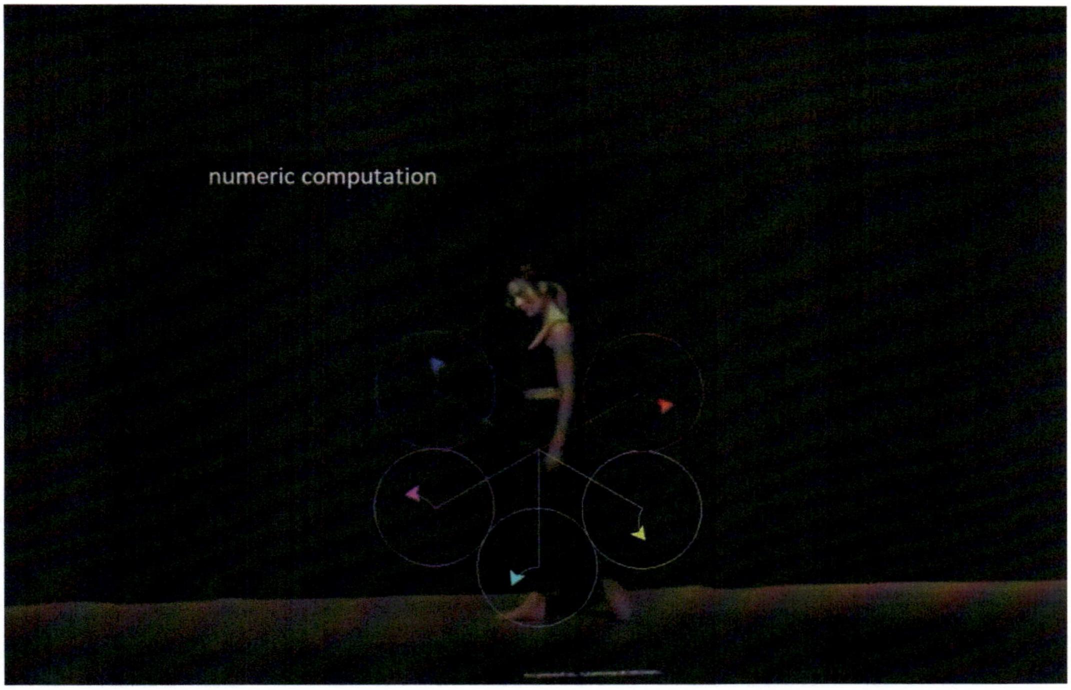
numeric computation

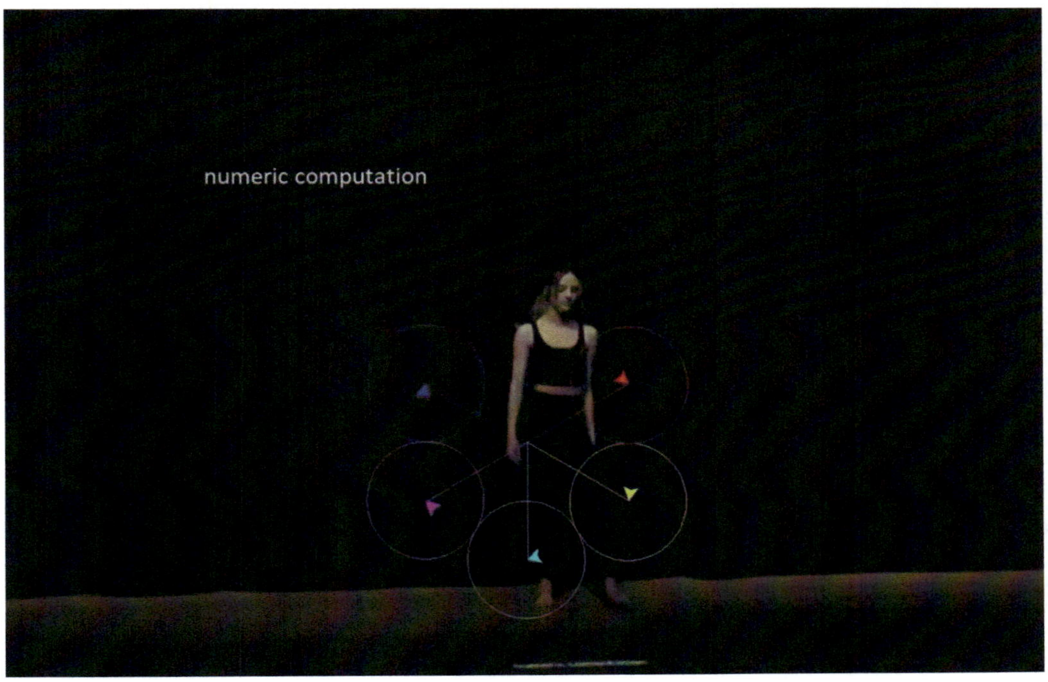
numeric computation

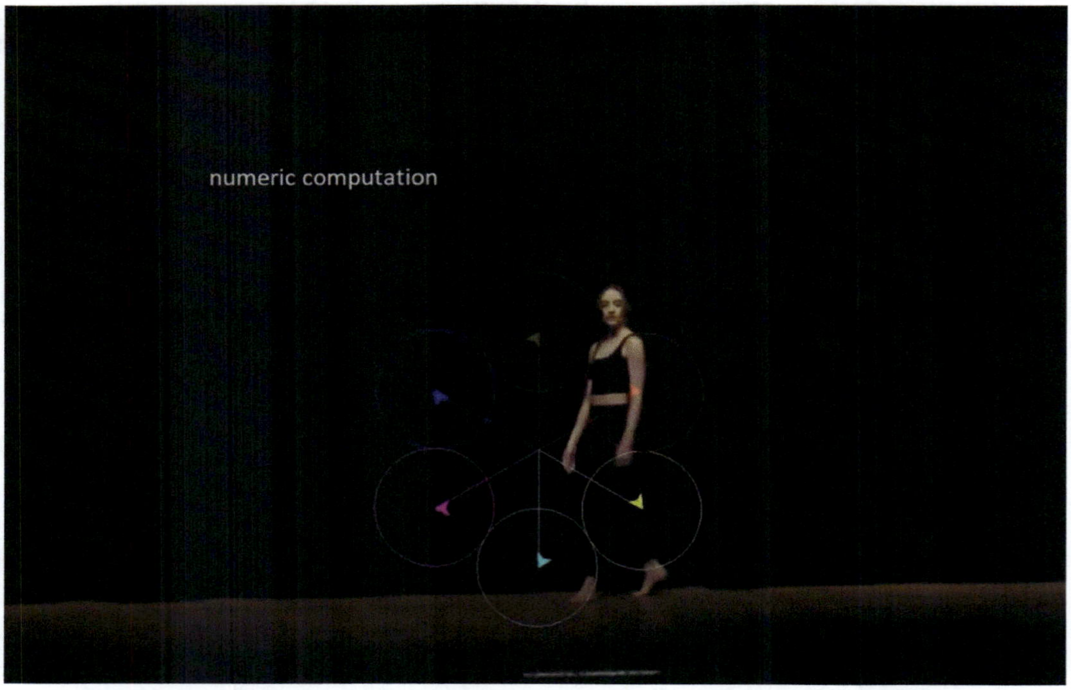
numeric computation

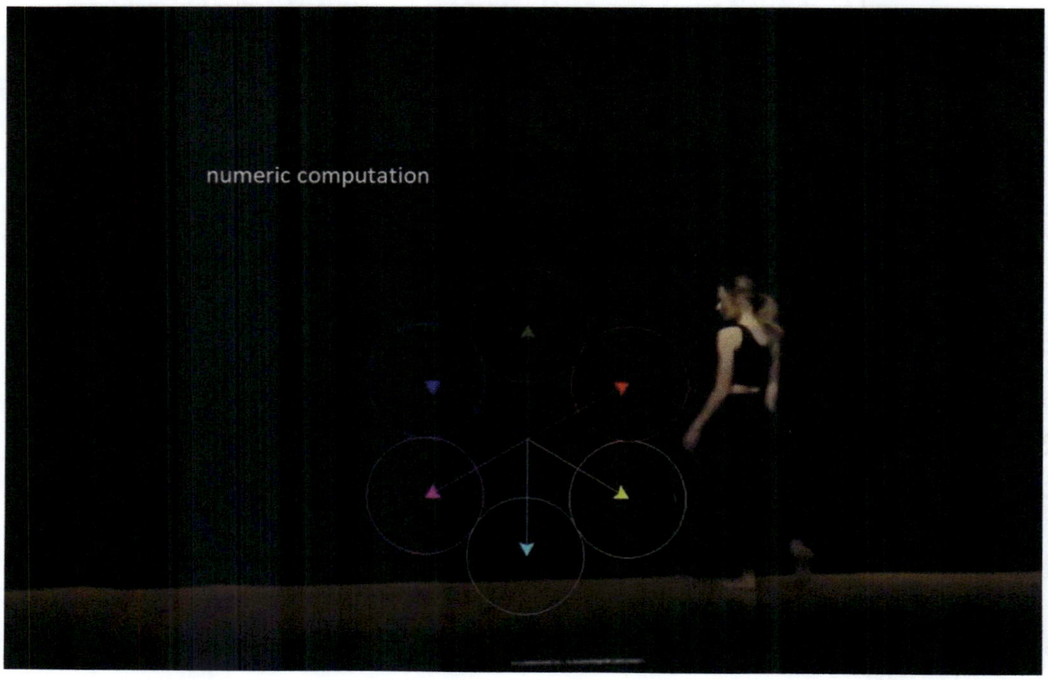
numeric computation

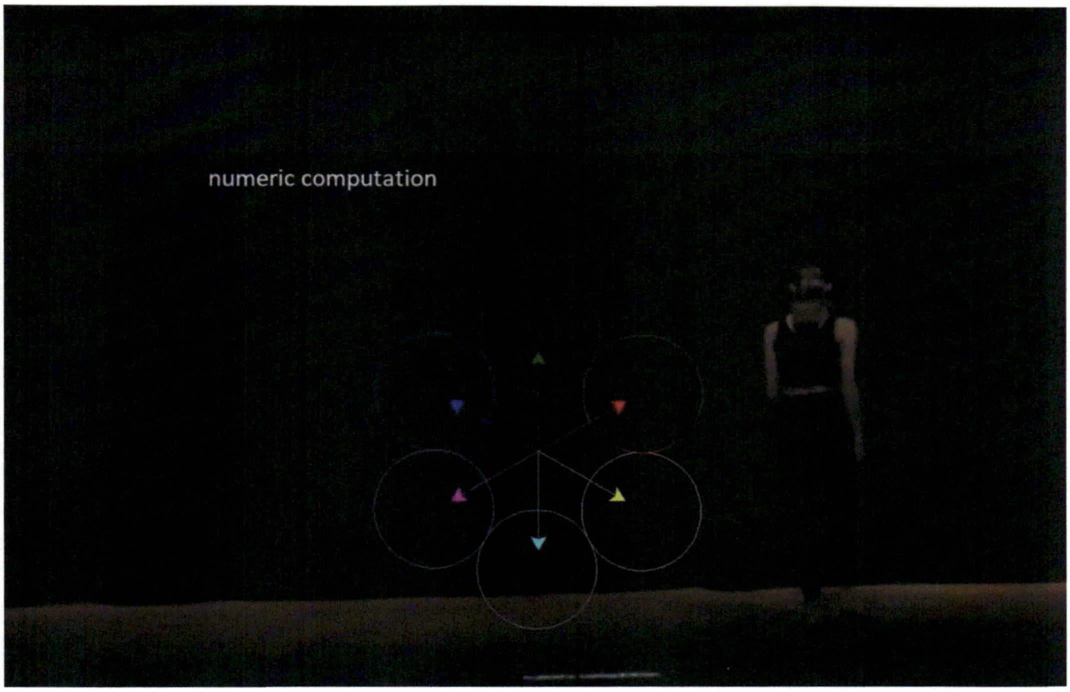
numeric computation

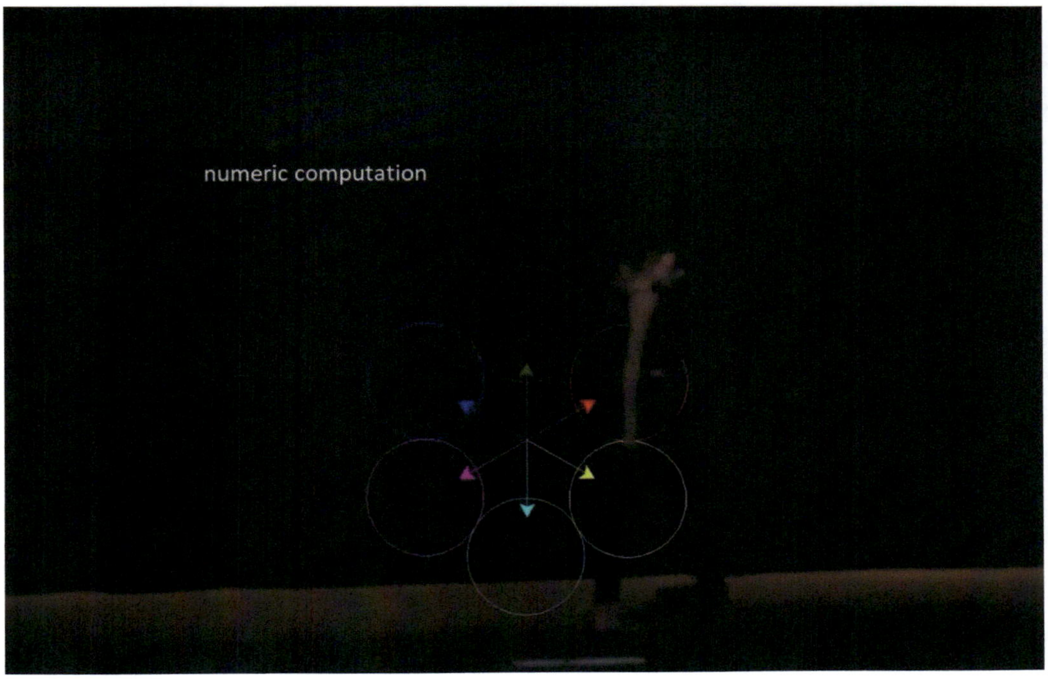
numeric computation

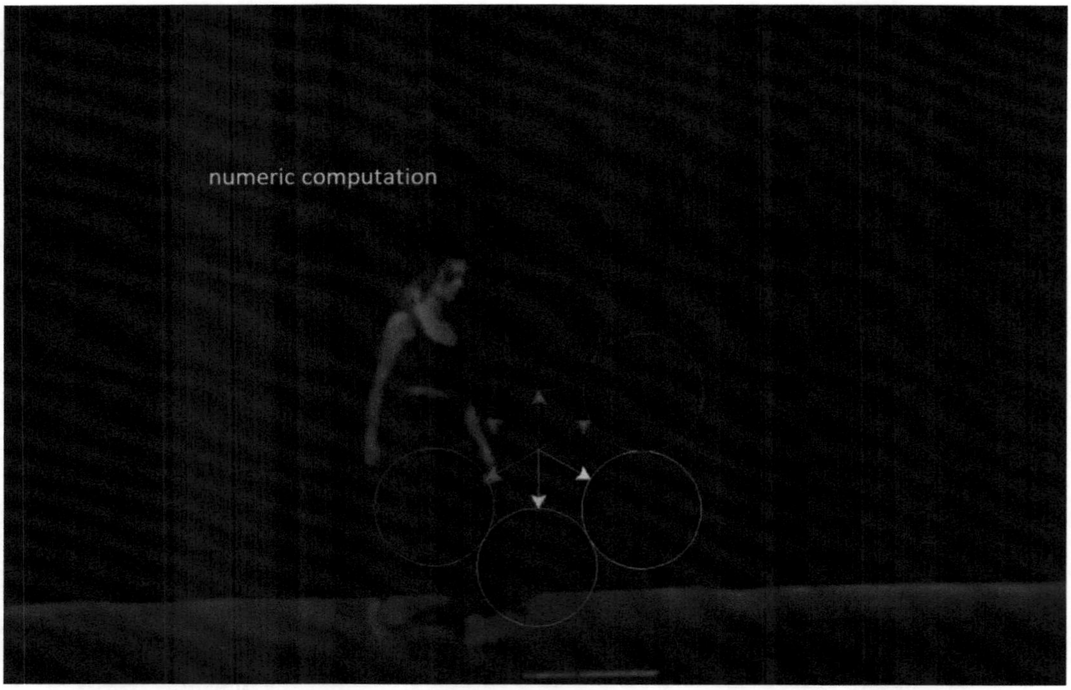
numeric computation

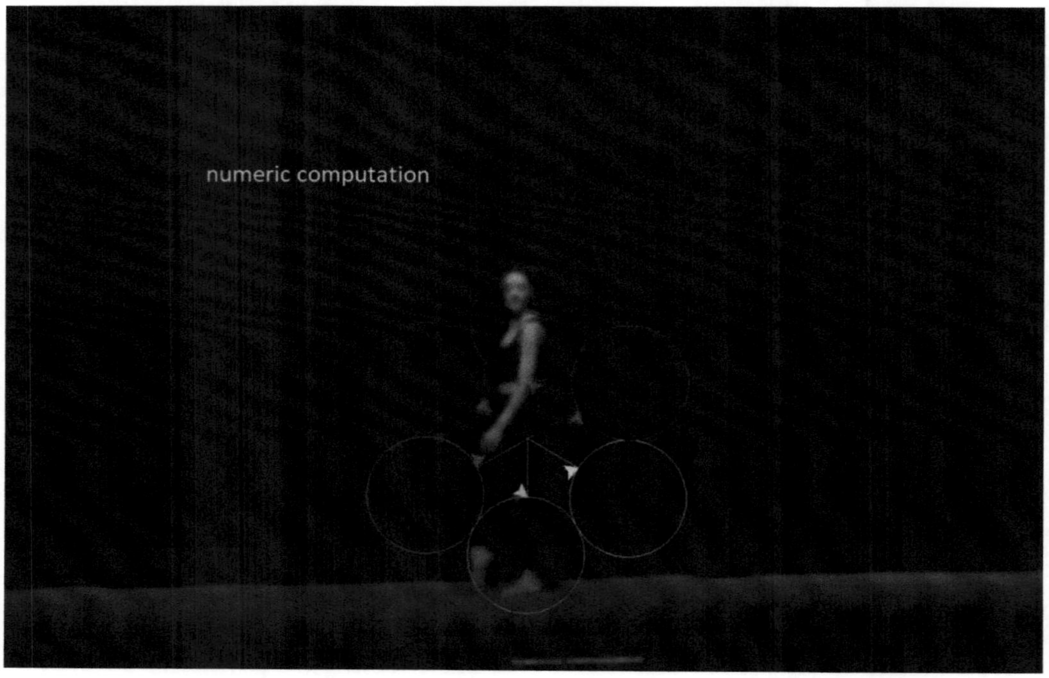
numeric computation

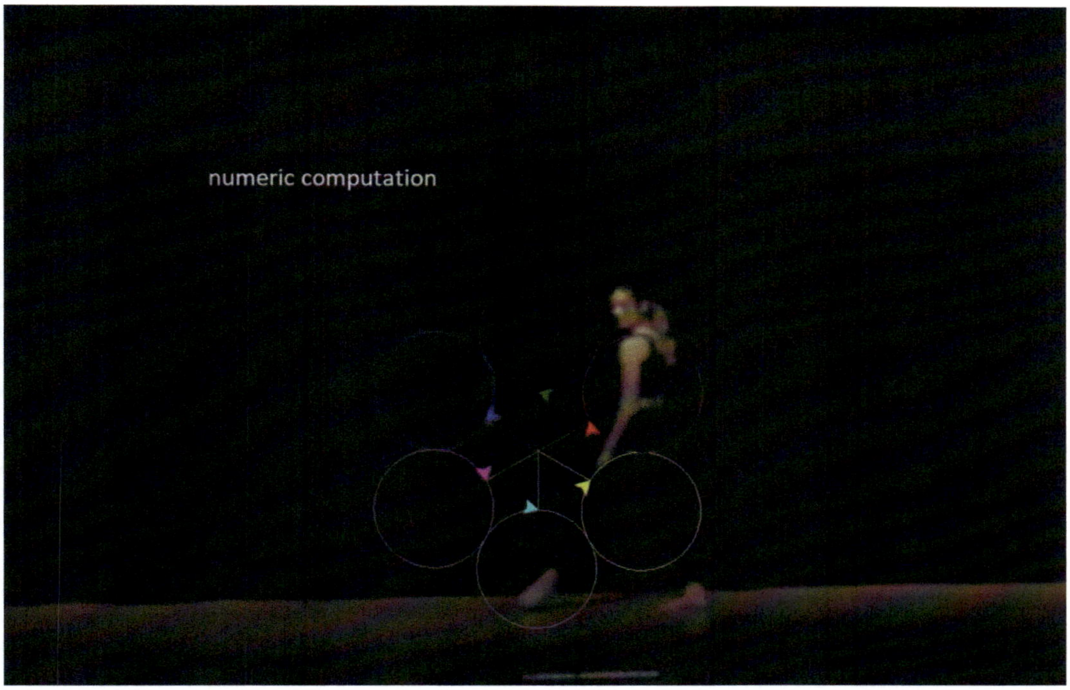
numeric computation

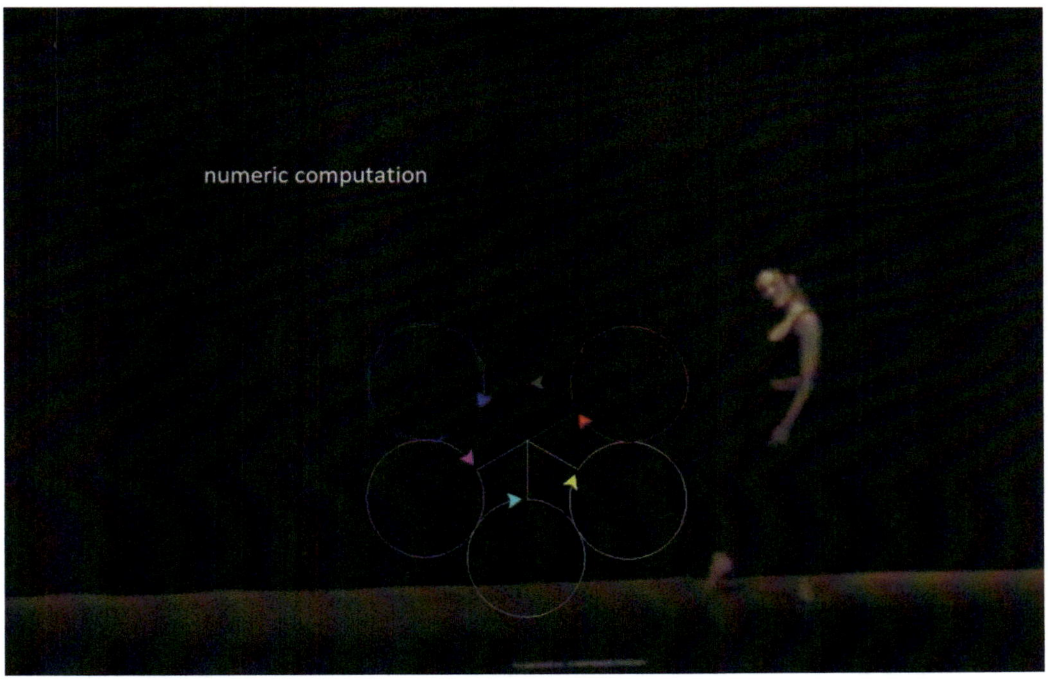
numeric computation

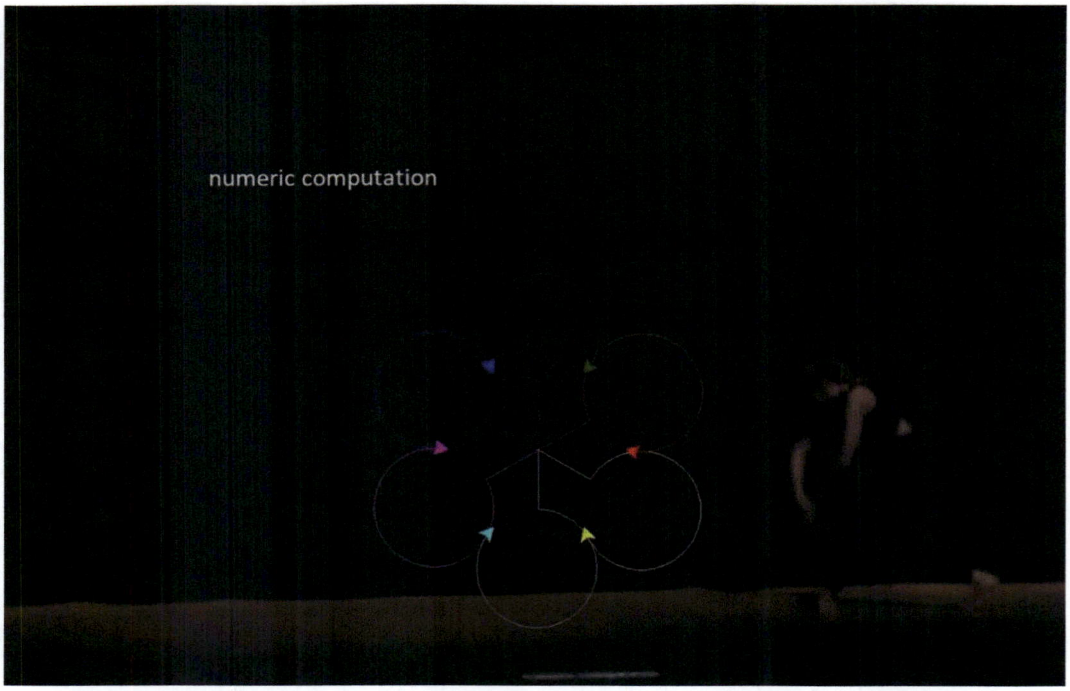
numeric computation

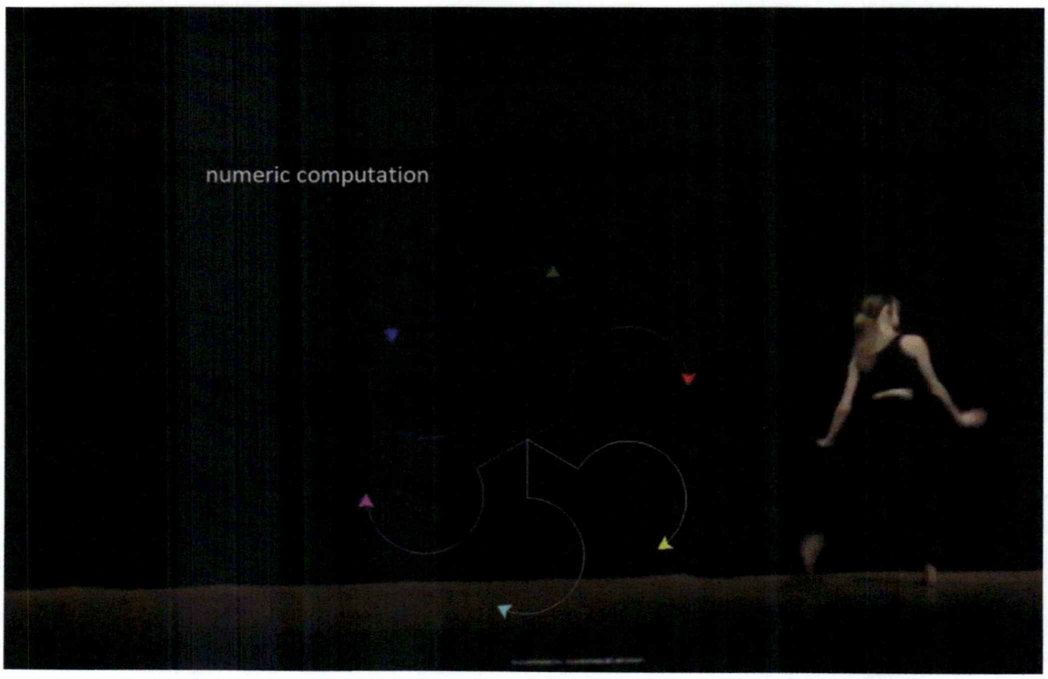
numeric computation

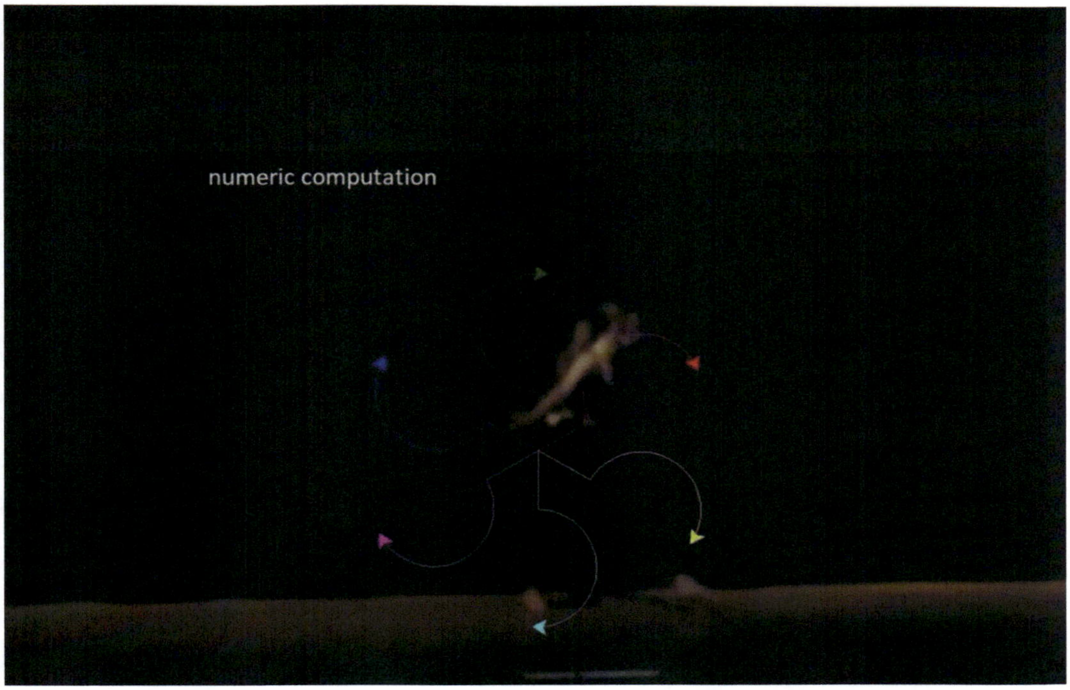
numeric computation

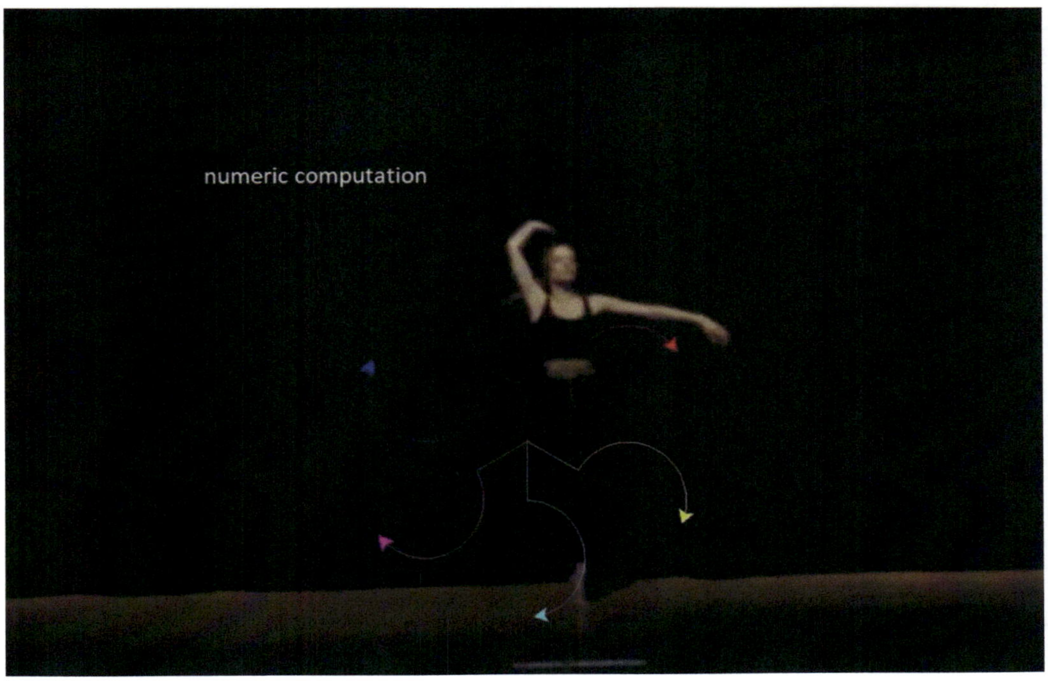
numeric computation

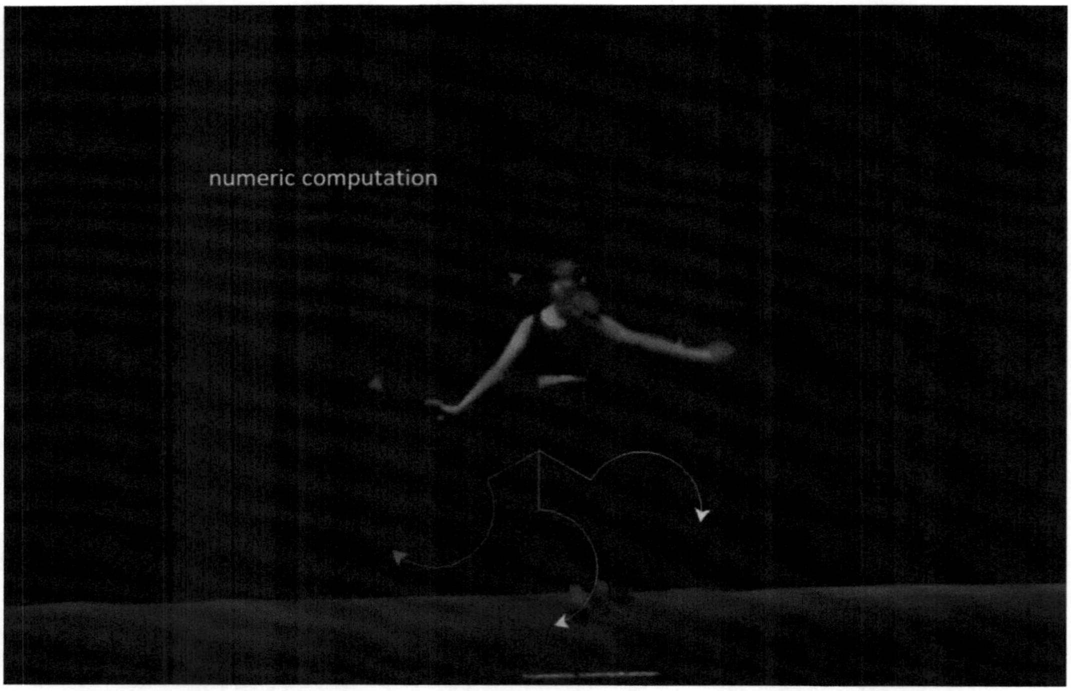
numeric computation

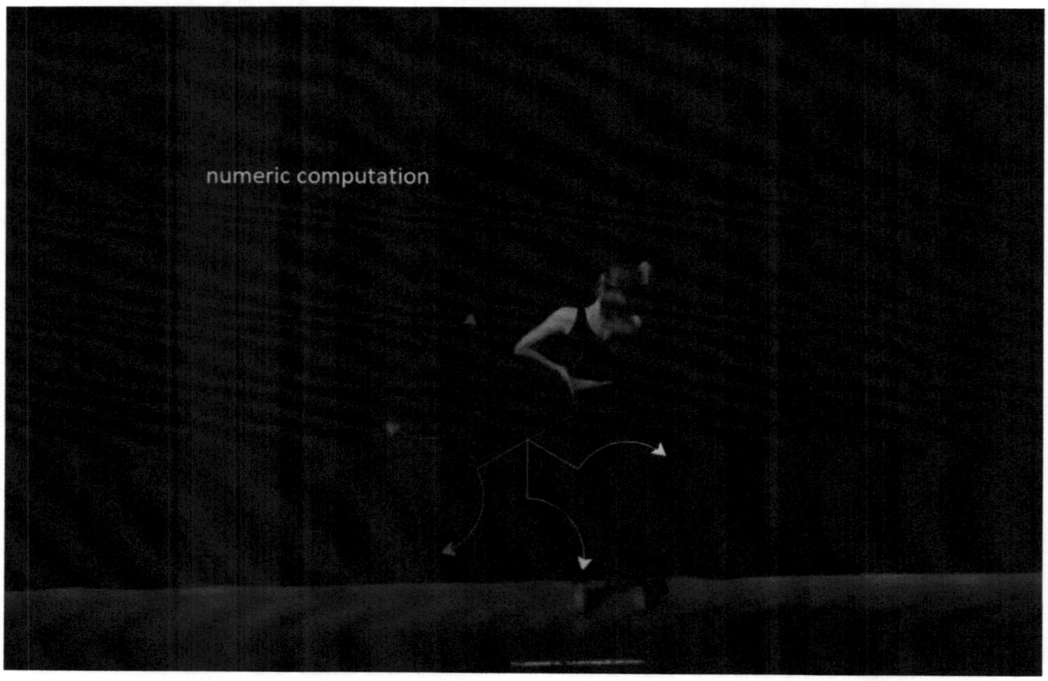
numeric computation

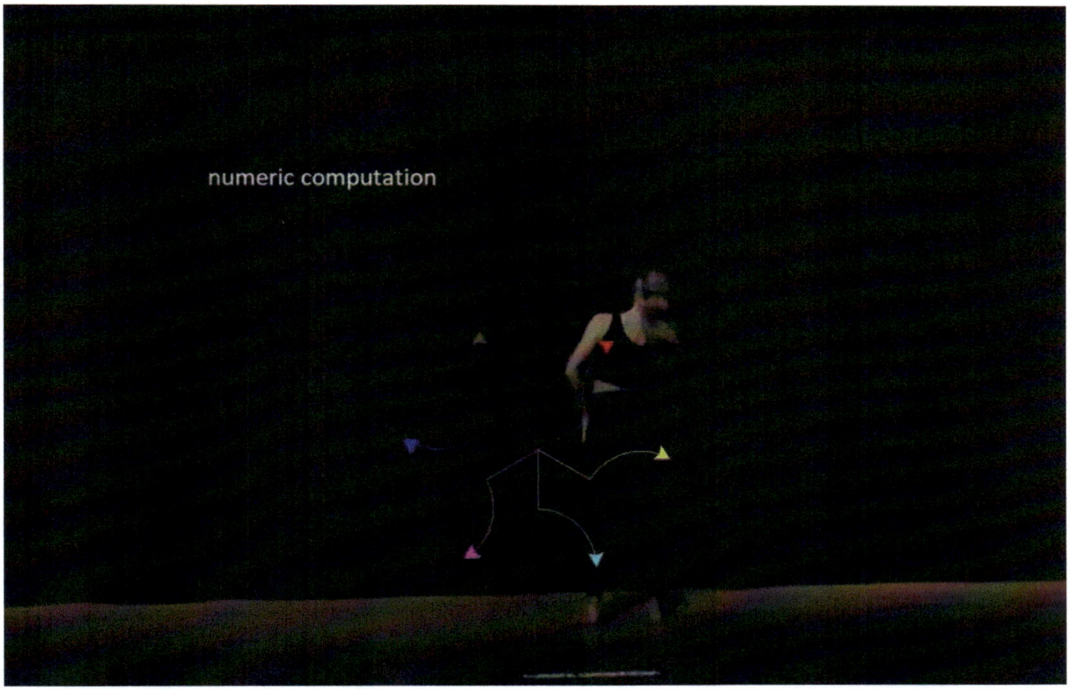
numeric computation

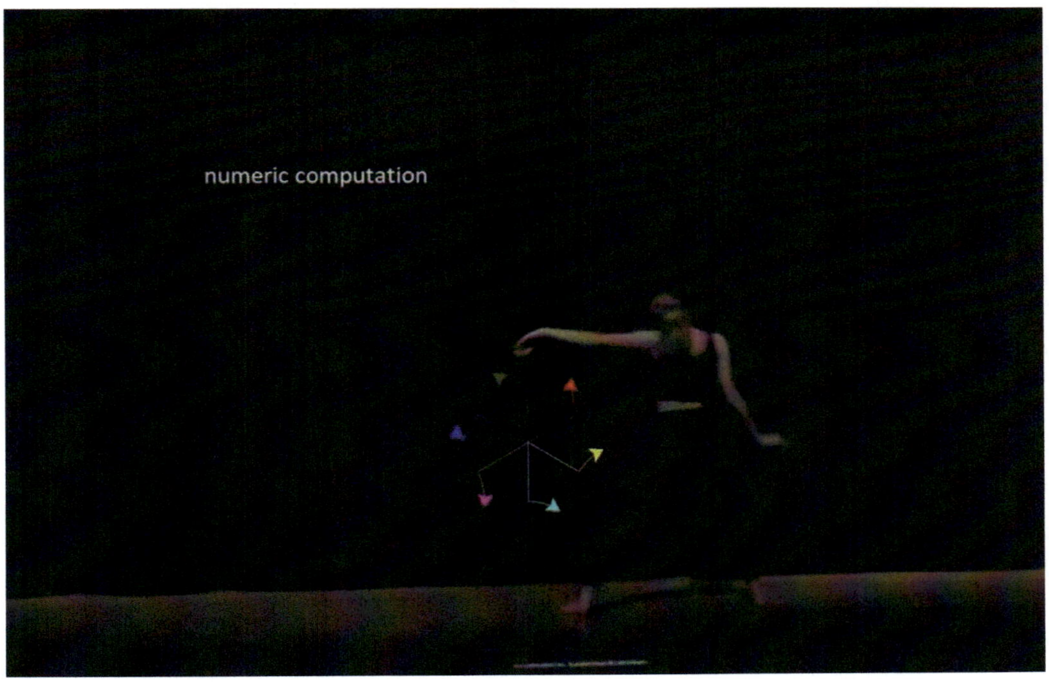
numeric computation

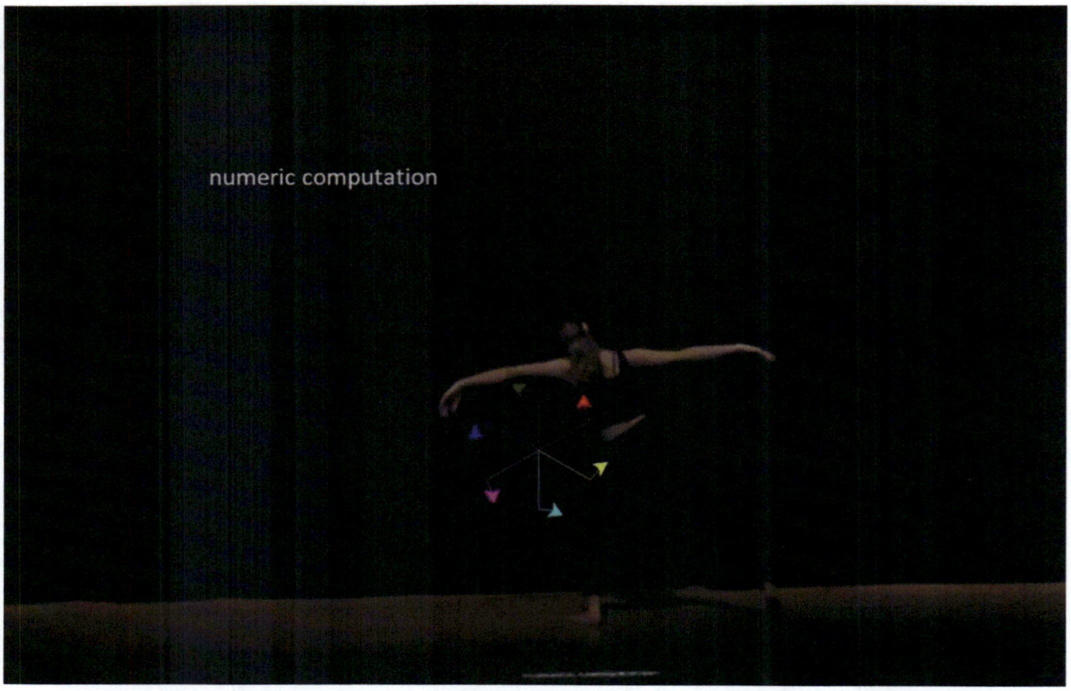
numeric computation

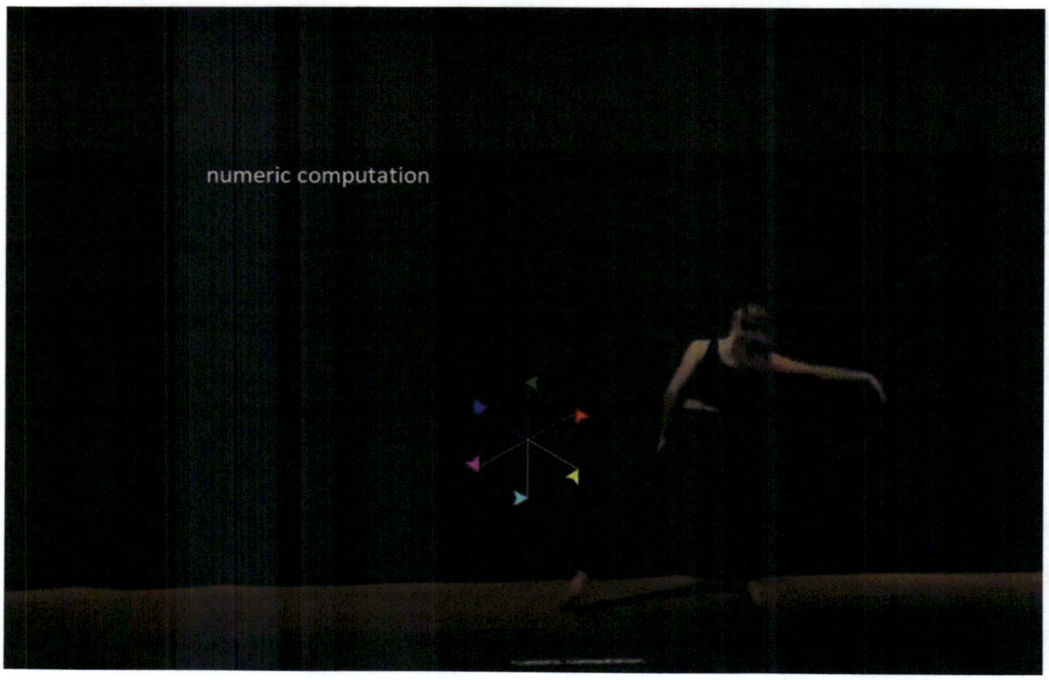
numeric computation

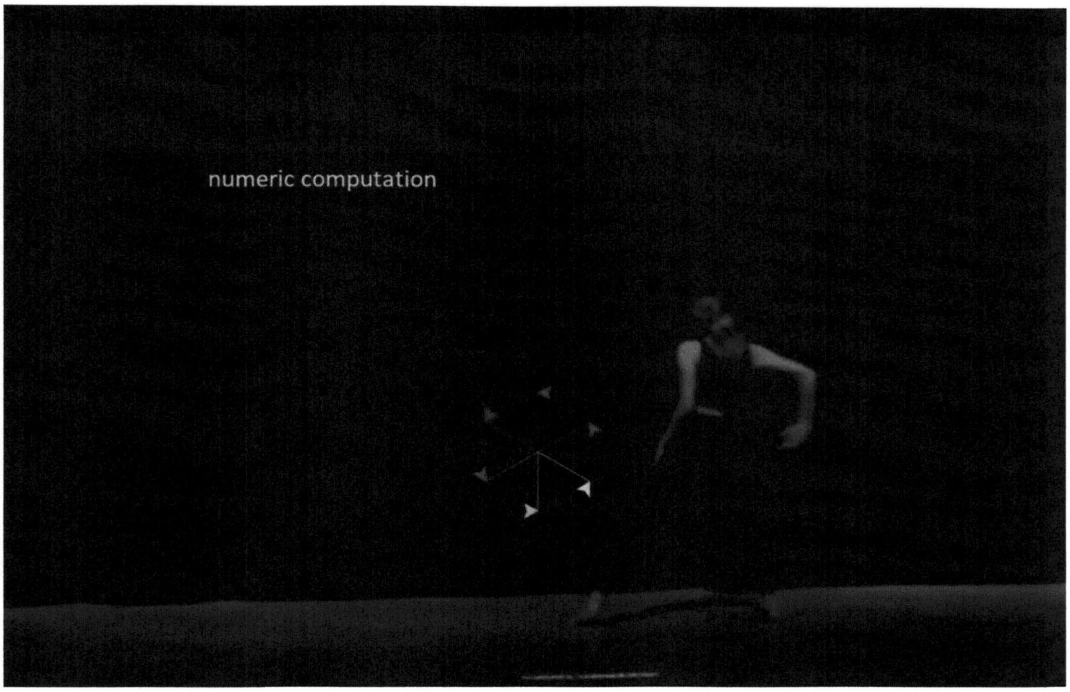
numeric computation

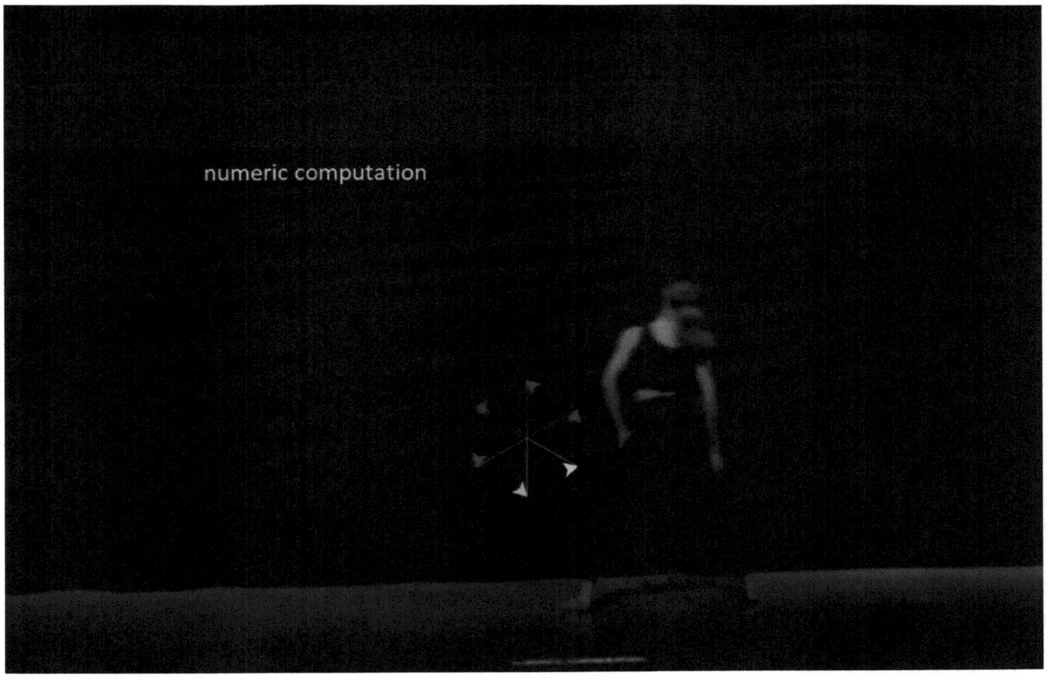
numeric computation

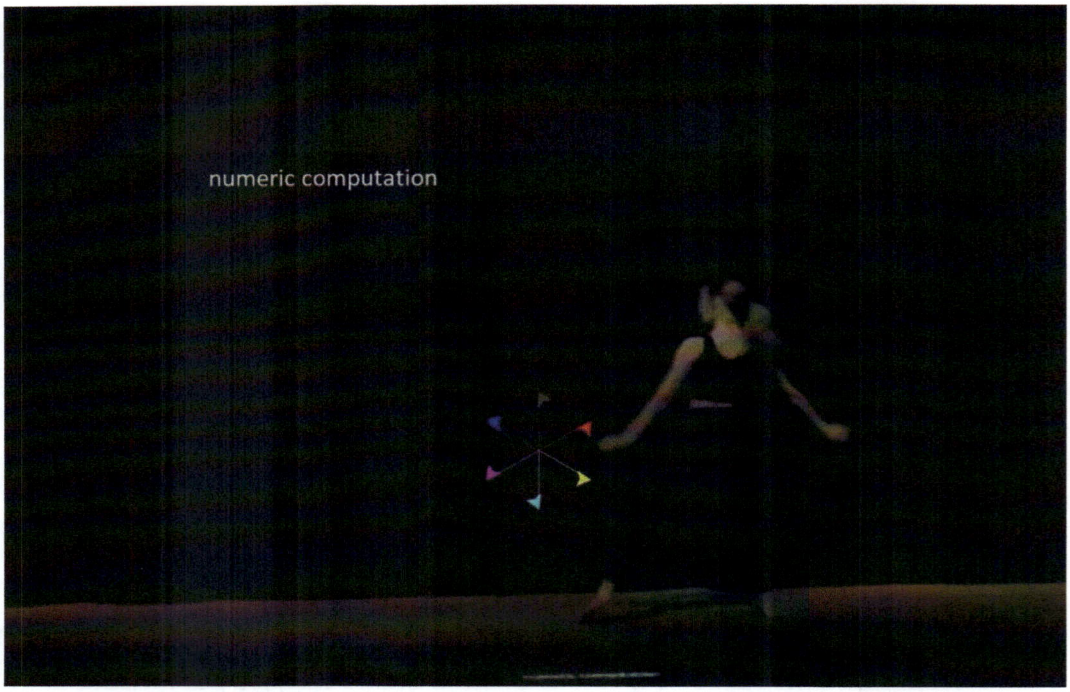
numeric computation

Sequence

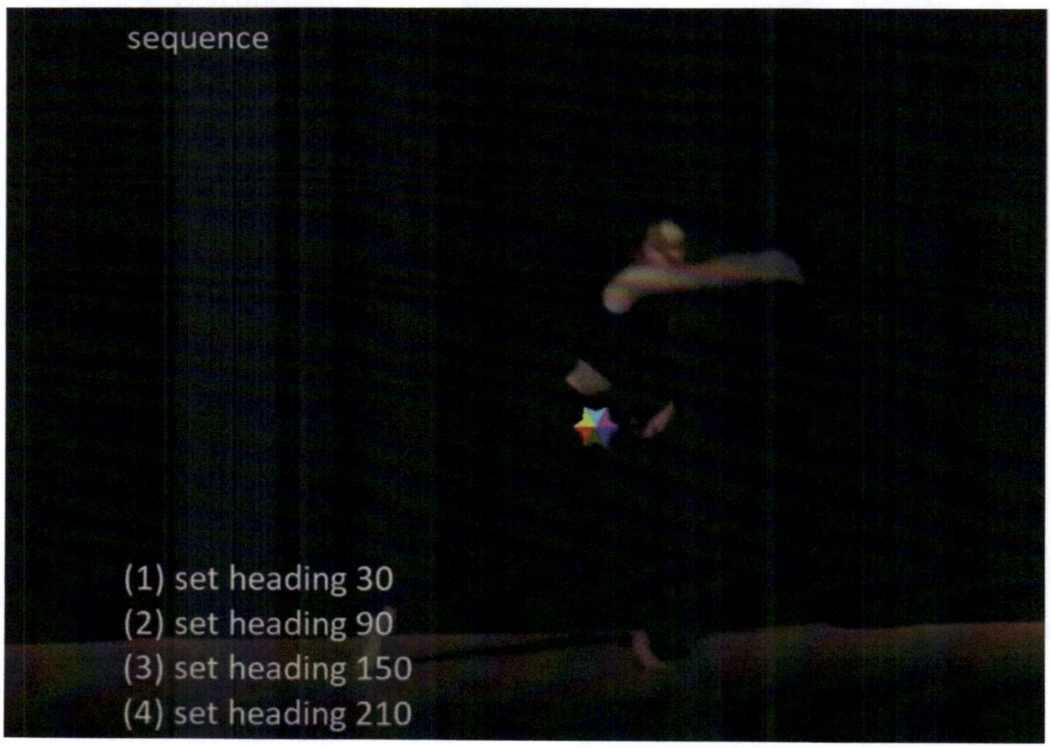
sequence
(1) set heading 30
(2) set heading 90
(3) set heading 150
(4) set heading 210

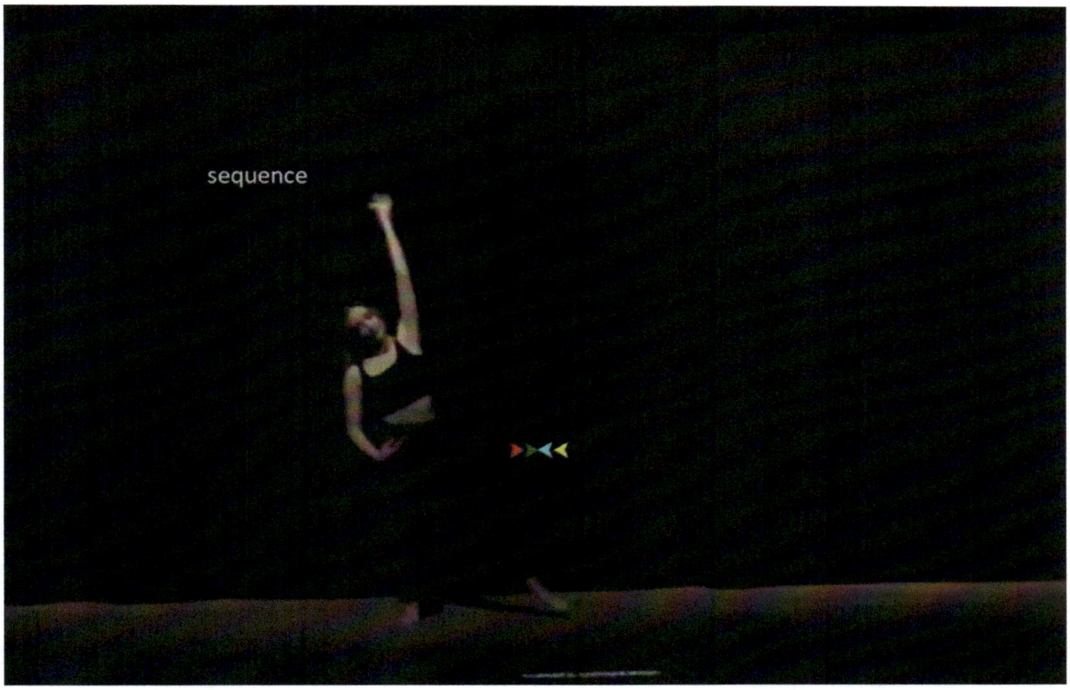
sequence

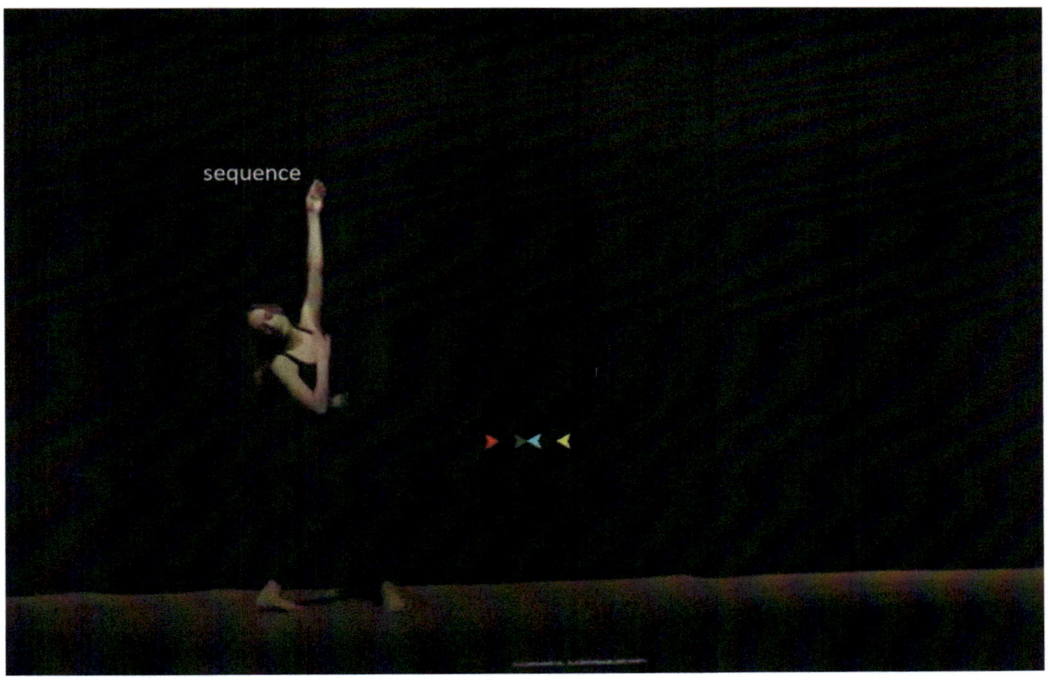
sequence

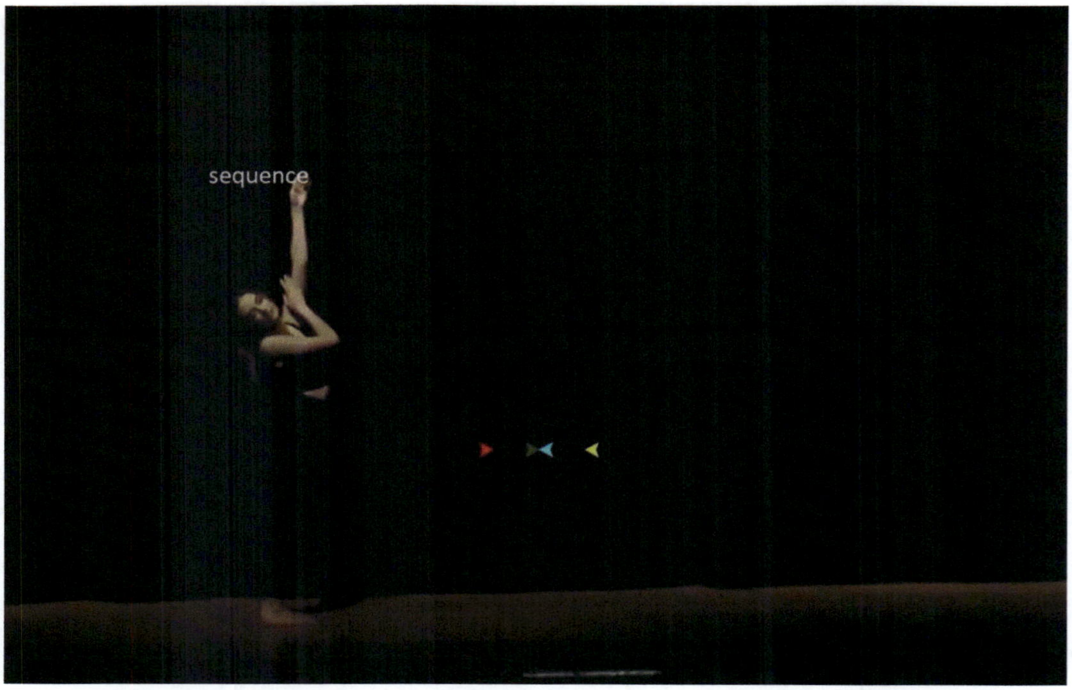
sequence

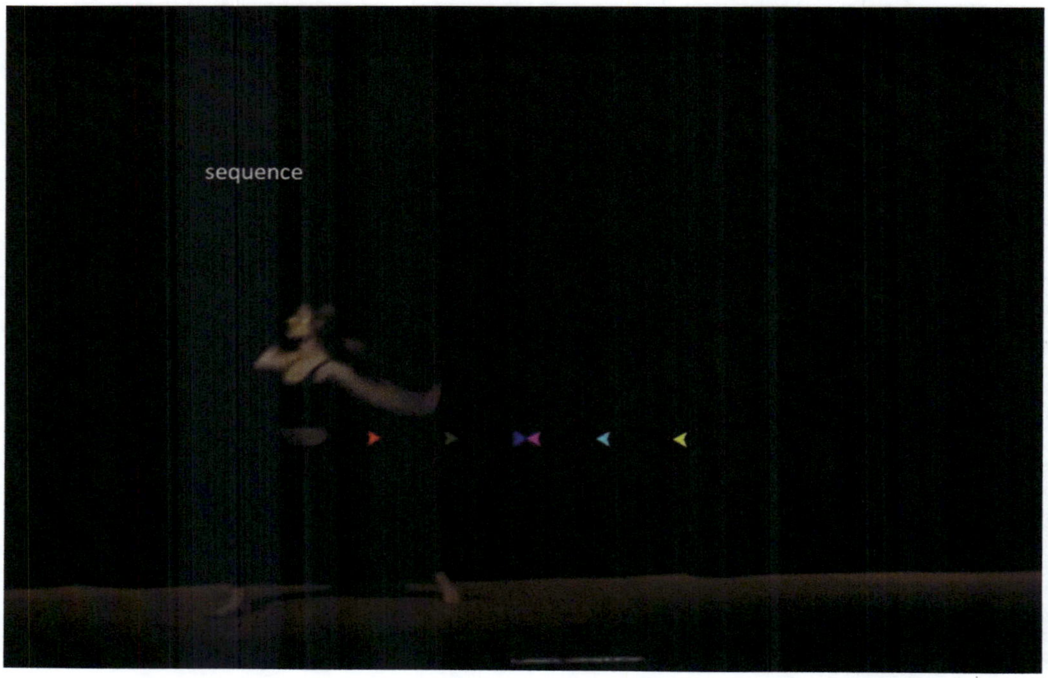
sequence

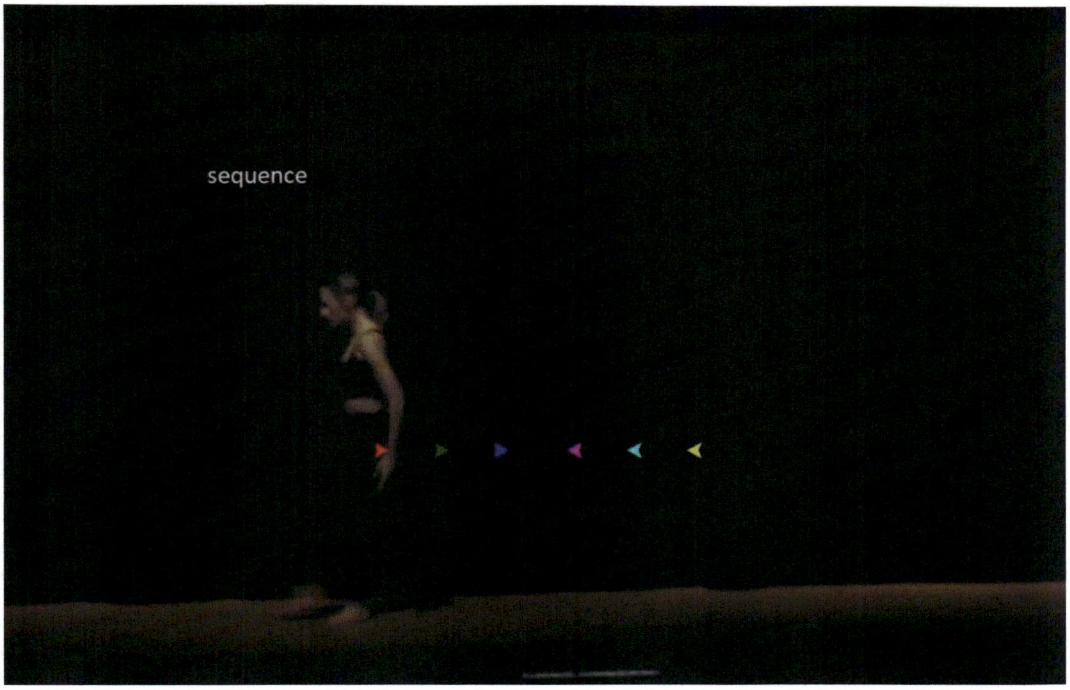
sequence

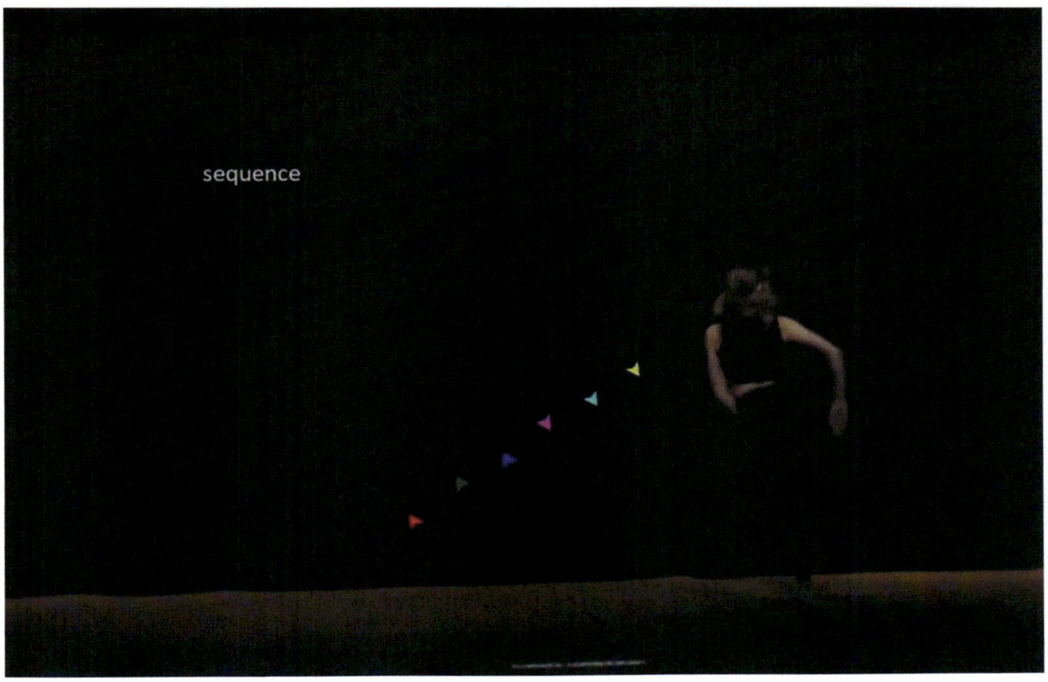
sequence

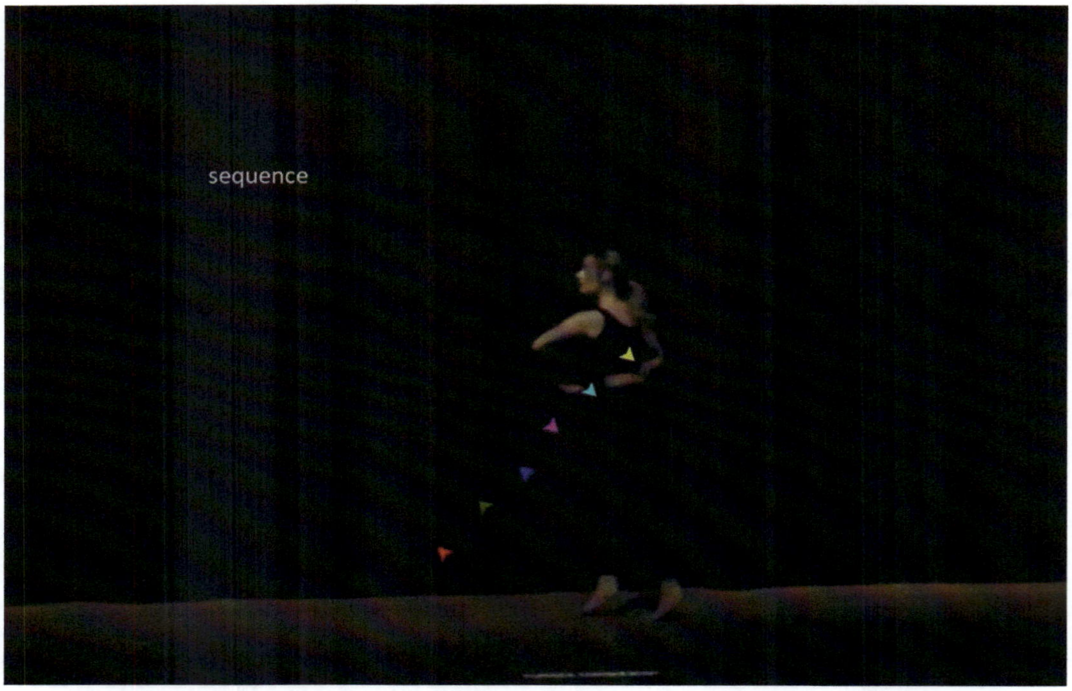
sequence

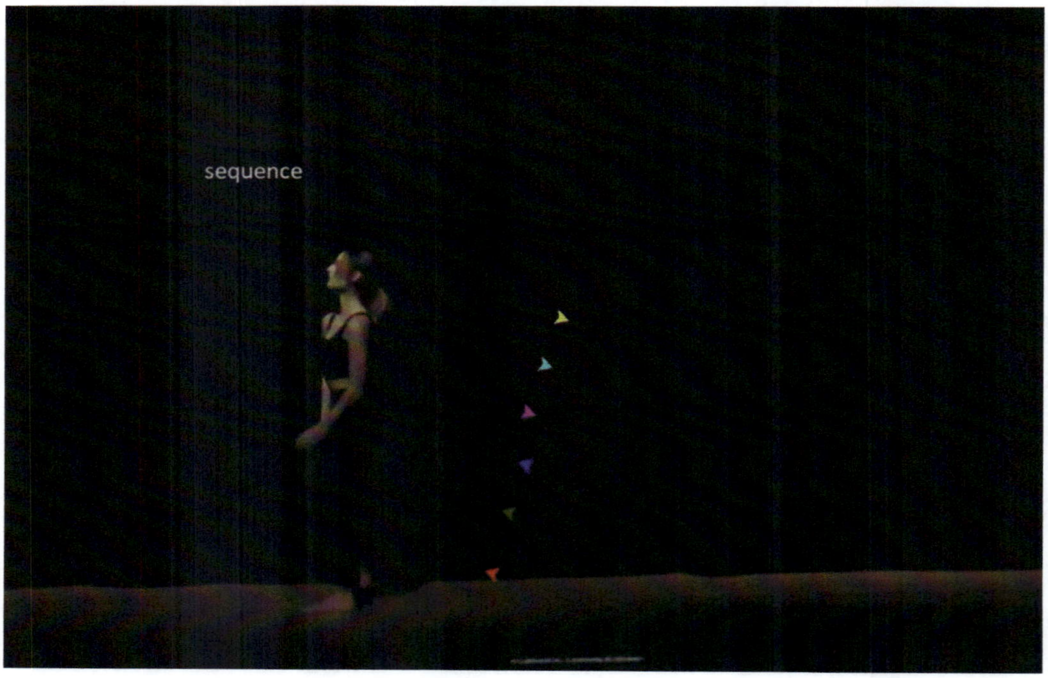
sequence

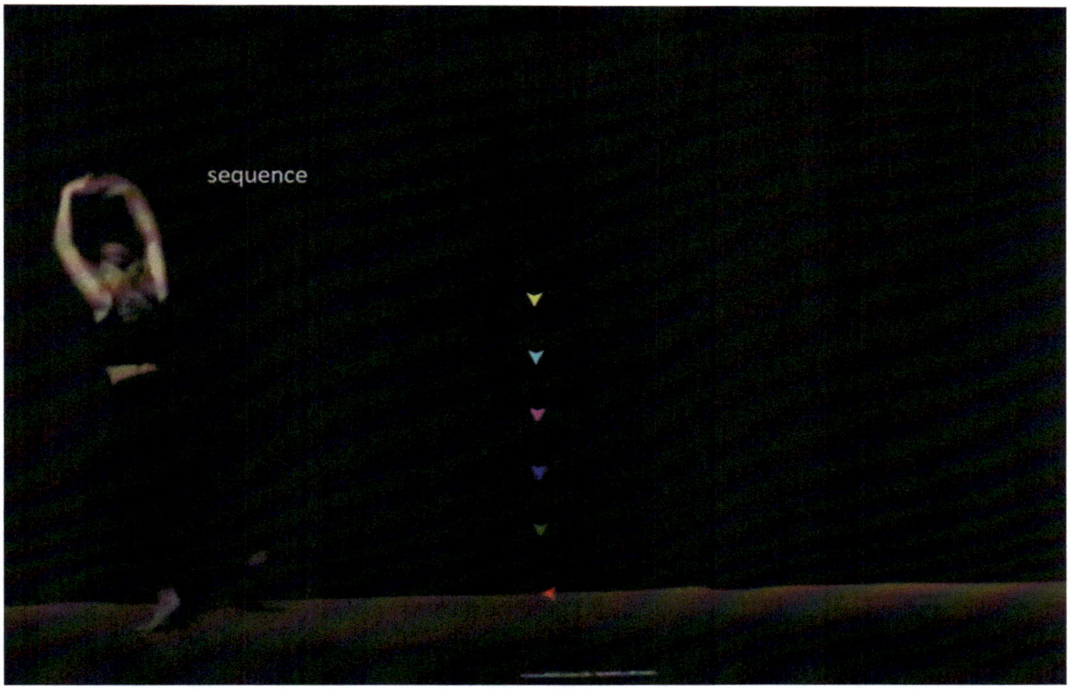
sequence

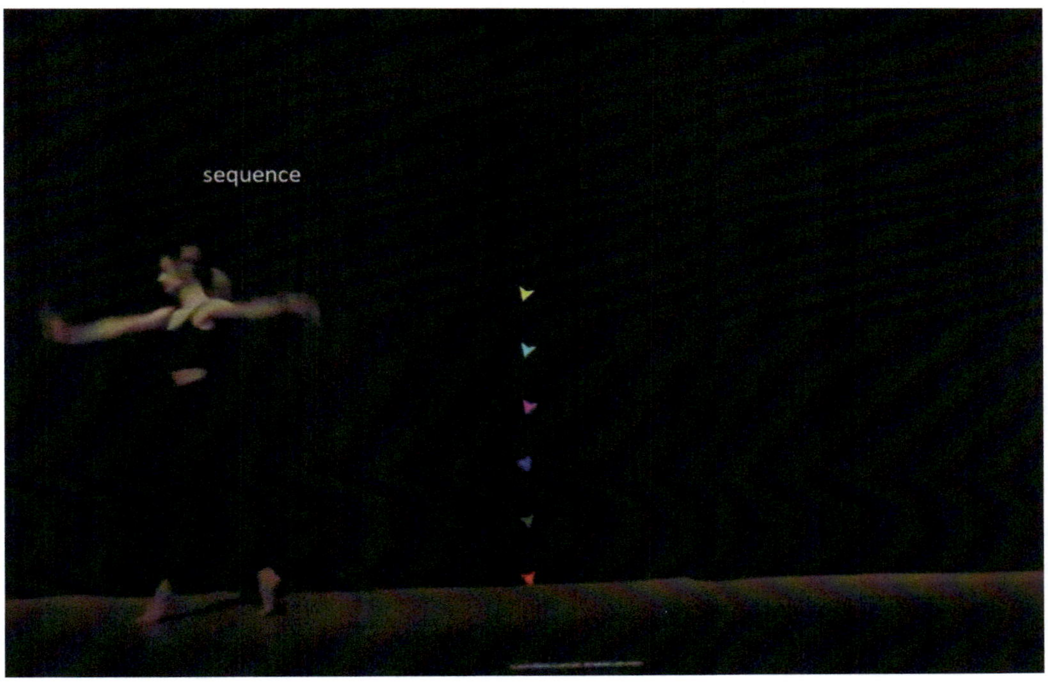
sequence

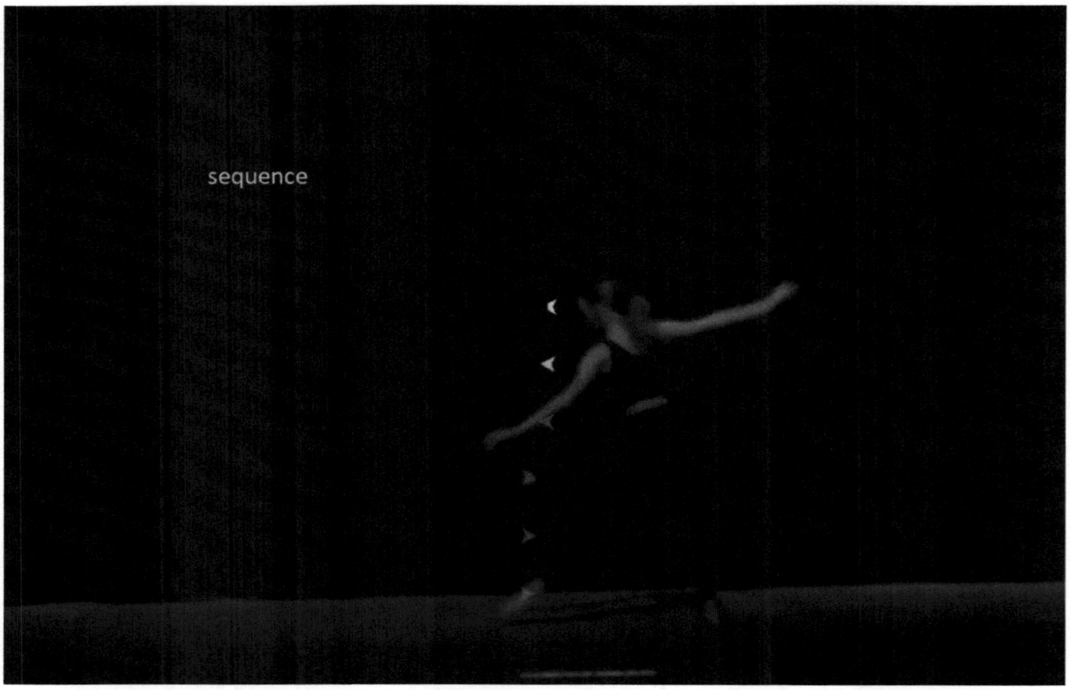
sequence

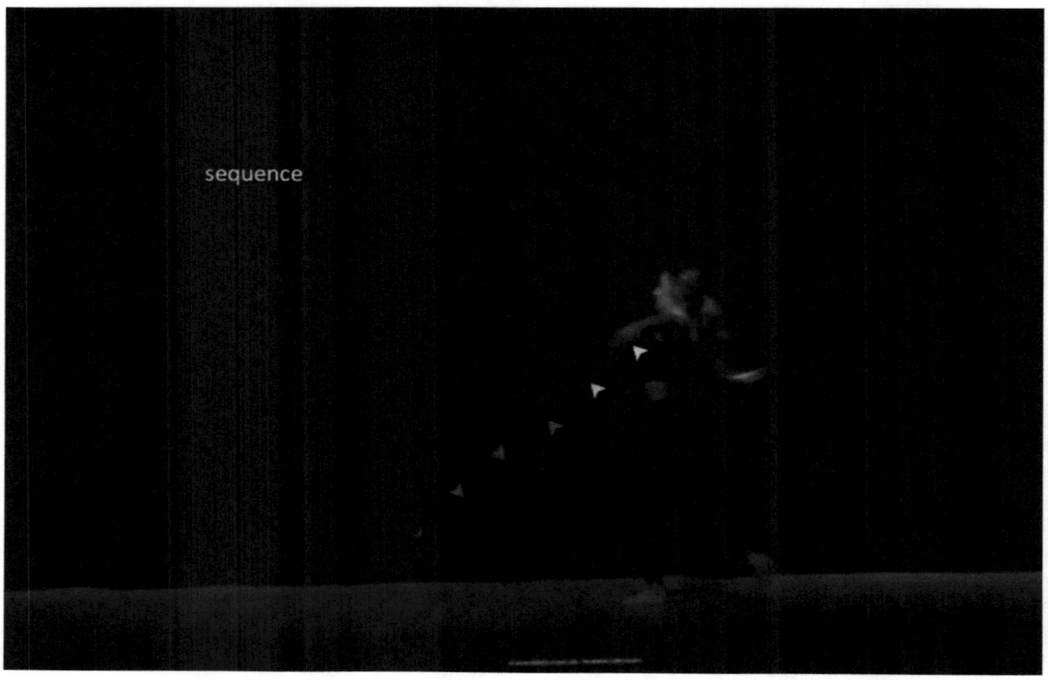
sequence

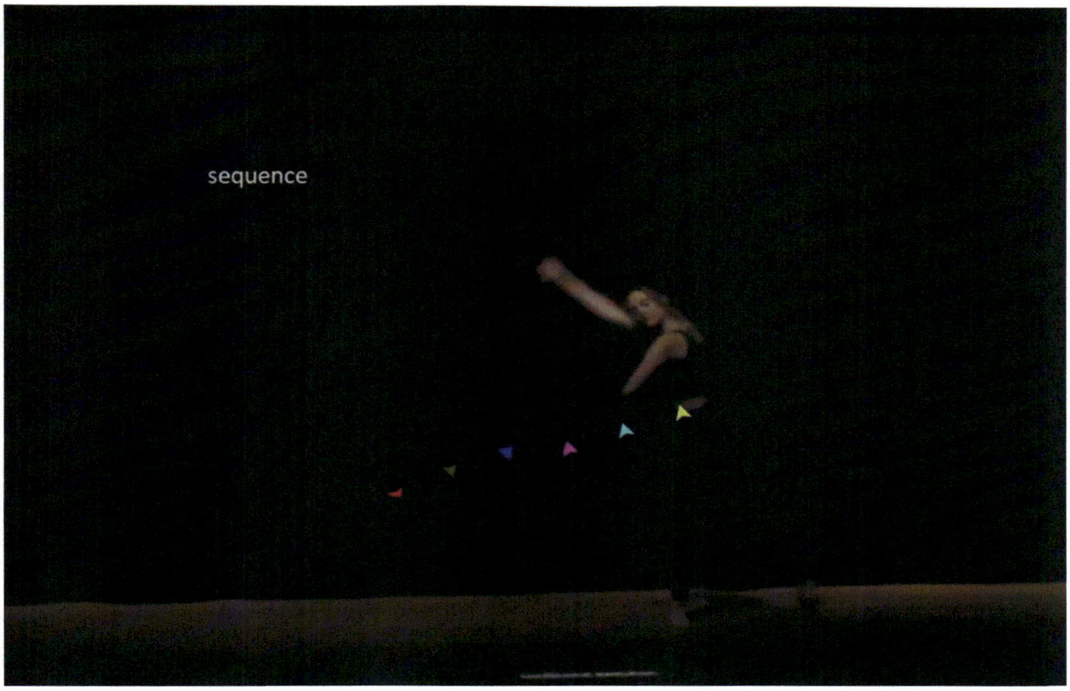
sequence

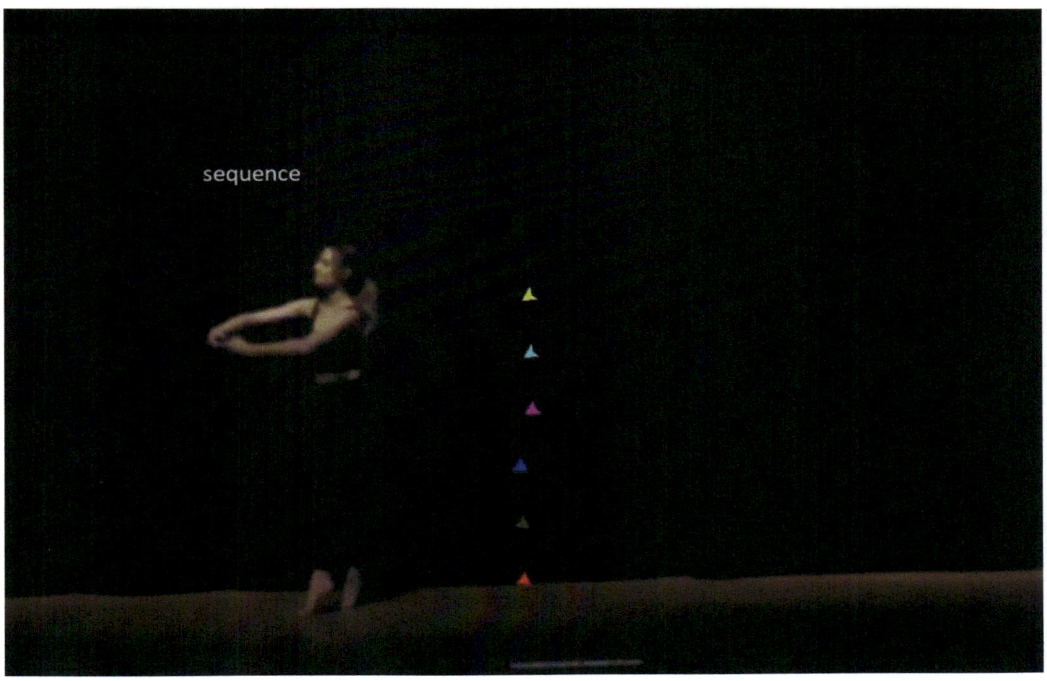
sequence

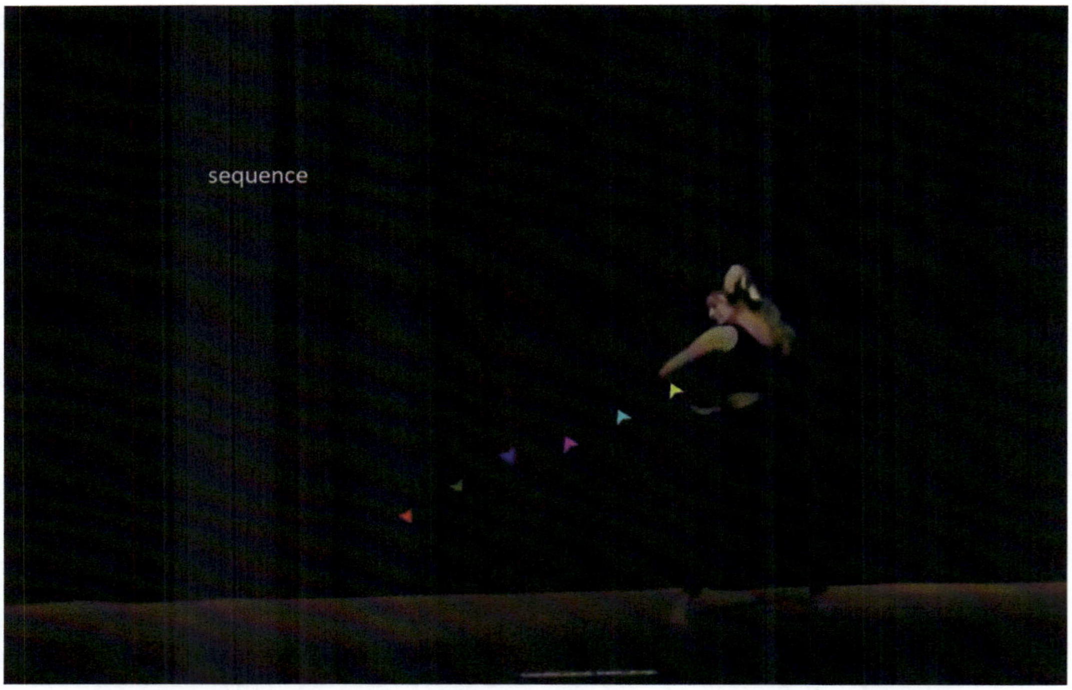
sequence

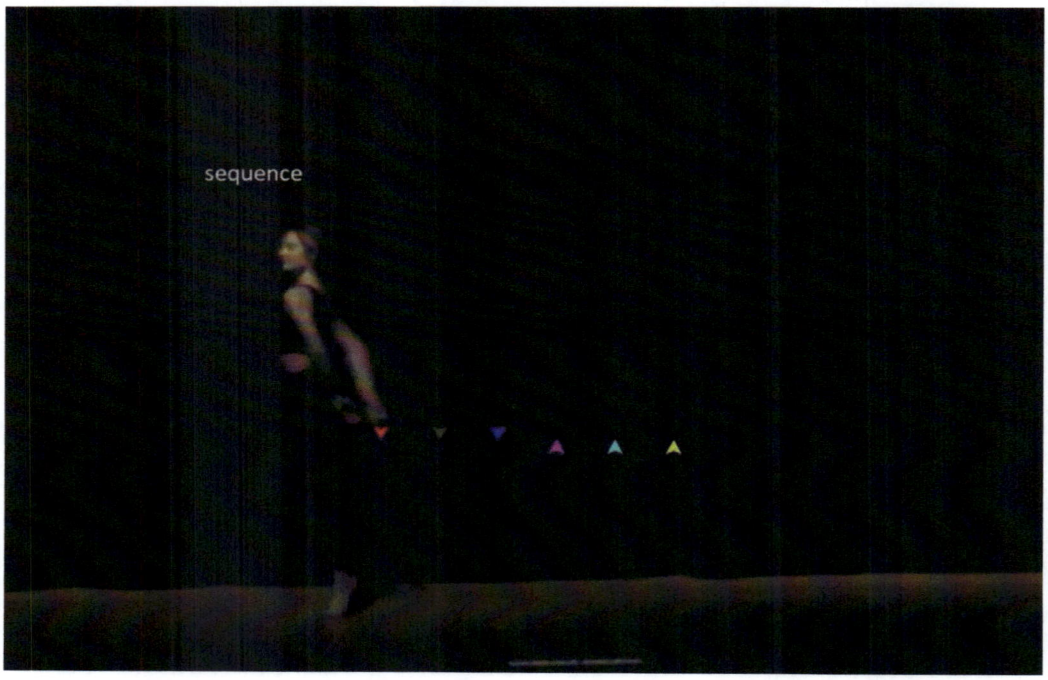
sequence

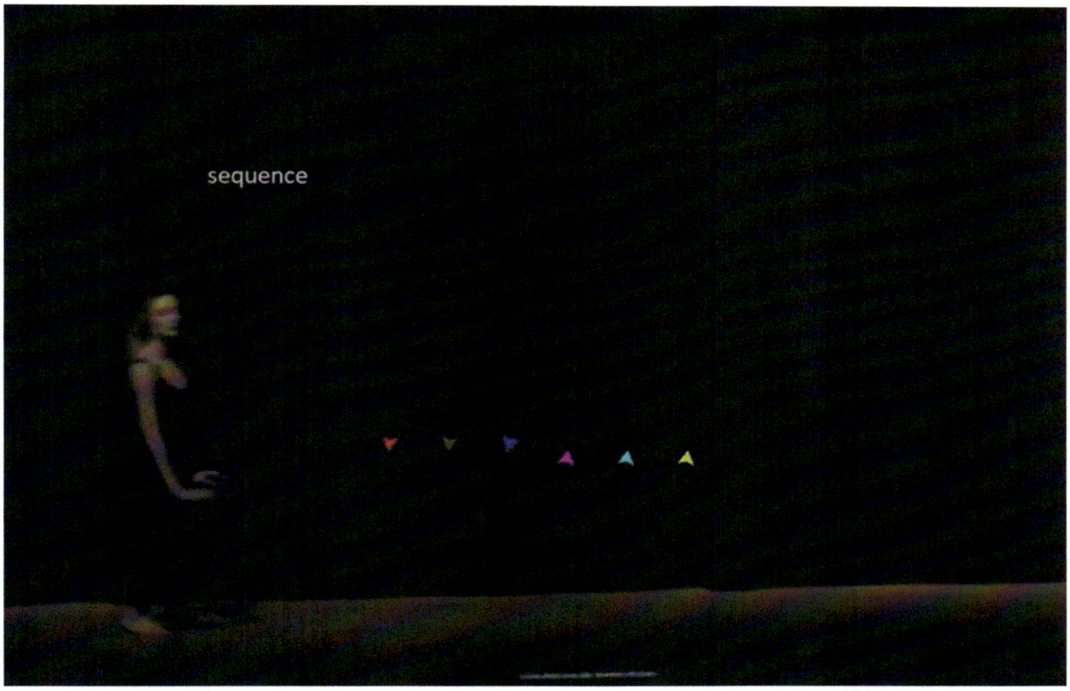
sequence

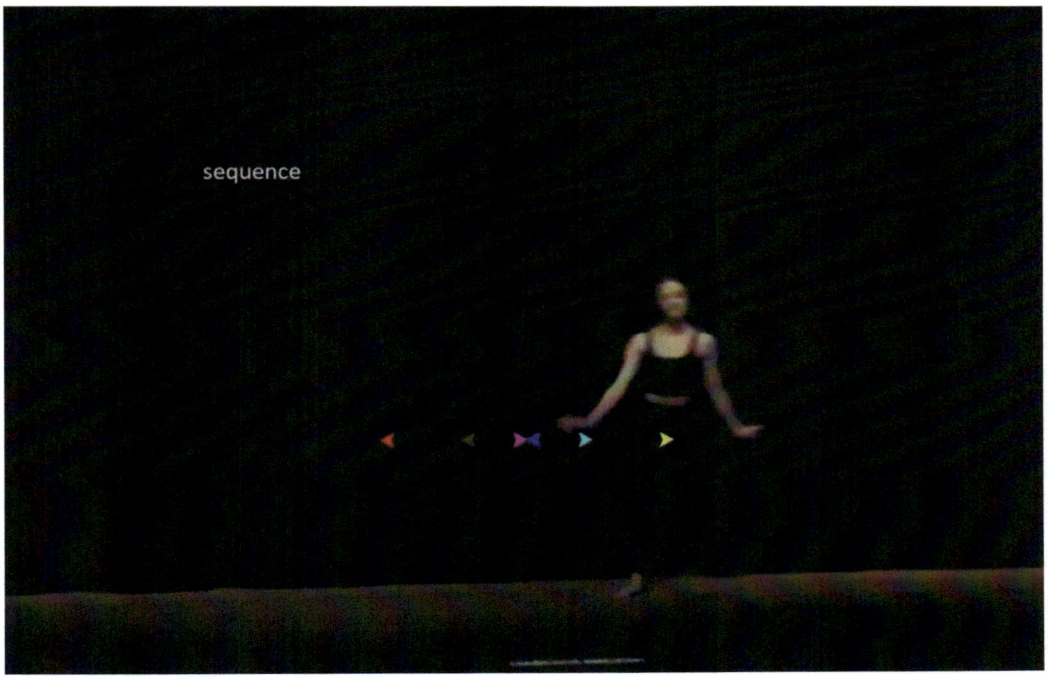
sequence

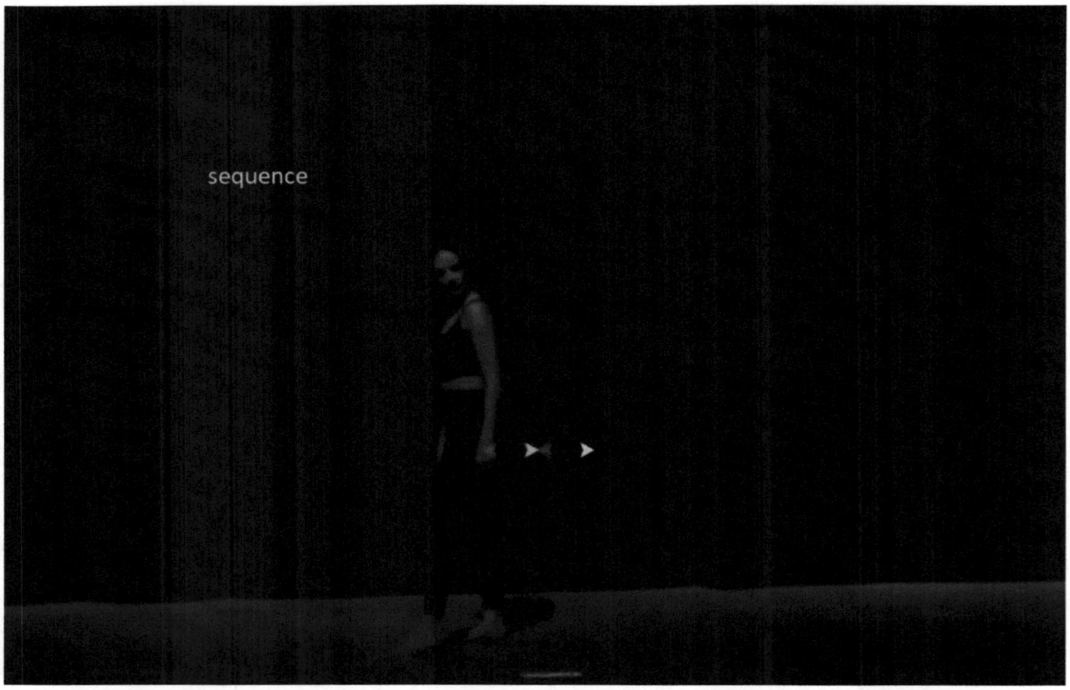
sequence

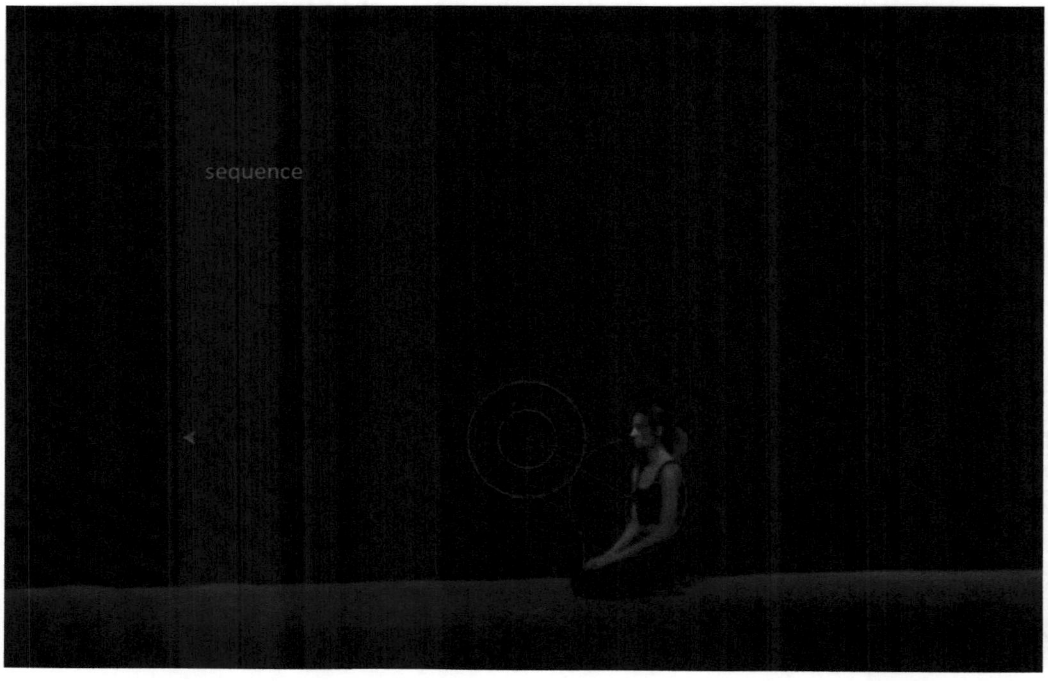
sequence

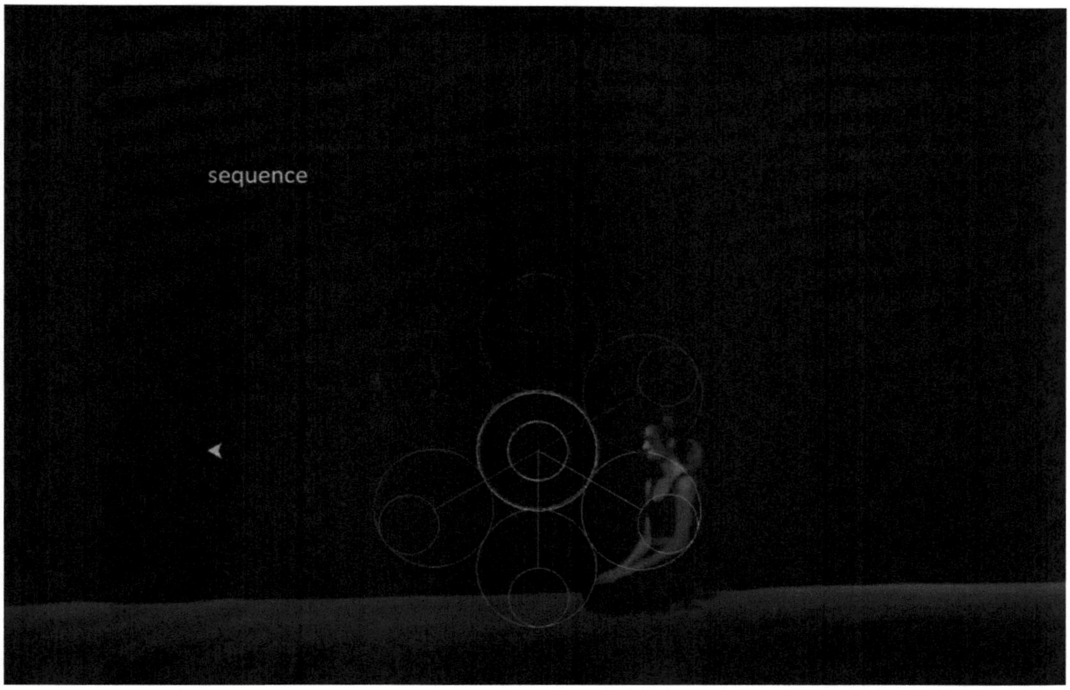
sequence

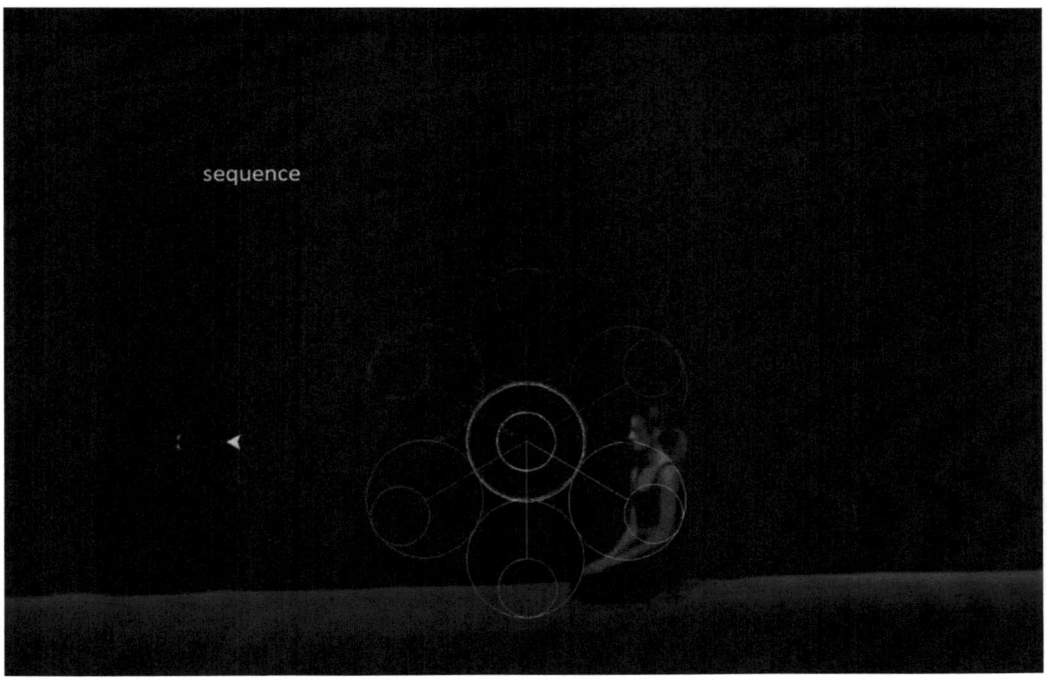
sequence

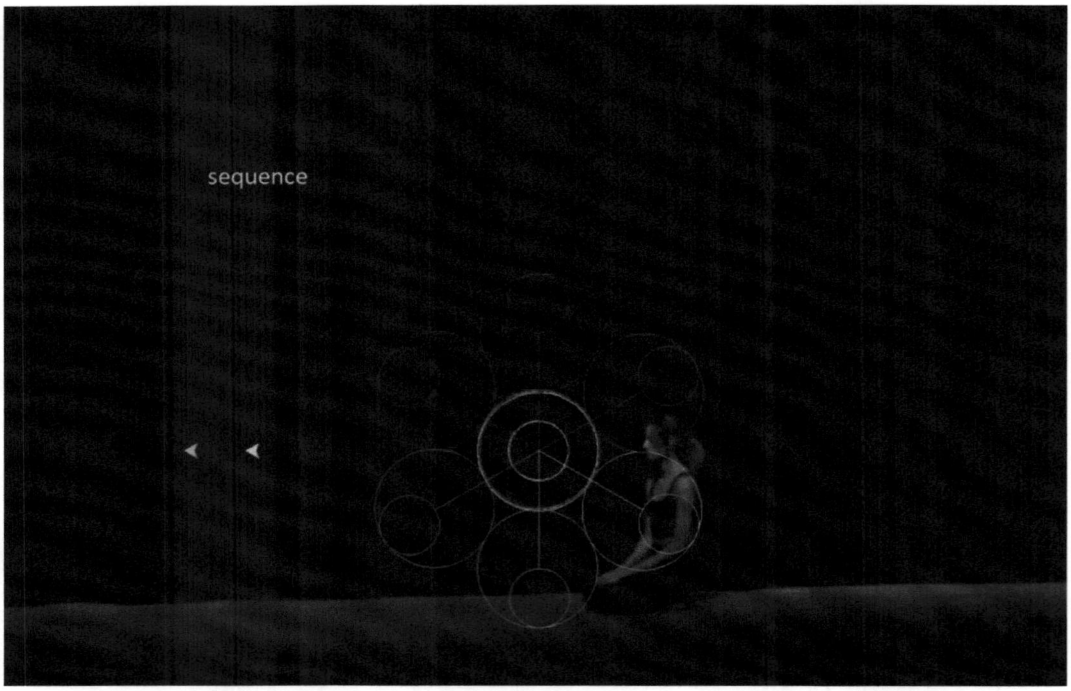
sequence

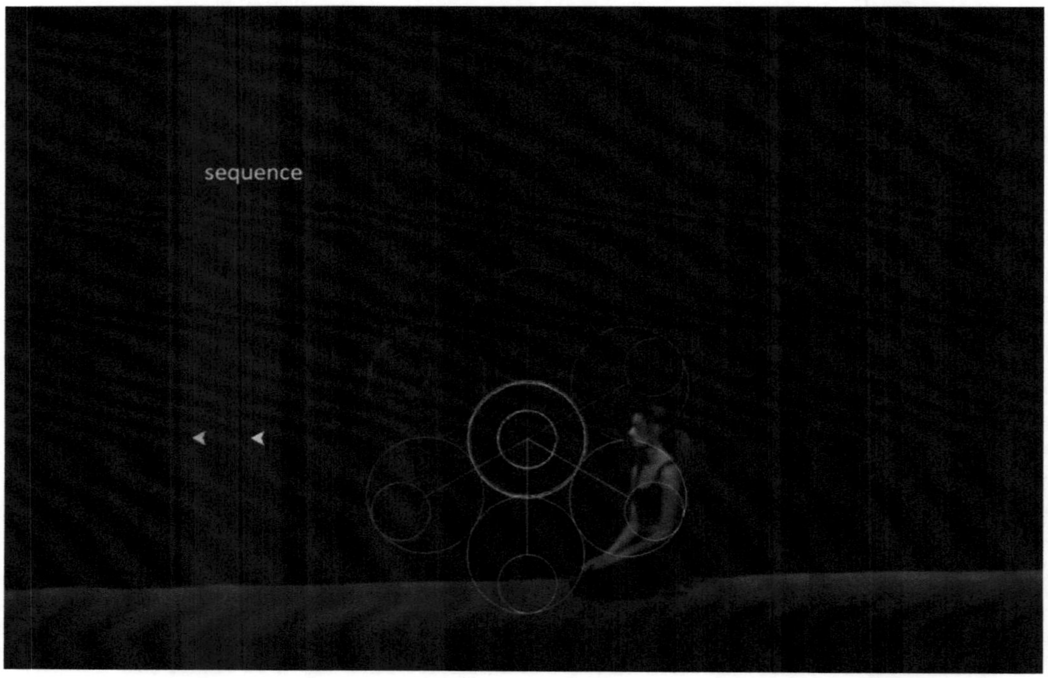
sequence

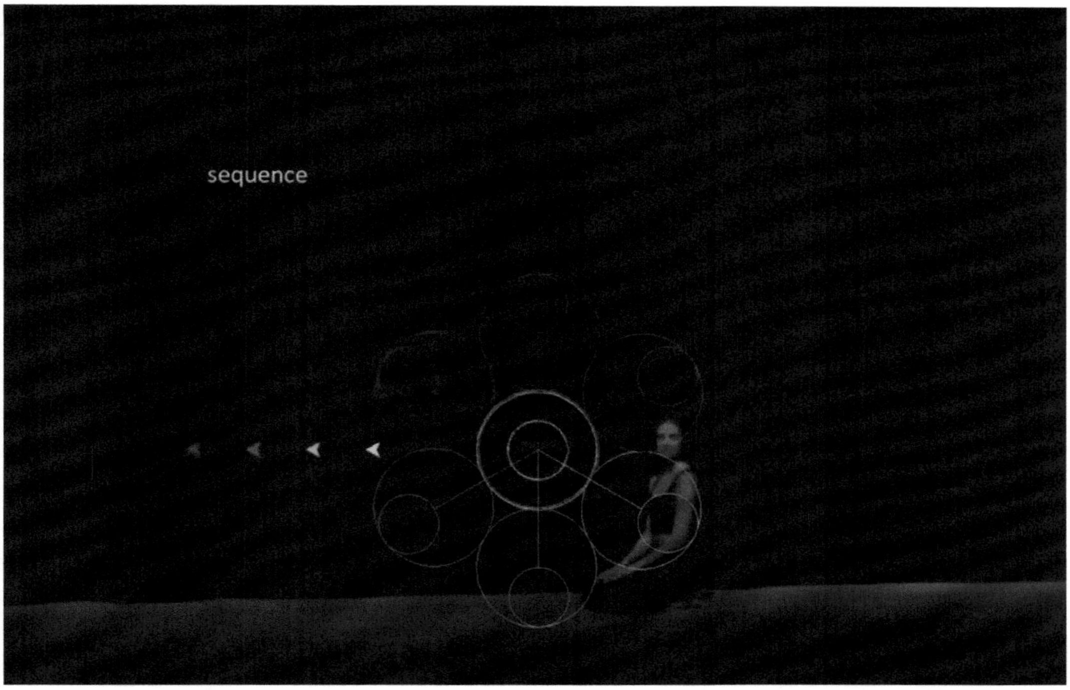
sequence

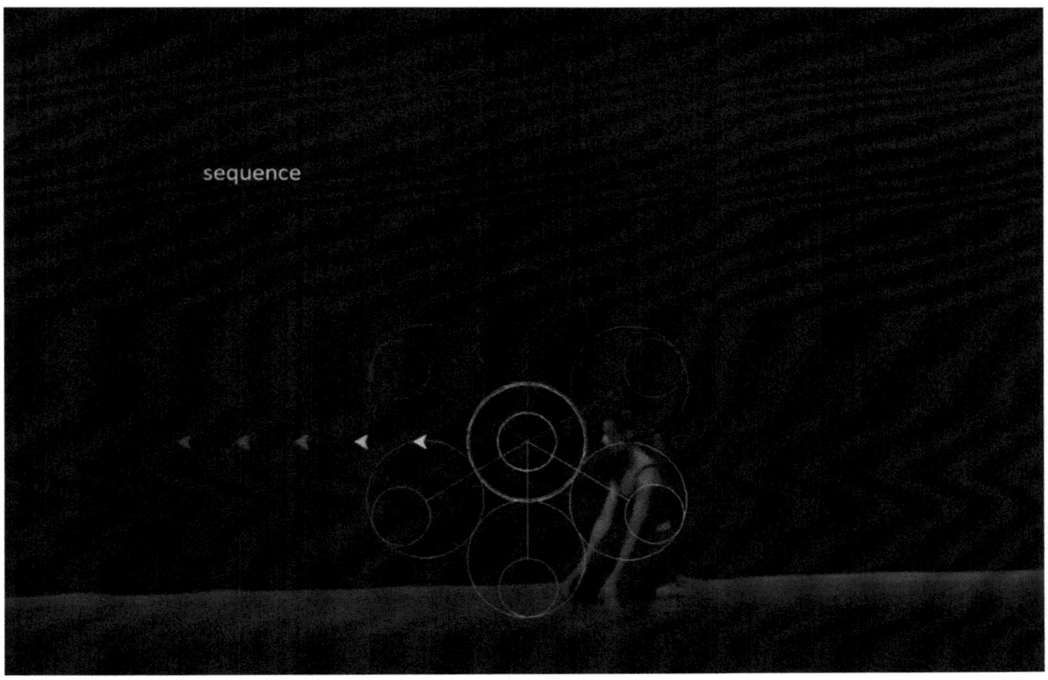
sequence

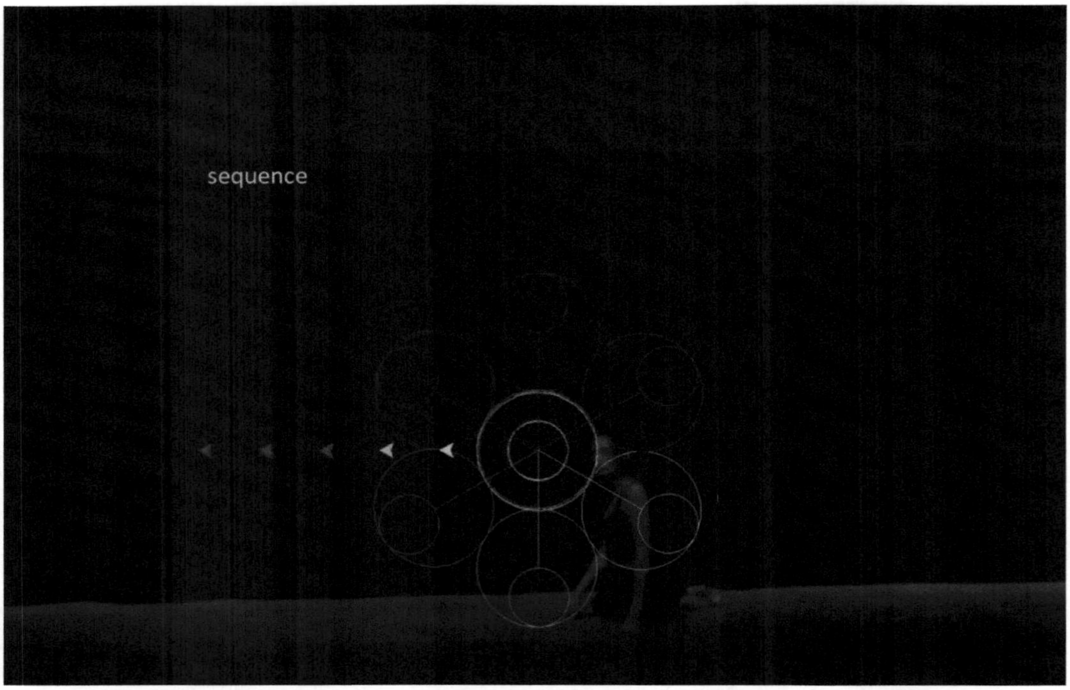
sequence

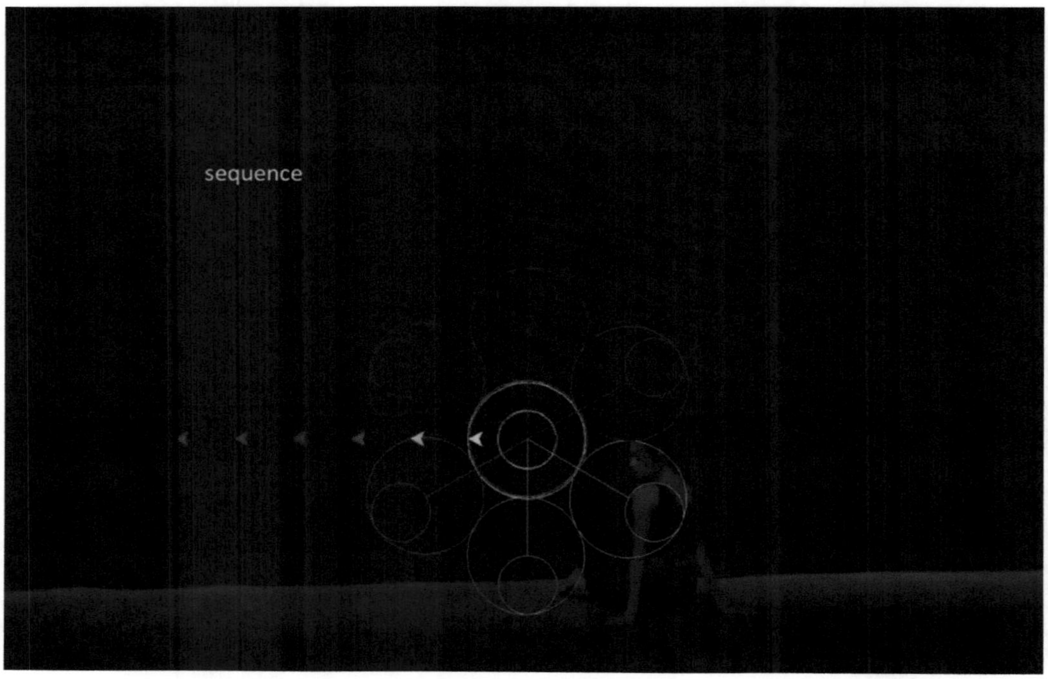
sequence

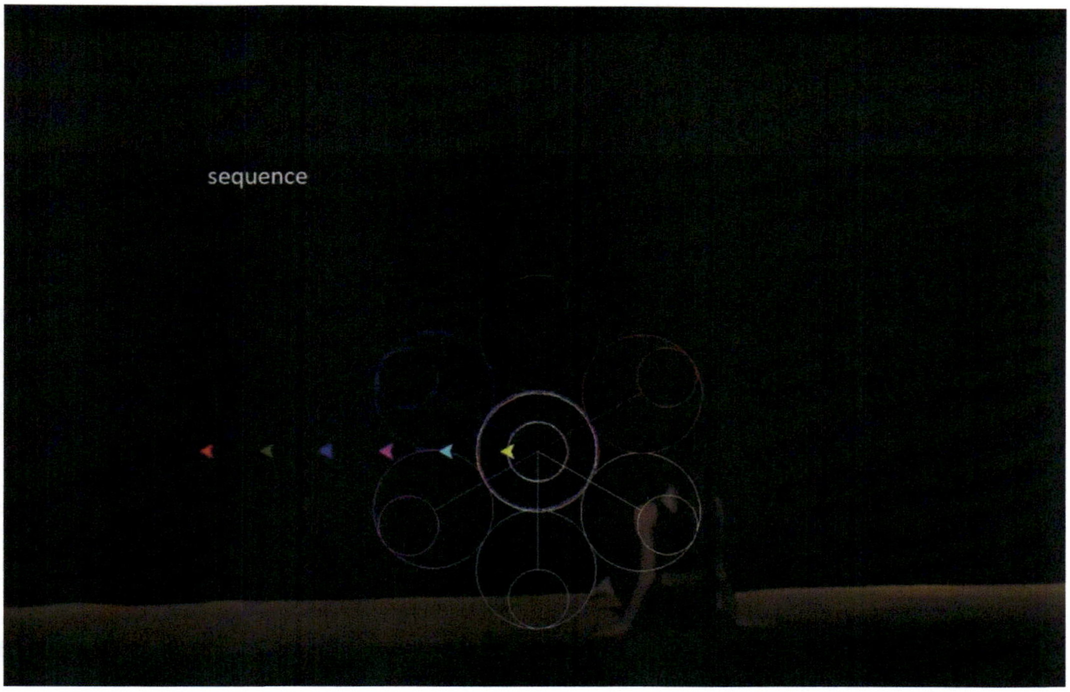
sequence

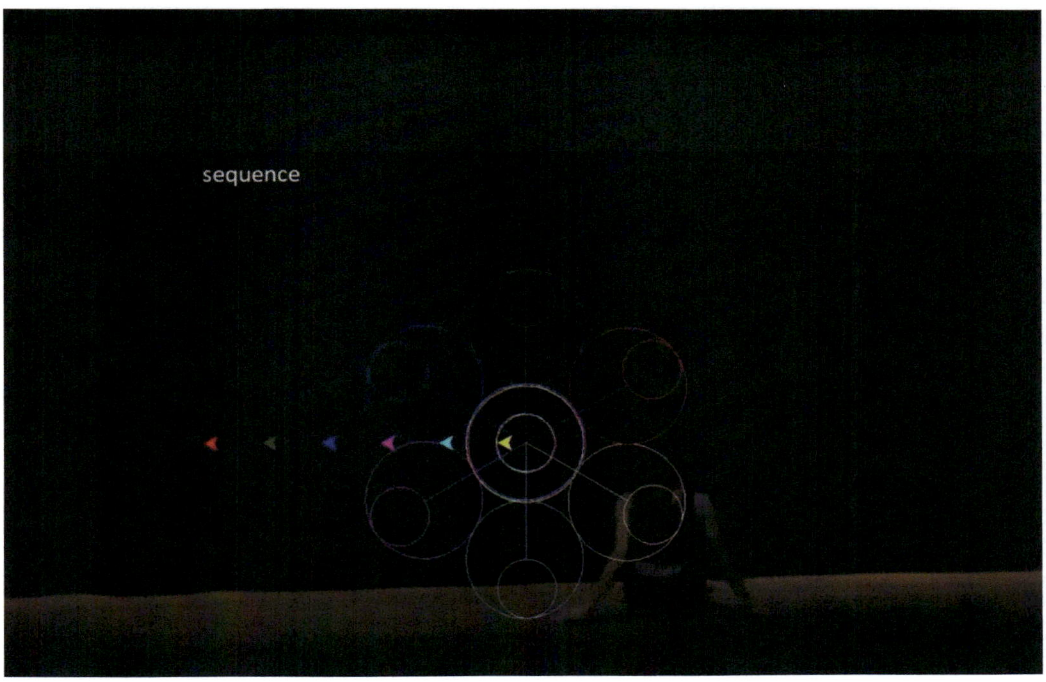
sequence

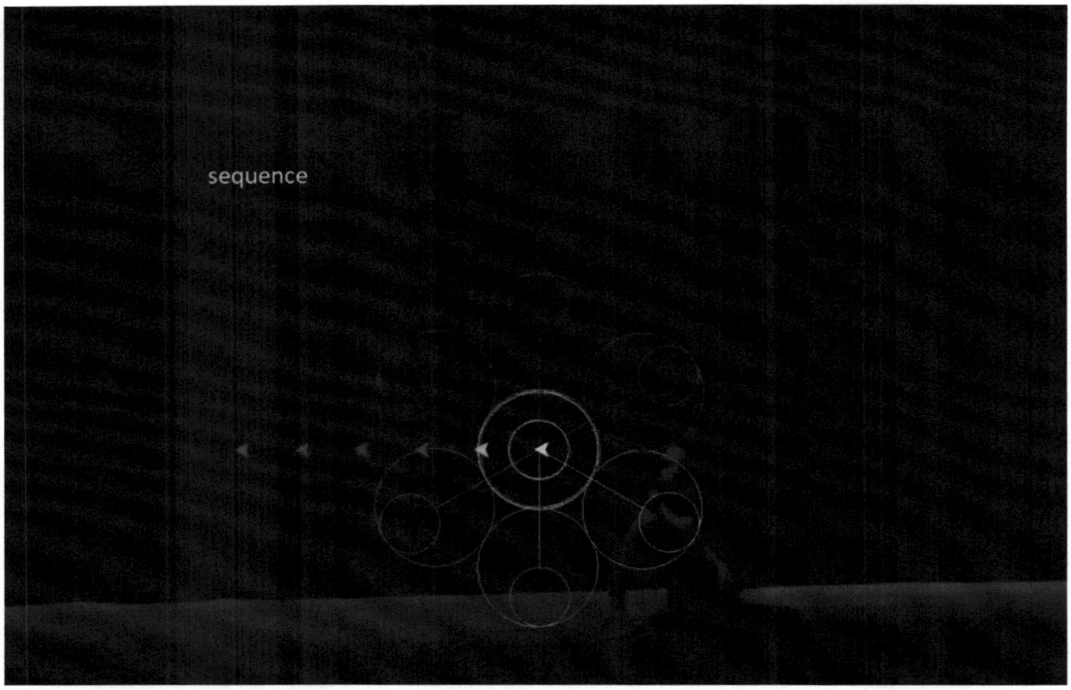
sequence

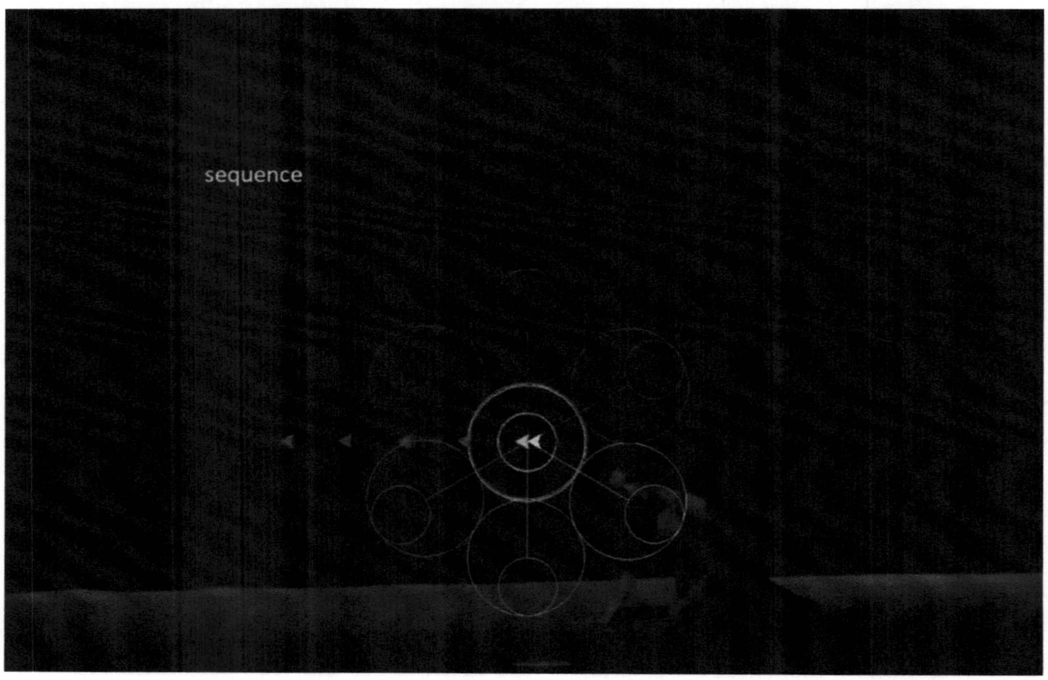
sequence

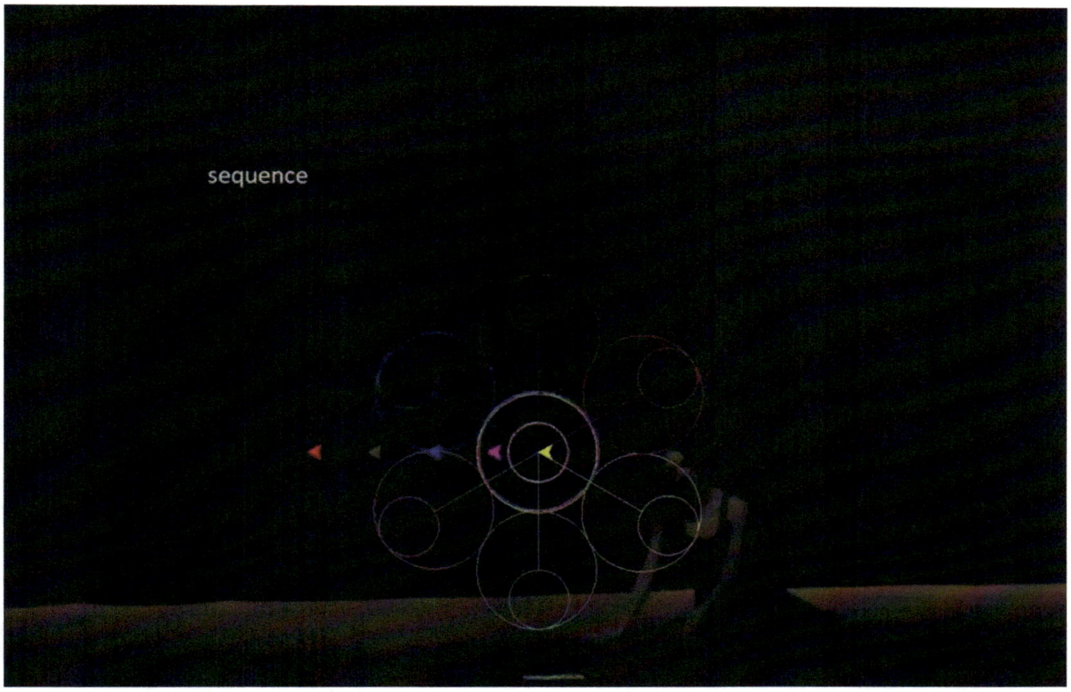
sequence

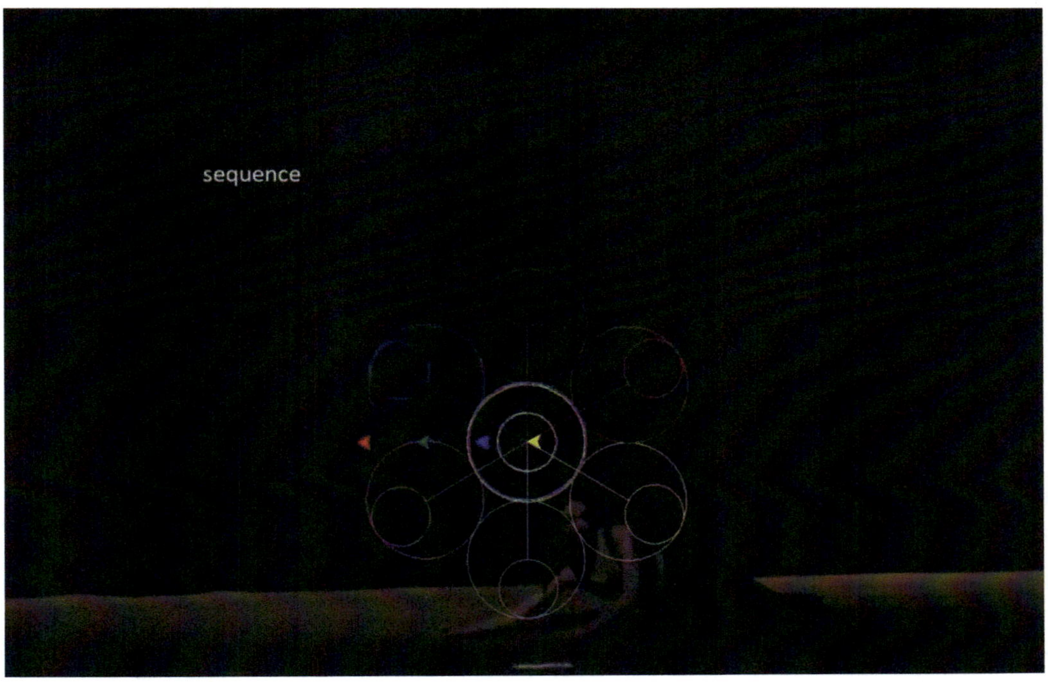
sequence

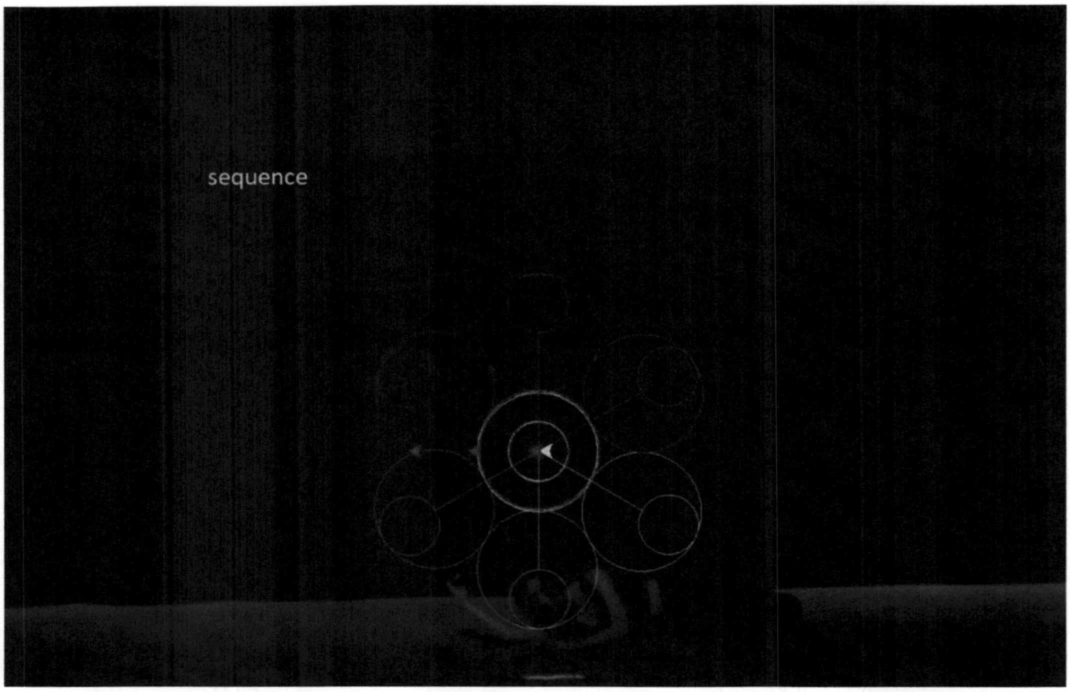
sequence

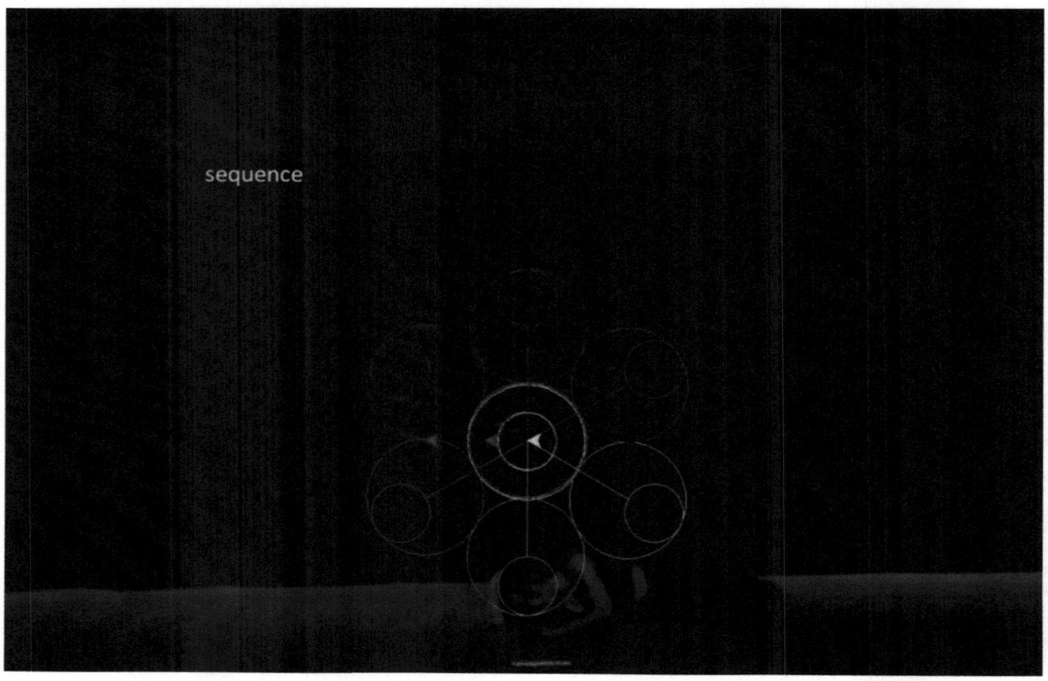
sequence

RULE BREAKING VIDEO:

Mary Jo Lasky dancing to a poem by Anna Ursyn,: recited by Izzy Sajdak:

https://www.youtube.com/watch?v=dsb-p7fFo_o

Based on:

RULE-BREAKING POEM:

He introduced her to performing functions, accurate guiding, path planning and following trajectories, automatic avoidance of objects, version interface, autonomous navigation, and replication of movements. She used hard coding with messy formatting code, several things in one function, variables with magic numbers and strings, and no comments.

She told him about the principles: centering, gravity, alignment, contraction, release, breath use, falling, then recovering, balancing, and then off-balancing, opposition, succession, tension, relaxation, spiral, swing, and momentum. He crossed his arms and put them on his hips when listening.

He told her about safety, security, reliability, testability, maintainability, and portability. Those she applied to her work, respectively.

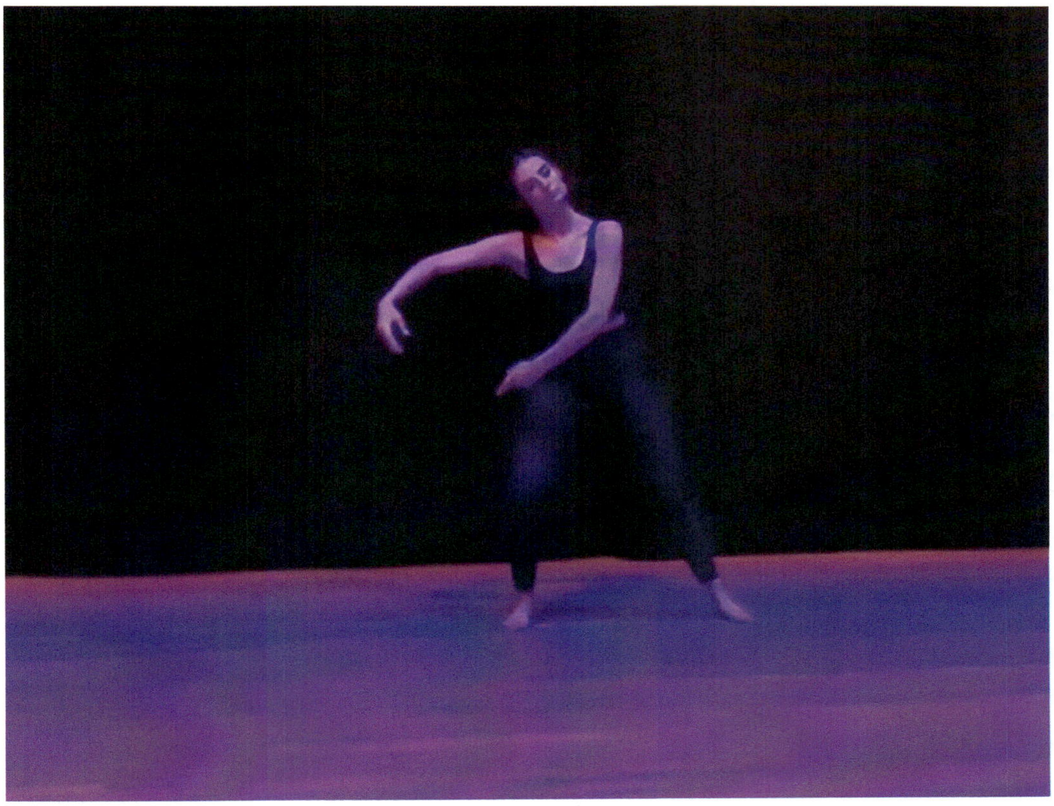

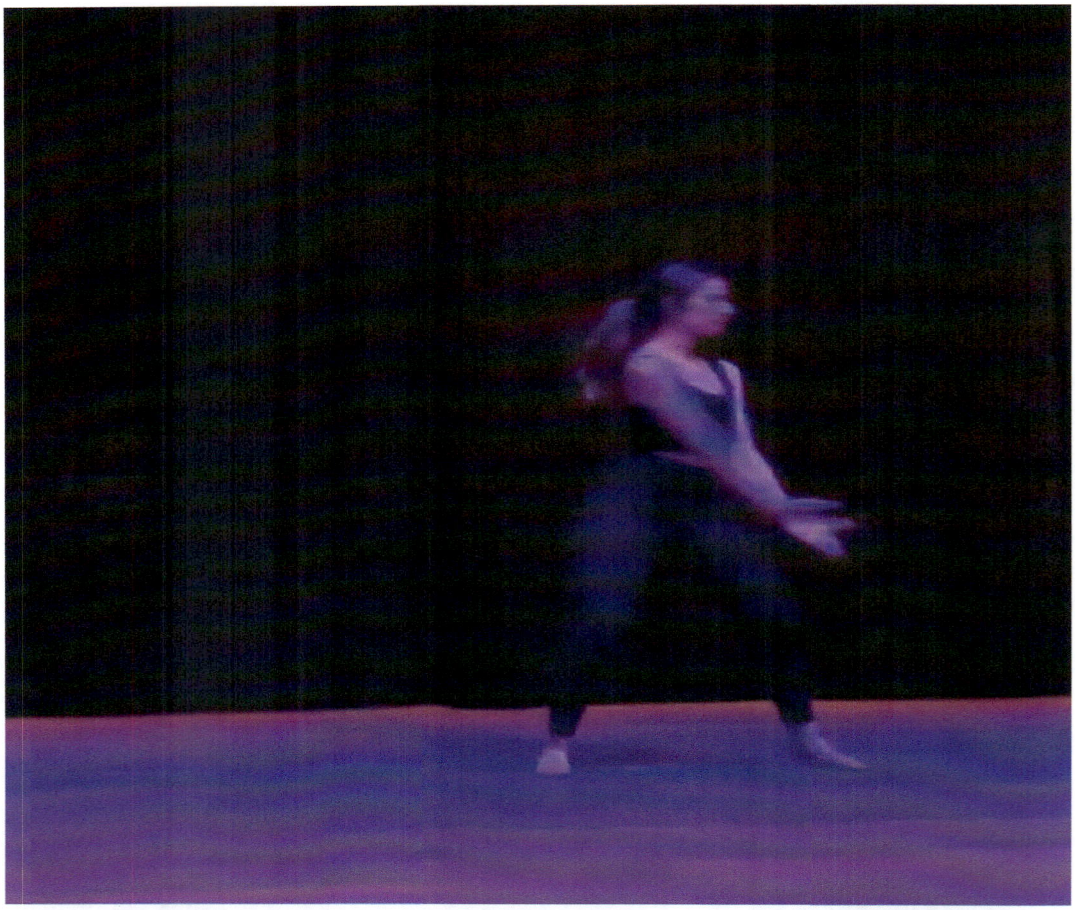

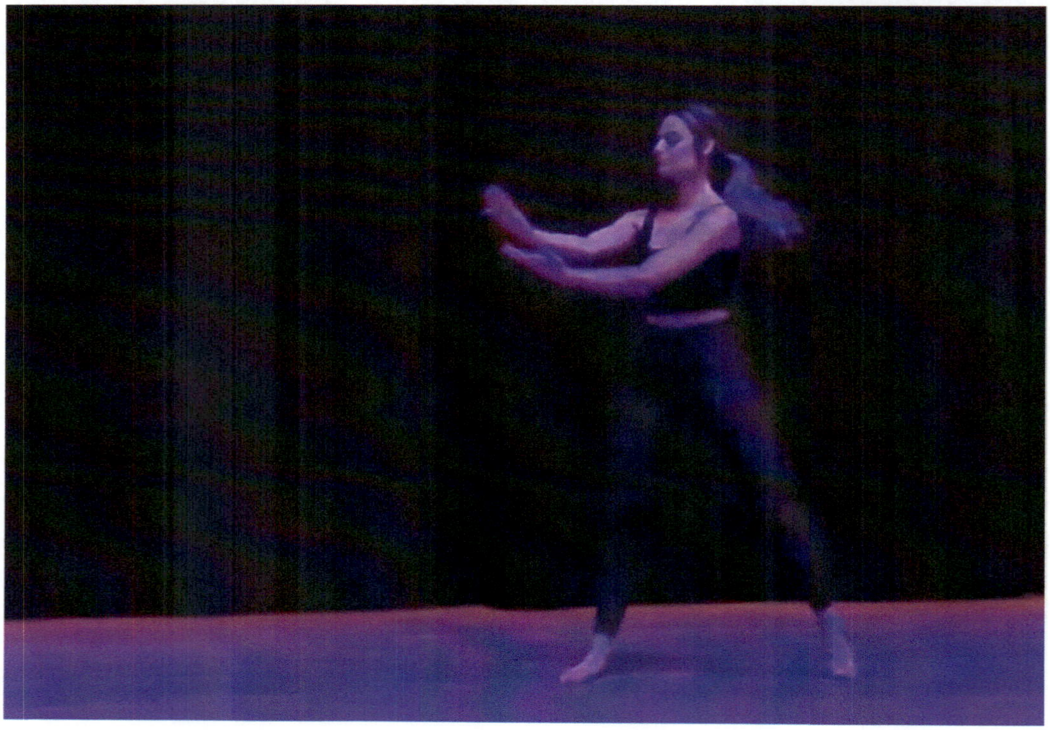

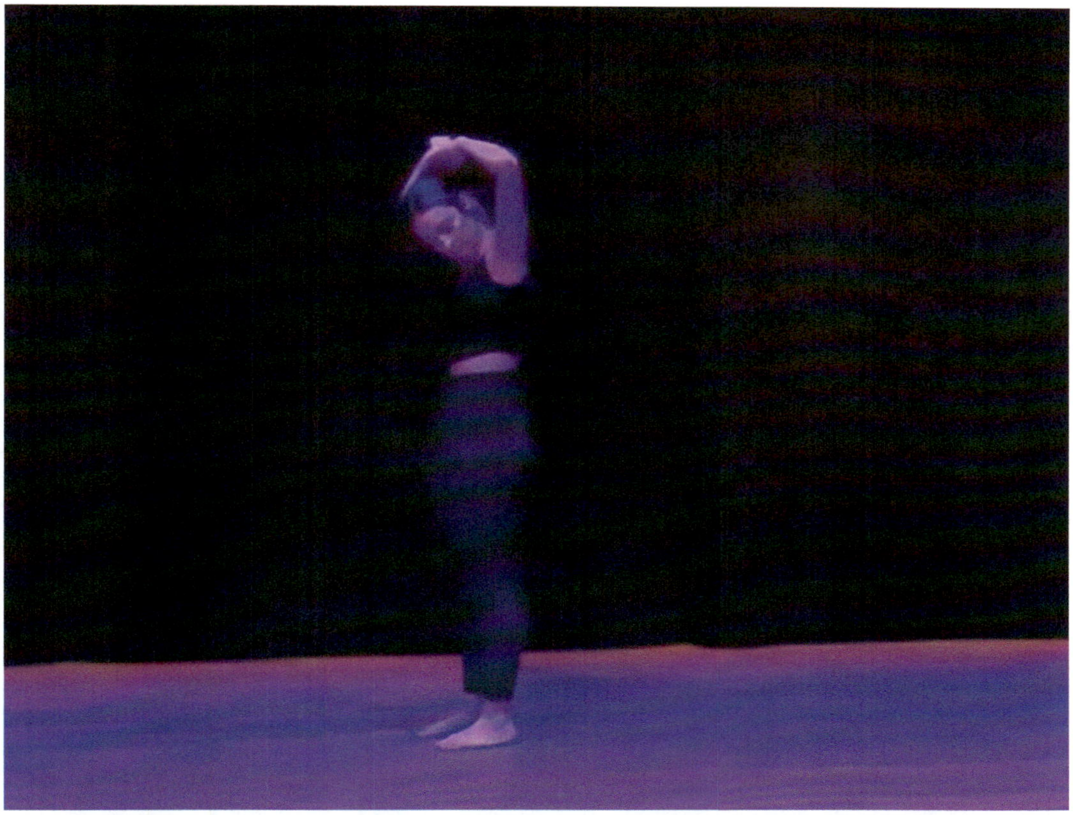

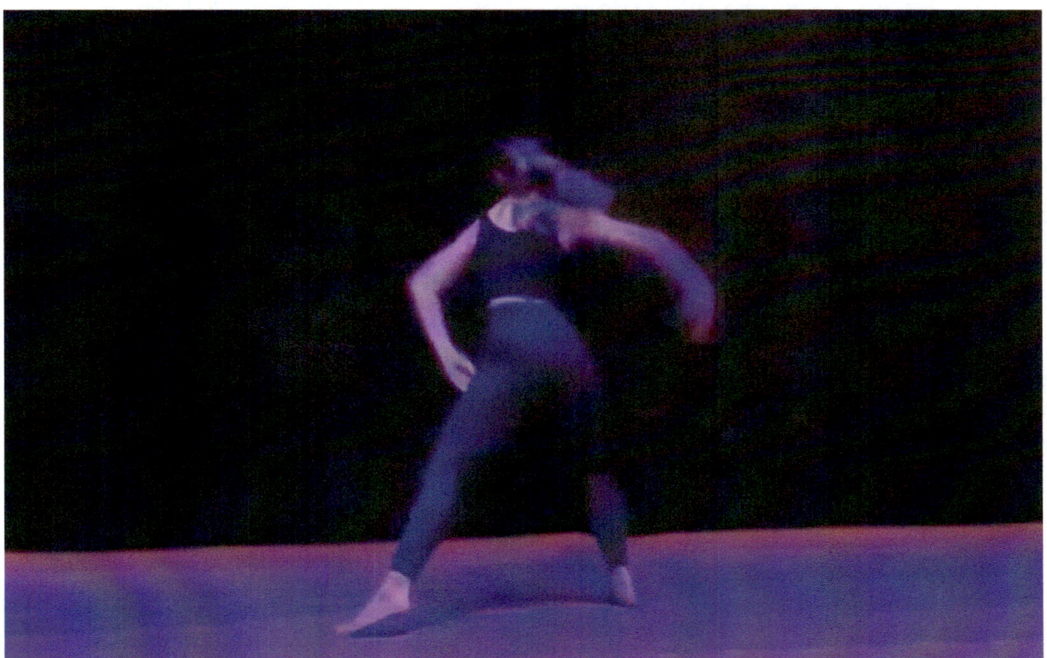

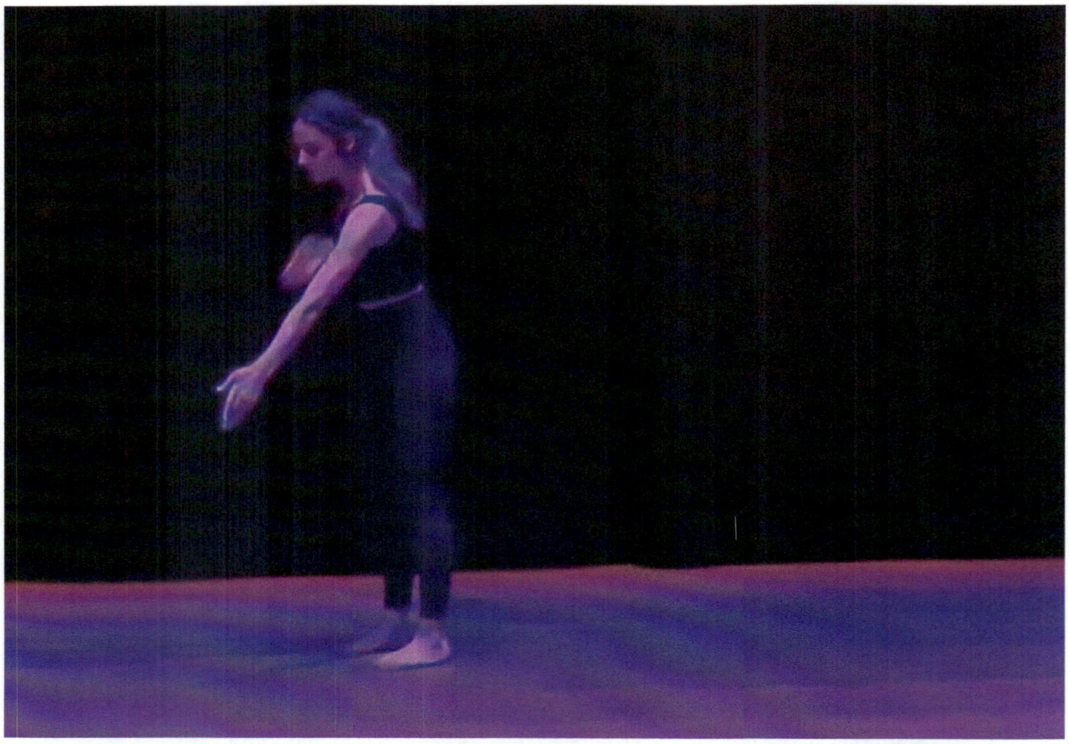

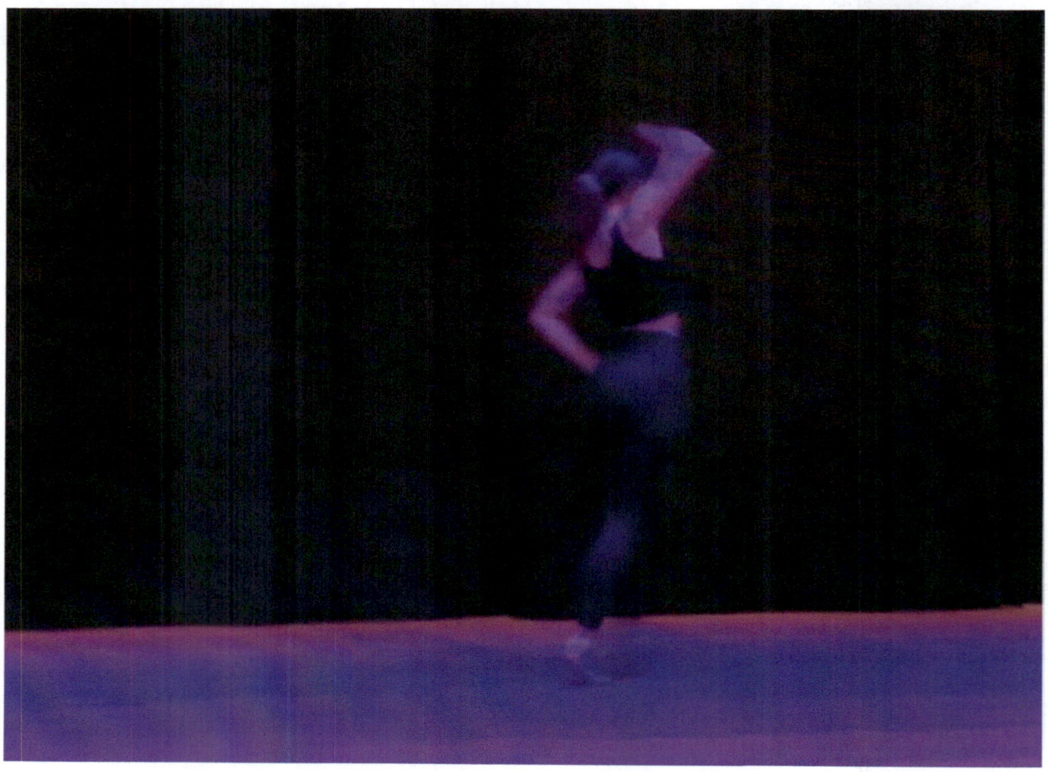

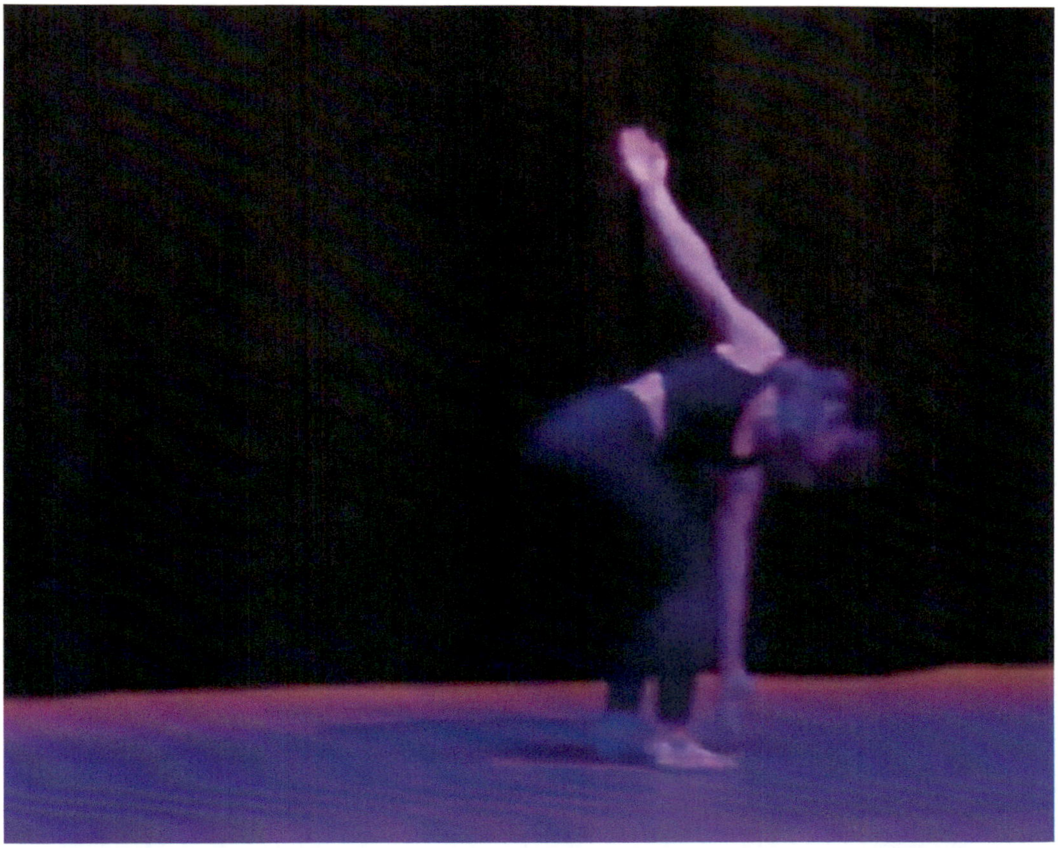

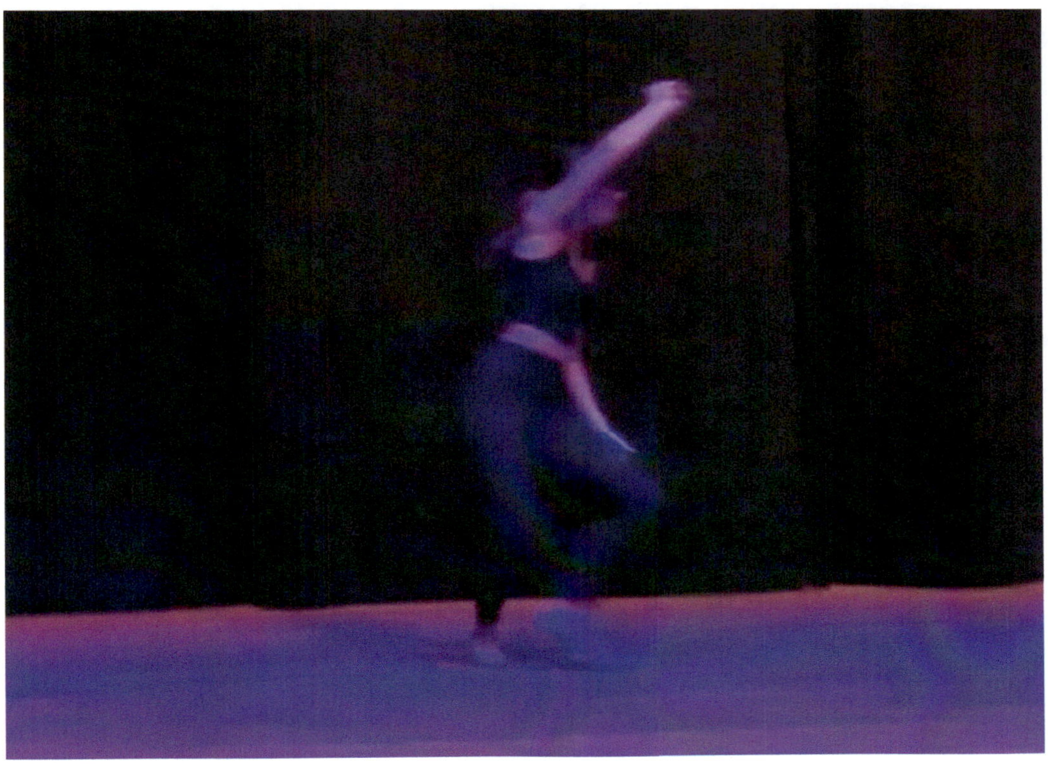

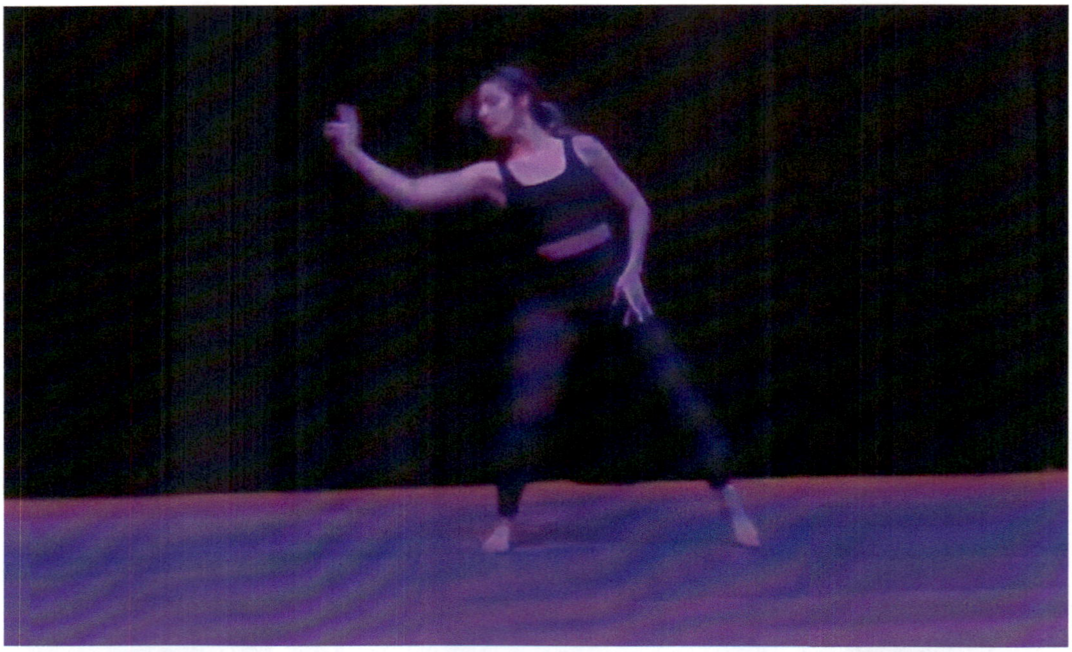

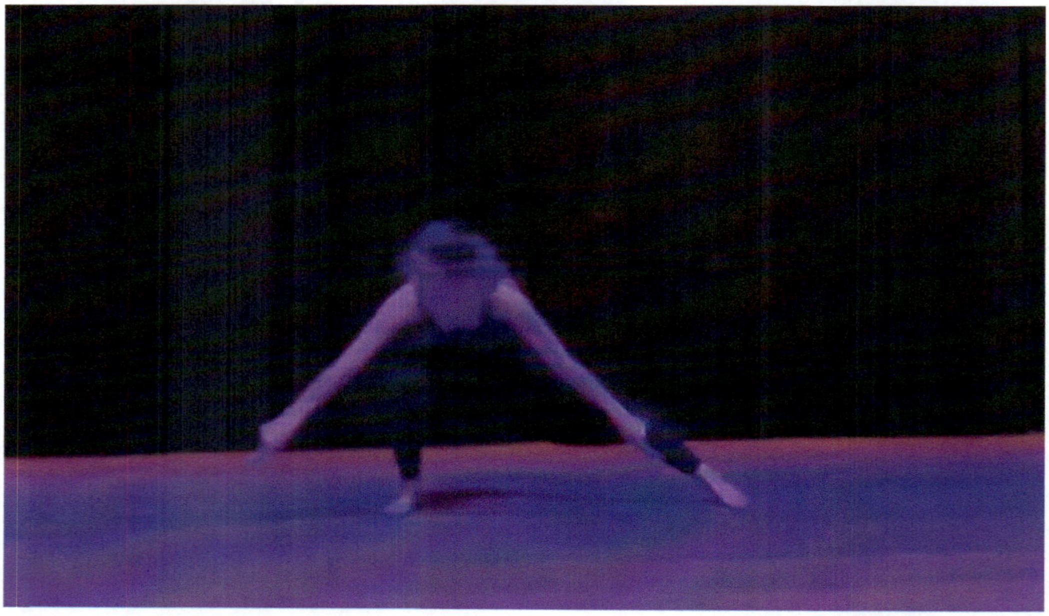

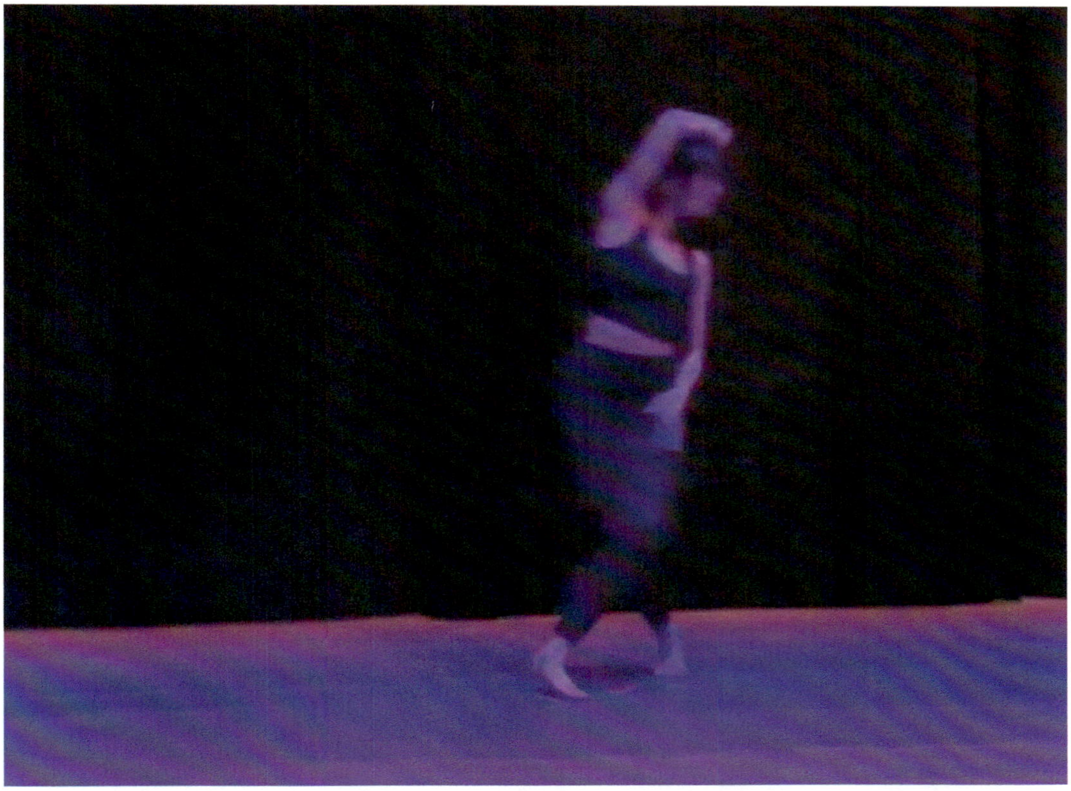

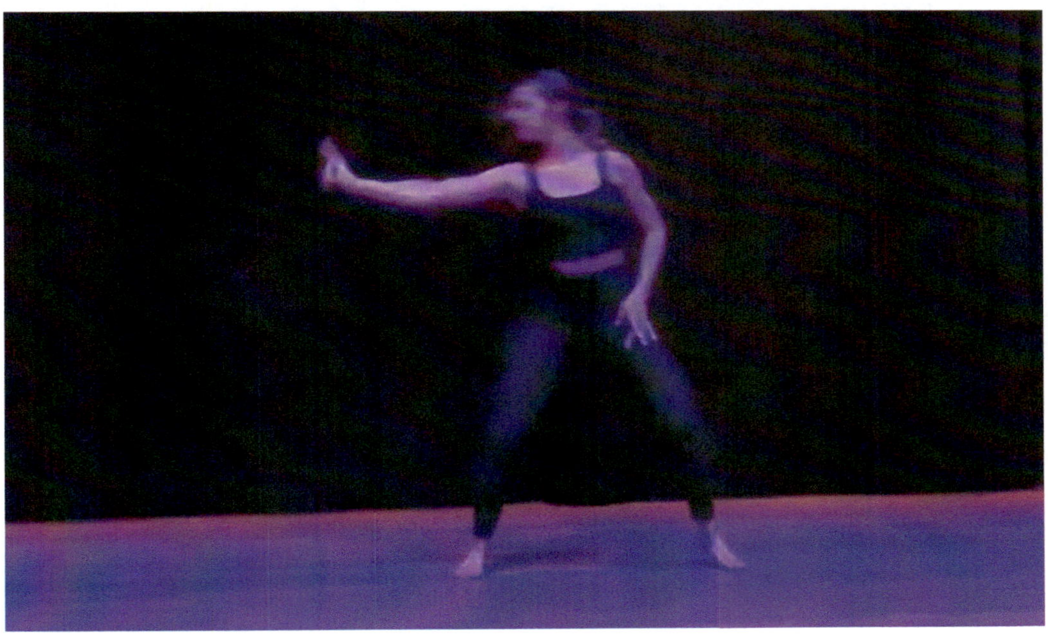

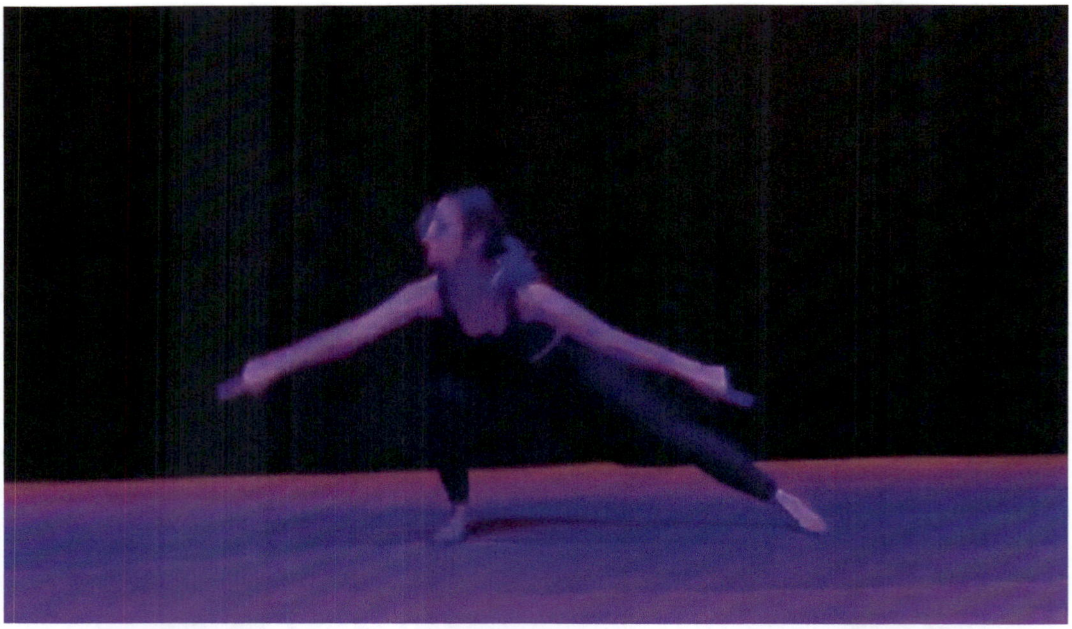

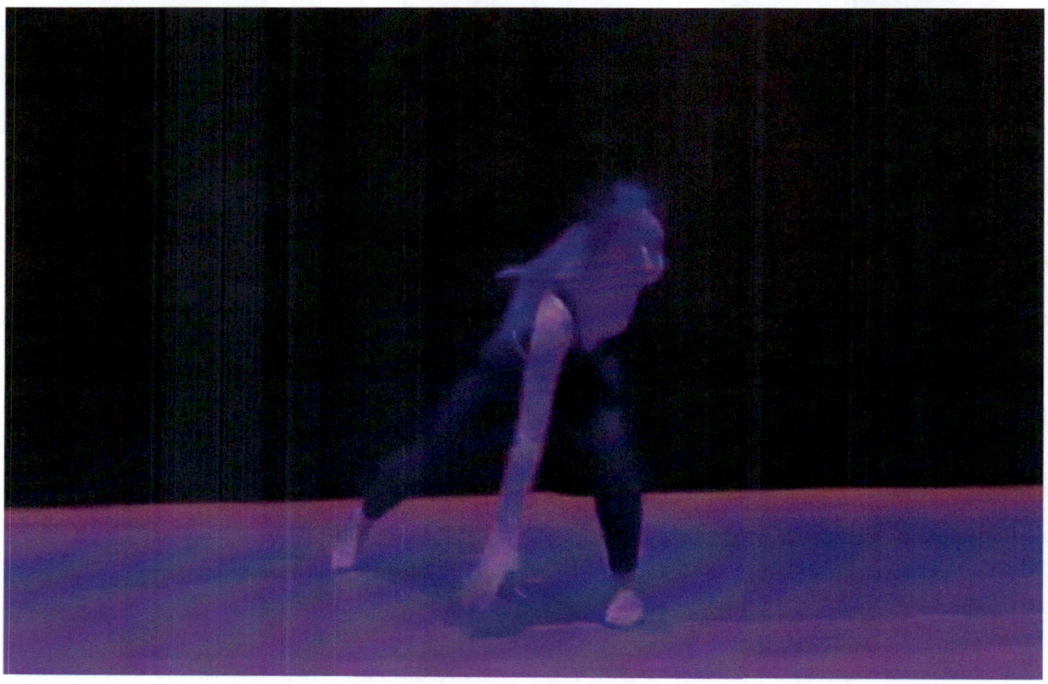

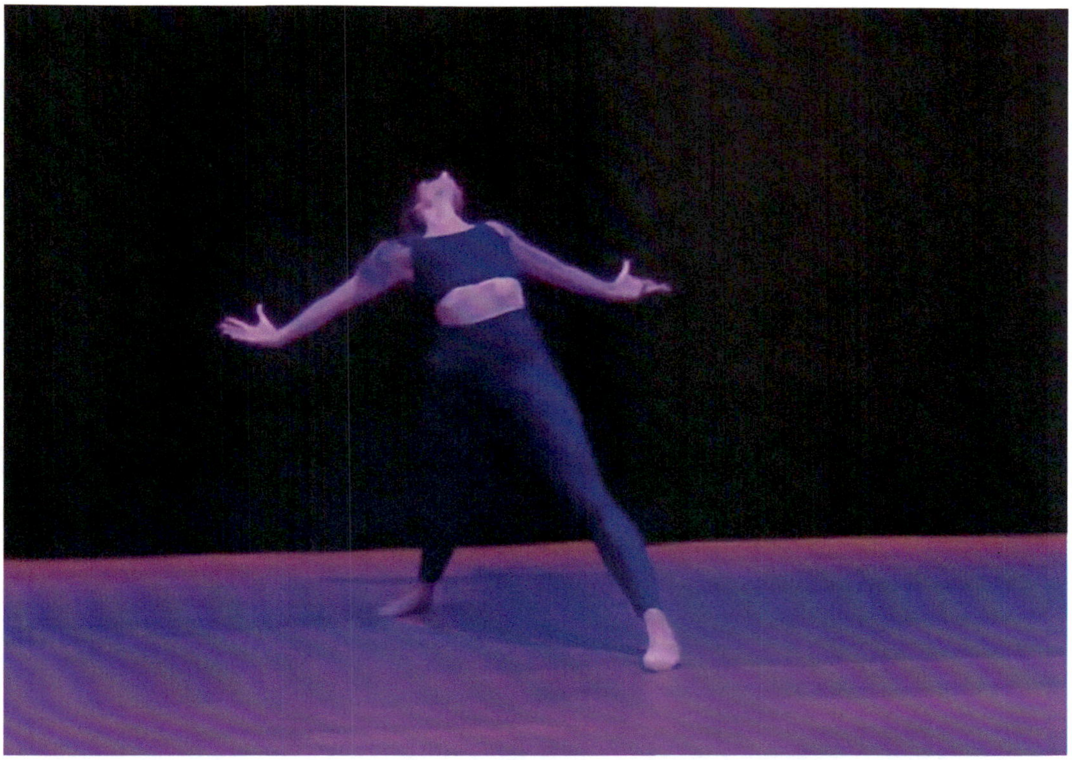

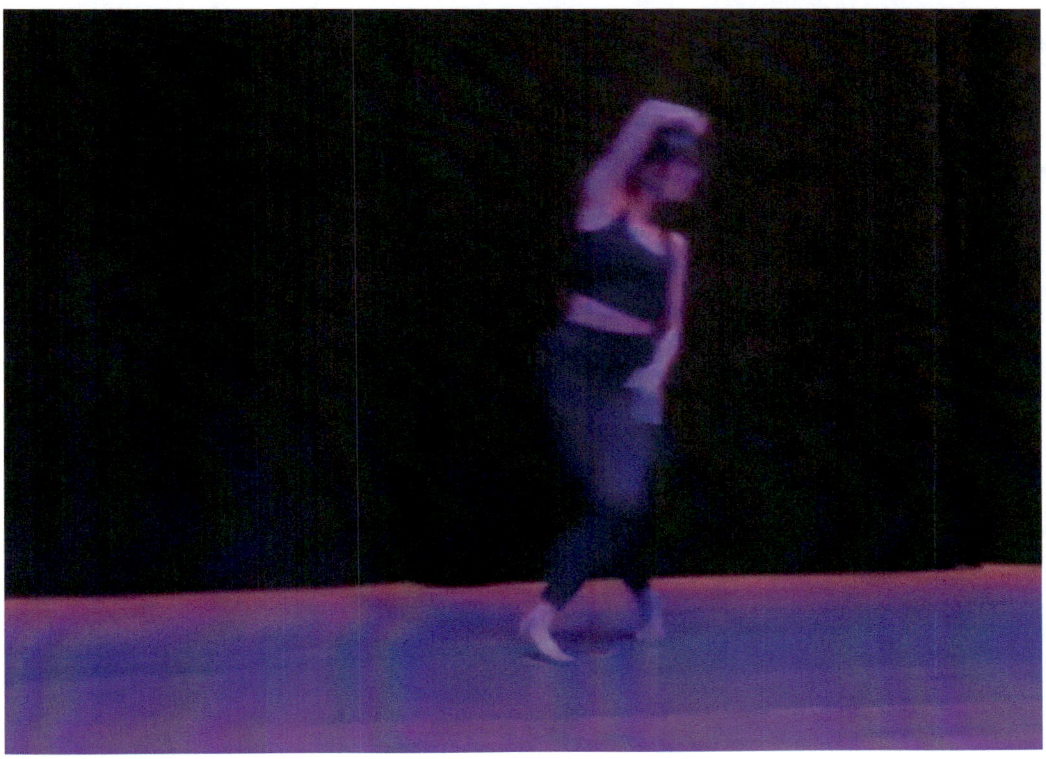

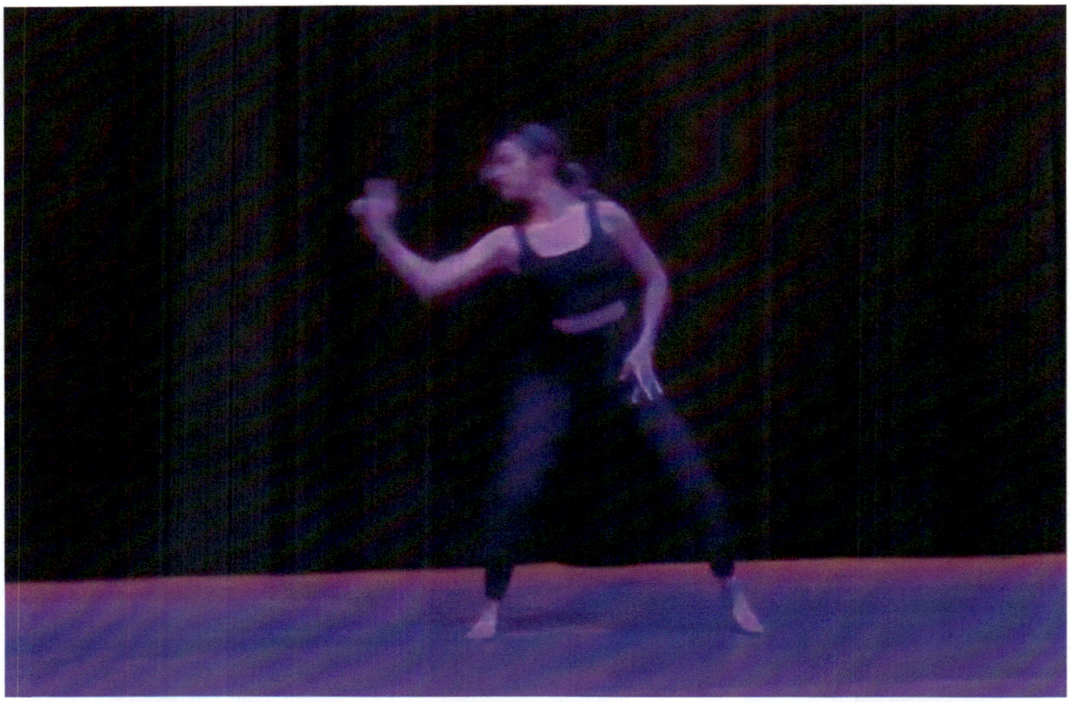

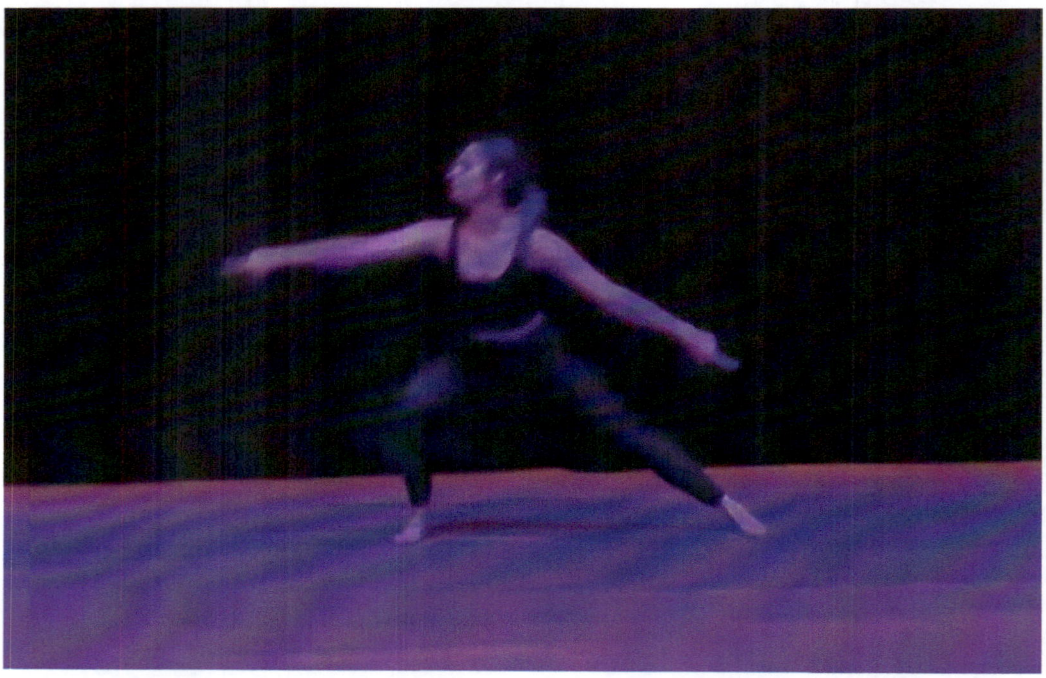

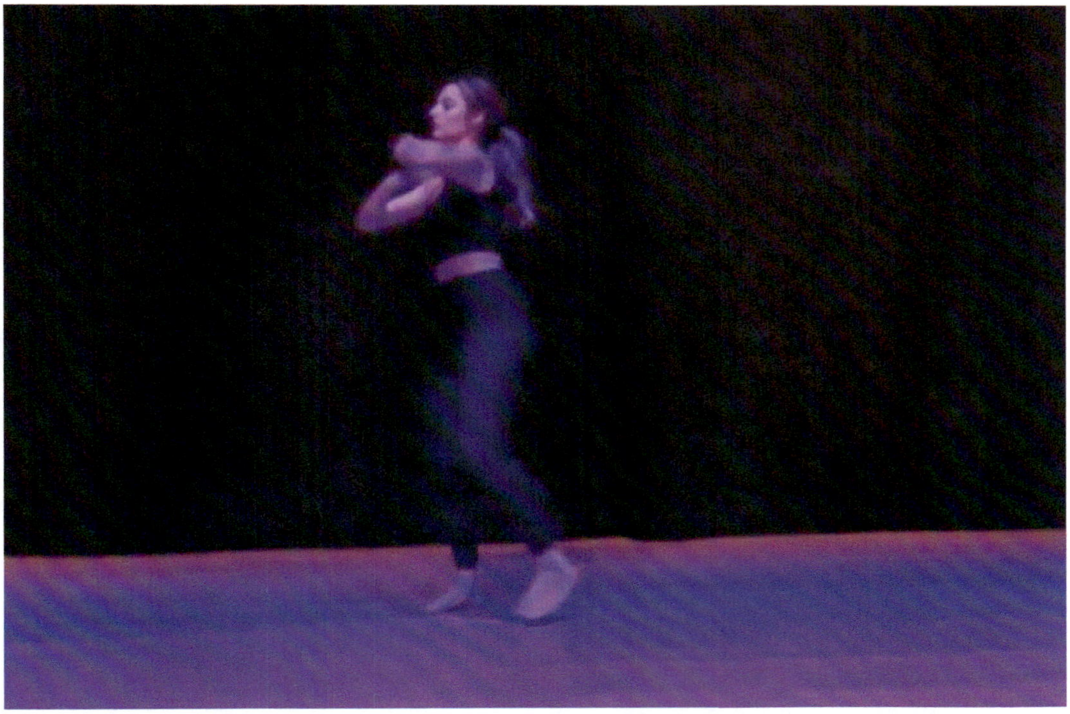

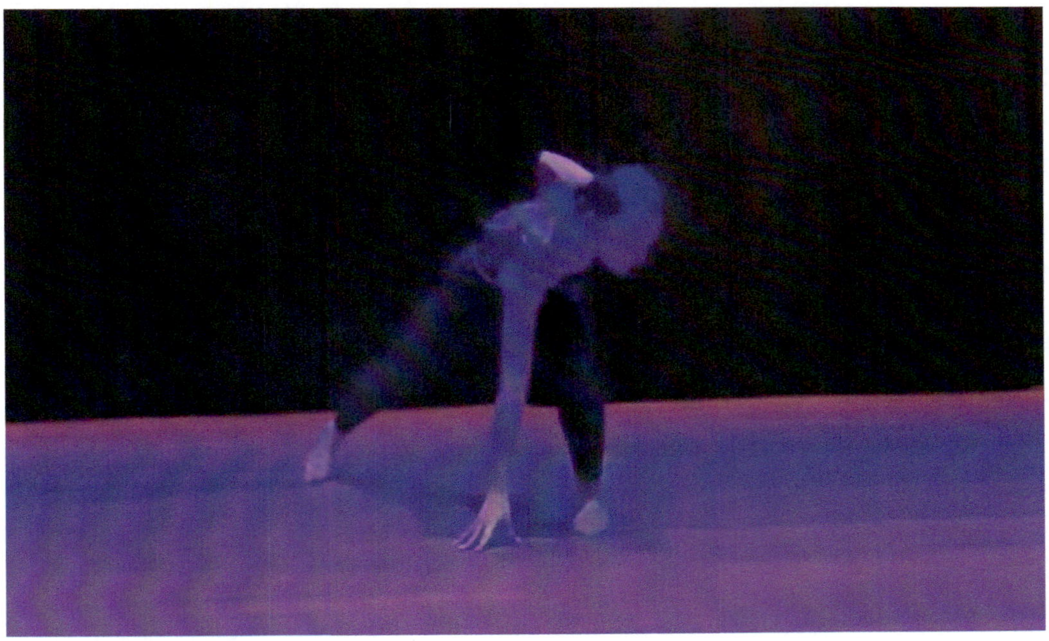

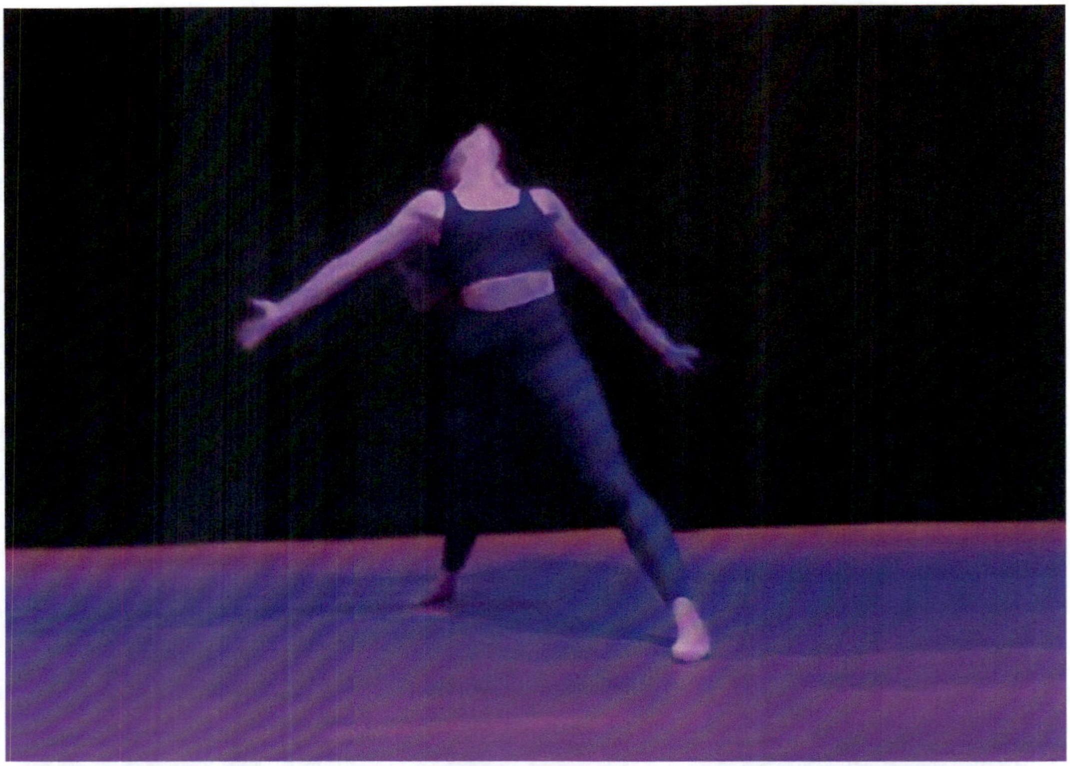

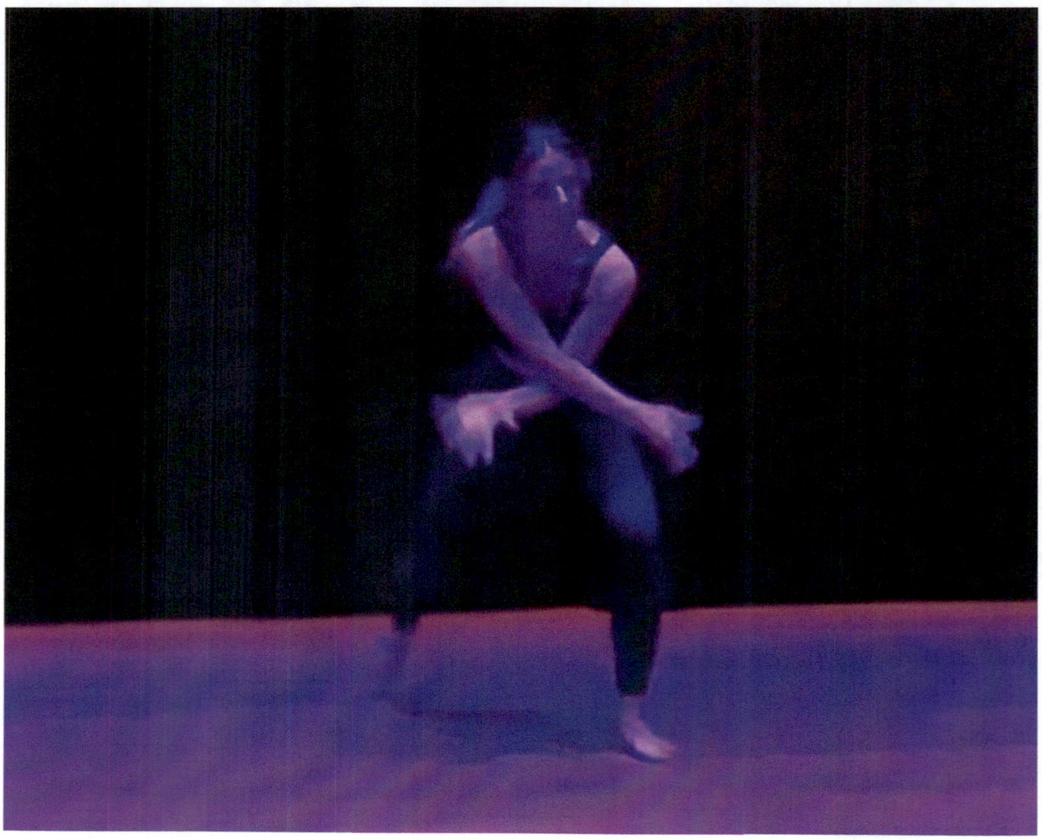

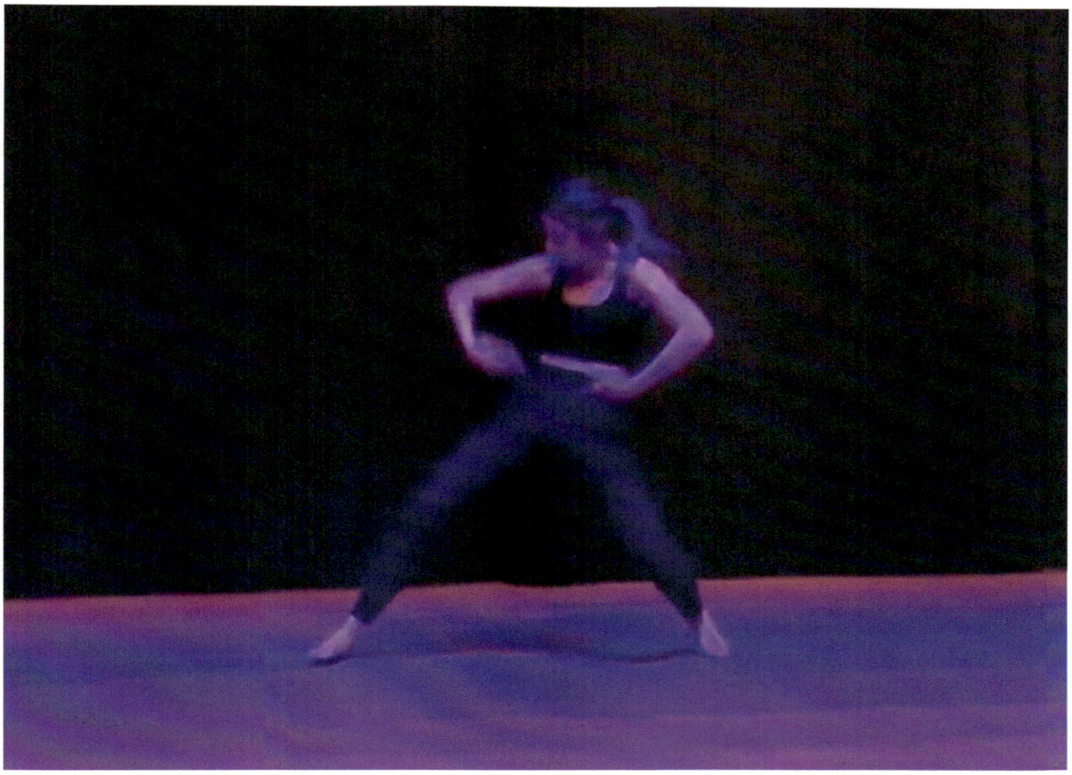

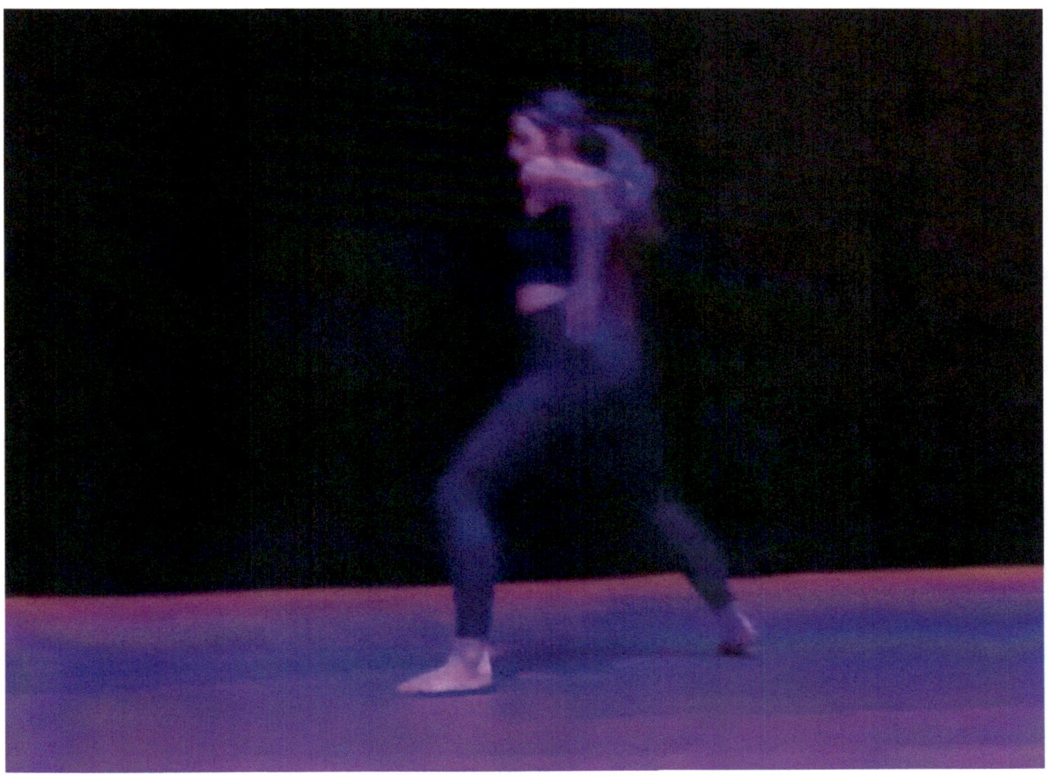

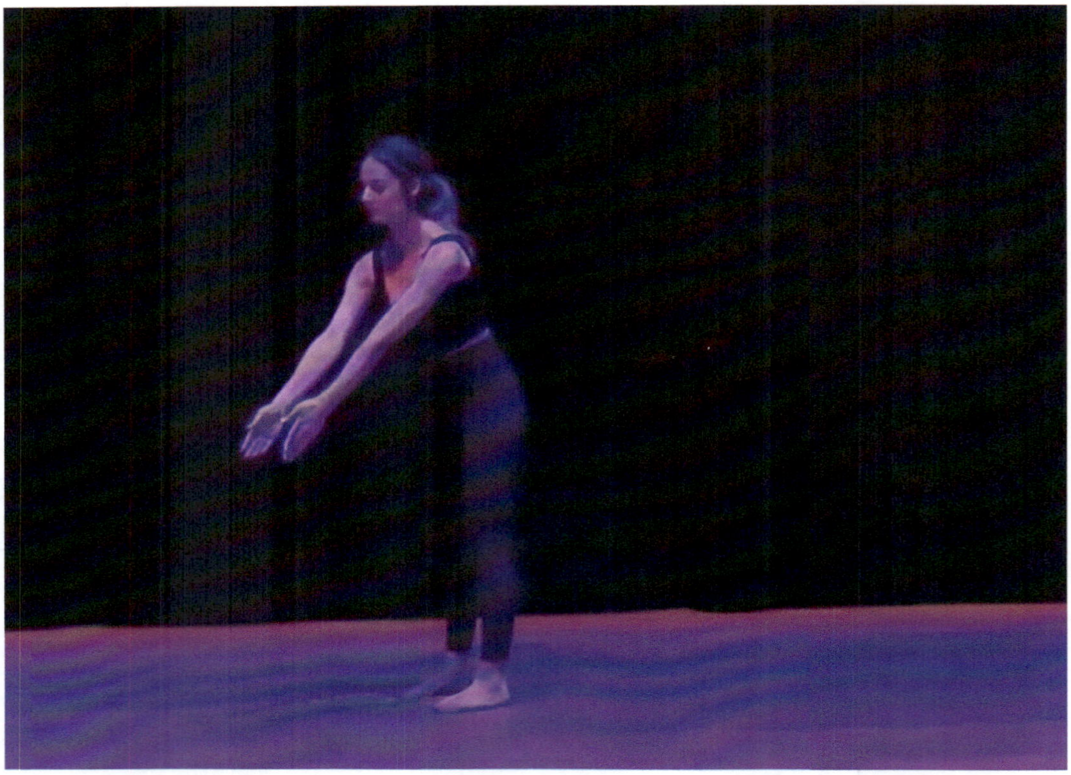

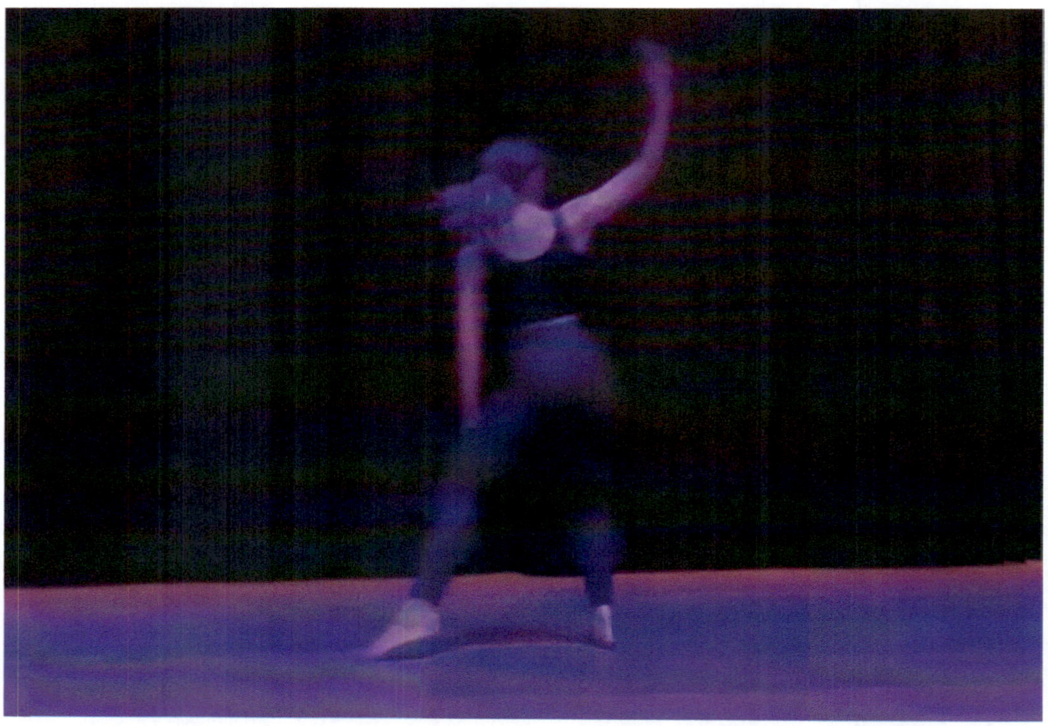

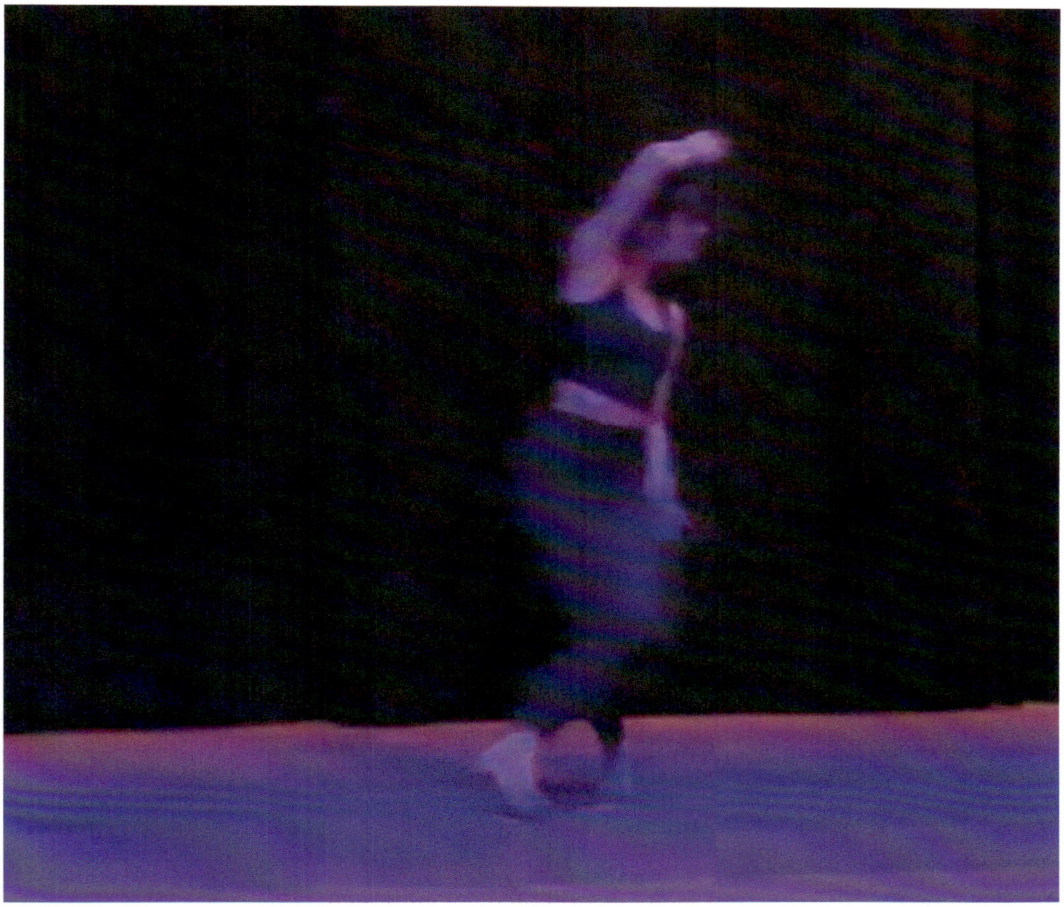

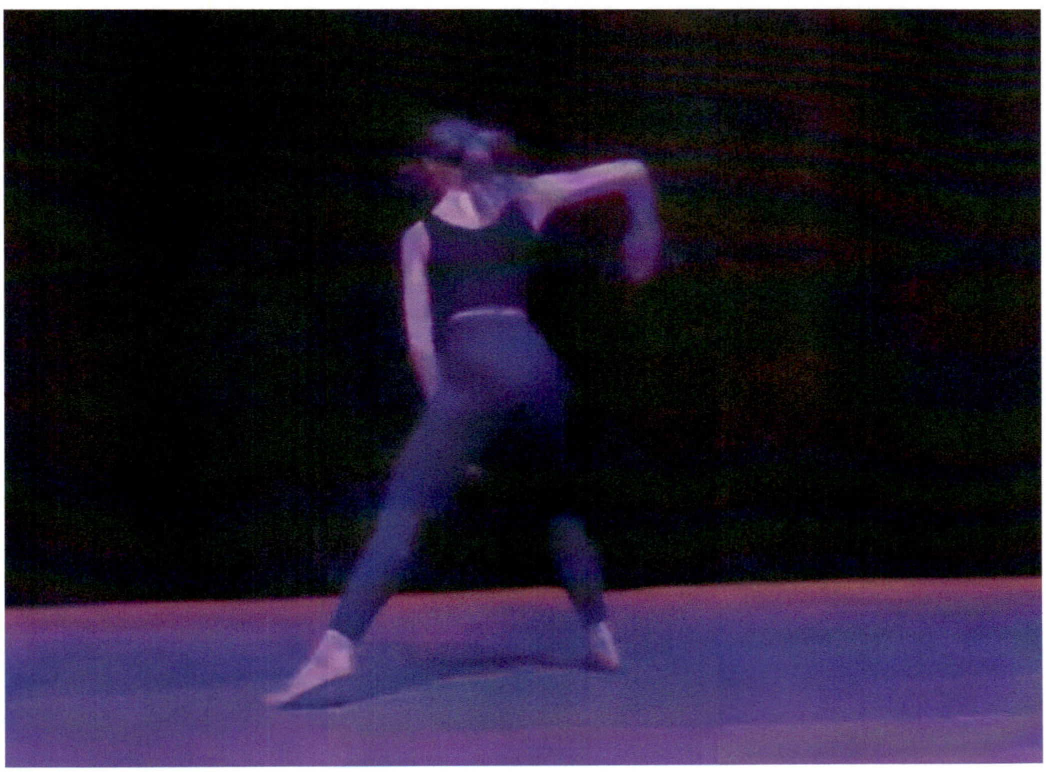

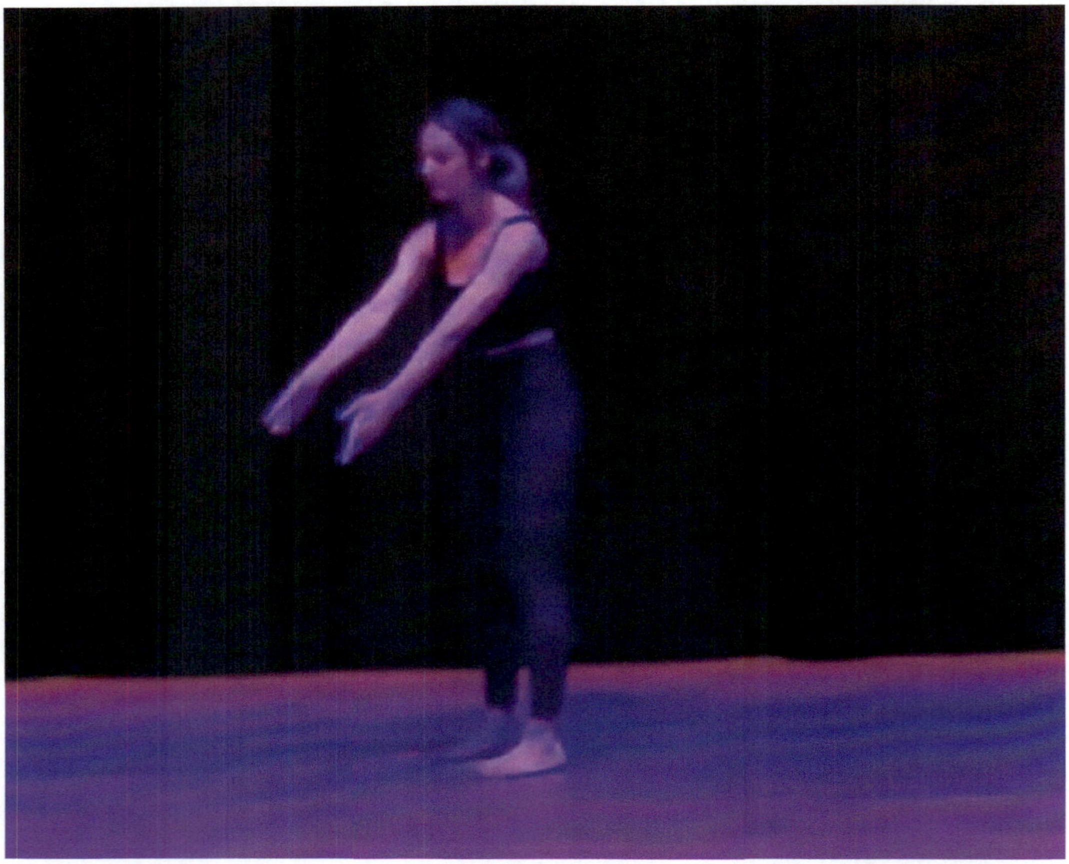

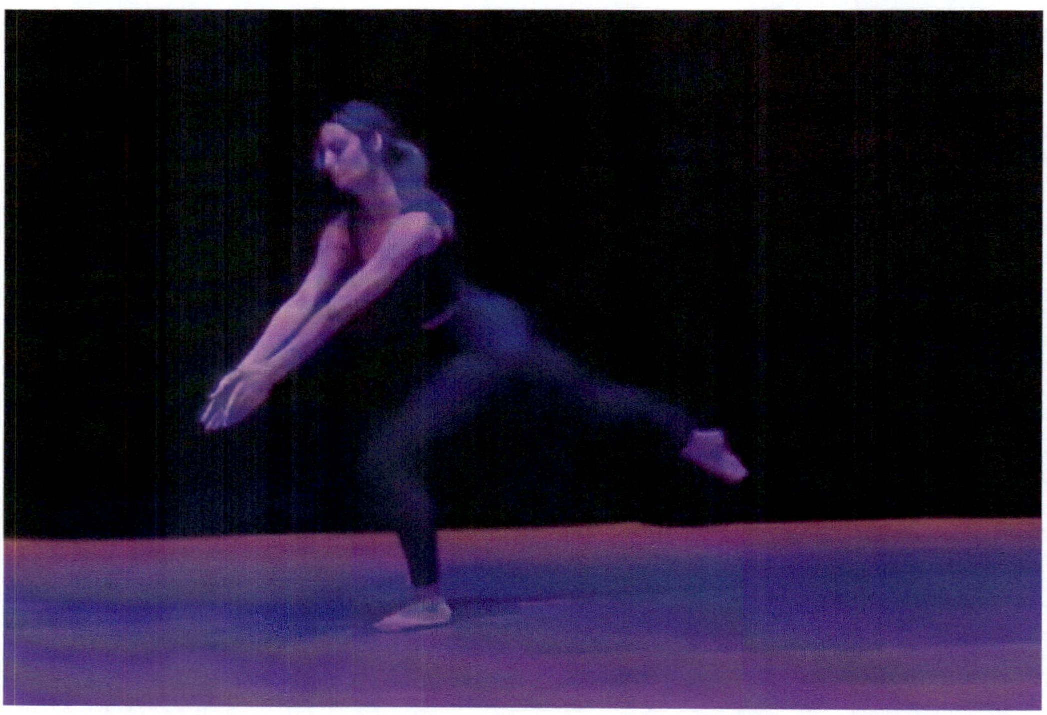

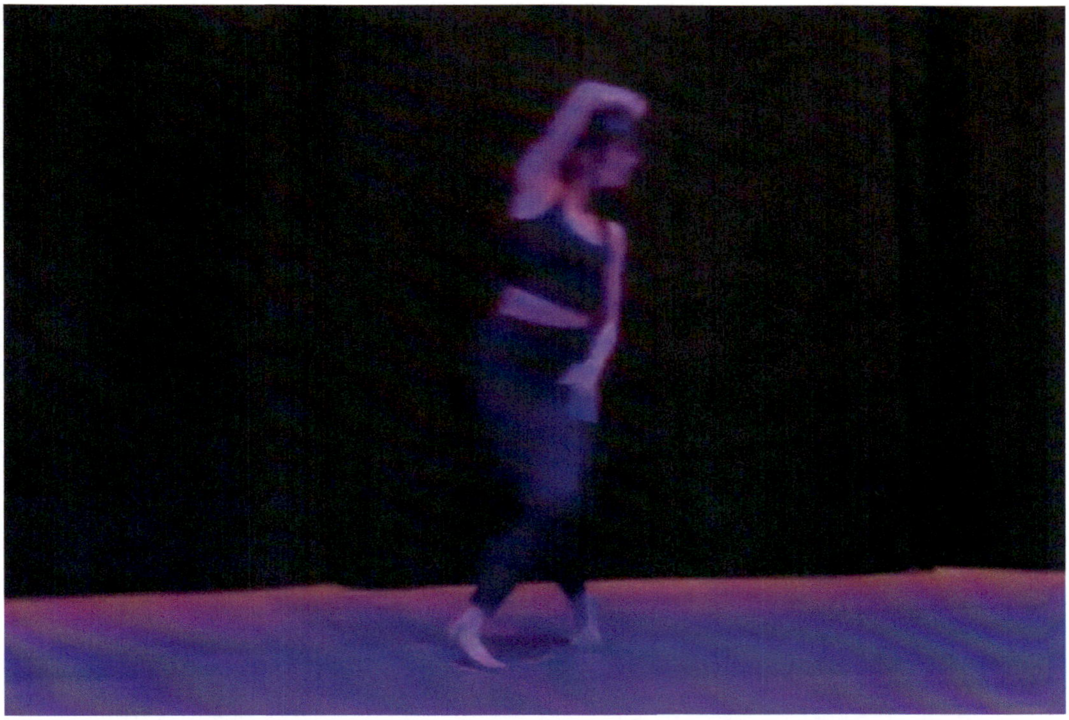

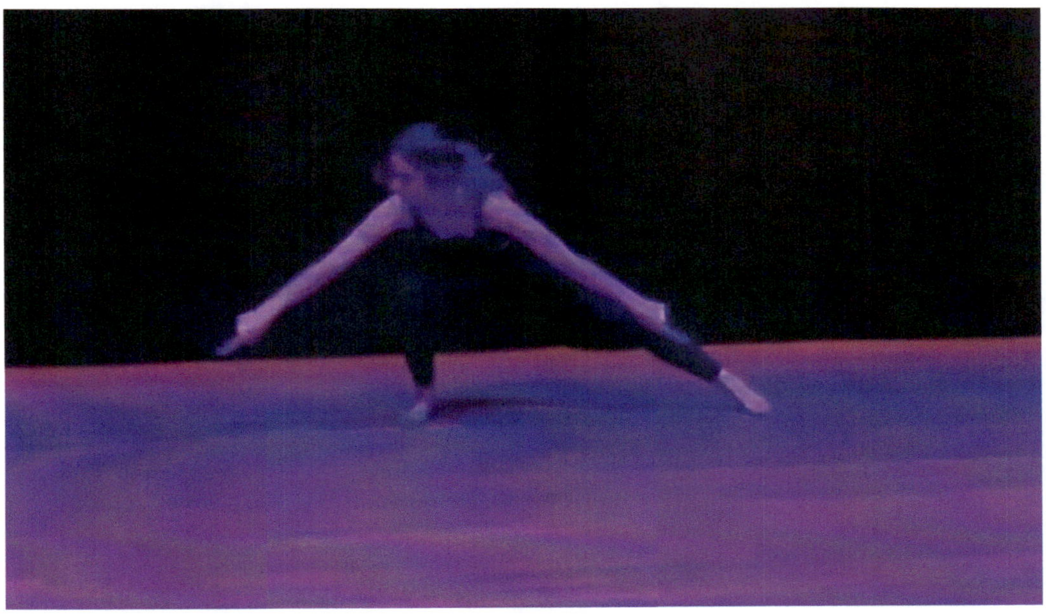

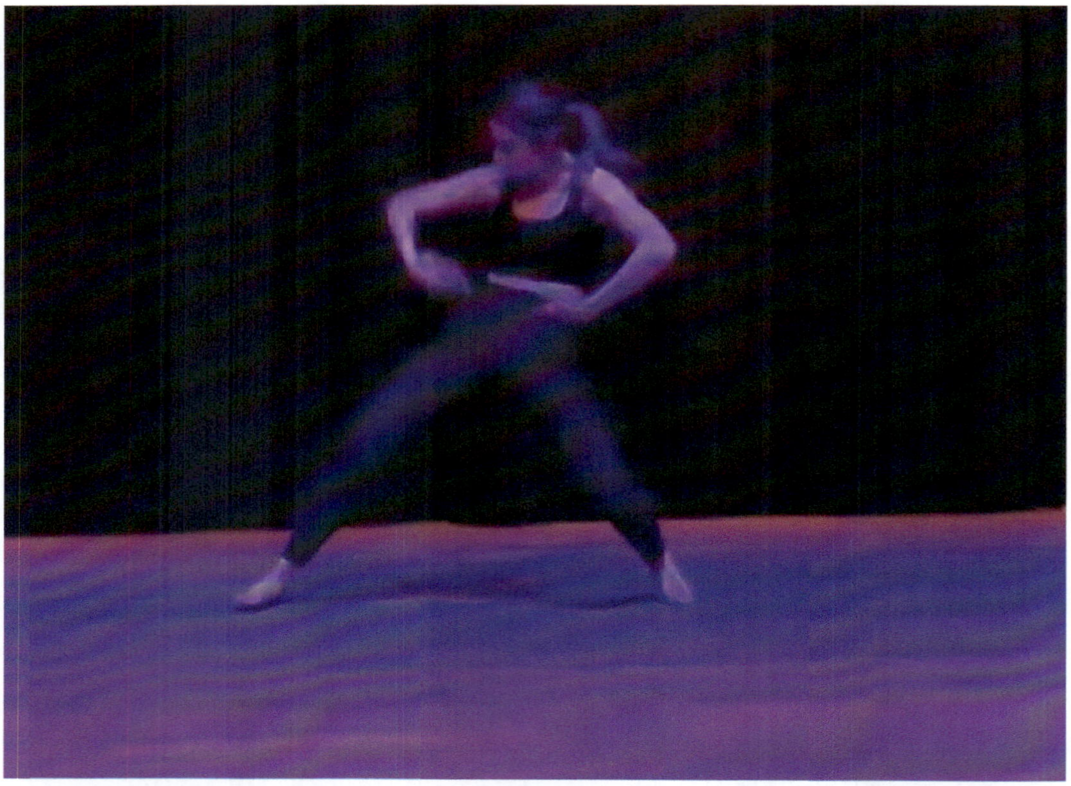

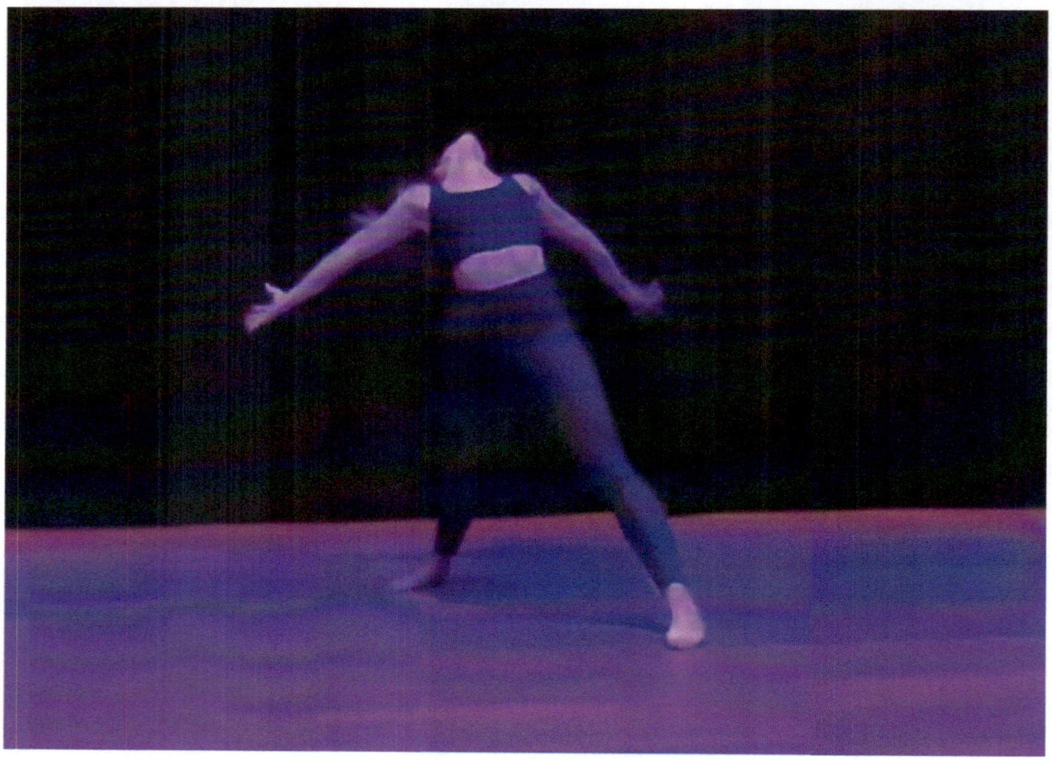

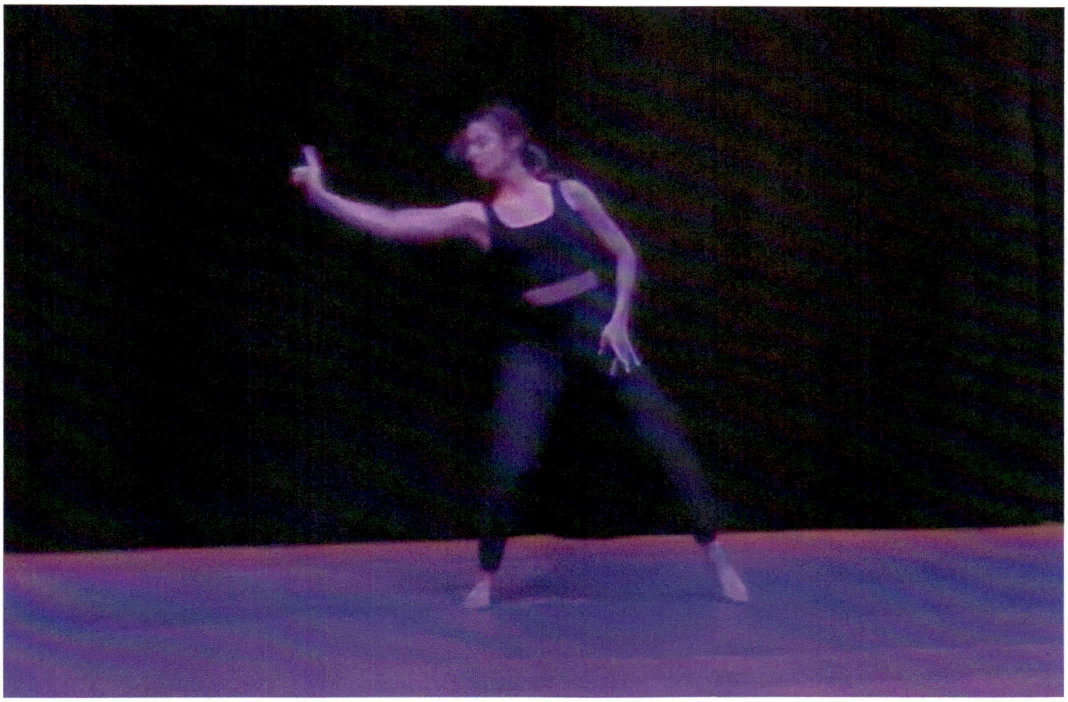

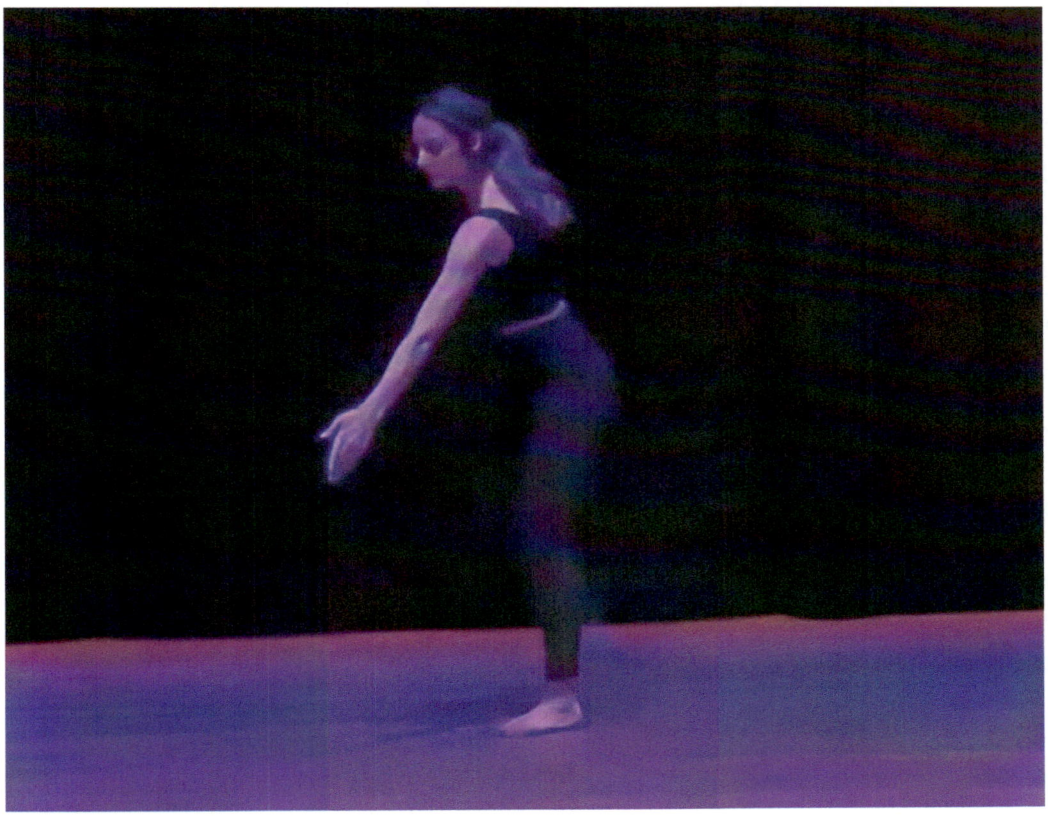

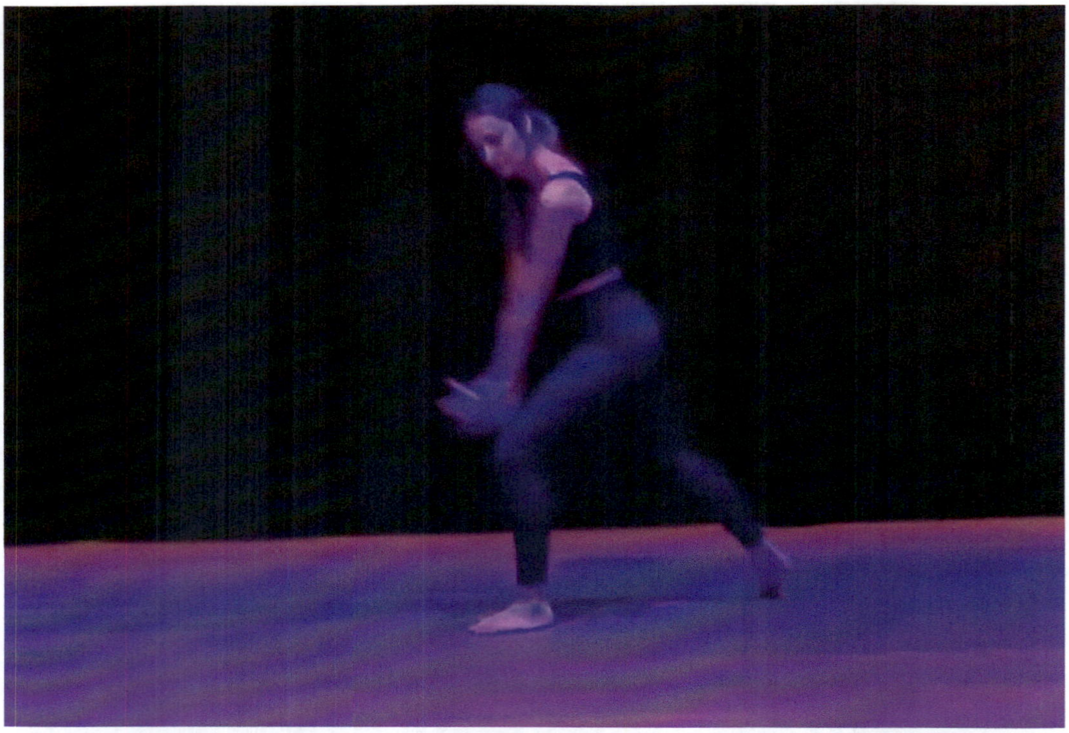

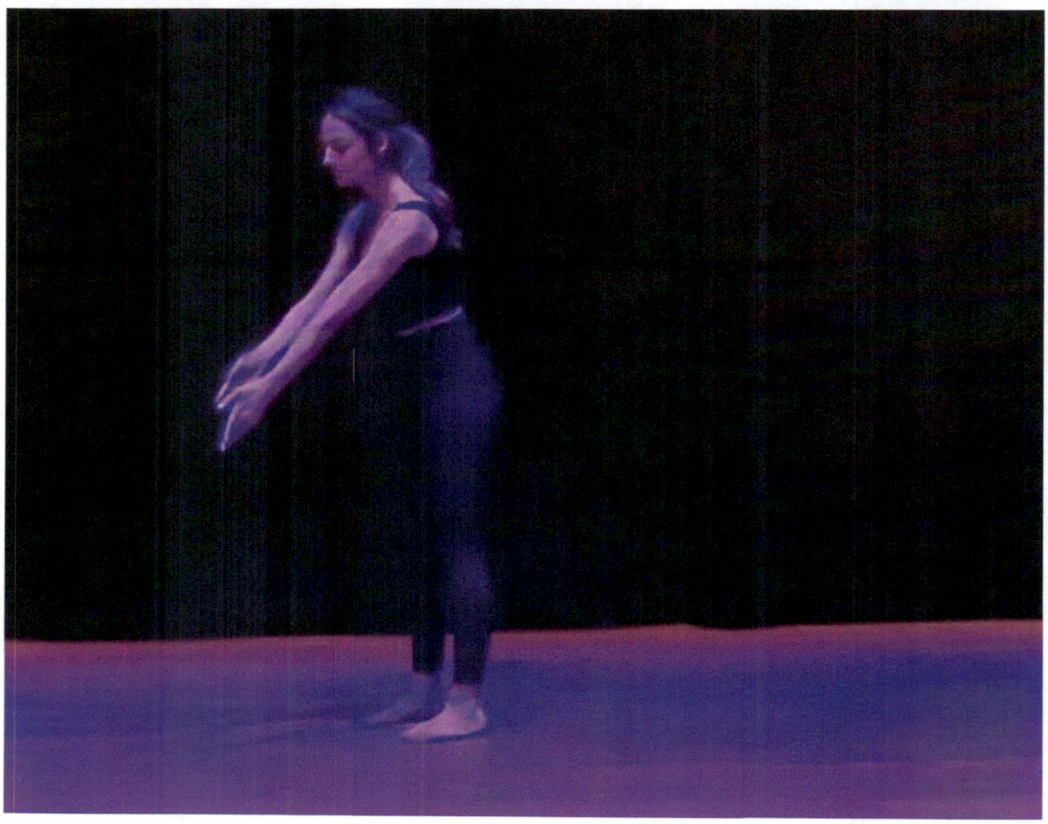

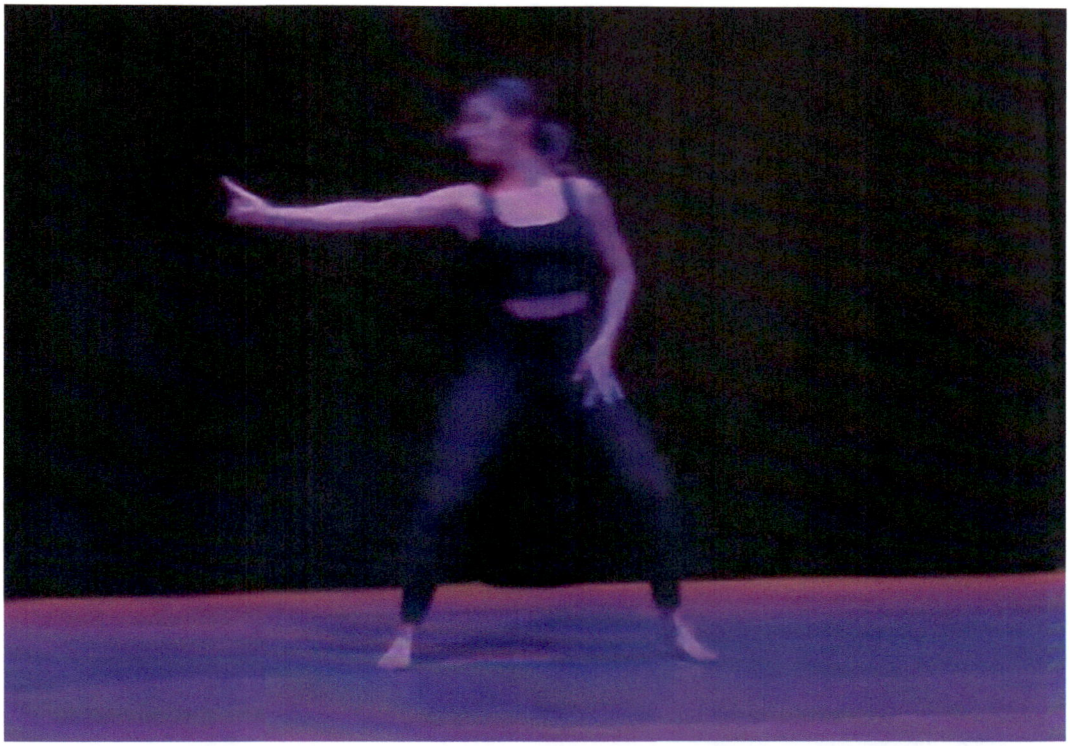

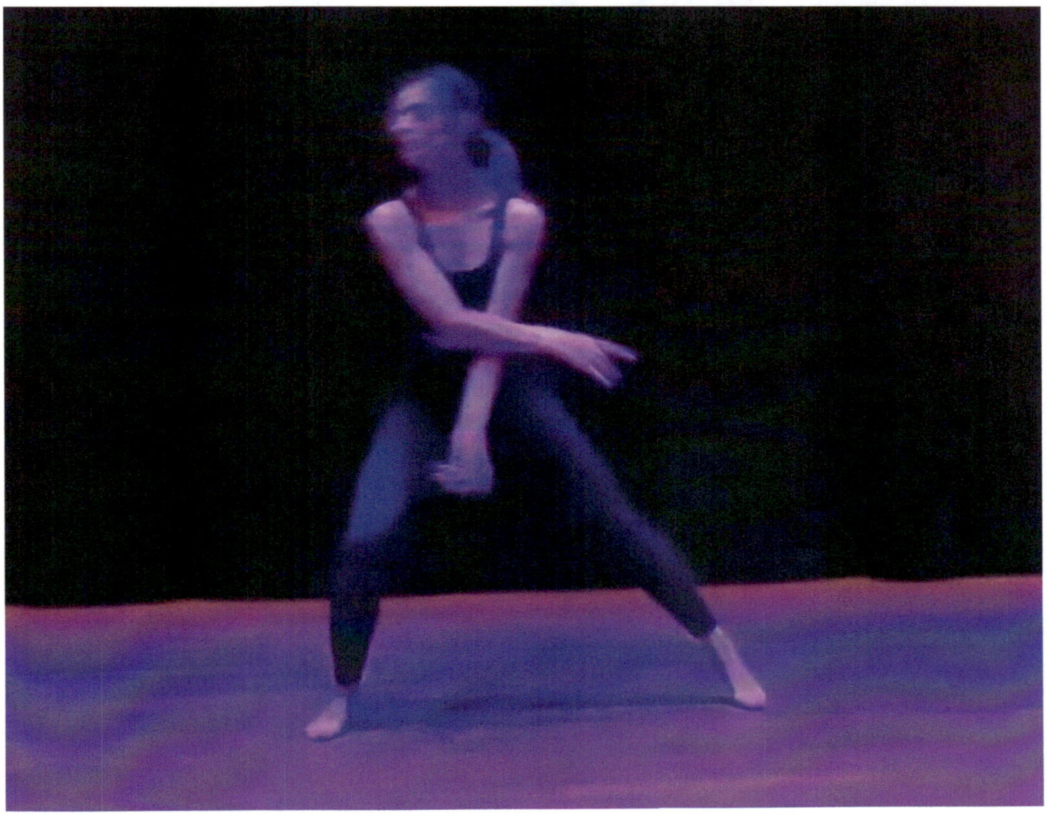

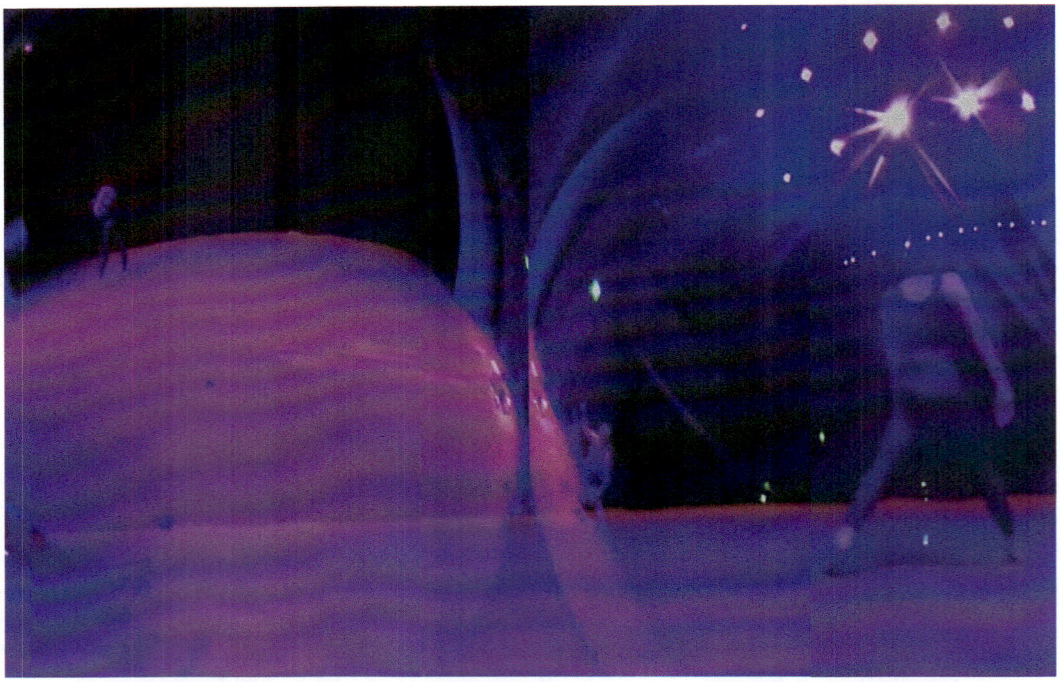

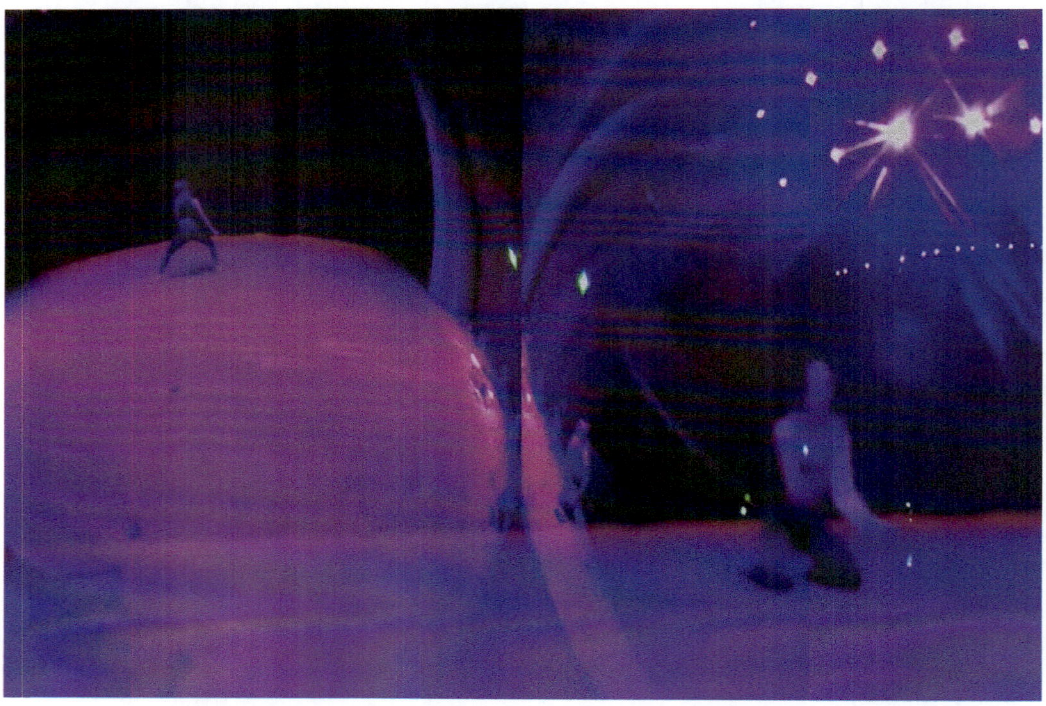

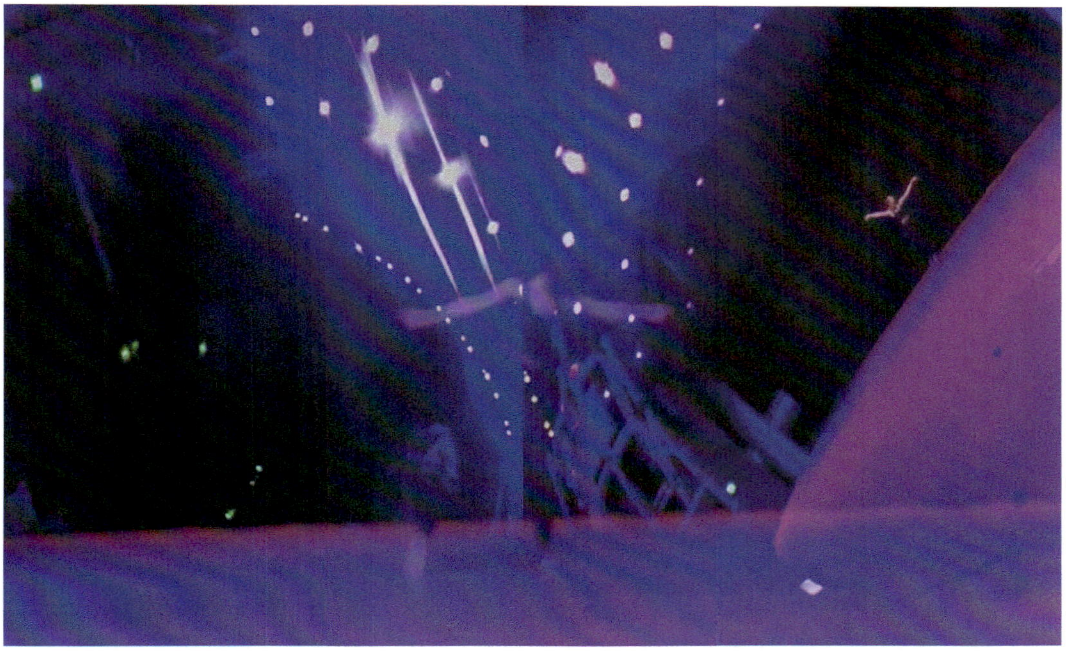

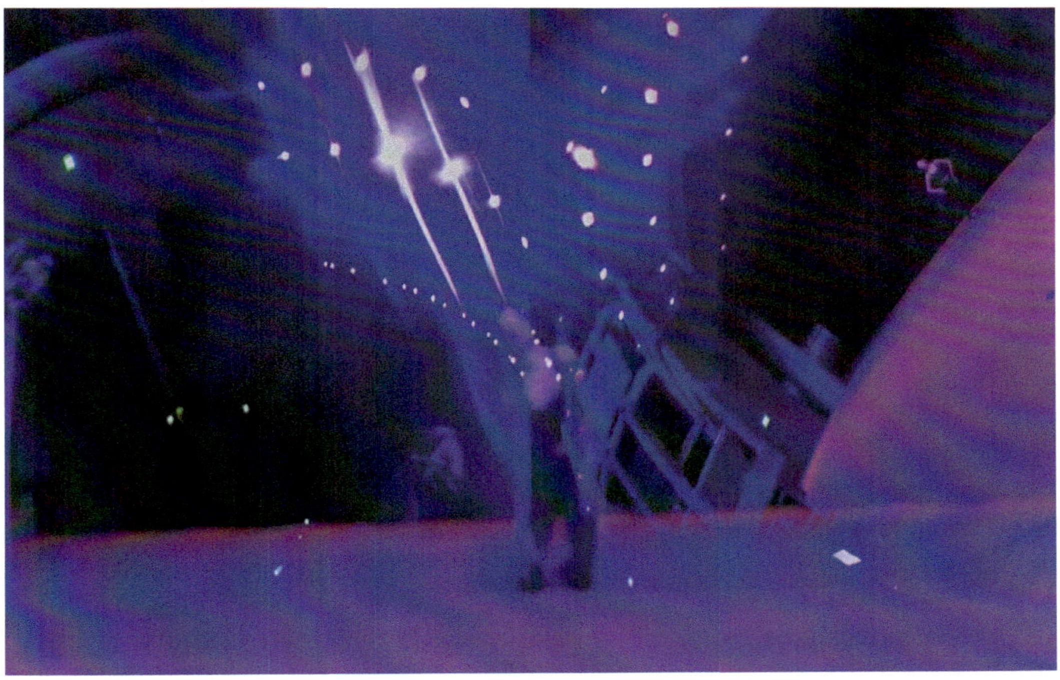

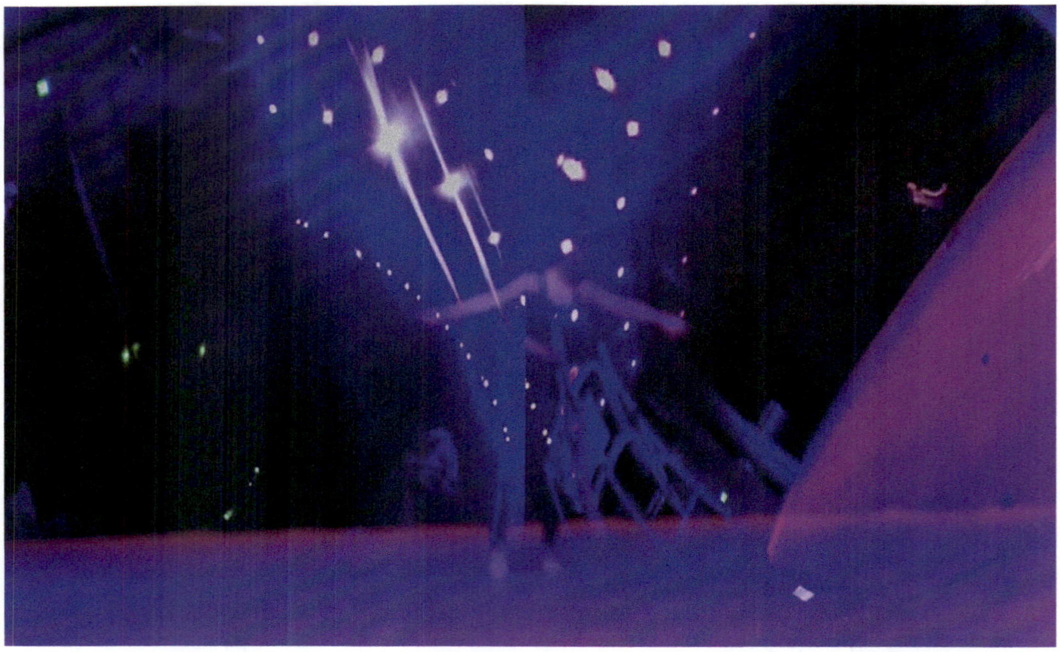

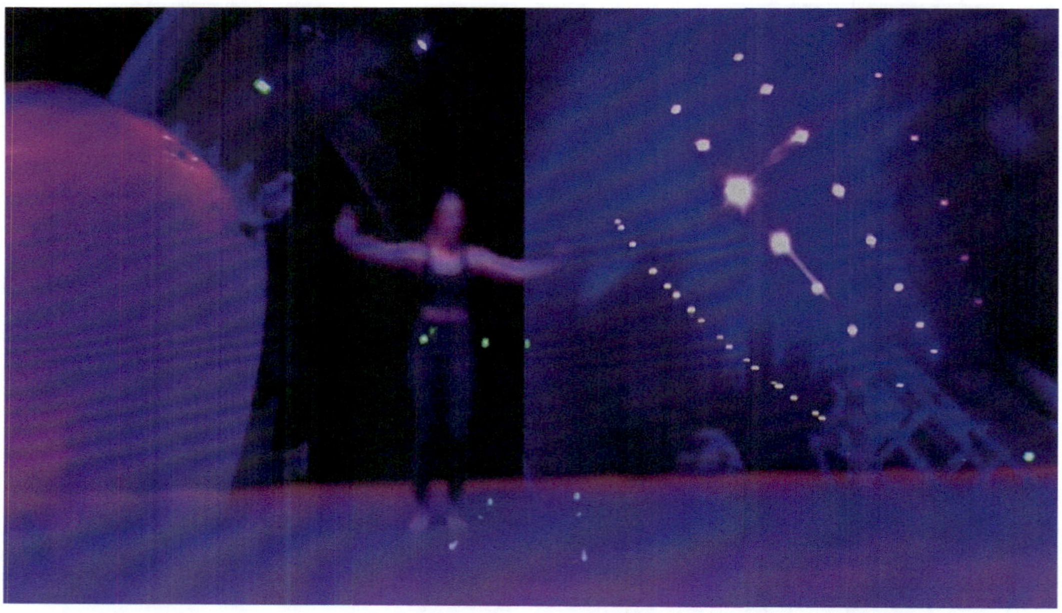

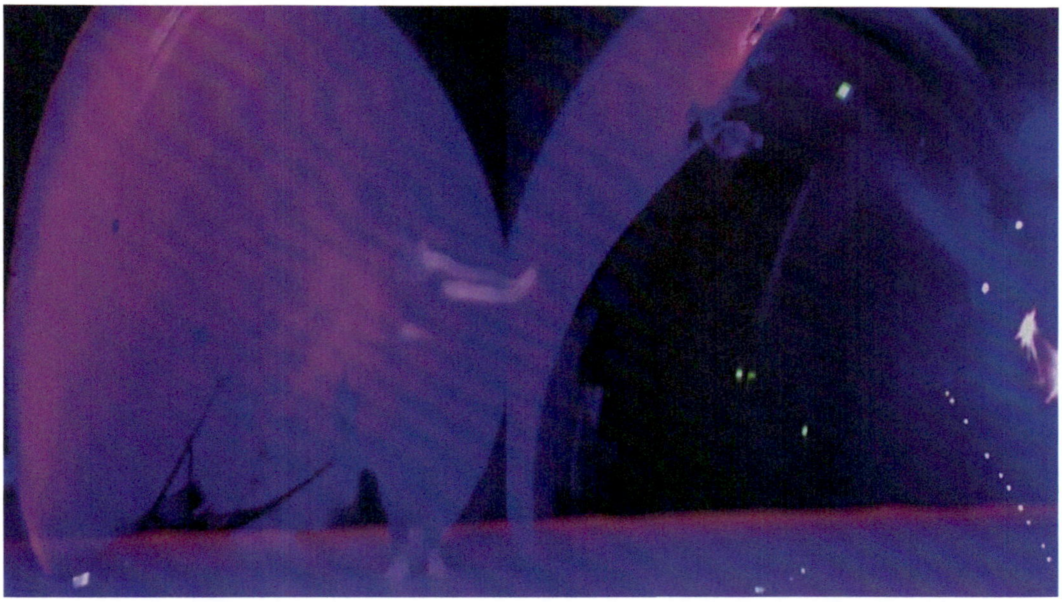

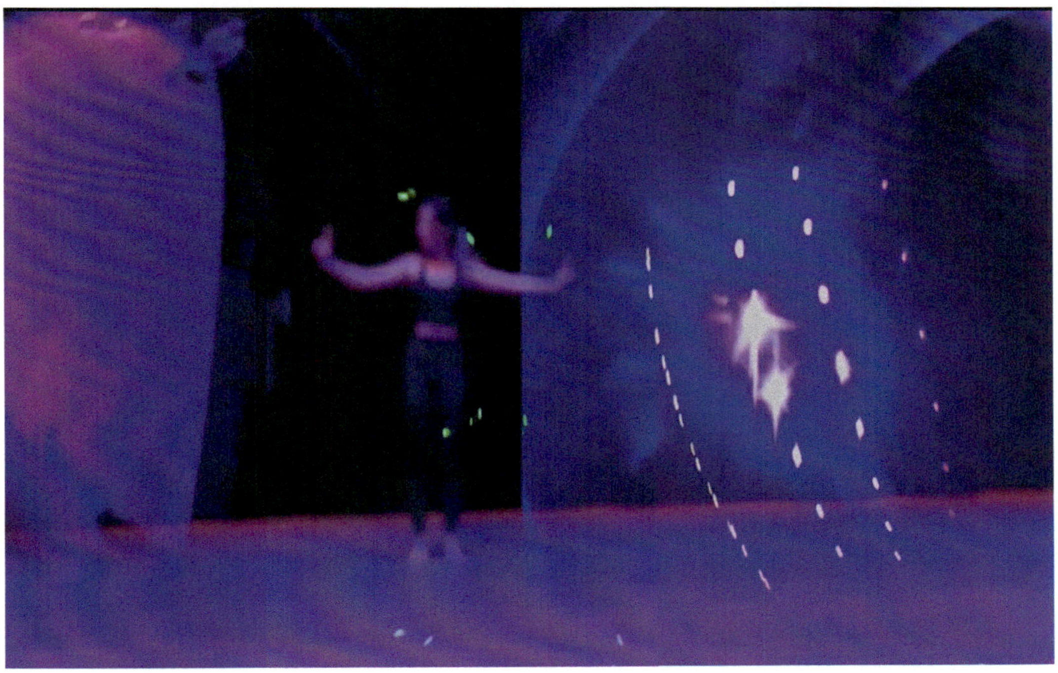

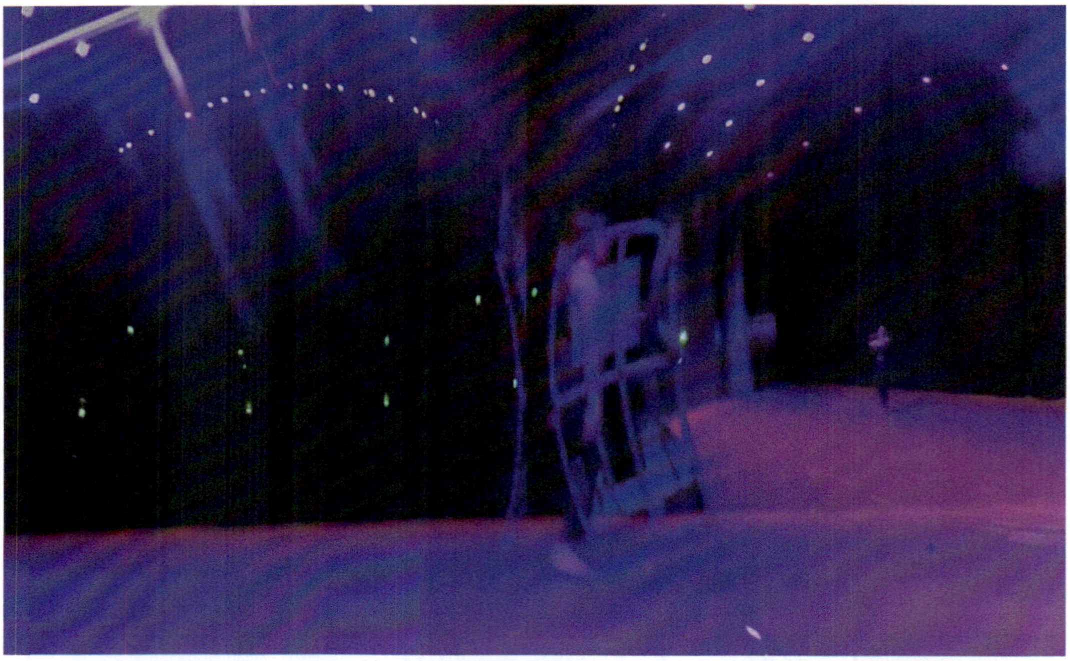

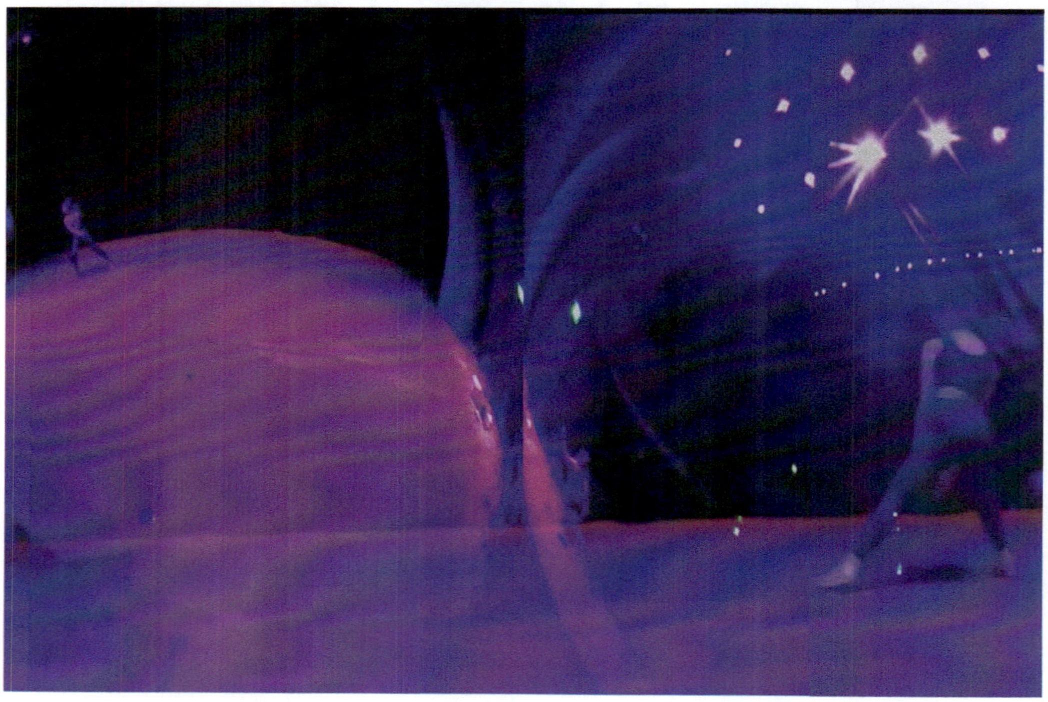

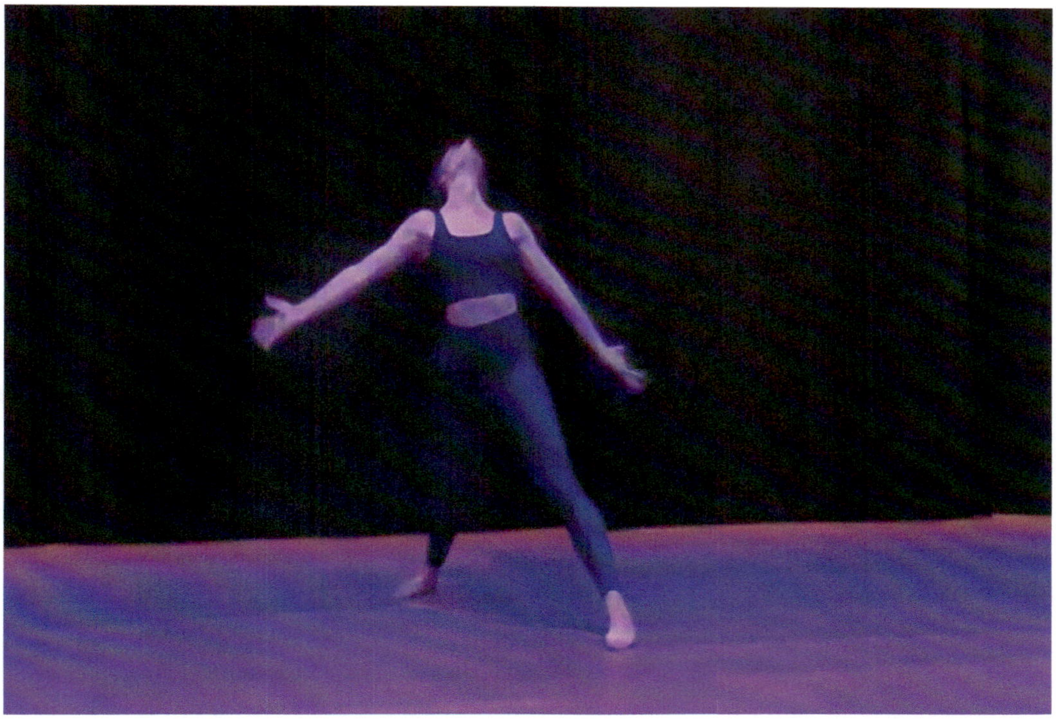

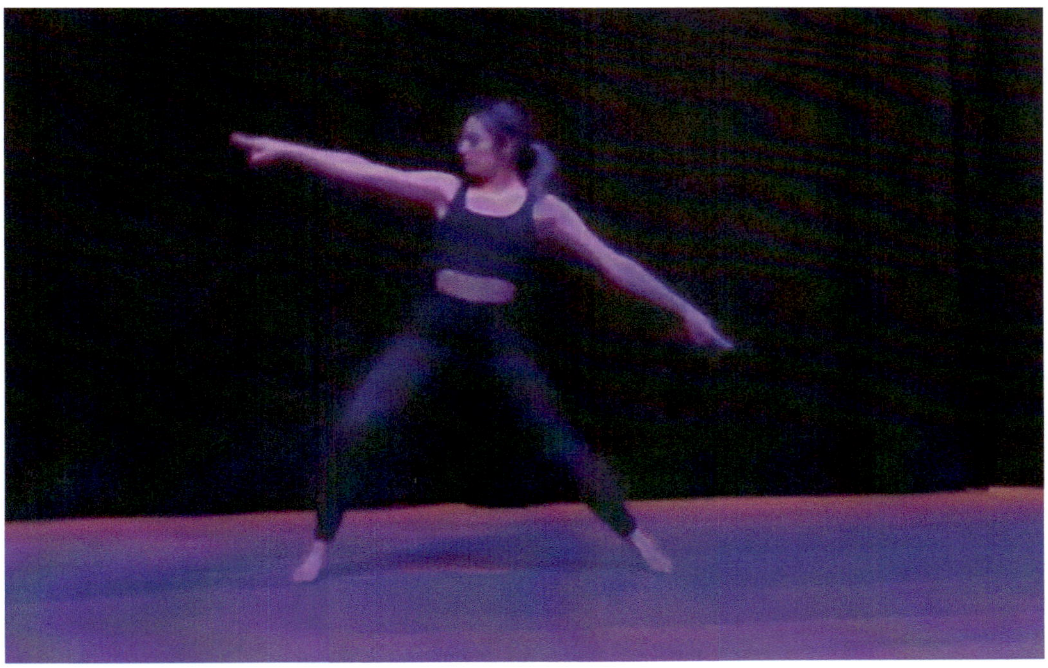

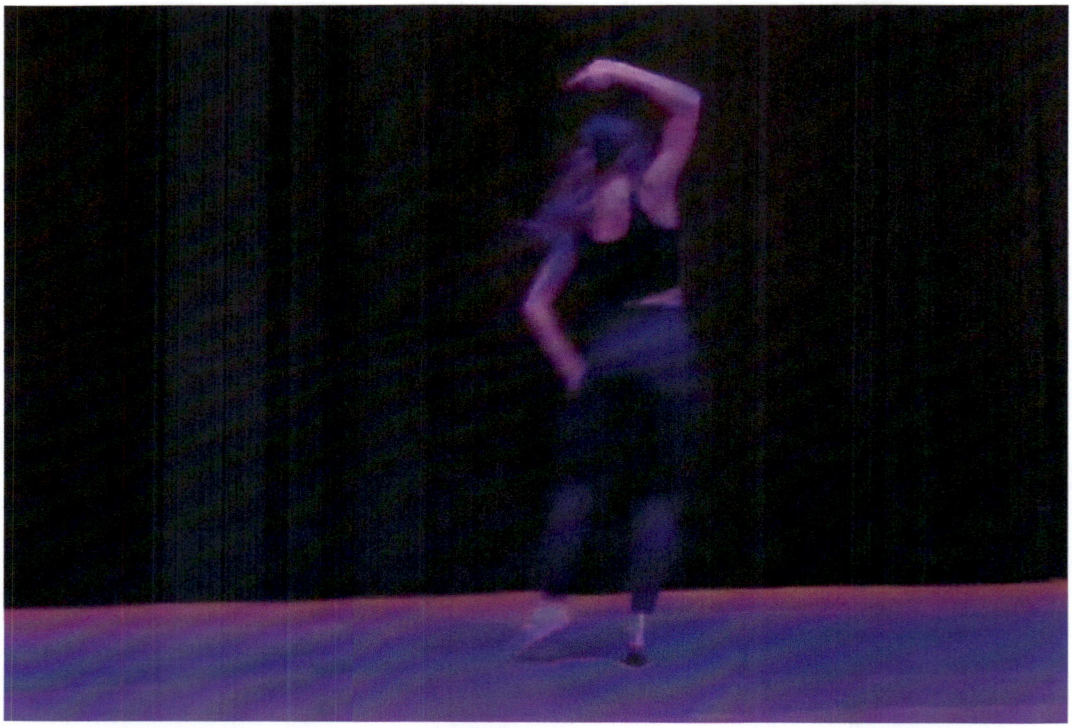

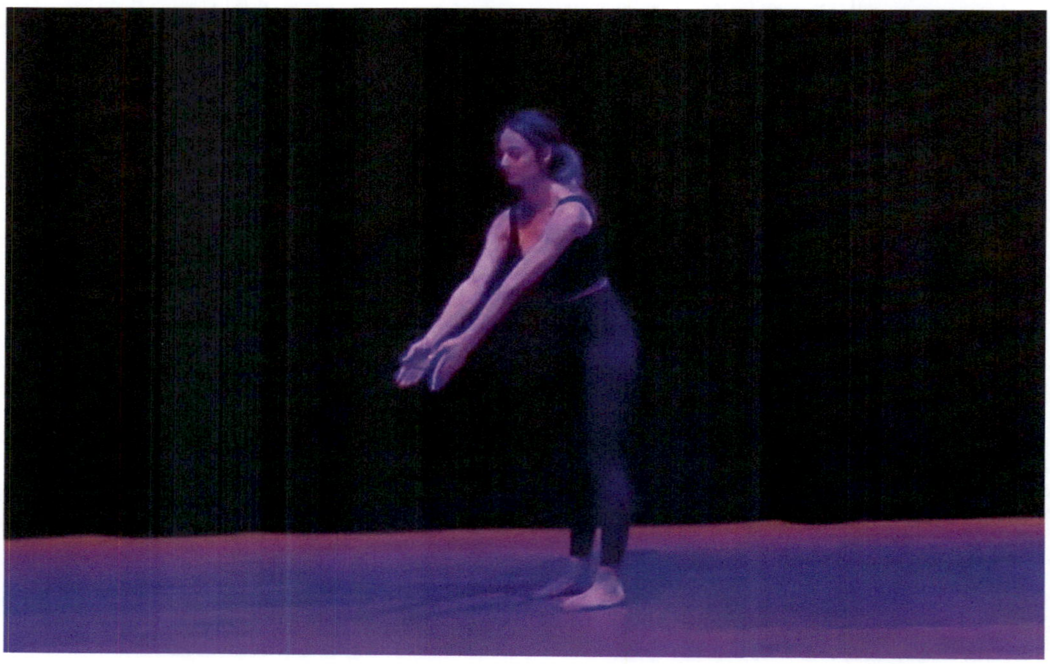

PREPARATORY ANIMATED VIDEO IMAGES

Izzy Sajdak dancing for the preparatory video.

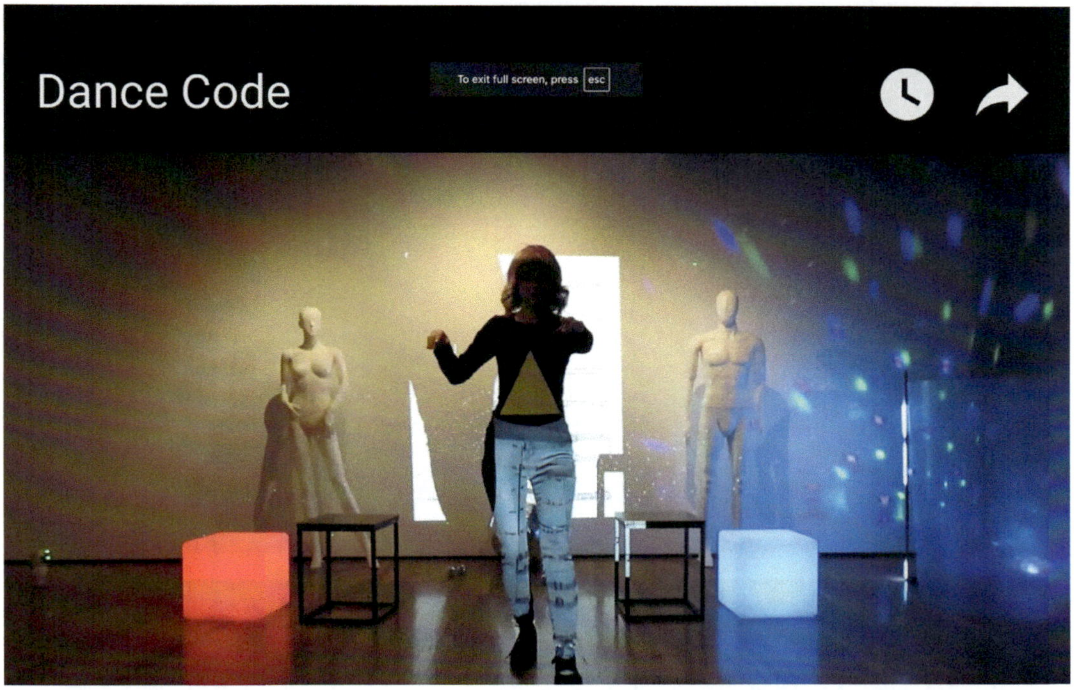

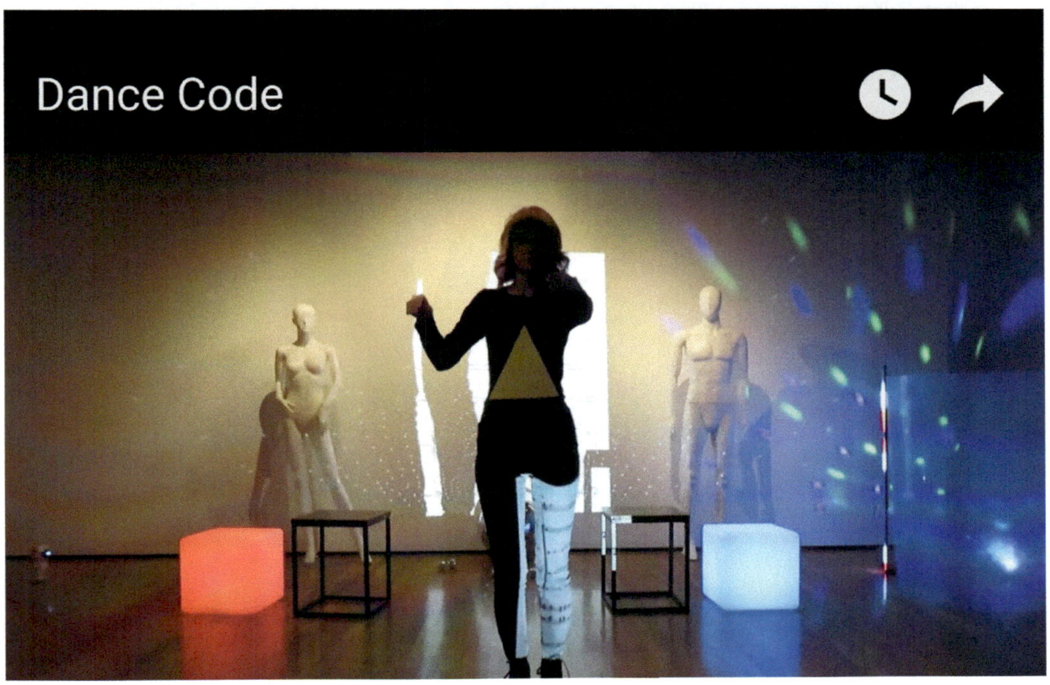

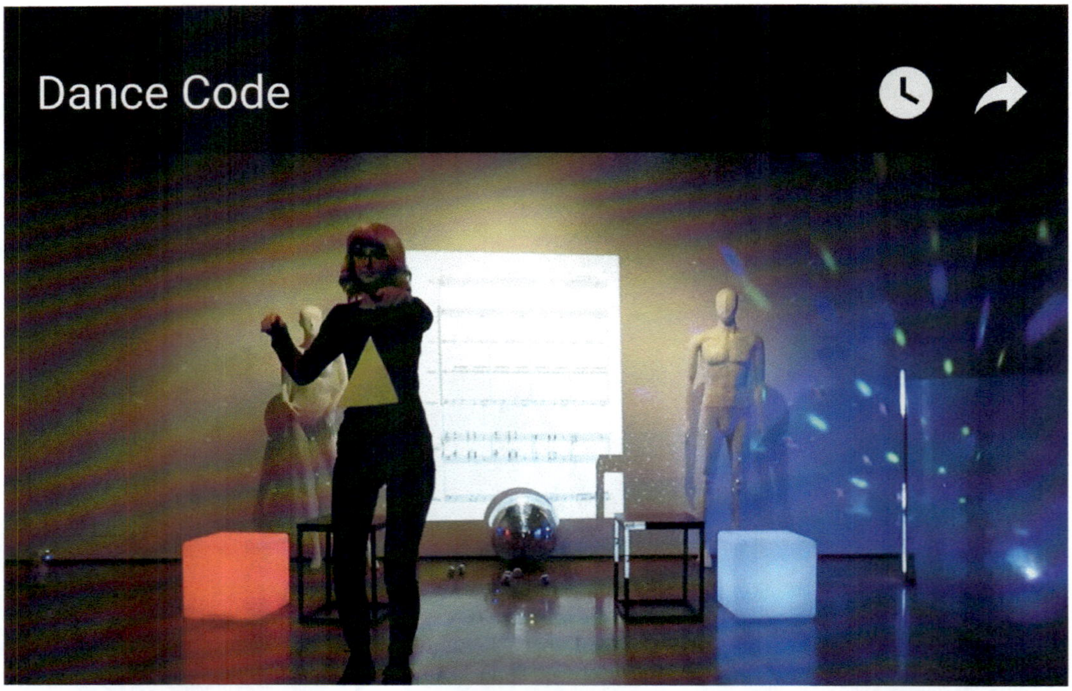
Dance Code

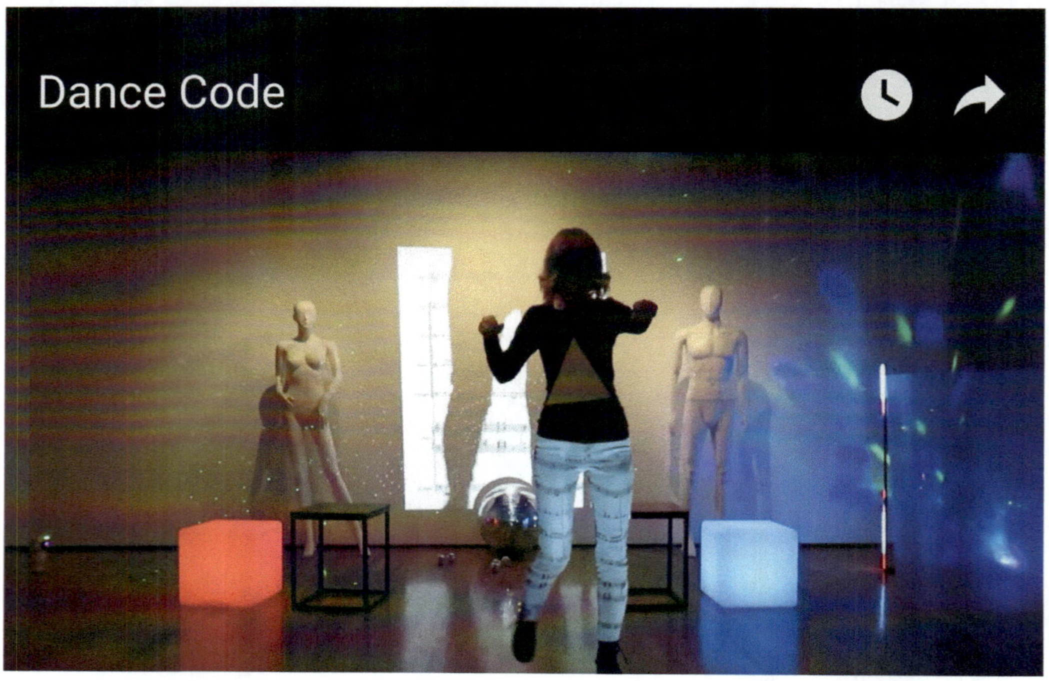
Dance Code

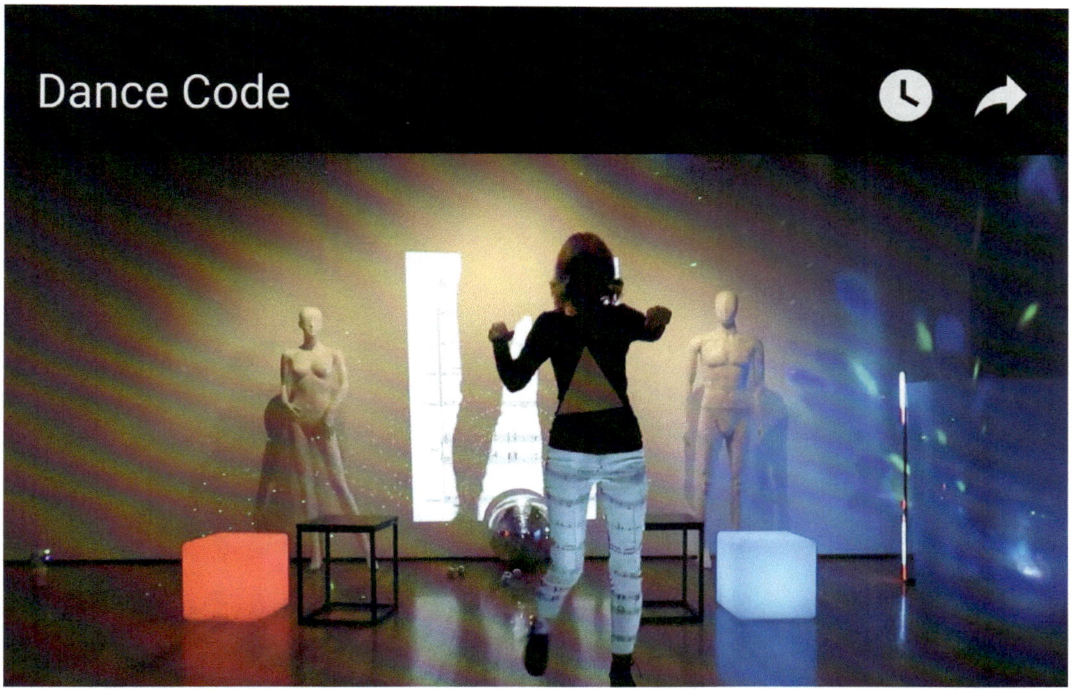
Dance Code

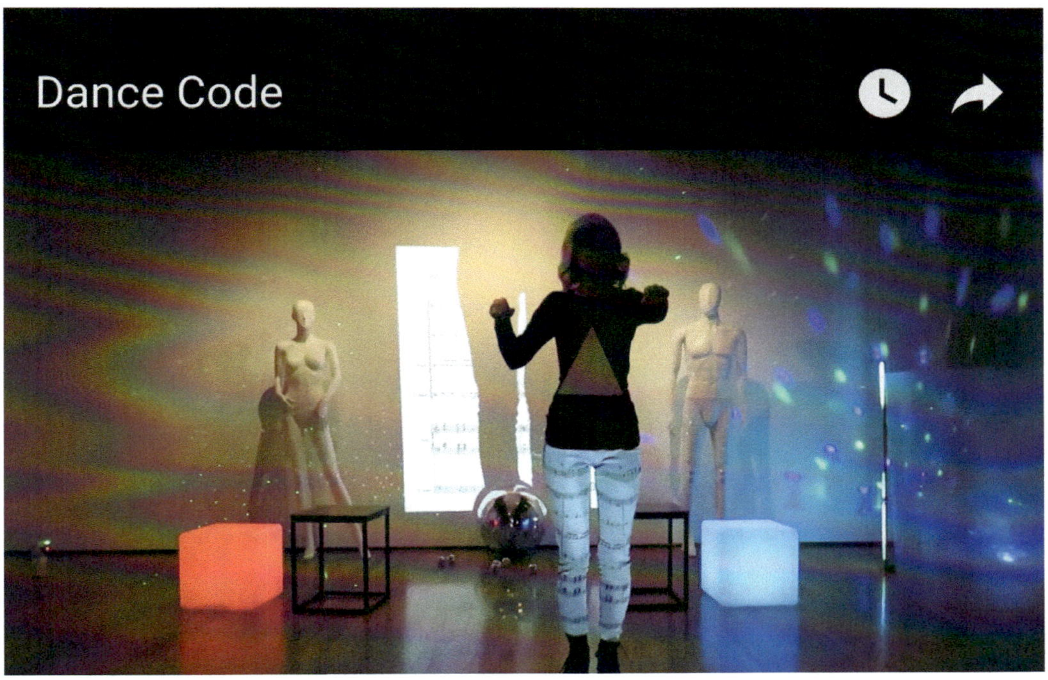
Dance Code

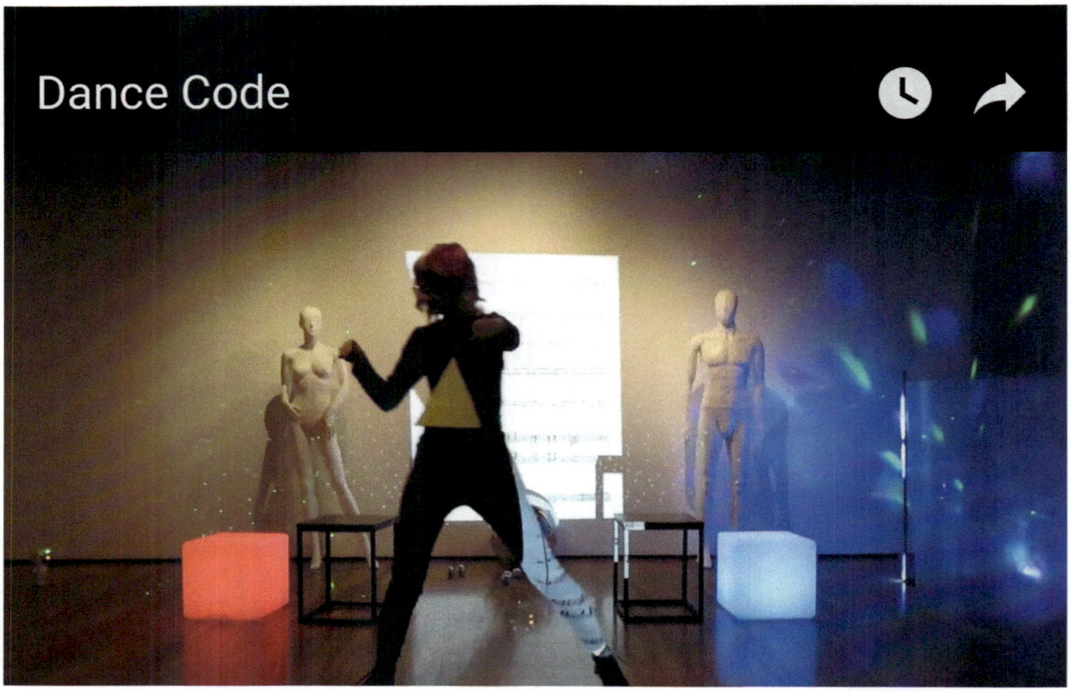
Dance Code

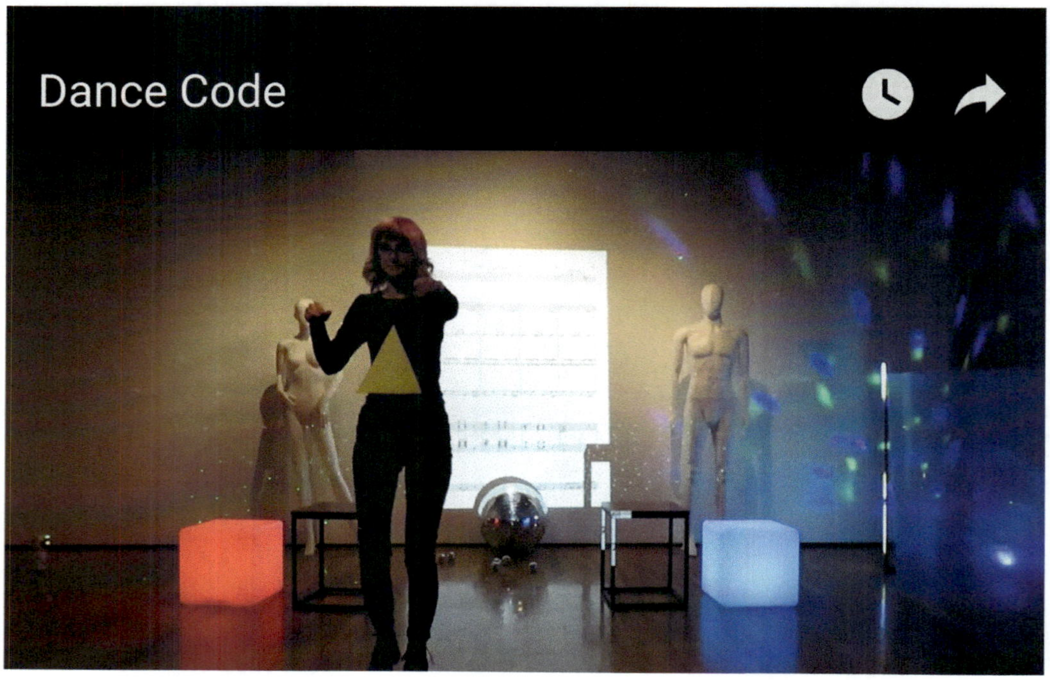
Dance Code

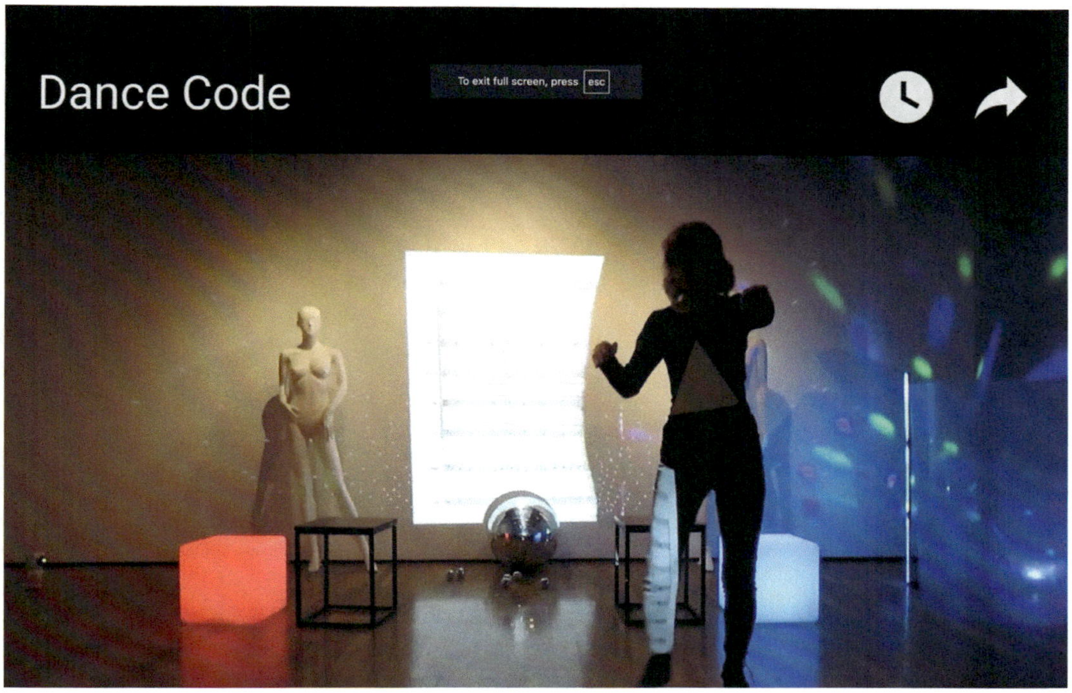
Dance Code
To exit full screen, press esc

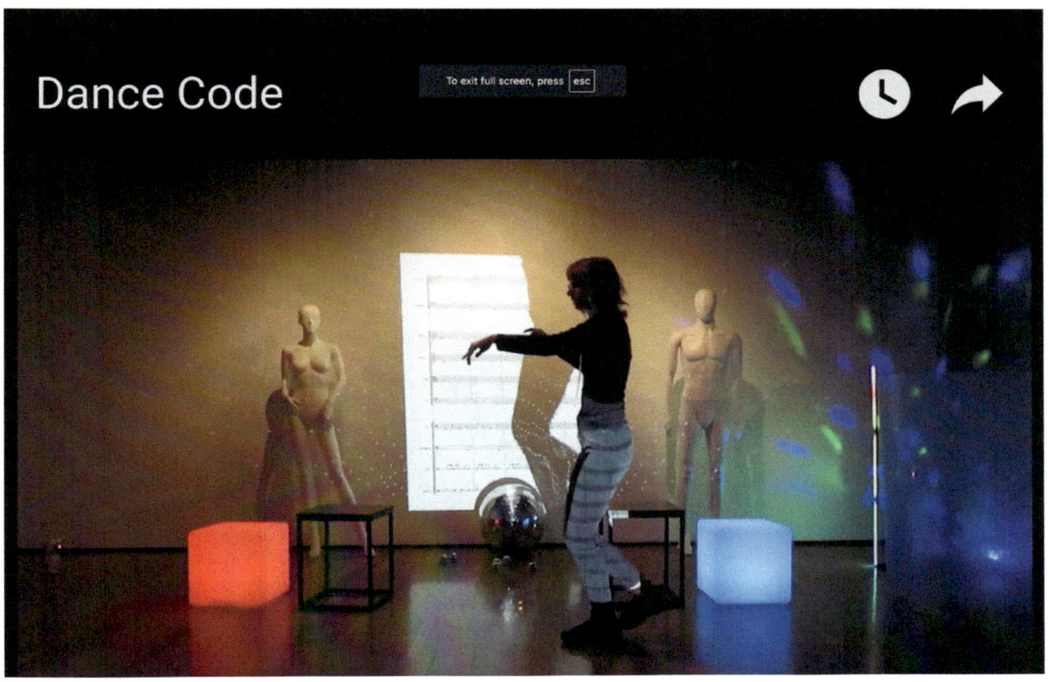
Dance Code
To exit full screen, press esc

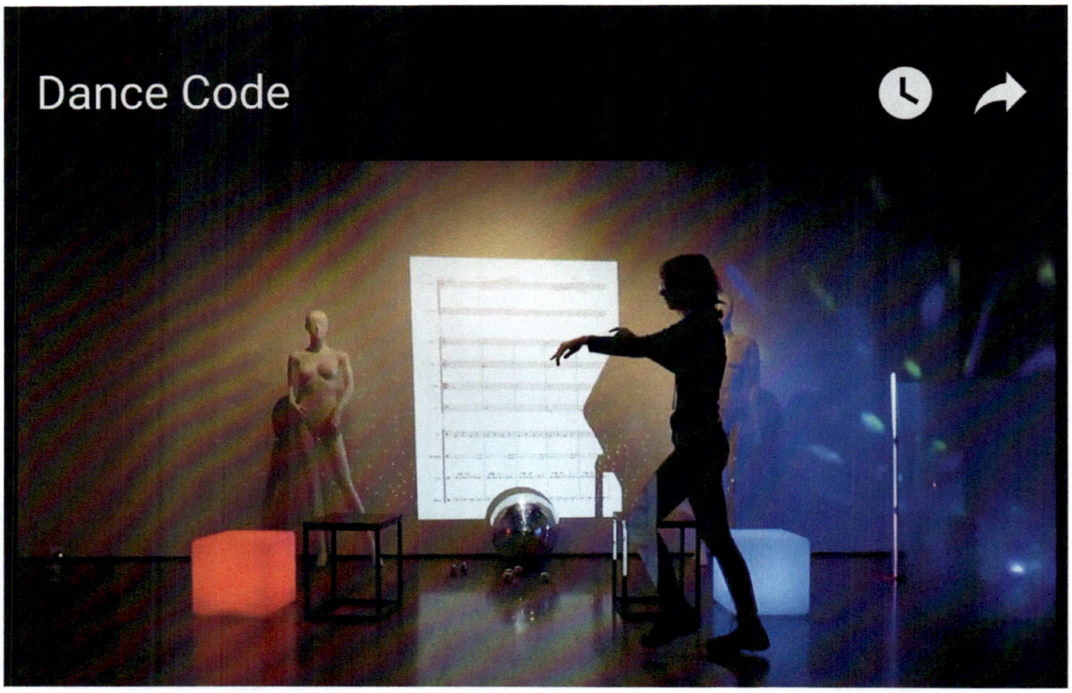
Dance Code

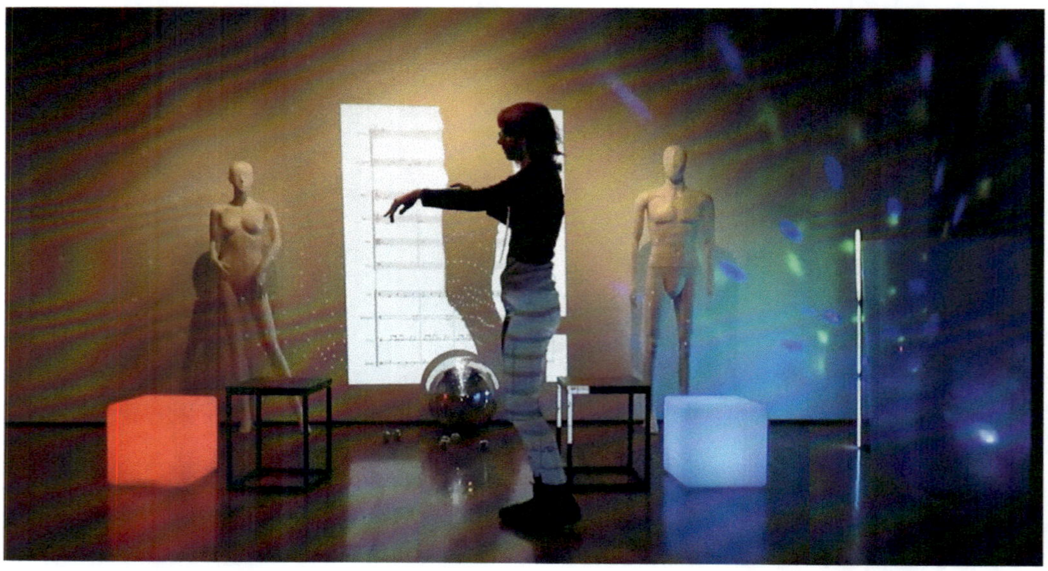

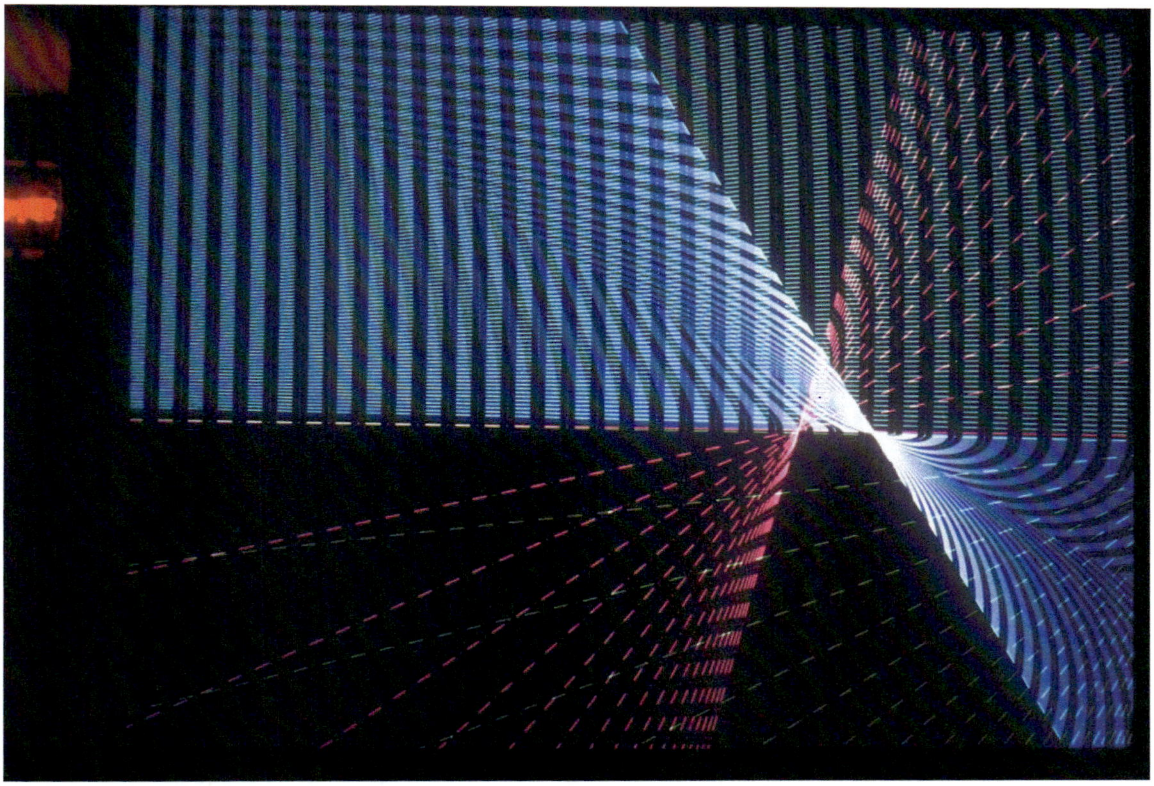

The Swirl. The output from the computer code. © *2025 by Anna Ursyn*

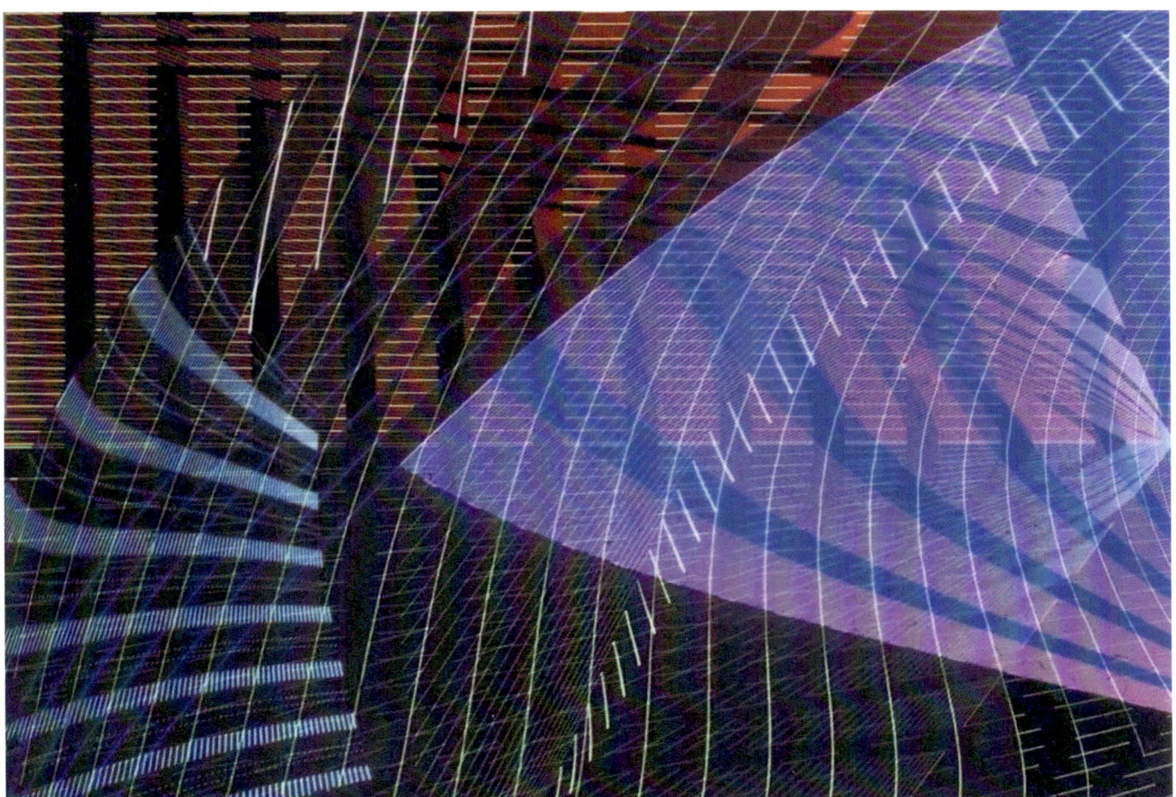

A Way Out. The output from computer codes. © *2025 by Anna Ursyn*

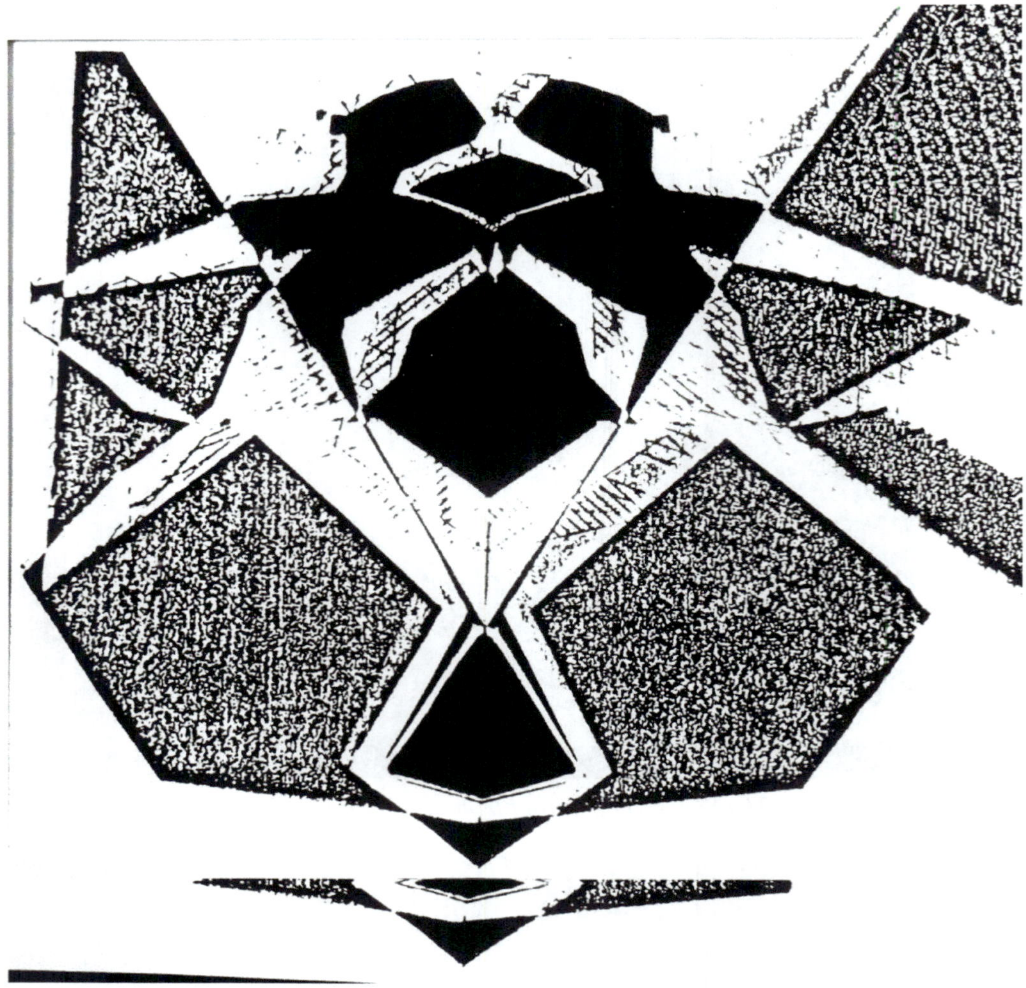

Dancers. The output from the computer code with translation. © 2025 by Anna Ursyn

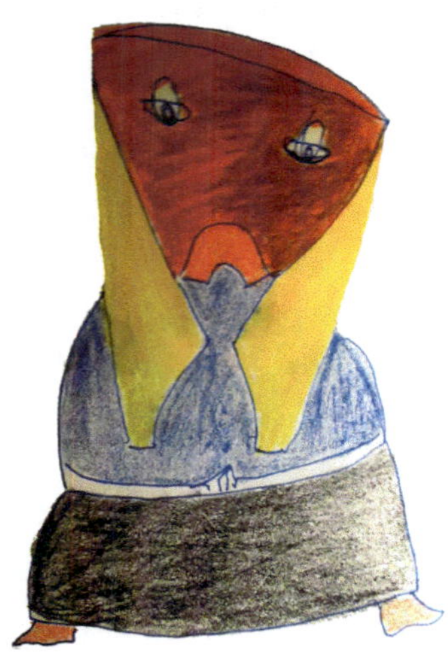

Dyran: always thinks about tomorrow.

Dyran: plans for tomorrow.

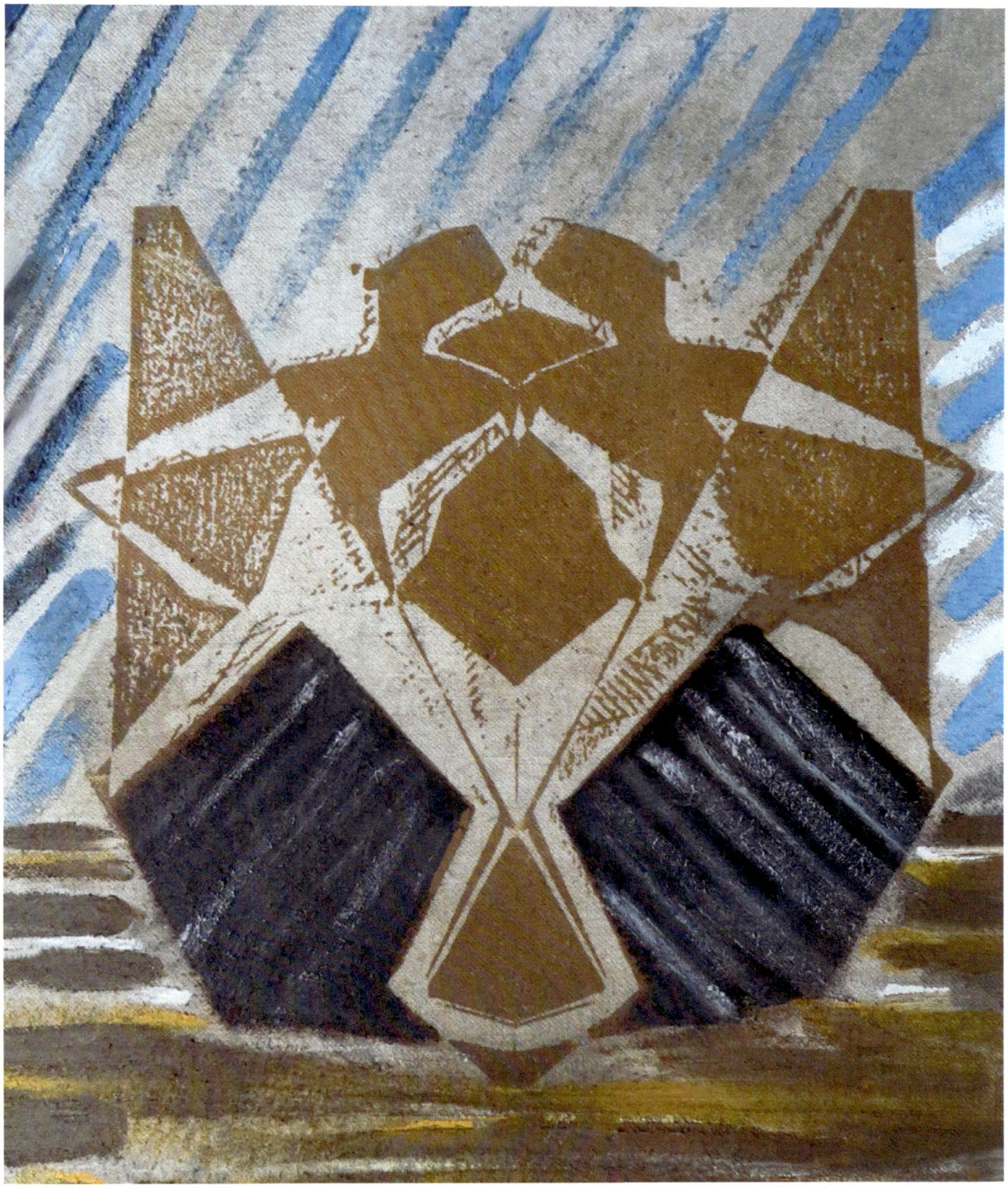

Dancers. Photosilkscreen after the output of the computer code © 2025 by Anna Ursyn

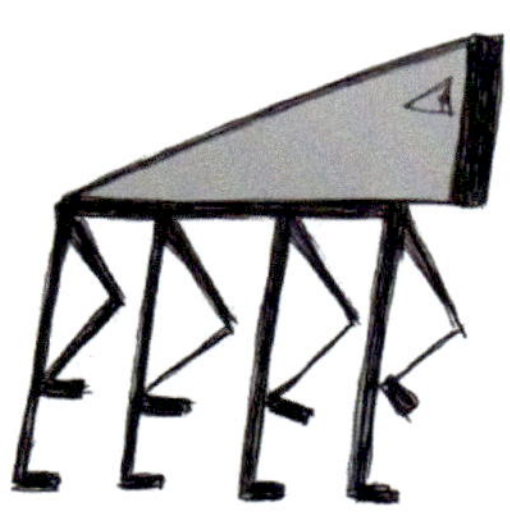

Yasala: goes rhythmically whenever he goes.

Yasala: moves rhythmically whatever he does

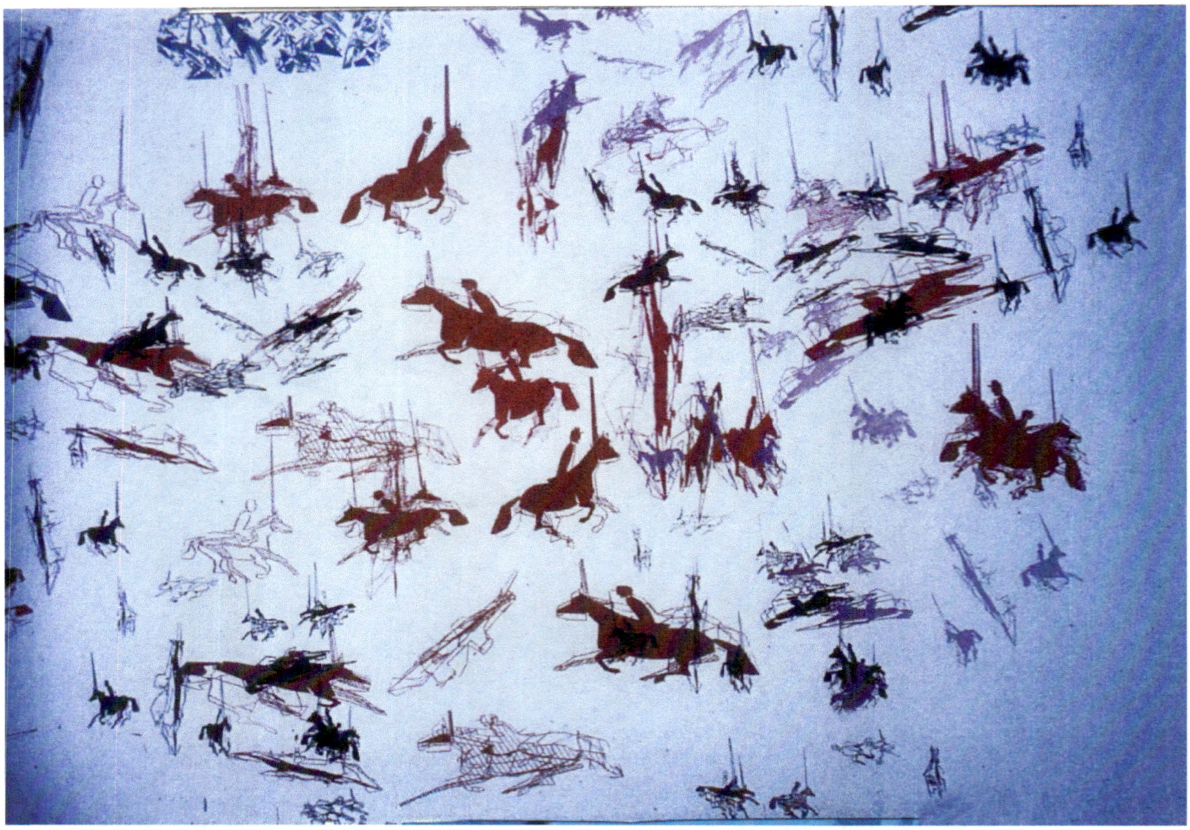

Passing Time. Photosilkscreen after computer codes. © 2025 by Anna Ursyn

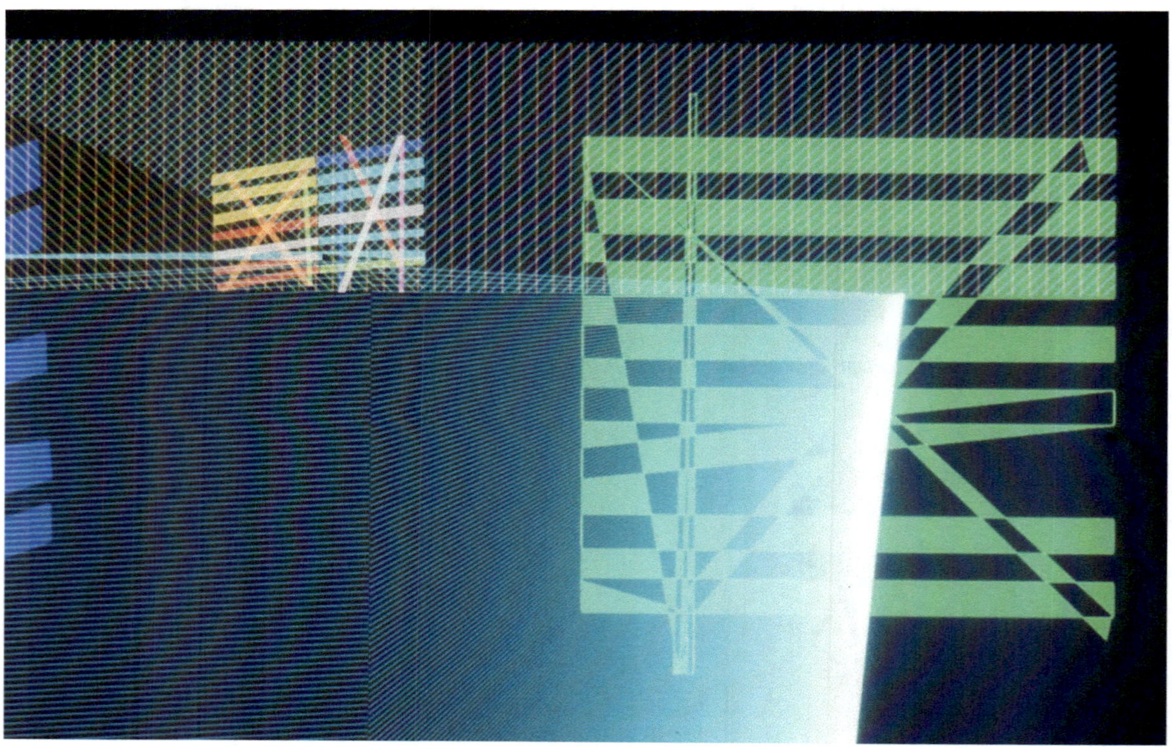

The Fences. The output from the computer code. © 2025 by Anna Ursyn

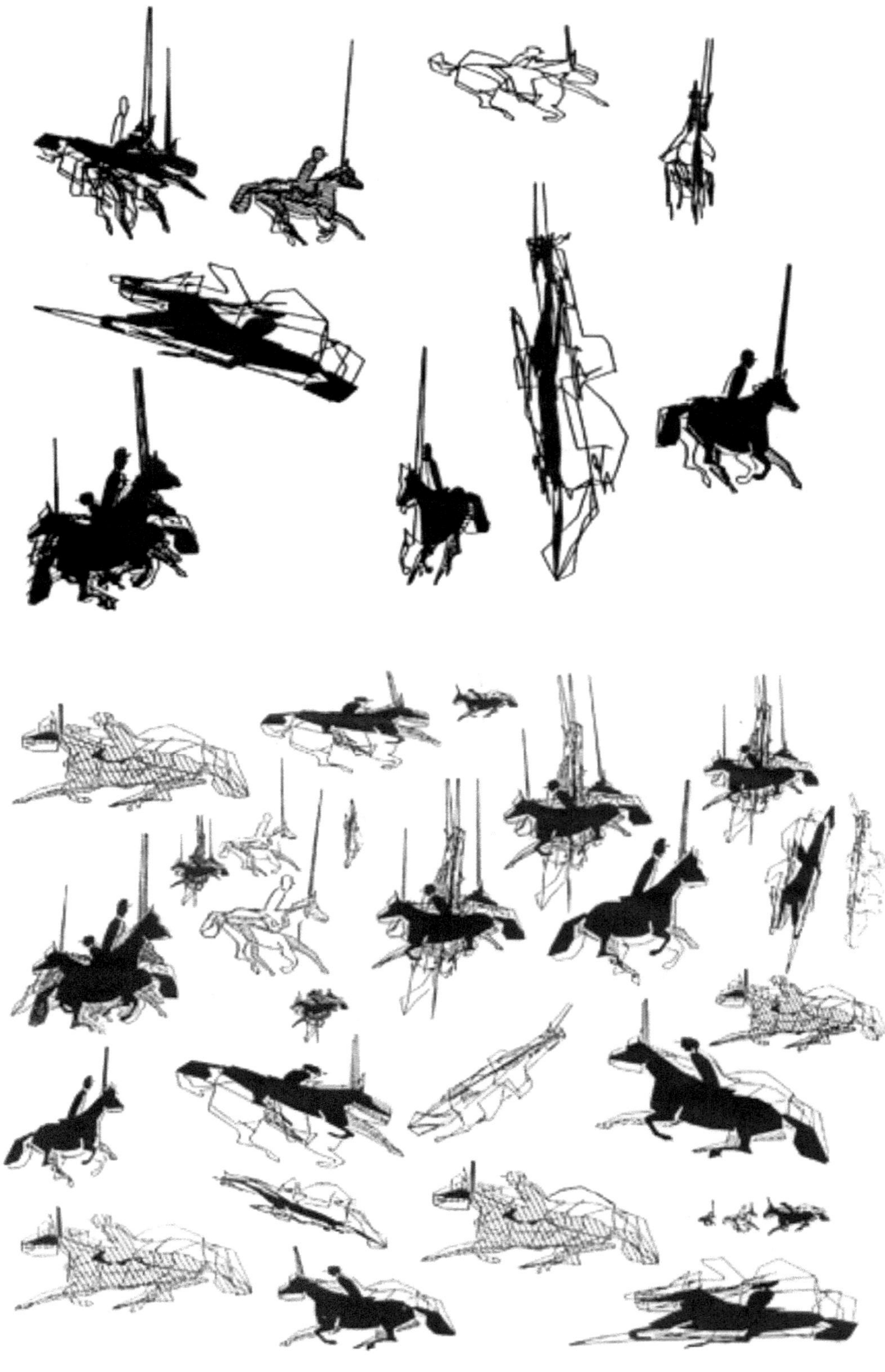

Hierarchy. The plot after computer codes. © *2025 by Anna Ursyn*

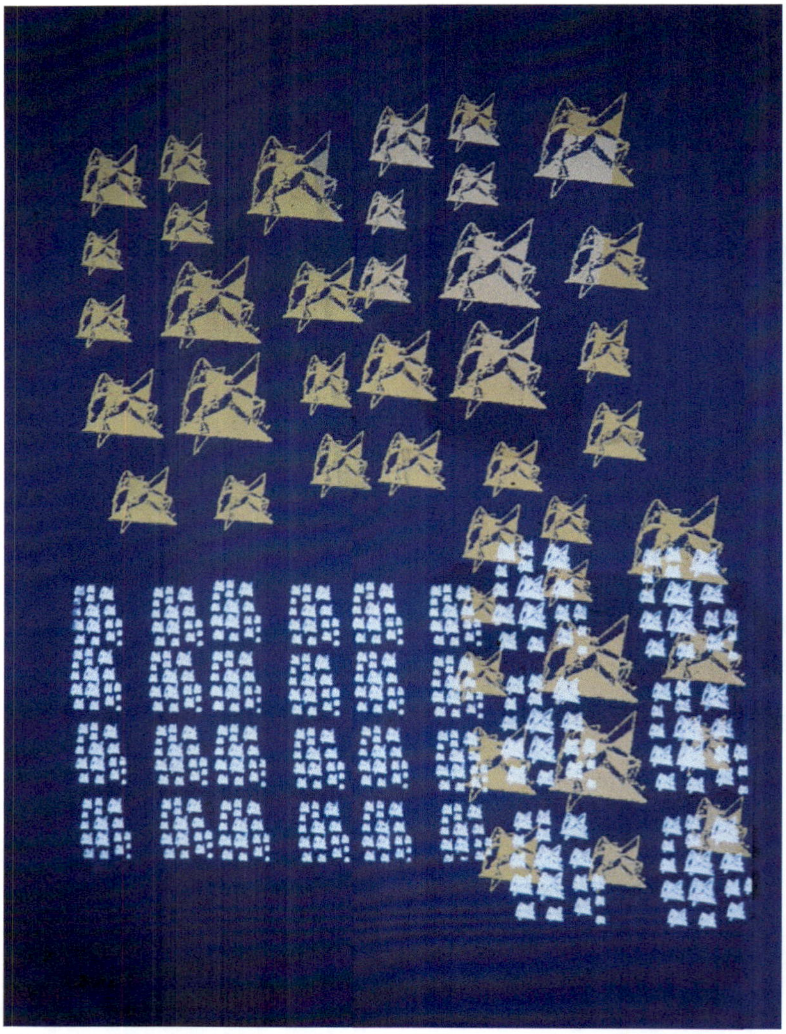

Divided Thoughts, Photosilscreens after 3D codes on fabric. © 2025 by Anna Ursyn

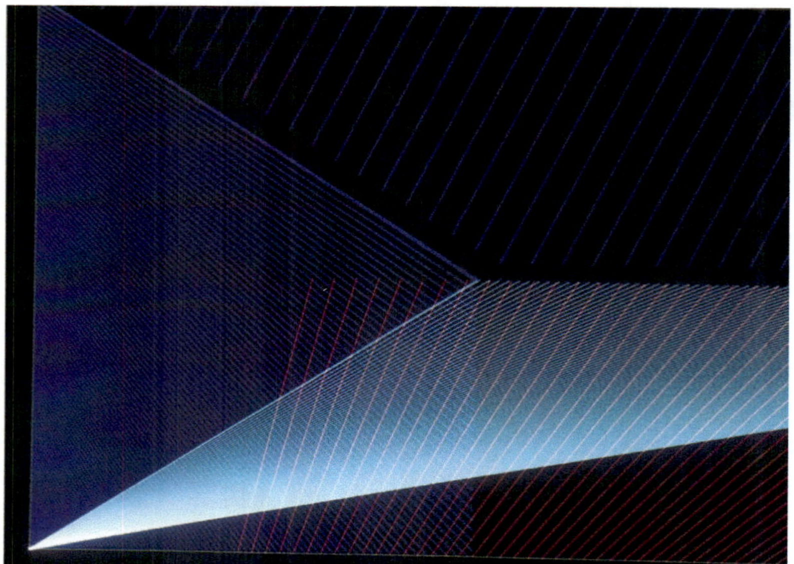

A Free Flow. The output after 3D computer code. © 2025 by Anna Ursyn

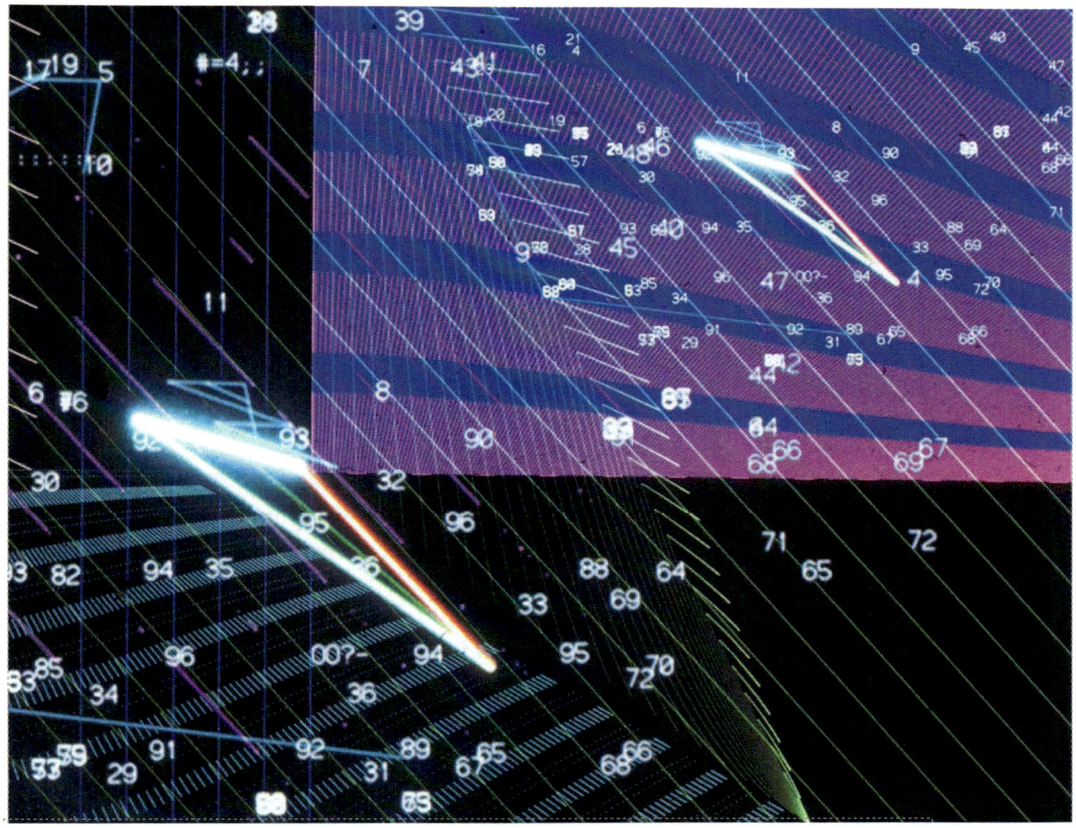

Stellar Messages. The output from the computer code. © *2025 by Anna Ursyn*

APPENDIX

Dance Code Video – Python User Notes

© 2021 by Stuart Smith.

Introduction

The dance project video was constructed from many parts of three different kinds: video clips, sound files, and still images, with the completed video as well as the video clips used to make it; the sound files, the still images used for headings, captions, and code summaries; and the Python program that created the animation, if you are interested in the programming involved in creating the dance video, what follows will describe how to get Python, install it on your system, and run the program.

Getting Python and installing it

Python is available as a free download from *python.org/downloads/* Download the appropriate version for your system (Windows, macOS, Linux/Unix). As of this writing, the latest version is 3.9. The Python installer will automatically be downloaded into your **Downloads** folder. Click on the filename to start the installation and follow the on-screen instructions. When the installation is complete, you will find that you have the following:

- IDLE(Python 3.9 64-bit)
- Python 3.9 (64-bit)

- Python 3.9 Manuals (64-bit)
- Python 3.9 Module Docs (64-bit)

For the dance project, you will be using only the IDLE "shell," the interactive interface to Python.

Using IDLE

To start IDLE, click on its icon. In Windows, the icon will be found in the **Start** menu under **Python**. You should get a screen that looks like Fig. 1. The ">>>" on the left is the Python command prompt. Every time it appears, Python is waiting for you to type or click something.
To see the Python program create the animation portion of the video, click on **File**. You should now get a menu of file options as shown in Fig. 2. I clicked on **Open**. I navigated to **Dance > Experiments** and clicked on *Dance28.py***.** This brought up a new window over the current one, as shown in Fig. 3. The new window contains the dance code. Since the program is fairly long I need to scroll down to see it all.
To run the dance program, I clicked **Run** at the top of the new window. This brought up a menu as shown in Fig. 4. Then, I clicked on **Run Module**. After a short initial startup delay, Python should play the entire animation portion of the dance video. There is no text or music in the animation. These elements were added later in a video editor after the dance animation itself had been converted to video.

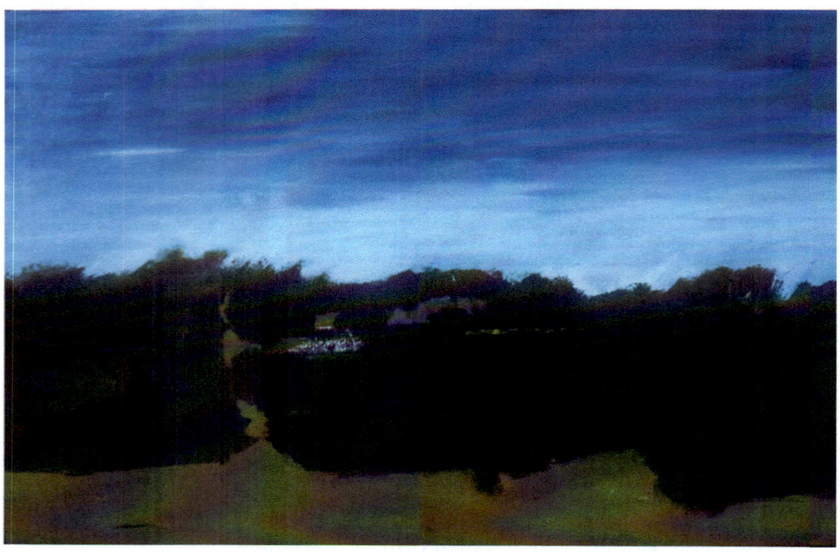

Evening Light. © 2025 by Anna Ursyn

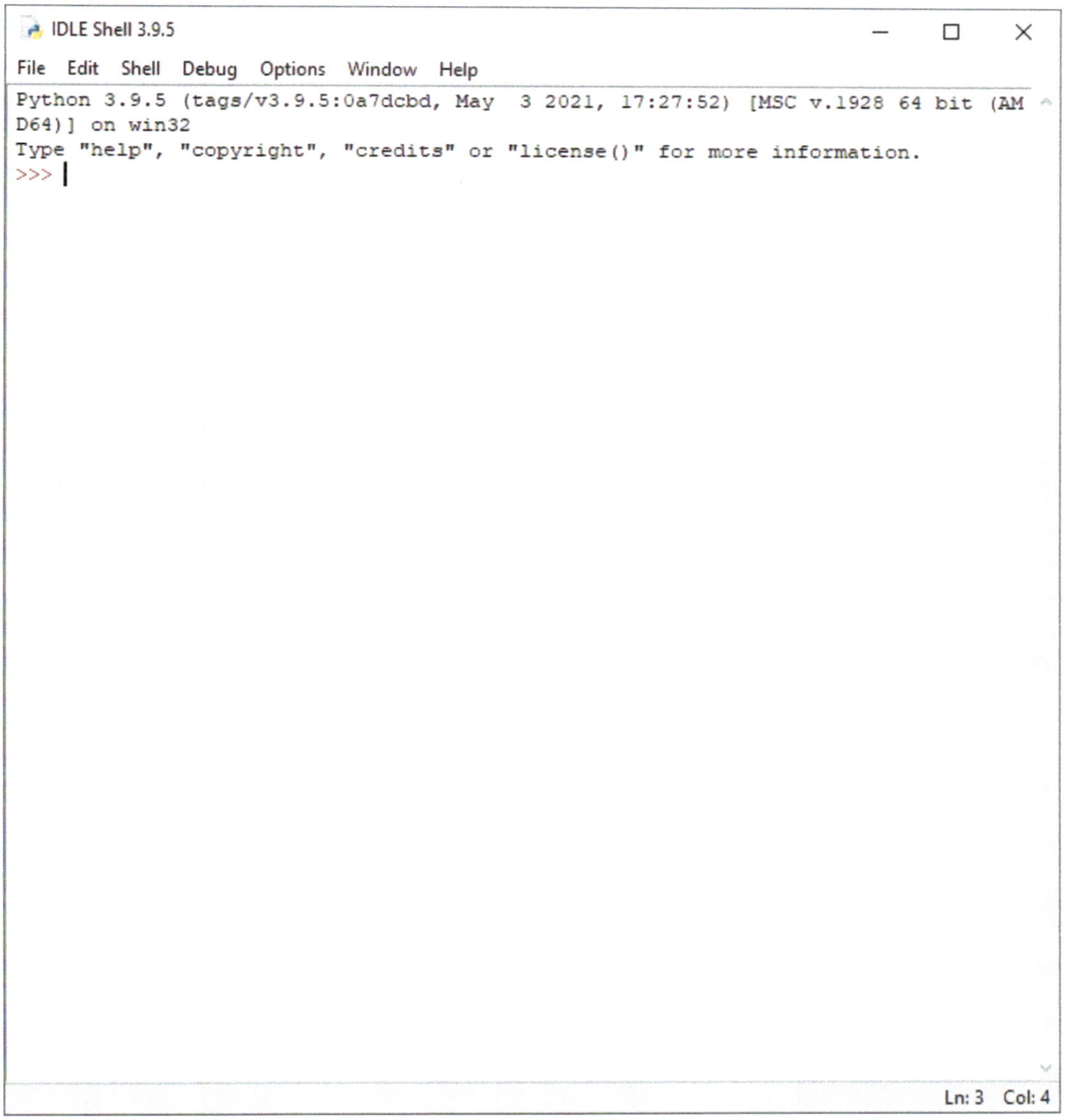

FIGURE 1

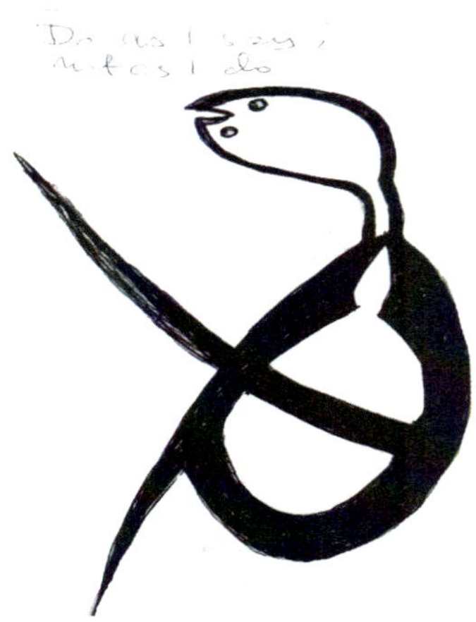

Pu: believes in "Do as I say, not as I do."

Pu: believes in "Do as I say, not as I do."

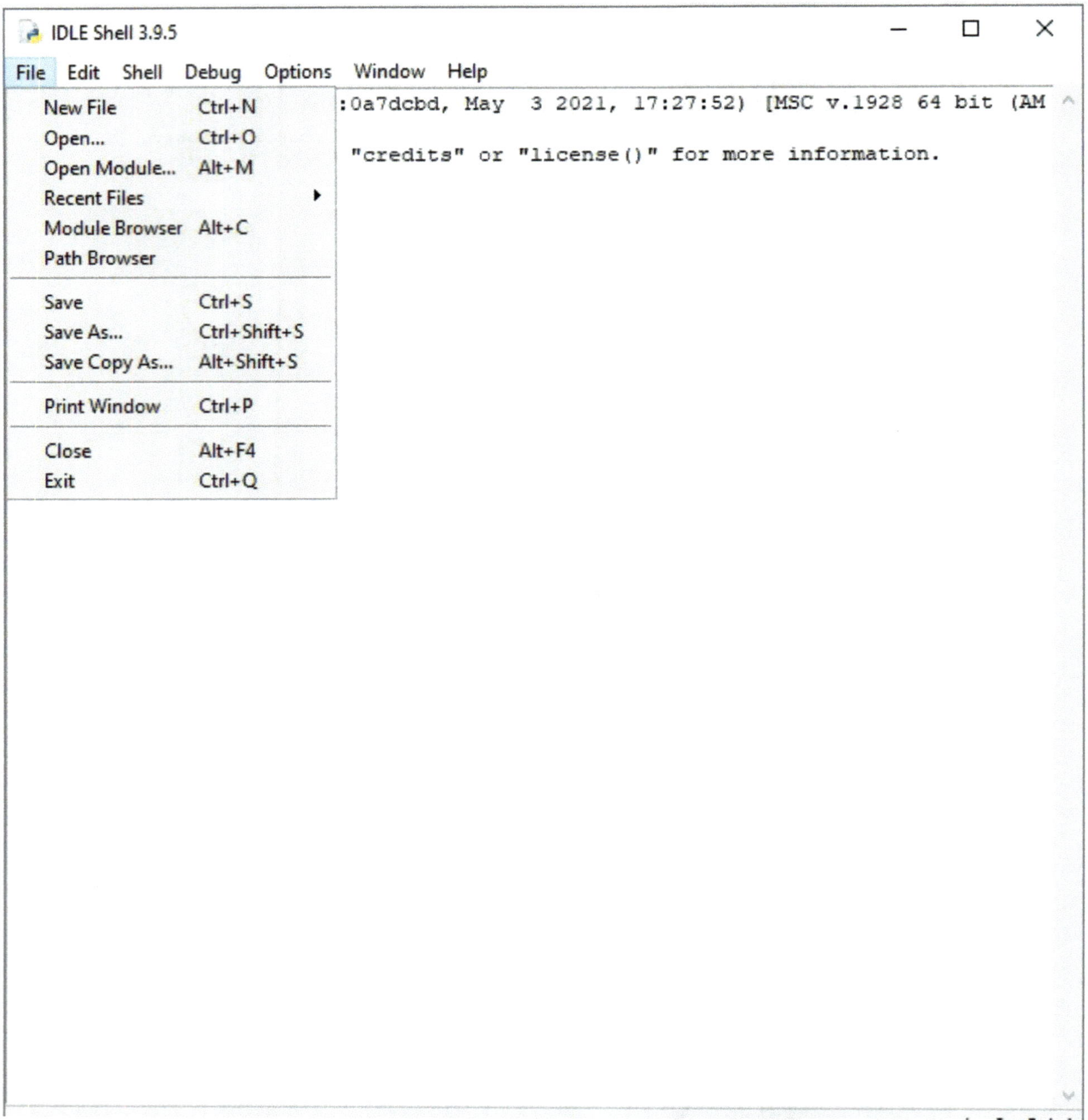

FIGURE 2

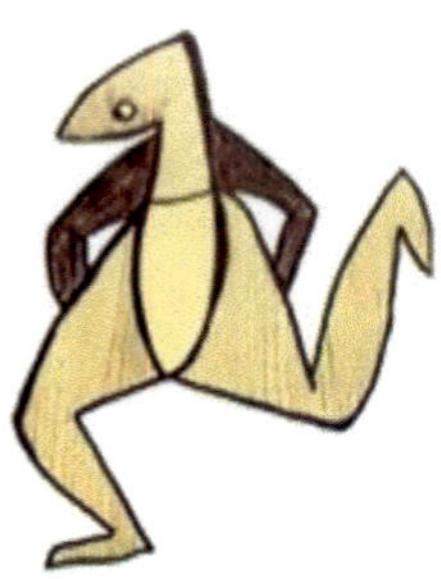

Yngus: whenever sad, he dances to his own humming.

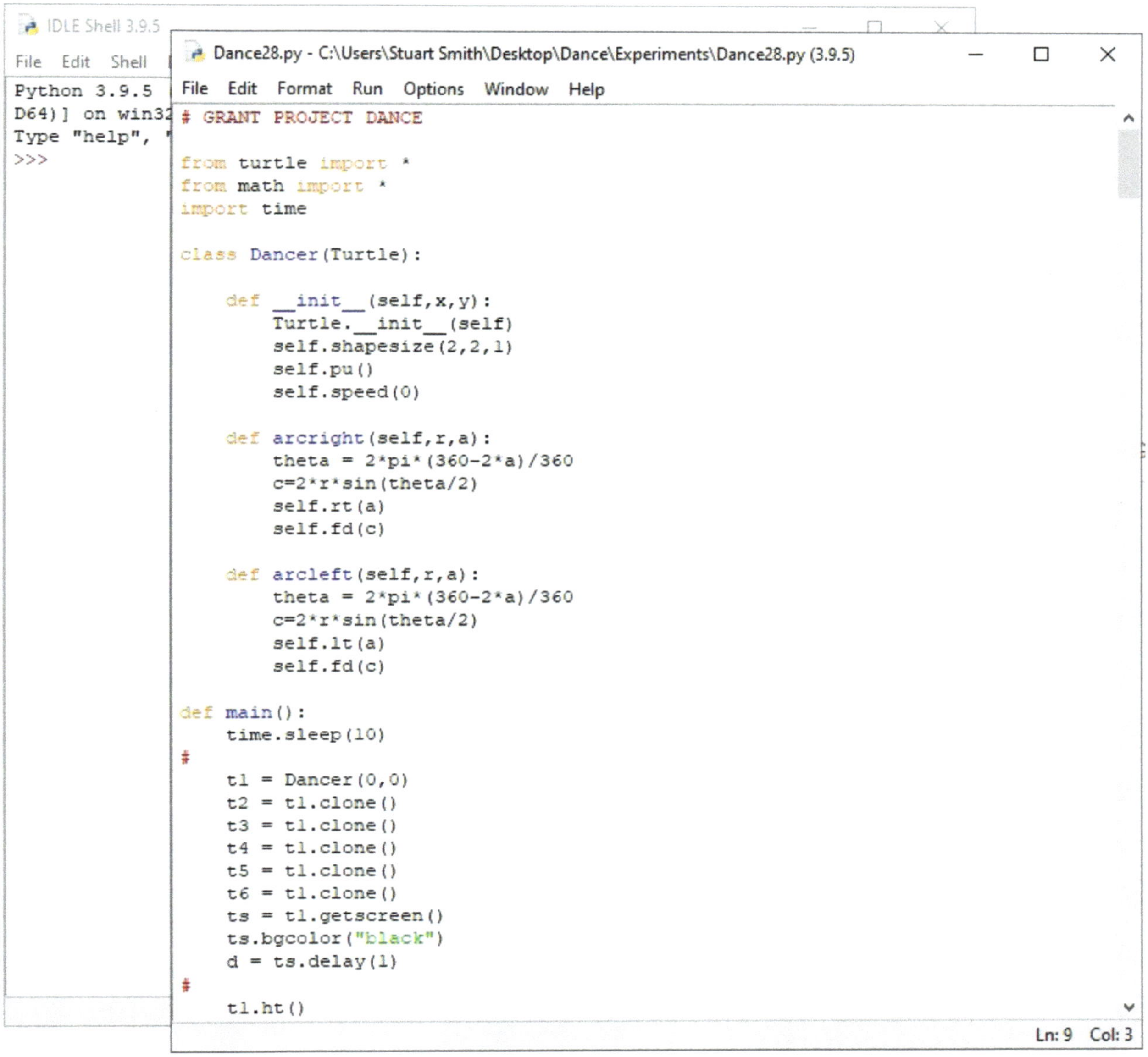

FIGURE 3

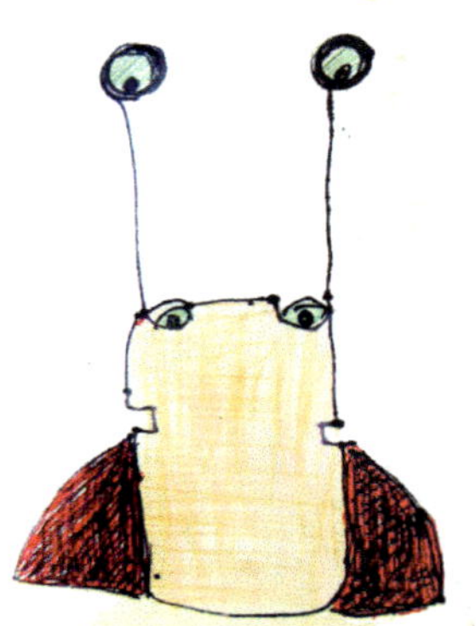

Usafegel: thinks twice before recording information.

Usafegel: careful but consistent.

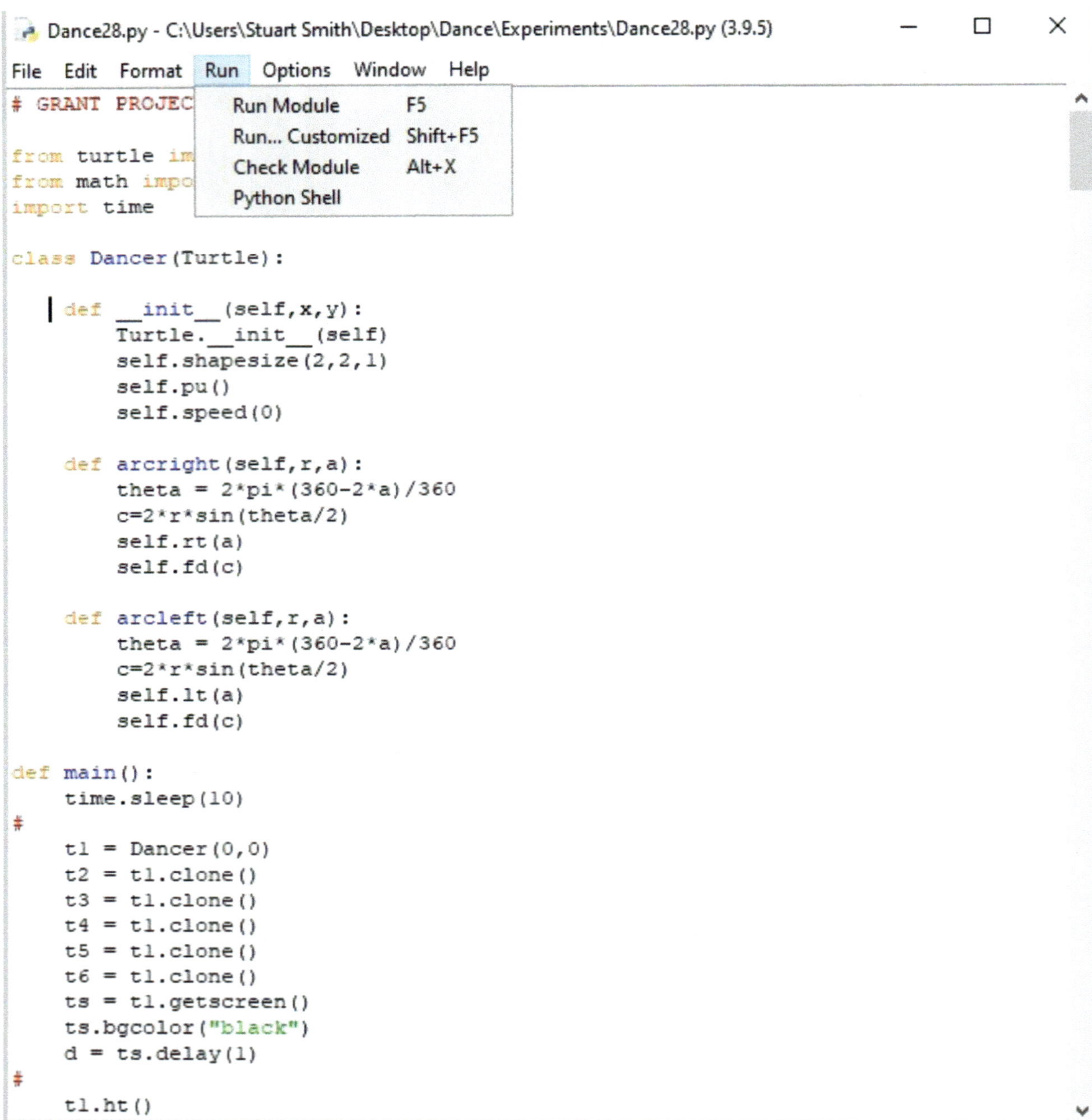

FIGURE 4

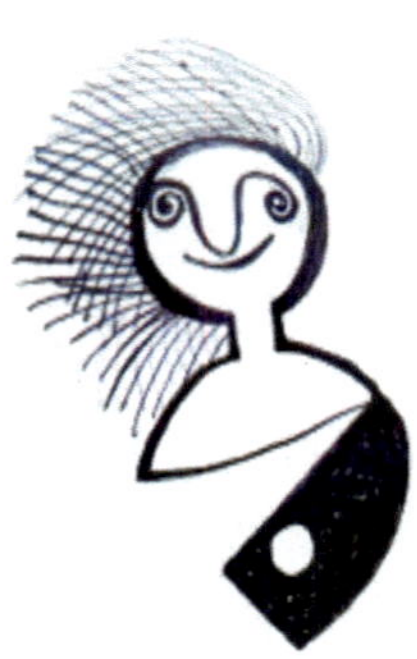

Lawade: composes music on a train.

Comments on the program

The program has two features that deserve some comment. First, the code is "object-oriented," which means that the beginning of the program is dedicated to creating the objects ("dancers" in this case) and the operations that can be performed on them, such as moving circularly through a given angle and with a given radius. Second, the dance video is an animation of *multiple* dancers. This means that when all the dancers are directed to make a particular move simultaneously, e.g., all turn right 90 degrees, the program has to move the dancers in round-robin order in equally small amounts several times until they have all completed the full 90-degree turn. This takes at least seven lines of code, as will be seen throughout the main section of the program. All was set in the folders:

Looping: Python code for the "Looping" section of the video
Music: The mp3 files that contain the music for the five major sections of the video
Numeric: Python code for the "numeric computation" section of the video
PNGs: Versions of the BMP and JPG files with Alpha channel added so that the images can be transparently layered one on top of the other in the video
Sequence: Python code for the "sequence" section of the video
Videos: The final video as well as no-audio versions of the video.

Ideas for Topics to Introduce Coding To The Novice:

Learning coding in a visual way
Our assumptions
To begin with, we have to rethink some basic questions.
First, how are we defining &quote coding? could mean either
(1) translating algorithms or specifications developed by somebody into computer programs or
(2) creating a program from scratch in a continuous cycle of design>code>test>repeat.
While writing codes for an initial video, we focused on sequences, conditions (If statements), and repetitions (or

s, FOR, or WHILE Statements).
Next, what is the relationship between dance to coding? Does the dance follow the code or does the code somehow evolve out of the dance? The same goes for the music: which comes first?
Finally, what do the dancers actually do? Classical ballet movements include basic moves that the professional dancer must master: arabesques, pirouettes, ronds de jambe (round of the leg – a circular movement of the leg that can be done either par terre or en air – on the ground or in the air, on or and several other moves. It is hard to envision dancers who are executing these movements accompanied by a dissonant modernist score; it would be a comedy performance.

Methods and proceedings

We have assembled several sets of coding for the selected dances. After collecting different sets of coding, we asked students from the UNC Dance Department to dance, so we could explain coding through dance movements. Sequences of dancers' steps were recorded and converted into computer programs—lists of instructions written to tell a computer what task it should perform. The results of this phase were translated back into

dance performances. Students were solo-performing traditional dances, which were recorded to obtain the time-based and graphical data. Videos and graphics served as visual tools for learning programming: to carry information with explanatory power. After the end of solo performing traditional dances, students from the UNC's Dance Department were asked to improvise to jazz-based music compositions. The students experienced different ways of recording their steps.
Anyone inclined to learn scare-free coding can use the website containing the results as a reference to learn about coding in a visual way. The codes can guide interested parties to perform and learn from. Visualizations of dances, created previously by Mohammad Majid al-Rifaie and Md. Fahimul Islam were published as a book chapter (Ursyn, 2015), and were used with permissions.

1. Conditions in programming by Stuart Smith

Programming styles can be seen as declarative vs. imperative. Declarative programming is like asking your friend to draw a landscape. You don't care how they draw it, that's up to them. Imperative programming is like your friend following instructions before painting a landscape. They are getting step-by-step directions to get the desired result.

These are very different, and one's choice determines which language or languages are chosen for the project. There are some obvious prosaic metaphors or similes for imperative programming: executing a sequence of steps is like following a recipe; looping is like the lather-rinse-repeat cycle on the shampoo bottle; branching is like filling out a tax return form that asks you to answer item 9 if you're filing jointly, otherwise skip to item 15. We can't think of a good metaphor for the subroutine jump-and-return, which is a key element of imperative programming. Perhaps, observing protons that are in rapid exchange with water. Metaphors for pointers and assignment is seen as a street address to a person's actual home.

It is hard to find metaphors for declarative language like, e.g., Prolog. And the examples used to teach such a language would be different from what is used for C, C++, Java, etc. For a dance piece, a declarative program would be like telling the dancers to "do a particular dance" for two minutes. The dancers would then consult their cellphones or laptops to find the definitions and rules for a specific dance, after which they would get together to see if the dance was possible. If so, they would go ahead and do it.

We discussed using the body to send signals. For example, an arm can signal some meaning, the angle of 90 degrees, like a biker signaling they're making a turn on the interstate, would mean turn left (LEGO), or a foot forward could mean available of one's choosing. Both classical and social movements have a set of meanings. This is where one could start or could look at coding rules and assign functions to them. It is very open-ended.

In terms of dance art – classical or modern – rather than social dancing, they both pose the same questions. There's a precise terminology for ballet, which we partly remember, and discovered that there's also a precise terminology for ballroom dancing. In both cases, the moves described by the terminology are only a framework to be elaborated by the dancers. We decided to focus on ballroom dances as perhaps closer to our audience. Great dancers have their styles, which can't be captured in a bare description according to the rules. To dance using only the prescribed moves would be like cooking without seasoning. The question is what the dancers would do with the rest of their bodies. We were concerned about how a dance consisting only of prescribed steps would look: robotic, stilted, or just plain boring. There is tension between dance as art and dance as a metaphor for coding. There are many dances guided by the rules and many coding languages with various principles. One can't optimize both aspects simultaneously. There's always going to be a tradeoff. It is hard to decide where to draw the line.

Python does have an integrated development environment, which would be easier to insist on for non-technical learners; however, the need for metaphors or similes diminishes very early in the learning process. Each new element of the language is self-justifying, i.e., we simply need it to carry out some useful computation. Each new element could be motivating with a problem to be solved. If one wants to find the class average on each of the homework and tests over the semester and find each student's average for the semester. To "use a spreadsheet" is not an acceptable answer in this class!

When teaching beginners, a standard imperative language that has an IDE, (Integrated development environment Software), first tell them that it's like (there's your simile) a calculator: just type in your calculation and hit ENTER. Once they're comfortable with the calculator simile it's no longer necessary to use a figure of speech to ease them further into programming. It's now a short step from here to introduce variables. After they've got variables, one can introduce the relational operators and IF-THEN-ELSE. Computing the absolute value of a number of ties these together. At this point, one can have them begin writing complete actual programs. This motivates the need for input/out operations like read and write. Put and get can be seen as some examples. Then, introduce arrays and loops with some simple problems like finding the average or the largest and smallest values in an array. Various useful algorithms, e.g., searching and sorting, can be presented. The next-to-last step is functions and subroutines, i.e., breaking a large problem into smaller sub-problems that are easier to solve. The last step is recursion, which many beginners have a lot of trouble understanding. Some coders explain it as finding a way to one's home. When one reaches it, simply stop searching for it, which can mean moving.

When using step-based dances, it could also involve what the dancers do with the rest of their bodies.

This use of variables, conditions, if, while, else, and more will serve the purpose and will enlighten the audience. We discussed the body parts of the dancers: to function-code them (say an arm means a condition, the head, a goal, a leg, etc.), and we can color-code them (yellow, move forward, orange, two steps forward, purple...etc.). Typically, they do the diagrams of the steps, coding man blue steps, and woman red steps, as an example. That would be a "bird's eye view" perspective.

Different colors for costumes (front and back), were discussed to convey a color-based message. Python is considered "easy to learn" and good for a novice.

We decided to use ballroom or social dancing. Instead of (or in addition to) colors, a sequence of basic dance steps could be represented in "turtle graphics" code (perfect for Python), e.g., Forward 10, left 90, etc. The question was how to represent branching and loops. Functions could be used to encapsulate the code for more complex moves, like the whisk or spin turn. Each function call would appear as a single step in the program. We discussed having a large screen monitor above the dancers that would show what the dancers were doing in the footprint notation or in some representation like what you can do in Processing, a language developed by Ben Fry and Casey Reas for visually motivated users. There is a need for some unusual or surprising element to all of this. There are many ways of assigning motions to conditions. But when looking from another angle, we could compare the 12 notes to the set of ballet terminology.

This project appeared to be not all that open-ended. Translating dance steps into program code is a quite specific task, imagining is new tricks, anecdotes, examples, etc. that work better for teaching the first thing was to come up with a way of representing the steps as program code, regardless of whether the representation is helpful to beginning learners.

As an example of an open-ended project: music was composed for wind orchestra and electronic sounds. Some type of visual spectacle was needed to accompany it. The only information was that the piece was "about circles." Hundreds of ping-pong balls were painted in fluorescent colors, as well as some 5 ft. diameter weather balloons. The piece was performed in semi-darkness. There was a powerful UV light that made the ping-pong balls glow intensely. Handfuls of the balls were thrown into the audience, who were also batting a weather balloon from one person to another like a beach ball.

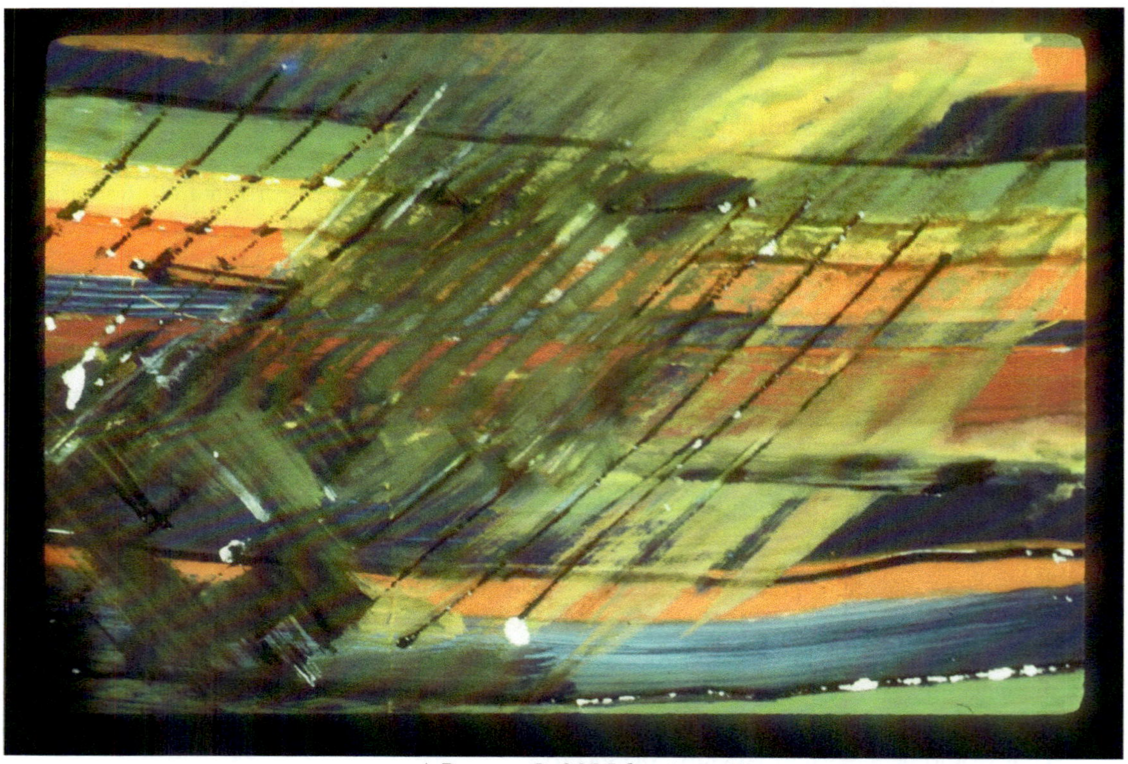

A Dance. © 2025 by Anna Ursyn

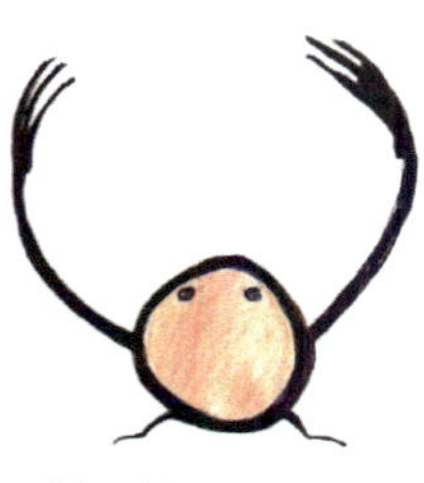

Pukaga: brings good wishes down.

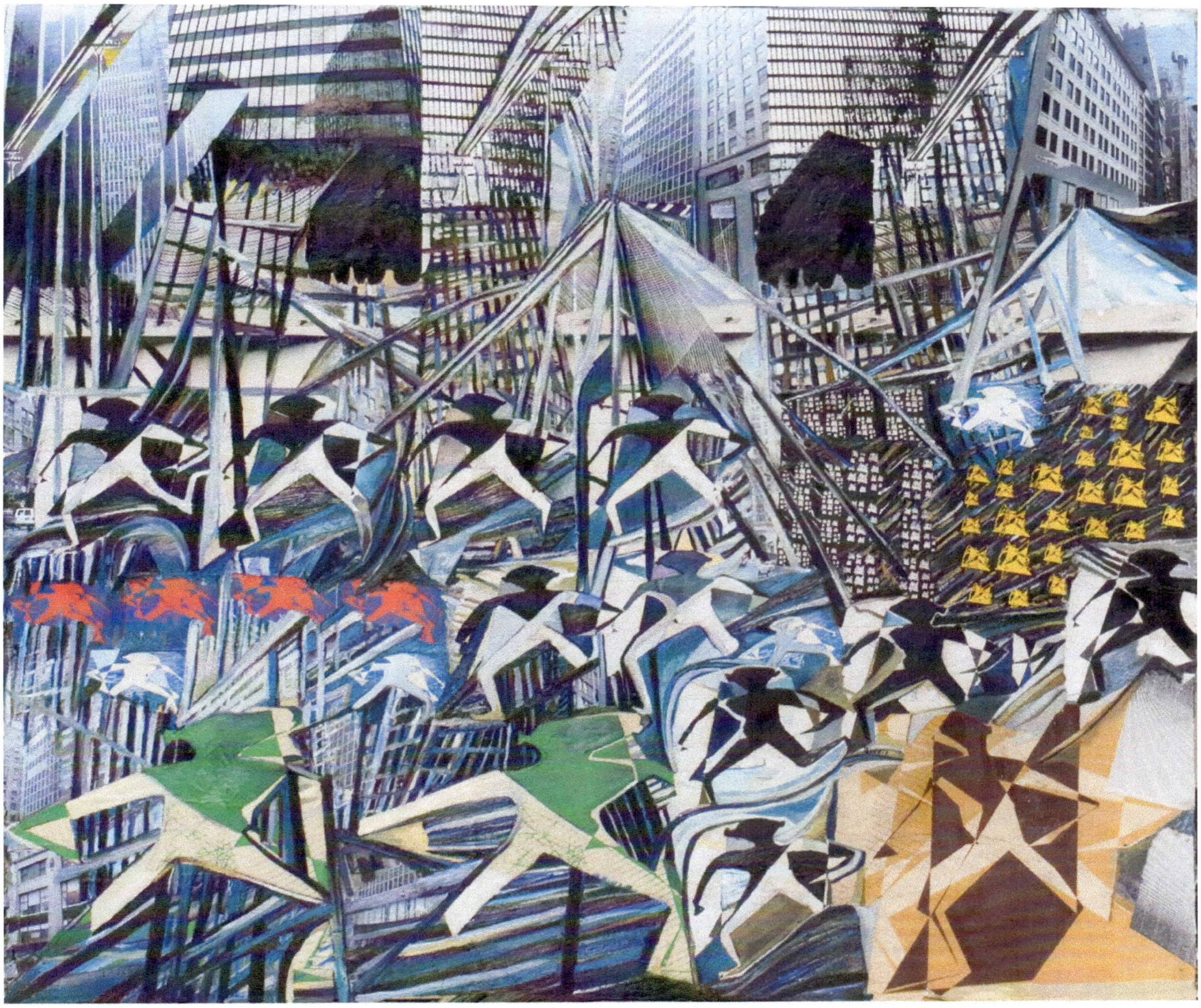

Urban Masquerade. © 2025 by Anna Ursyn

Rhythm and Pattern in Dance and Programming

In dance, rhythm involves the movements' timing related to music, while the pattern means the design of a steps' sequence and movements in the choreography. While rhythm refers to the timing (time), pattern is about the organization and sequence on the floor (space).

In programming, rhythm is about the timing and flow of code execution, while pattern refers to repeatable design structures that solve programming problems and structure the program. Thus, rhythm is about timing and flow, while pattern refers to the code's architecture.

This analogy allows for the application of the dance metaphor of coding.

Tango – 2/4 or 4/4 time with counting beats 1-2-3-4
Waltz – with a triple meter 3/4 time, with a count 1-2-3, 1-2-3
Cha-Cha – 4/4 time with a count pattern 2-3-cha-cha-cha

Rumba – 4/4 time, with counting beats slow, quick, quick
Foxtrot – 4/4 time, with counting beats slow, slow, quick, quick
Salsa – 4/4 time, with a count 1-2-3, 5-6-7” with pauses in the fourth and eighth beats
Swing (Lindy Hop) – 4/4 time with a count: 1-2, 3-4, 5-6, 7-8”

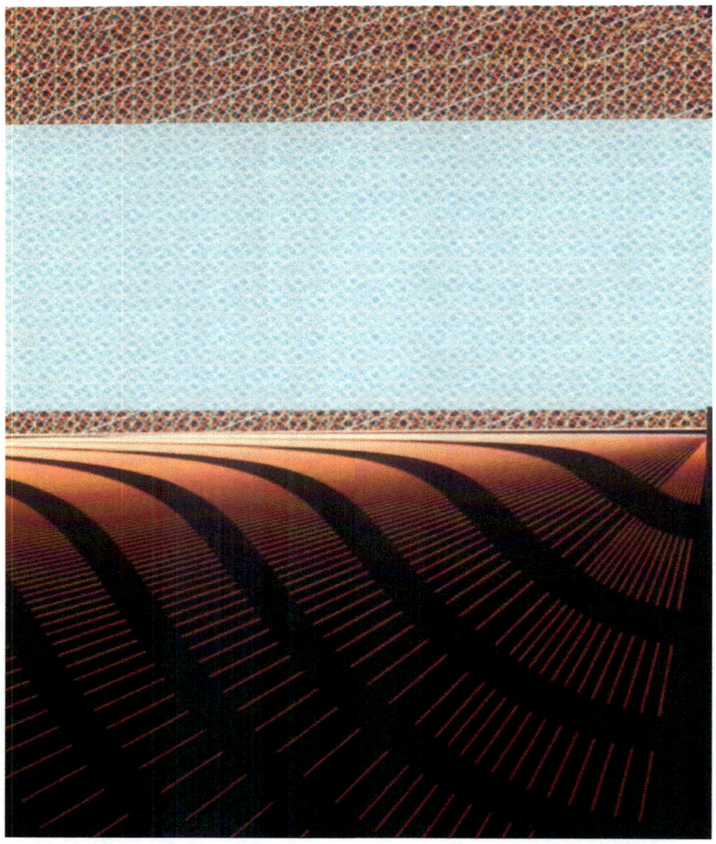

Fieldwork. The output from the computer code. © 2025 by Anna Ursyn

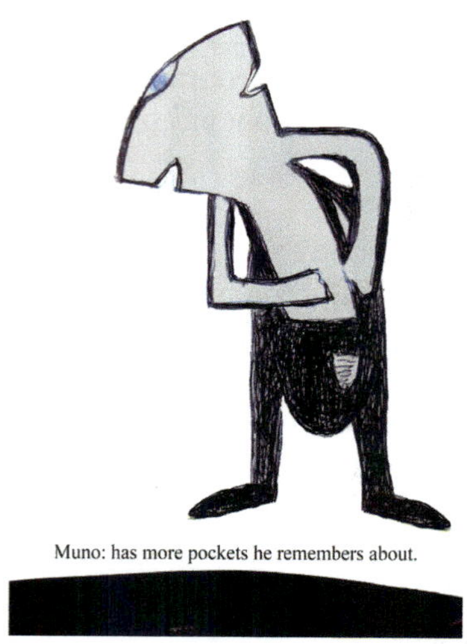

Muno: the more pockets, the better.

PUBLICATIONS RELATED TO THIS BOOK

al-Rifaie, M. M., Ursyn, A., & Wyeld, T. (2020). *The Art of Coding: The Language of Drawing, Graphics, and Animation.* CRC Press: Chapman and Hall/CRC. Taylor and Francis Group. ISBN 978-0-367-90037-3; 978-1-138-62964-6; 978-1-315-21033-9.

Arya, A. (2020). *Anyone Can Code.* Routledge. ISBN-10: 0367199696, ISBN-13: 978-0367199692

Holoworld, *The Basics of STREAM Education: A Comprehensive Guide* (2024). Retrieved from https://www.blogs.holoworld.one/what-is-stream-education/

Beginner's Step-by-Step Coding Course: Learn Computer Programming the Easy Way. (DK Complete Courses), Hardcover –2020. ISBN 1465482210

Erwig, M. (2022). *Once Upon an Algorithm: How Stories Explain Computing*. MIT Press. ISBN-10: 0262545292, ISBN-13: 978-0262545297.

Isaacson, W. (2021*). The Code Breaker: Jennifer Doudna, Gene Editing, and the Future of Human Race.* ISBN 978-1-9821-1585-2.

Kassing, G., & Jay-Kirschenbaum, D., (2021). *Dance Teaching Methods A Curriculum Design: Comprehensive K-12 Dance Education,* 2021Human Kinetics, ISBN: 978-1-4925-7239-8 (print).

Montfort, N. (2021). *Exploratory Programming for the Arts and Humanities, second edition* MIT Press. ISBN-10: 0262044609.

Rahman, A. (2024) A video: https://youtu.be/zhuMGF8_Whw

Smith, S. (2021). Jazz and Jazz Theory. In *Describing Nature Through Visual Data,* A. Ursyn, Editor. IGI Global. https://www.igi-global.com/chapter/jazz-and-jazz-theory/259689

STEAM (2023). Rhode Island School of Design. Retrieved from https://www.risd.edu/steam. VisionShiftStudios (2024).

STEM to STEAM: Integrated Studies STEM to STEAM Resources Toolkit (2024). Edutopia, George Lucas Educational Foundation. Retrieved from https://www.edutopia.org/article/STEM-resources-downloads

Ursyn, A. (2024). *The Transfer of Knowledge through Art and Visualization: Novel Technologies, Collaboration, and Visual Communication of Science-based Projects.* Anna Ursyn (Author). Taylor & Francis Group. Routledge. ISBN 9781032624228.

Ursyn, A., & Mostowfi, M. (2015). Visualization by Coding: Drawing Simple Shapes and Forms in Various Programming Languages. In *Handbook of Research on Maximizing Cognitive Learning through Knowledge Visualization* A. Ursyn, Editor.
DOI: 10.4018/978-1-4666-8142-2.ch008. https://www.igi-global.com/chapter/visualization-by-coding/127482

Ursyn, A., al-Rifaie, M. M., & Md Islam, F. (2015). Metaphors for Dance and Programming: Rules, Restrictions, and Conditions for Learning and Visual Outcome. In Anna Ursyn, Author/Editor (2015), *Knowledge Visualization and Visual Literacy in Science Education,* 2015, IGI Global, 572 pages, ISBN-13: 9781466681422

What is Coding: Coding 101 tutorial (2024). Retrieved from https://www.bitdegree.org/tutorials/what-is-coding/

Bruce Wands, Art of the Digital Age, The first major illustrated survey of this exciting, new, and experimental field. © 2006. Pages: 224. Thames & Hudson, ISBN-13 : 978-0500286296

(Used with permissions.) * – Denotes content reprinted from Ursyn, A., al-Rifaie, M. M., & Md Islam, F. (2015). Metaphors for Dance and Programming: Rules, Restrictions, and Conditions for Learning and Visual Outcome. In Anna Ursyn, Author/Editor (2015), *Knowledge Visualization and Visual Literacy in Science Education,* 2015, IGI Global, 572 pages, ISBN-13: 9781466681422.

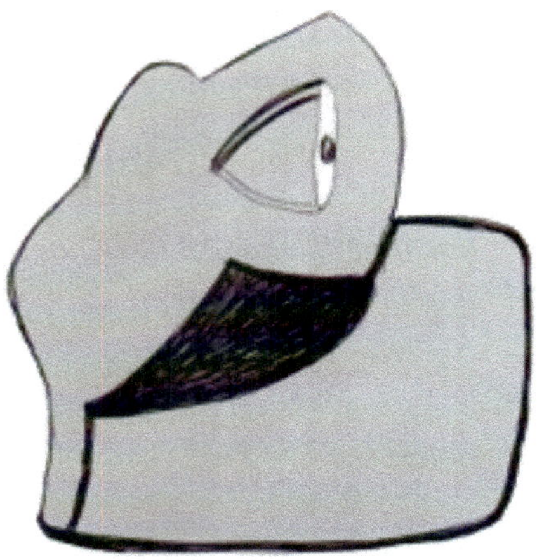

Hadura: always waits for her followers

Hadura: always waits for all her followers.

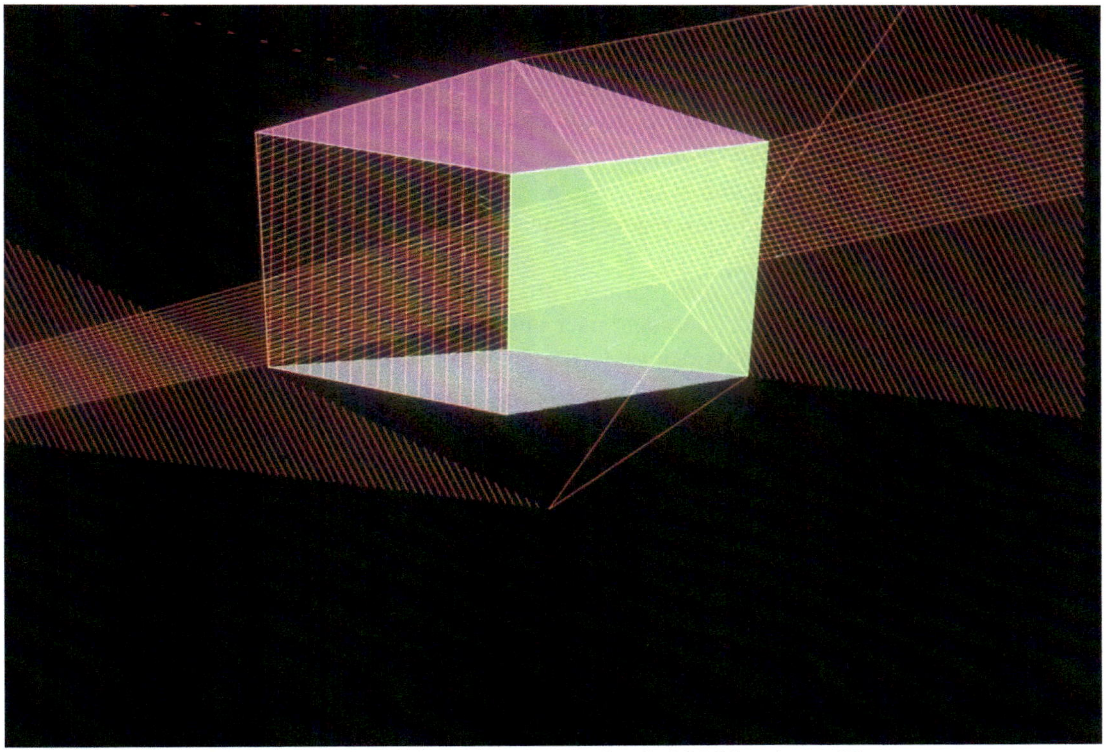

The Stage. The output from the computer code. © 2025 by Anna Ursyn

TERMS WITH SHORT DESCRIPTIONS

Abstract – a concept or thought that does not have a physical, concrete existence. Also, an artistic object that has a meaning independent of narrative or pictorial representation. Also, a summary of a larger text.

Aesthetics – analysis of principles related to beauty in terms of arts or philosophy

Algorithm – a sequence of instructions telling a computer how to perform computation or write programs. In a recursive process, algorithms can be written many times to create repetition. Algorithms help to solve a complex problem by writing a sequence of simple steps.

Anthropology means the study of human characteristics. Physical anthropology studies human biological and physiological features and their evolution. Cultural anthropology studies societies, cultures, and their development.

API – application programming interface, is a software interface that allows communication between computers and/or programs.

Artifact – an object produced by a human being, examined in artistic or historical terms (it may be an artwork, a craft object, a tool, a weapon, or an ornament of archaeological or historical interest).

Artificial intelligence (AI) uses computer systems and applies machine learning for tasks, involving big collections of data, that would require human intelligence.

Binary numeral system uses two symbols: typically, "0" (zero) and "1" (one).

Biointerface is a place of contact between a biomolecule, living tissue, or organism and a living or another material. Nanostructure interfaces are extensively studied.

ChatGPT is a computer program that uses artificial intelligence (AI) and natural language processing (NLP) to understand user questions, providing automatic responses, and simulating human conversation (ibm.com/topics/chatbots)

Cognitive abilities enable to think abstractly (not only on concrete objects), learn from experience, reason, comprehend ideas and concepts, and solve problems. They support gaining knowledge, utilizing information, creative and critical thinking.

Cobots are robots that collaborate with people.

DALL·E 2 is a version of GPT-3 that can create realistic images and art from text descriptions.

Data – factual information specially organized for analysis, reasoning, or making decisions.

Data Visualization: information abstracted in a schematic form to provide visual insights into data sets. Data visualization enables us to go from the abstract numbers in a computer program (ones and zeros) to the visual interpretation of data. Text visualization means converting textual information into graphic representation, so we can see the information without reading the data, as tables, histograms, pie or bar charts, or Cartesian coordinates.

Dimension relates to spatial measures such as width, height, or length. Dimension means how many numbers (coordinates) are needed to determine a position of a point in space. In the Cartesian three-dimensional space, three coordinates, x, y, and z, describe the position of a point: x-horizontal axis, y-vertical axis, and z-depth. René Descartes (Cartesius) revolutionized mathematics by providing the first systematic link between geometry described by Euclides and algebra). A point has no length, area, volume, or other dimensional attributes, so some mathematicians claim a point is 0-dimensional. A line has one dimension because only one coordinate is needed to describe a point in space, while a surface has two dimensions. Data may be linear or multidimensional.

Empathy means understanding and sharing another person's or creature's feelings. One may apply cognitive (by knowing how to take somebody's perspective), emotional (by feeling somebody's emotions), and compassionate (by combining intellect and feelings) empathy.

Extended Reality (XR) comprises real-and-virtual environments generated by computer technology and used in wearables.

Virtual Reality (VR) refers to immersive experiences created with the use of real-world content (e.g., 360° Video), purely synthetic content (for instance, computer-generated), or a hybrid of both.

Augmented Reality (AR) is an overlay of computer-generated content on the real world that can superficially interact with the environment in real time. With AR, the CG content and the real world do not block themselves.

Mixed Reality (MR) – in MR, the synthetic content is overlaid and anchored to interact with objects in the real world in real-time.

Fiber lasers use fiber optic cables with added rare earth elements and other materials to provide and transmit light and light signals.

Form – mass that occupies space, giving the weight, density, and thickness of a 3D object or is implied in a picture of this object.

Fractal geometry describes self-similar or scale-symmetric objects called fractals. When magnified or reduced in size, fractal things are self-similar: their parts are similar to the whole, and the likeness continues when the parts are magnified more and more. They are present in nature, designs, and artworks. Fractal geometry can explain the origination of patterns in living organisms.

GPT (generative pre-trained transformer) is a deep learning model originated by Google in 2017.

GPT-4 is a multimodal model that accepts prompts of images and texts as inputs and gives out text outputs, such as natural language or code.

Graphic is an image represented by a graph or relating to graphics. Computers often generate graphic displays.

Haptic – relates to the sense of touch; the senses of touch and proprioception enable the perception and manipulation of objects.

Higher-order thinking goes beyond observation of facts and data memorization, enabling one to build data-based knowledge. Higher-order thinking makes critical thinking possible in an evaluative, creative, and innovative way.

HTML – Hypertext Markup Language. This language codes formatting, links, and other things on Web pages. XML – eXtensible Markup Language. A markup language like HTML; it lets individuals define and use their tags.

Hyperlink – a link from a hypertext file or document to another file activated by clicking on a highlighted word or image on the screen.

Icon, iconic object or image. An icon represents a thing or refers to something by resembling or imitating it; thus, a picture, a photograph, a mathematical expression, or an old-style telephone may be regarded as an iconic object. Therefore, an iconic object or image has qualities in common with things it represents by looking, sounding, feeling, tasting, or smelling alike.

Iconography means visual elements such as images and symbols used in a specific culture, society, or artwork. It also may indicate a study about these elements.

Immersive experience happens when a 3D or 4D (including time) image (maybe plus sound or other stimuli) surrounds you creating extended reality. Platforms that support immersive experiences include augmented reality (AR), virtual reality (VR), Mixed reality (MR), 360° video, wearable devices, the Internet of Things (IoT), and more.

Infographics: tools and techniques involved in the graphical representation of data, mostly in journalism, art, and storytelling.

Information – knowledge derived from study, experience, or instruction based on facts or data.

Information visualization – representation plus interaction; it means using the computer-supported, visual-spatial representation of abstract data to amplify cognition and derive new insights. Data presented as information visualization are often interactive, numerical, verbal, and graphical (such as the text of geographic representations).

Ions: cations and anions. An atom or molecule with an electric charge is called an ion. An electric charge results from the presence of single, double, triple, or even higher negative electrons, unequal to the number of positive protons in the nucleus of an atom. Removing or adding one or more electrons changes a neutral atom into an ion. A cation is an ion or group of ions with a positive charge. An anion is a negatively charged ion. In an electrolyte solution, cations move toward the cathode (negative electrode) after applying an electric current, and anions migrate to an anode.

Interface denotes a connection or interaction between two units (which may be subjects, organizations, physical or electronic systems), for example, social interface, or interface between phases of matter such as solid, liquid, and gas, and also interaction of hardware and software. The user and computer interface allows communication with the operating system. Interface with the internet may go through the Internet protocol.

Internet of Things (IoT) means a set of objects connected over a wireless network and acting without human actions. It serves for personal use in smart homes (with everyday physical devices embedded with sensors and software), for elder with disabilities care, as an internet of medical things (IoMT), for business, industrial, vehicular communication, and many kinds of infrastructure systems.

Kinesthesia means knowing about the position and movement of the parts of the body. Sensory organs (proprioceptors) in the muscles and joints signal changes in position.

Knowledge visualization uses visual representations to transfer insights and create new knowledge in communicating different visual formats.

Latitude and longitude are parts of a grid system giving us horizontal and vertical coordinates in the geographic coordinate system. Latitude lines run east-west, and longitude lines run north-south. Every location on Earth has a universaladdress described by two numbers called latitude and longitude coordinates; they say where the coordinates intersect.

LED, the light-emitting diode, is a semiconductor diode that glows when a voltage is applied.

Mandala is a geometric circular figure depicting the universe according to Hindu and Buddhist traditional symbols.

Manifesto is a public declaration of opinions, statements, and programs mostly of political or social aims. An art manifesto expresses aesthetic and philosophical thoughts about the art defined by an artist or a group of artists. An individual art manifesto facilitates the translation of artistic, visual messages into literary form and conveys it to the viewers, critics, and jurors.

Meander is a geometric pattern made by a line bending along and repeatedly crossing a straight line but not crossing itself. In the arts, it is a decorative pattern constructed from a continuous line shaped into a repeated, ornamental motif.

Metaphor – indicates one thing as representing another, thus making an implied mental comparison. For example, we use metaphors every day when we talk about desktop items, such as folders, documents, and in and out boxes, using familiar objects for organizing elements related to the computer. Also, a figure of speech where we use one word or phrase to designate another one, thus suggesting comparison, as in 'a sea of trouble.'

Mime – a performance art involving body motions without the use of speech.

Mythology is a collection of stories, myths, legends, and beliefs typical of a given culture or religion; for example, Christian, Jewish, Egyptian, Hindu, or Muslim mythologies.

NASA is the *National Aeronautics and Space Administration* civil space program and space exploration. This agency works with contractors, academia, and international and commercial partners to explore, discover, and expand knowledge for the benefit of humanity.

Negative space is the space around the objects depicted in a picture. A figure-ground reversal is an optical illusion that happens when we perceive an image formed by the negative space.

Network is a set of connected (wired or wireless) devices, such as computers, servers, mainframes, peripherals, firewalls, smartphones, and more, which allows people to share data. Devices may be linked through cables, telephone lines, radio waves, satellites, or infrared light beams.

Networking means exchanging ideas and information among people for social or business contact.

Neurotransmitters are chemical substances released by nerve cells that transfer signals from neurons to other cells.

OpenAI is a research unit that derives its solutions from deep learning and natural language processing.

Pantomime – a musical, theatrical production using gestures and movements but not words

Paradigm is a pattern of thought: a set of concepts or a way of thinking characteristic of a specific field in science (research methods, postulates, and standards) or philosophy. Paradigm shifts – transitioning to a new approach to the field under study is how the developments come about. Paradigm also means a model, a typical example.

Pattern means the regular order existing in nature and a manmade design. We see patterns everywhere in nature, mathematics, art, architecture, and design. In nature, patterns can be seen as symmetries (e.g., snowflakes) and/or structures having fractal dimensions such as spirals, meanders, or surface waves. In computer science, design patterns serve in creating computer programs. In the arts, the pattern is an artistic or decorative design made of recurring lines or any repeated elements.

Perspective is the appearance of depth, picturing the world on a two-dimensional, flat surface of a piece of paper or a computer screen. Perspective represents 3D objects and depth relationships on a 2D surface.

Photonics is the science of light (photons) generation, detection, and manipulation. It can go through emission, transmission, modulation, sensing, signal processing, switching, and amplification.

Scale is a range of values that forms a standard system for measuring the size of something. Units for measuring in science are known as the International System of Units, which is based on the meter. Very small physical and biological structures are measuring fractions of a meter: micrometer (one-millionth of a meter – 1×10^{-6} m), nanometer one billionth of a meter – 1×10^{-9} m), and picometer (one trillionth of a meter – 1×10^{-12} m) thus, the picometer is one-thousandth of a nanometer. Below are examples of objects of various sizes.

Scientific visualization presents real, abstract, or model-based objects in a digital way directly from the data. It may deliver art-science cooperative learning projects and make knowledge understandable to a broad audience. Visualization as storytelling comprises narratives, interactive graphics, explanatory and animated graphics, and multimedia.

Semiconductor material has a conductivity between a conductor (most metals, for example, copper) and an insulator (such as glass). The properties of a semiconductor, its moving electrons, and electron holes in a crystal lattice can be explained by quantum physics. Silicon crystals are semiconductive materials; they are used in microelectronics and photovoltaics used for direct conversion of sunlight to electricity with the use of solar panels.

Semiotics studies the meaningful use of signs, symbols, codes, and conventions for communication. The name 'semiotics' is derived from the Greek word 'semeion,' which means "sign." "Meaning" is always the result of social conventions, even when we think something is natural or characteristic and use signs for those meanings. Therefore, culture and art are a series of sign systems. Semioticians analyze such sign systems in various cultures; linguists study language as a system of signs, and some even examine film as a system of signs. The semiotic content of visual design is essential for non-verbal communication applied to practice, especially for visualizing knowledge.

Sign tells about a fact, an idea, or information; it is a distinct thing that signifies another thing. Natural signs signify events caused by nature, while conventional signs may signal art, social interactions, fashion, food, interaction with technology, machines, and practically everything else. Signs take conventional shapes or forms to tell about facts, ideas, or information. Icons and symbols help visually compress information. An icon represents a thing or refers to something by resembling or imitating it; thus, a picture, a photograph, a mathematical expression, or an old-style telephone may be regarded as an iconic object. Therefore, an iconic object has qualities in common with things it represents by looking, sounding, feeling, tasting, or smelling alike. Natural signs signify events caused by nature, while conventional signs may signal art, social interactions, fashion, food, interactions with technology, machines, and practically everything else.

Signage is a visual graphic that displays information, for example, street signs, room identification signs, or any informational or regulatory signs. Signs, symbols, and icons are collectively called signage. Effective design of a complicated product may help memorize and learn how to use this product (for example, 'Where is the switch?' or 'How to open this thing?').

Social media include social networking sites, image-sharing sites, video hosting sites, discussion sites, community blogs, and economic networks.

Social networking means using Internet-based social media websites and applications to interact with other users for a social or business purpose and stay connected with friends, family, colleagues, customers, or clients. Social networking acts through sites like Facebook, Twitter, LinkedIn, and Instagram.

Space is a three-dimensional expanse in the physical universe where we can measure the position and direction of objects and events. A Cartesian coordinate system (with the x, y, and z axes) serves this purpose in Euclidean geometry.

Spectroscopy records how matter interacts with or emits electromagnetic radiation. A spectroscope serves to study matter by measuring properties of specific frequencies and wavelengths of electromagnetic radiation.

Symbol does not resemble things it represents but refers to something by convention; for example, the word "red" represents red. We must learn the relationship between symbols and what they represent, such as letters, numbers, words, codes, traffic lights, and national flags. A symbol represents an abstract concept, not just a thing, and is comparable to an abstract term. Highly abstracted drawings that show no

realistic graphic representation become symbols. Symbols are omnipresent in our life, for example:
– An electric diagram that uses abstract symbols for a light bulb, wire, connector, resistor, and switch.
– An apple for a teacher or a bitten apple for a Macintosh computer.
– A map – a typical abstract graphic device.
– A 'slippery when wet' sign.

Symmetry is present when similar parts of an object are arranged around a line, point, or plane. A crystal shows symmetry when it has a center of symmetry, rotation axes, or mirror planes (imaginary planes that divide it into halves). There are several types of symmetry: for example, line or mirror symmetry, radial, cylindrical, or spherical symmetry. A figure with a line, bilateral symmetry has two identical halves when folded along its line of symmetry, and these halves are congruent, meaning they are the same size and shape. An object has a radial symmetry when it can be rotated around the rotation axis. For example, with a fourfold rotation axis, the crystal repeats itself each 90°. Angles of rotational symmetry possible for crystals are: 60 degrees, 90 degrees, 120 degrees, 180 degrees, and 360 degrees. The halves of the bilaterally symmetrical animals, for example, butterflies, form each other's mirror images when seen along the axis. A circle is symmetrical about its center. Optical isomers are 'mirror images' of each other and are symmetrical around a plane. For example, molecules of some sugar isomers deflect the rays of light in the right or left direction.

Synesthesia means the merging of two or more senses. It happens when a sensory stimulus is received by one sense but also triggers perception in another sense.

Synapse: a structure that links (through a synaptic cleft) a nerve cell with another nerve or some other cell type. In an electrical synapse, an activated neuron develops an electric signal on its presynaptic cell membrane, which induces a voltage change on the postsynaptic cell. In a chemical synapse, electrical activity on the presynaptic cell membrane initiates the release of a neurotransmitter and its diffusion to a postsynaptic cell.

Texture: a characteristic of a substance or a material. Texture exists all around us. It can be actual (natural, invented, or manufactured) or simulated (made to look rough, smooth, hard or soft, or like a natural texture). Simulated textures are made to represent real textures, such as smooth arms or rough rock formations. But they are not actual textures; if you touch the picture, you feel only the paint or the pen or pencil marks.

URL means Uniform Resource Locator. Also, resources are called well-known codes, such as http://www.sciam.com/, used in hyperlinks (activated by clicking).

Variable is a feature or a factor that can take on different numerical values.

Vernacular is an adjective to describe the local, indigenous, folkloristic, common, or simple form of a specific culture, architecture, language, or other human activities or products.

Vernacular language is typical of a particular country or region or a dialect that can be regional, professional, or otherwise linking a group of people. It may also denote the architecture of houses in a particular region.

Visuals are pictures, diagrams, graphics, film clips, or various forms of display.

Visual media are primarily images, videos, and infographics. PowerPoint presentations may serve as an example.

Visualization means the communication of information with graphical representations. Interactive visual representations of abstract data use easy-to-recognize objects connected through well-defined relations.

Wireframe is a skeletal three-dimensional model in which only lines and vertices are represented. It can also be a way to design a website at the early structural level, a 2D outline of a web page or app that displays the functional elements of a website.

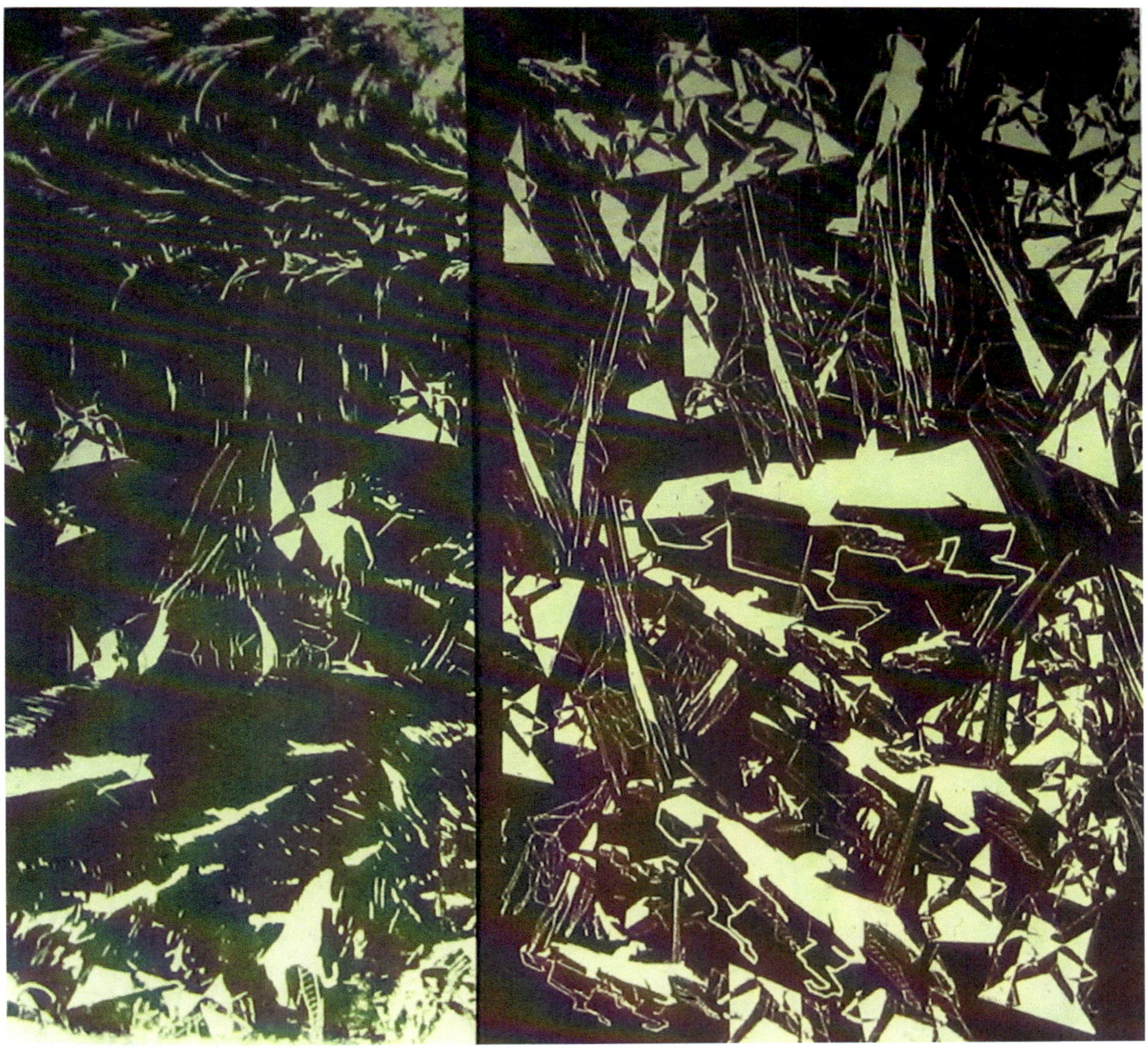

Different Ways. Photolithographs after computer codes. © 2025 by Anna Ursyn

Cosmic Numbers. Photolithograph after computer codes. © 2025 by Anna Ursyn

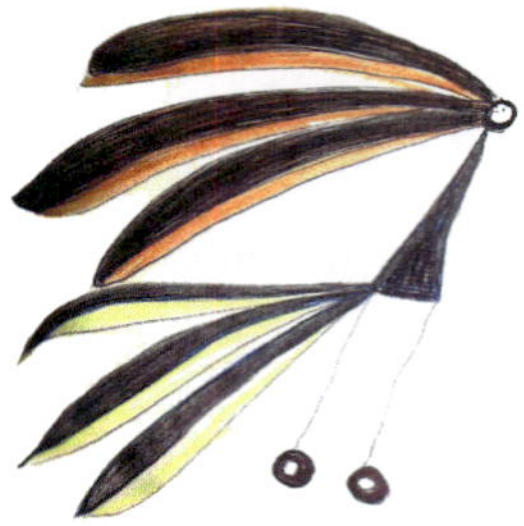

Fuftu: makes only right turns.

Fuftu: makes only right turns.

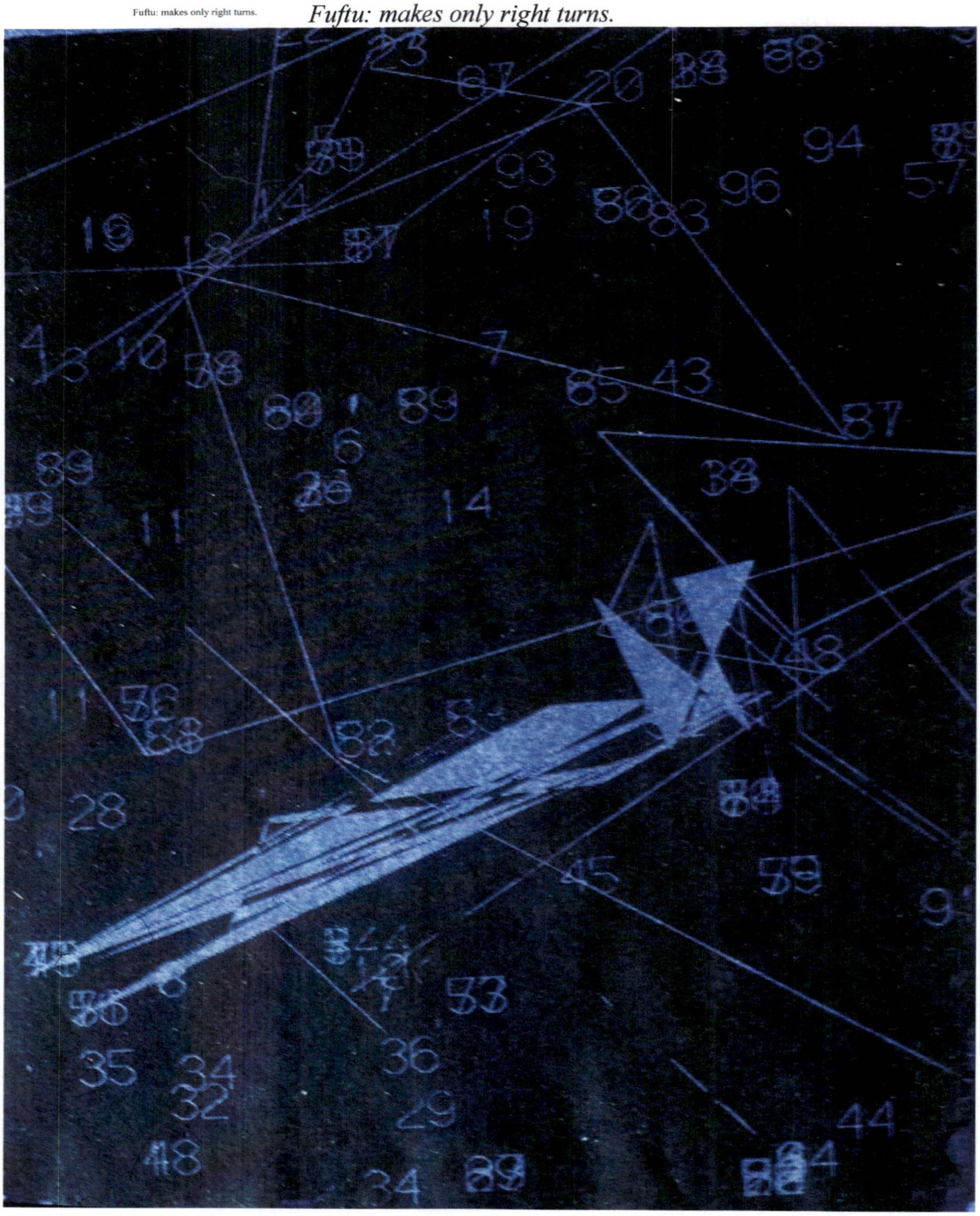

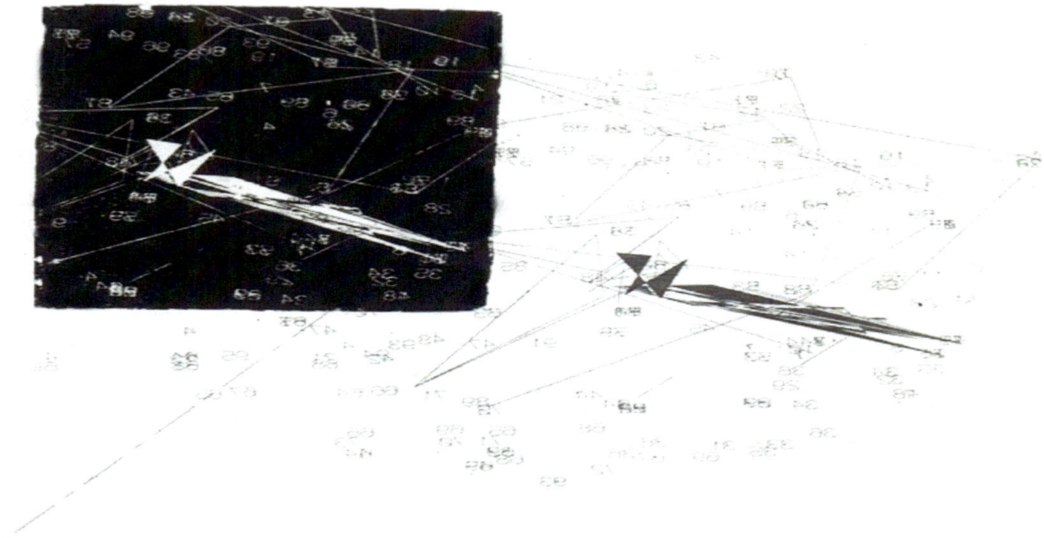

Kepler_438b_Kepler_442b, Photolithograph on the code's outcome. © 2024 by Anna

CREDITS FOR THE PERFORMANCE

Campus Commons Performance Hall (CCPH) stage,
https://arts.unco.edu/campus-commons/

Anna Ursyn, Computer Graphics, UNC - Concept, Story, Directing, Video recording and Editing, Stage Design, and Codes with Outputs
Stuart Smith, Music and Computer Science, University of Massachusetts Lowell - Music, and Codes with Outputs
Brian Luedloff, Professor of Music, Director of Opera Theater, School of Music, UNC, and Advising, Organizing, Consulting, and Managing
Mohammad Majid al-Rifaie, Associate Professor in Artificial Intelligence, University of Greenwich London, United Kingdom
Md Fahim Islam JPMorgan Chase, Chicago, IL. - Codes with Outputs
Md. Asifur Rahman, Infozillion Teletech BD, Dhaka, Bangladesh, Codes with Outputs
Kim Clay, Assistant Professor of Dance, School of Theater Arts and Dance, UNC – dancing, leading ballet students
Mary Jo Lasky, student of Digital Media, UNC - Dancing at the Campus Commons Performance Hall (CCPH) stage.
Isabella Sajdak, Student of Digital Media, Art&Design, at UNC - Dancing, Drawing, Animation, Voice Recording, Singing, Reciting In the Music Studios labs, developed animations, videos, and recorded voices.
Adam Alduaij, Student of Graphic Design, Art&Design, UNC - Filming, Voice Recording
Jennifer Wagner, graduate student (MM) of Vocal Performance, UNC - reading script, acting, singing
Silas Vasquez, student of Musical Theater Department, UNC - reading script, acting, dancing
Brandon Ingold, Technical Director, PVA, CCPH), UNC - Stage Technical Assistance
Students from the Department of Dance - Dancing at the CCPH with their professor:

Spencer Bowman, Musical Theater and Dance Minor
Lauren Greiner, Musical Theater and Dance Minor
Joslyn Joseph, Dance and Theater major
Sabrina Patten, Musical Theater
Jeremiah Ulufanua, Musical Theater

Many thanks to everyone who participated in those collaborative endeavors.

The script written by the author of this book served our international, UNC's faculty, staff, and student collaborations, resulting in grants supporting videos and performances. The text was written during summertime with images by the author, Matylda Czarnecka, and Joohyun Pyune with copyright-free resources, AI, Grammarly, and Wordtune.

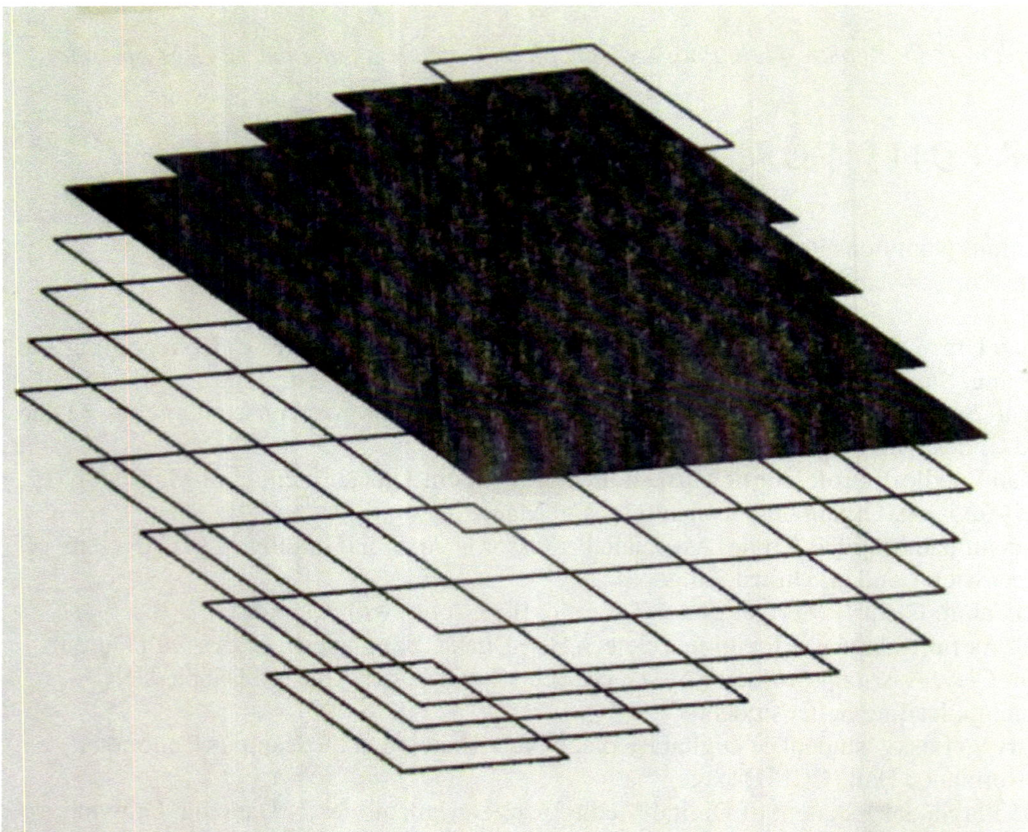

Free Fall. The output after the computer code. © 2025 by Anna Ursyn

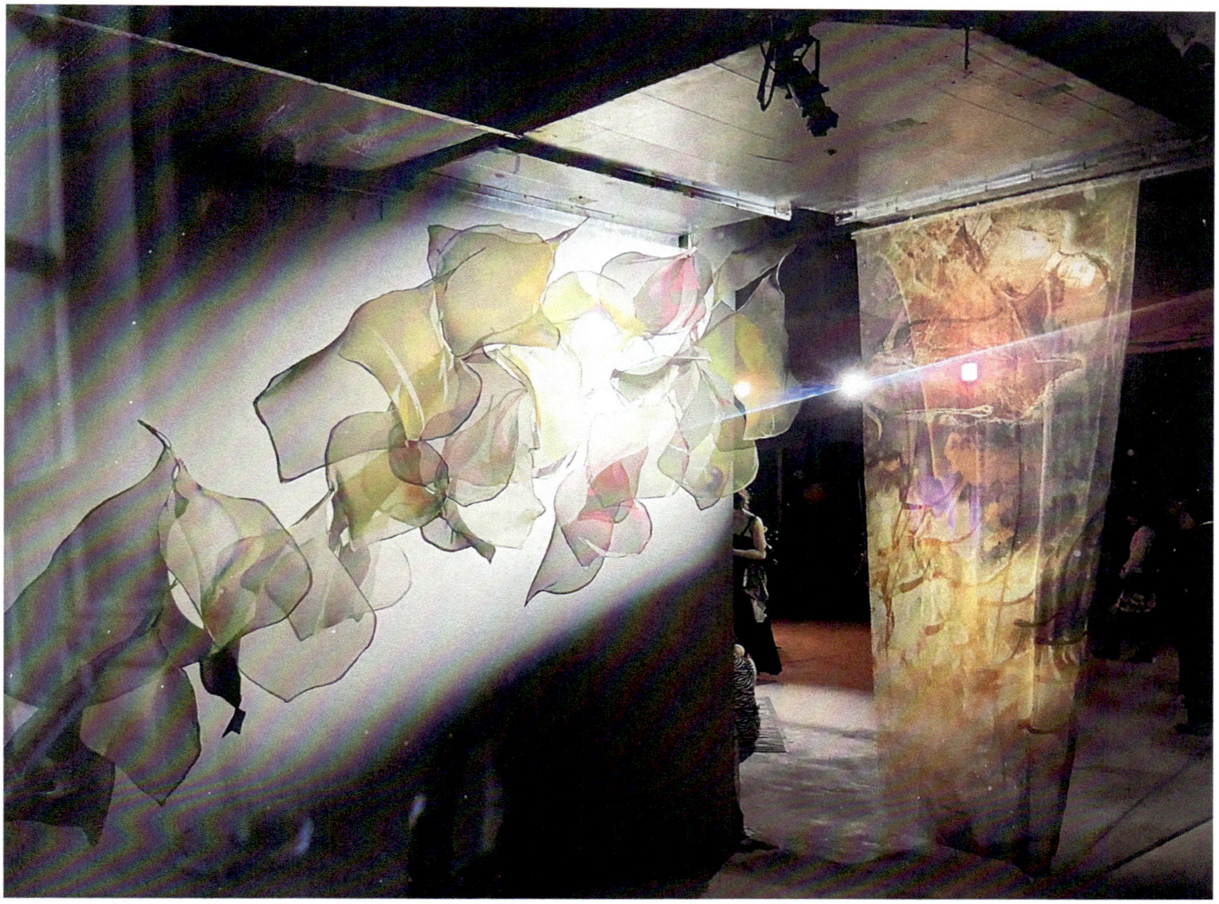

Artwork by Joohyun Pyune. (Used with permission).

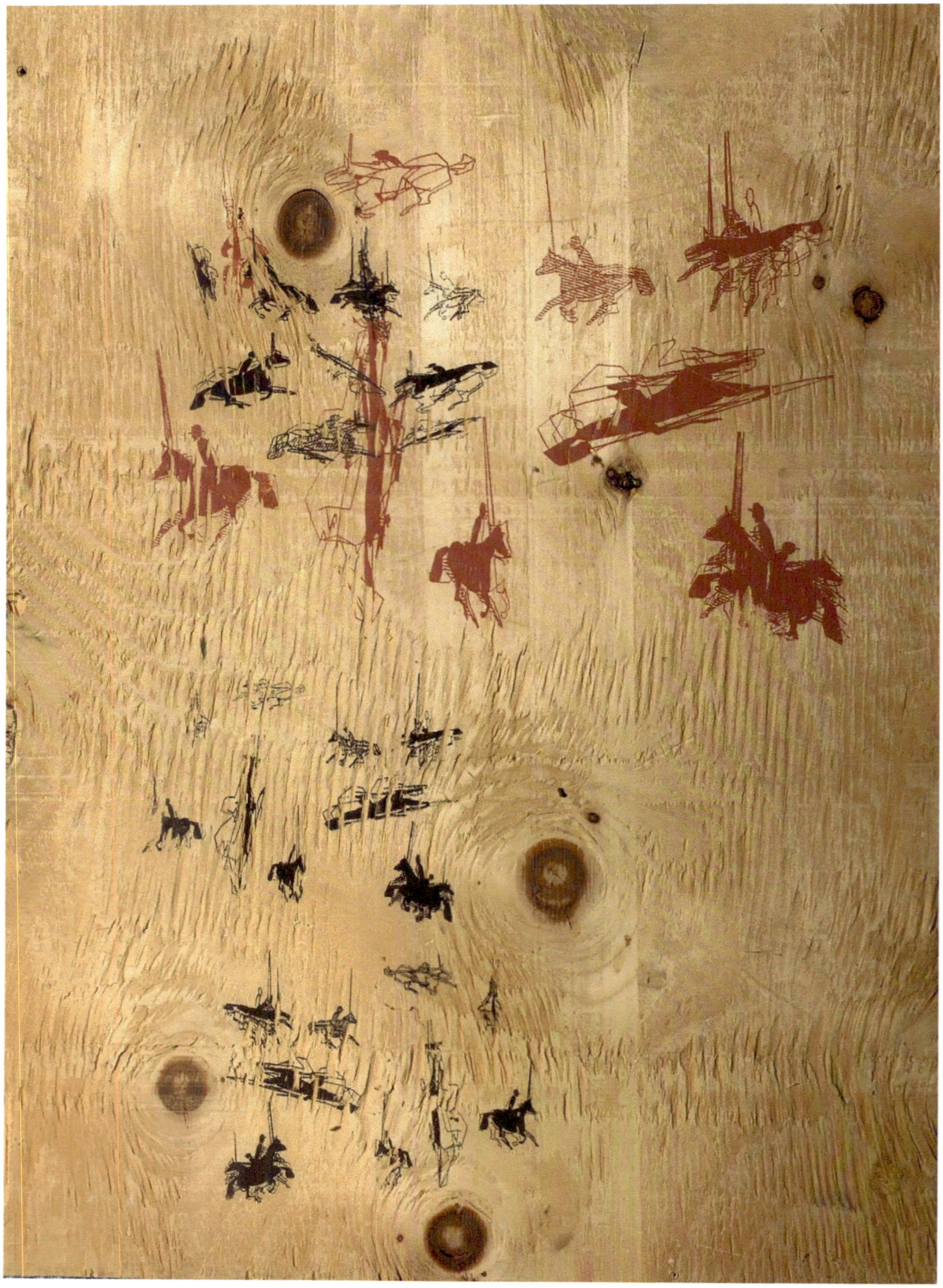

Horse Dance. Photosilscreens after computer codes on wood. © 2025 by Anna Ursyn.

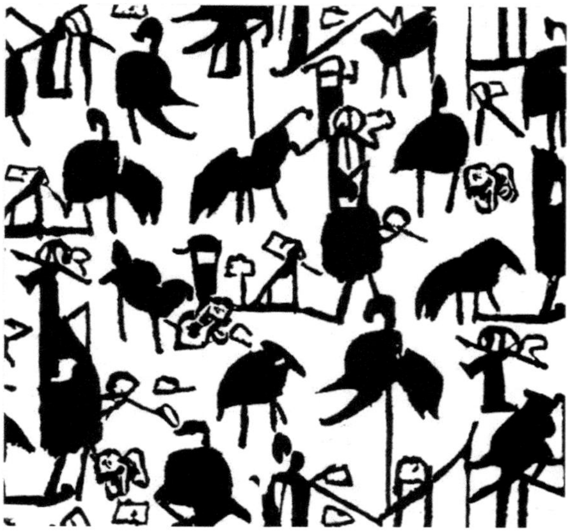

A Colony. © 2025 by Anna Ursyn

ABOUT THE AUTHOR

Anna Ursyn, Ph.D., is a professor of Computer Graphics at the University of Northern Colorado. She combines programming with software and printmaking media to unify computer-generated and painted images and sculptures.
She had over 50 single juried and invitational art shows and participated in over 200 juried and invitational fine art exhibitions, such as over a dozen times in ACM SIGGRAPH Art Galleries and traveling shows, including Louvre, Paris, Der Prix Ars Electronica Linz, Austria; Eurographics (Artistic Merit Award), Denver International Airport Show, Denver Capitol shows. Her work was selected to be sent to the Moon by NASA as a part of Carnegie Mellon's MoonArk project and traveling shows, including Centre Pompidou in Paris. Animations projected at the Victory Media Network, Dallas, TX, ISEAs, New York Digital Salons, 20th Century Matrix for NTT Museum, Tokyo, Japan (2000 images and 5000 texts describing the 20th Century), Winner of BauBoden, Deutsche Bau-und Bodenbank AG, cover and 18 color plates, Computerkunst/Computer Art Gladbeck, Germany + traveling shows. Salon de la Culture et des Jeux Mathematiques, CIJM, University Pierre et Marie Curie, Paris, Honorable Mention; Art of the Animal, Australia (award), GRAPHITEs, Art of Photography Show San Diego, CA, Common Ground the Haun Tie Art Museum in Beijing, M.I.A.D. Torento Argentina (Second Place), Immaginando, Digital Visions of Etruria (Honorable Mention), Italy, Bridges Mathematical Art Galleries, Joint Mathematics Meetings Art Galleries, and Generative Art galleries. Her work in ABADs is in the permanent collection of the Museum of Modern Art MOMA in New York City, the Los Angeles County Museum of Art, CA, the Smithsonian Institution, Washington, DC, and more. Anna published over a dozen books, chapters, and articles with her students' artwork included in her book.

Research and pedagogy interests include integrated art, science, and computer art graphics instruction. Chaired the SpaceTime international competitions for students by ACM SIGGRAPH, the International Conference on Computer Graphics and Interactive Techniques. Since 1997, Anna has served as a Liaison, Organizing, and Program Committee member of the International IEEE Conferences on Information Visualization (iV) in London, UK, and has chaired the Symposium and Digital Art Gallery D-ART iV.

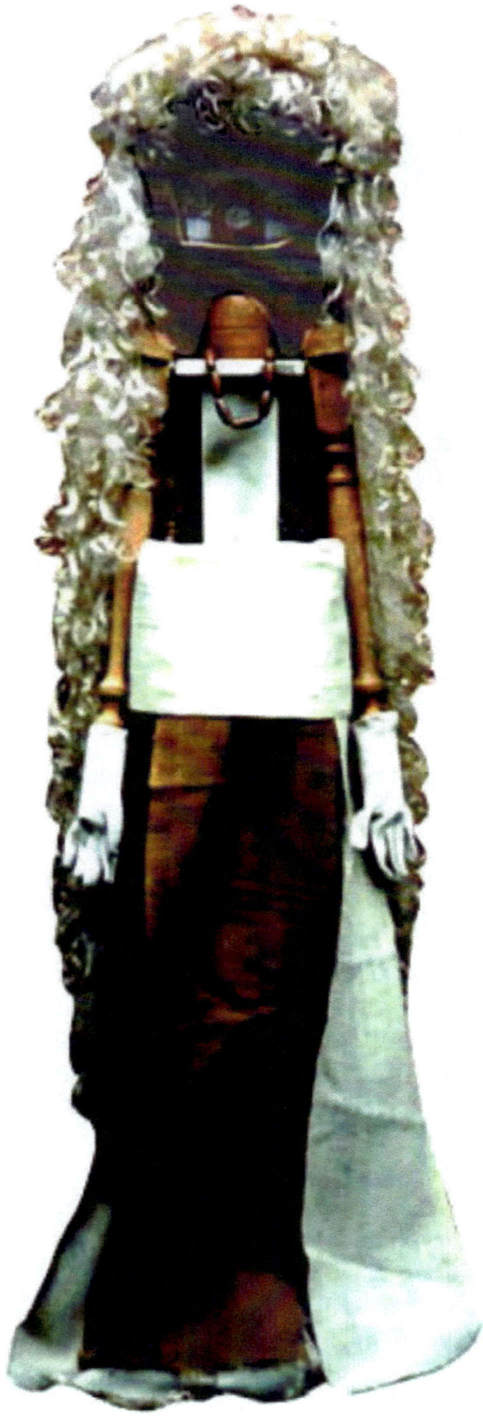

The Opera Singer. © 2025 by Anna Ursyn

ABOUT THE CODERS-COLLABORATORS

Dr. Stuart Smith, Department of Music, and Miner School of Computer and Information Sciences, UMass, Lowell, is a musician, coder, and visual artist. His expertise is in Parallel and Distributed Processing, Data Visualization, and music composition. Stu develops his software to create art. Besides having his band, Stu has a long history of flying planes for recreation.

Dr. Mohammad Majid al-Rifaie, Associate Professor in Artificial Intelligence, Goldsmiths College, University of London, is a coder, passionate about swarm intelligence, a writer, and an artist. He was the editor-in-chief of the oldest AI society(the Society for the Study of Artificial Intelligence and the Simulation of Behaviour or AISB) and has been interviewed multiple times by the British Broadcasting Corporation (BBC) and the New Scientist magazine for his work on swarms, AI, creativity, robotics, and arts.

Md. Fahimul Islam is a Lead Software Engineer and maintains cutting-edge enterprise solutions in a US-based Corporate Arena for JPMorgan Chase & Co., Chicago, IL Fahim completed a Bachelor of Science and Engineering in Computer Science and Engineering in 2004 in Bangladesh. Then, he completed his Masters in Telecommunications Engineering in 2007. As a student, he used to participate in programming contests for the ACM International Programming Contest. It involved thinking through complex programming-oriented problems which involves Back Tracking all kinds of Graph Traversing based on memory and time limitation. His final project was funded by the Ministry of Bangladesh as a young scientist in 2003-2004. The project was titled "Design and Implementation of Computer Based Security System," which was totally designed and developed using discreet Digital and Analog Electronics and capable of sending password store in Windows-based PC. The software was developed in C++ and Visual Basic. It was able to play sound when a Door-open or closed. When unauthorized access is detected an embedded Alarm is initiated and unauthorized data is sent to the computer program. It was awarded First prize in the Computer and Electronics category of the inter-university fair in 2004.

Md. Asifur Rahman is leading the Network Operations Center and Service Assurance teams at Infozillion Teletech BD, Dhaka, Bangladesh, the first service provider to introduce Mobile Number Portability in Bangladesh. Asif is a passionate mentor and IT Professional. He has a proven track record of training, building, and administering high-performance technical teams in diverse environments, and enabling a company to reach its financial and marketing goals. He completed his Bachelor of Science in Computer Science & Engineering in 2009 in Bangladesh.
He is a polyglot programmer who practiced different programming languages like BASIC, C, PHP, Python, Ruby, JavaScript, SQL, etc. Having a keen interest in Electronics from an early age, he loves to tinker with analog and digital circuits and work on practical solutions using Microcontrollers and IoT. He has a keen interest in street photography.

BOOKS – AUTHORED, CO-AUTHORED, AND EDITED BY ANNA URSYN

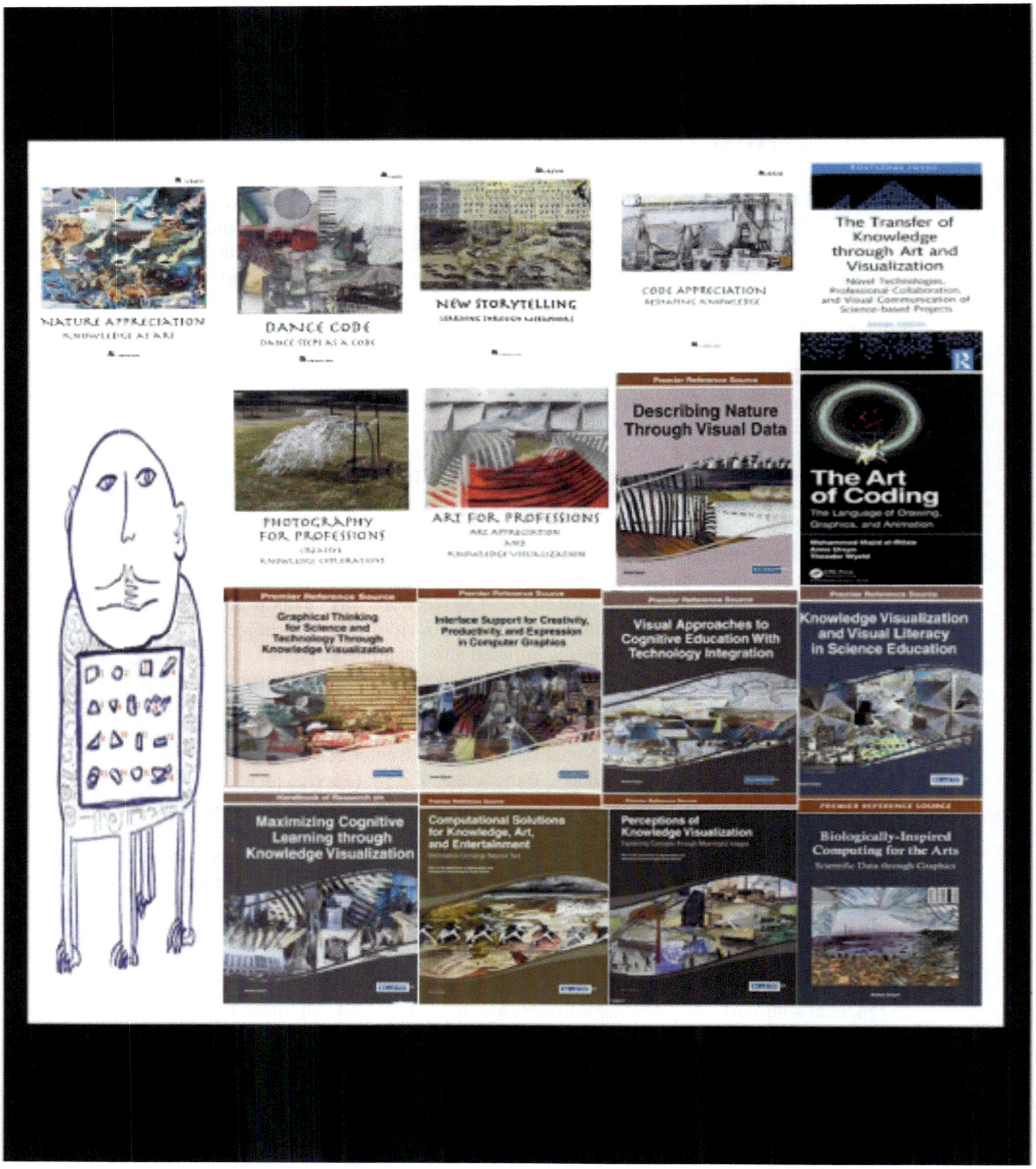

A LIST OF TITLES FROM THE PREVIOUS PAGE

1. *Dance Code, Dance Steps as a Code,* Anna Ursyn (Author). Taylor & Francis Group.
2. *New Storytelling: Learning through Metaphors.* Anna Ursyn (Author). Taylor & Francis Group.
3. *Code Appreciation, Reshaping Knowledge.* Anna Ursyn, (Author). Taylor & Francis Group.
4. *Nature Appreciation: Knowledge as Art,* Anna Ursyn (Author).
5. *The Transfer of Knowledge through Art and Visualization: Novel Technologies, Collaboration, and Visual Communication of Science-based Projects*. Anna Ursyn (Author). Taylor & Francis Group. Routledge. 2024, 143 pages, ISBN 9781032624228.
6. *Photography for Professions: Creative Knowledge Explorations*. Anna Ursyn (Author), Kendall Hunt Publishing Company, 2022, 177 pages, ISBN: 979-8-7657-4703-2.
7. *Art for Professions: Art Appreciation and Knowledge Visualization*. Anna Ursyn (Author), Kendall Hunt Publishing Company. 2021, 538 pages, ISBN-10: 1792466609, ISBN-13: 978-1792466601.
8. *Describing Nature Through Visual Data*. Anna Ursyn (Author/Editor), IGI Global, 2020, 367 pages, ISBN-10: 1799857530, ISBN-13: 978-1799857532.
9. *The Art of Coding: The Language of Drawing, Graphics, and Animation*. by al-Rifaie, M. M., Ursyn, A., Wyeld, T. CRC Press, Taylor & Francis Group, A Chapman & Hall Book. 2020, 312 pages. ISBN: 9780-0-367-90037-3 (hbk), 978-1-138-62964-6 (pbk), 978-1-315-21033-9 (ebk).
10. *Graphical Thinking for Science and Technology Through Knowledge,* Visualization (Advances in Multimedia and Interactive Technologies) (Anna Ursyn, Author//Editor), IGI Global. 2020, 374 pages, ISBN-10: 1799816516, ISBN-13: 978-1799816515.
11. *Interface Support for Creativity, Productivity, and Expression in Computer Graphics*. Anna Ursyn (Author/Editor) IGI Global. 2019, 355 pages, ISBN13: 9781522573715, DOI: 10.4018/978-1-5225-7371-5.
12. *Visual Approaches to Cognitive Education with Technology Integration*. Anna Ursyn (Author/Editor), IGI Global. 2018, 387 pages, ISBN-13: 9781522553328.
13. *Knowledge Visualization and Visual Literacy in Science Education.* Anna Ursyn (Author/Editor). IGI Global. 2016. 431 pages, ISBN-13: 9781522504801.
14. *Maximizing Cognitive Learning through Knowledge Visualization*. Handbook of Research. Anna Ursyn (Author/Editor). IGI Global, 2015, 572 pages, ISBN-13: 9781466681422.
15. *Computational Solutions for Knowledge, Art, and Entertainment: Information Exchange Beyond Text.* Anna Ursyn (Author), IGI Global, 2014, 511 pages, ISB-13: 9781466646278.
16. *Perceptions of Knowledge Visualization: Explaining Concepts through Meaningful Images*. Anna Ursyn (Author), IGI Global, 2014, 418 pages, ISBN-13: 9781466647039.
17. *Biologically Inspired Computing for the Arts: Scientific Data Through Graphics.* Anna Ursyn (Author/Editor), IGI Global, 2012, 442 pages, ISBN-13: 9781466609426.

Codes and videos can be found at:

https://ursyn.com

then select the Art & Programming

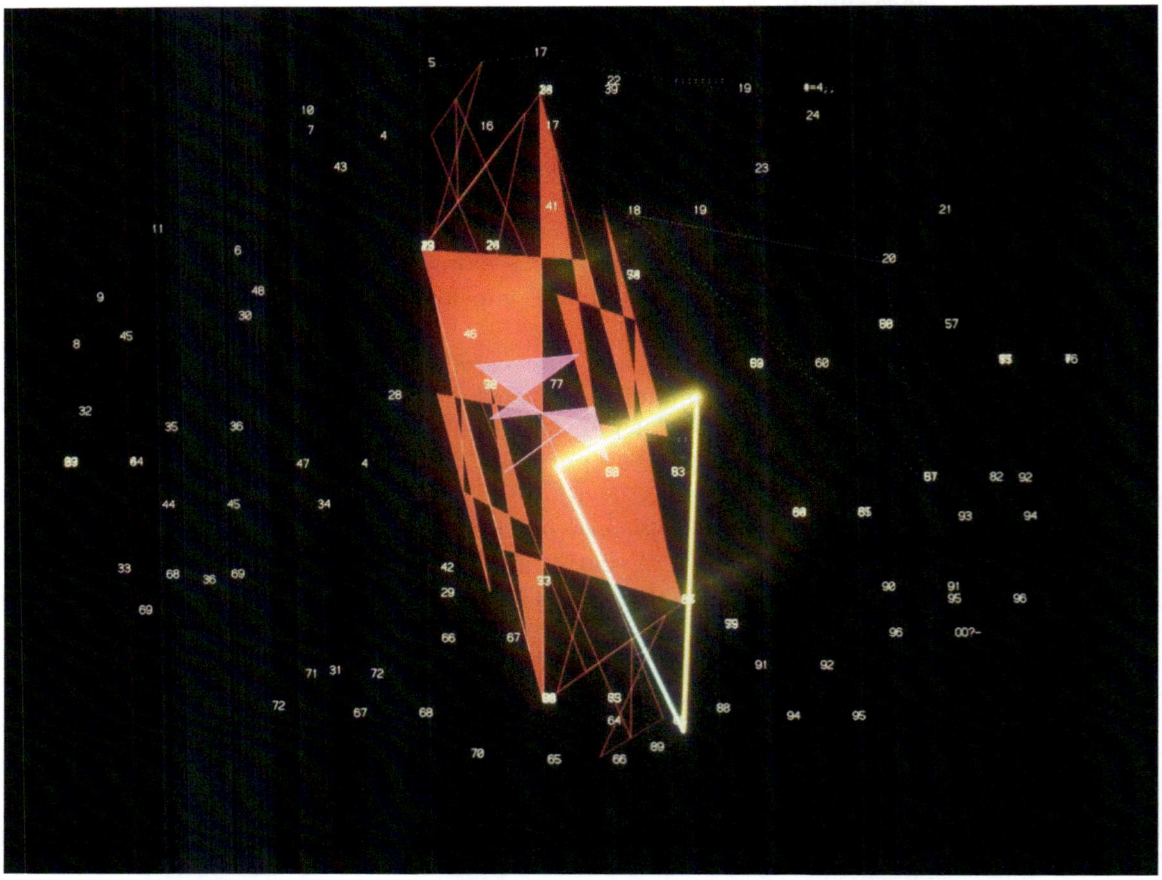

Stellar Numbers. The output from the computer code. © 2025 by Anna Ursyn

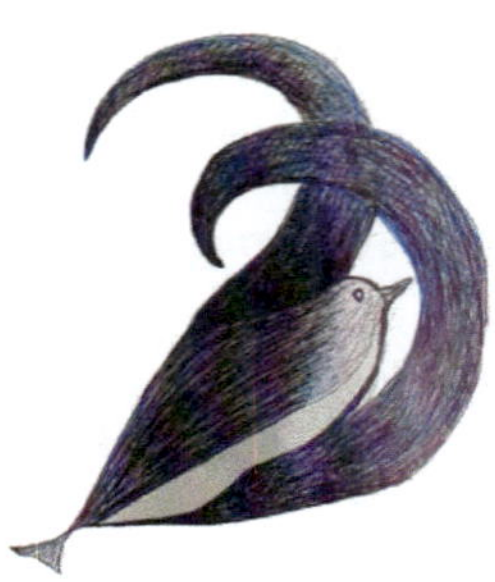

Lurti: dances in the air.

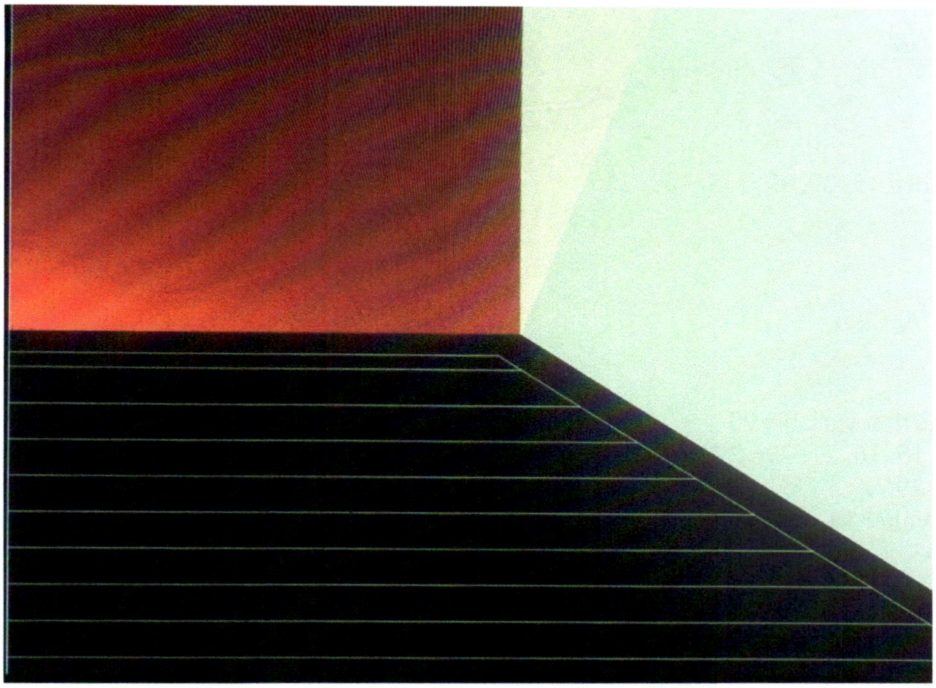

A Scene. The output from the computer code. © 2025 by Anna Ursyn

INDEX

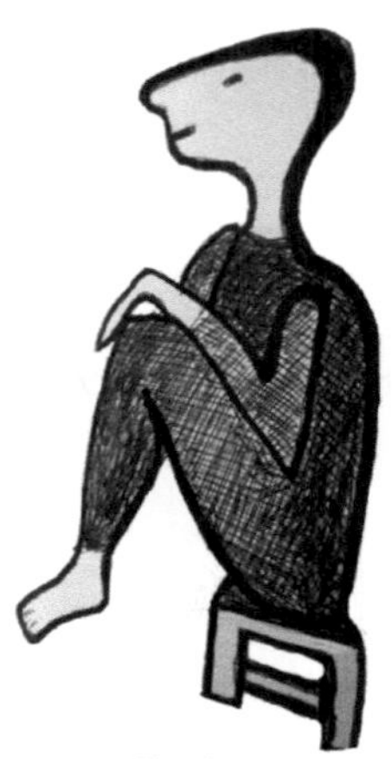

Ziz: nobody knows why.

Ziz: nobody knows why.

P

Q

R

S

T

V

W

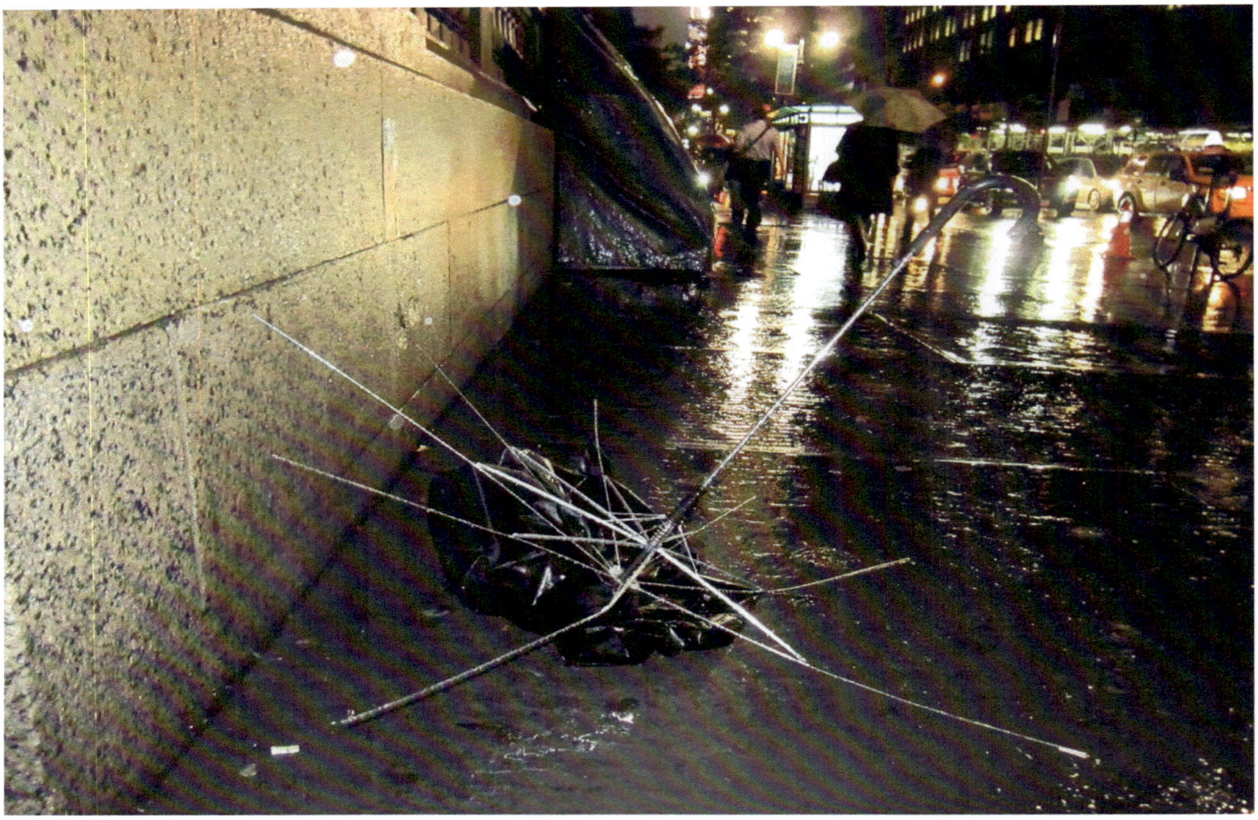

October in New York City. © 2025 by Anna Ursyn

For Product Safety Concerns and Information please contact our EU representative GPSR@taylorandfrancis.com Taylor & Francis Verlag GmbH, Kaufingerstraße 24, 80331 München, Germany

9781032971629

T - #0174 - 130625 - C316 - 279/216/15 - PB - 9781032971629 - Gloss Lamination